# Advances in Intelligent Systems and Computing

Volume 1381

The series "Advances in Intelligent Systems and Computing" contains publications on theory, applications, and design methods of Intelligent Systems and Intelligent Computing. Virtually all disciplines such as engineering, natural sciences, computer and information science, ICT, economics, business, e-commerce, environment, healthcare, life science are covered. The list of topics spans all the areas of modern intelligent systems and computing such as: computational intelligence, soft computing including neural networks, fuzzy systems, evolutionary computing and the fusion of these paradigms, social intelligence, ambient intelligence, computational neuroscience, artificial life, virtual worlds and society, cognitive science and systems, Perception and Vision, DNA and immune based systems, self-organizing and adaptive systems, e-Learning and teaching, human-centered and human-centric computing, recommender systems, intelligent control, robotics and mechatronics including human-machine teaming, knowledge-based paradigms, learning paradigms, machine ethics, intelligent data analysis, knowledge management, intelligent agents, intelligent decision making and support, intelligent network security, trust management, interactive entertainment, Web intelligence and multimedia.

The publications within "Advances in Intelligent Systems and Computing" are primarily proceedings of important conferences, symposia and congresses. They cover significant recent developments in the field, both of a foundational and applicable character. An important characteristic feature of the series is the short publication time and world-wide distribution. This permits a rapid and broad dissemination of research results.

Indexed by DBLP, EI Compendex, INSPEC, WTI Frankfurt eG, zbMATH, Japanese Science and Technology Agency (JST).

All books published in the series are submitted for consideration in Web of Science.

More information about this series at http://www.springer.com/series/11156

Tarun K. Sharma · Chang Wook Ahn ·
Om Prakash Verma · Bijaya Ketan Panigrahi
Editors

# Soft Computing: Theories and Applications

## Proceedings of SoCTA 2020, Volume 2

Springer

*Editors*
Tarun K. Sharma
Department of Computer Science
Shobhit University Gangoh
Gangoh, Uttar Pradesh, India

Chang Wook Ahn
Gwangju Institute of Science
and Technology
Gwangju, Korea (Republic of)

Om Prakash Verma
Department of Instrumentation and Control
Engineering
Dr. B. R. Ambedkar National Institute
of Technology
Jalandhar, Punjab, India

Bijaya Ketan Panigrahi
Department of Electrical Engineering
Indian Institute of Technology Delhi
New Delhi, Delhi, India

ISSN 2194-5357 ISSN 2194-5365 (electronic)
Advances in Intelligent Systems and Computing
ISBN 978-981-16-1695-2 ISBN 978-981-16-1696-9 (eBook)
https://doi.org/10.1007/978-981-16-1696-9



This Springer imprint is published by the registered company Springer Nature Singapore Pte Ltd.
The registered company address is: 152 Beach Road, #21-01/04 Gateway East, Singapore 189721, Singapore

# Preface

This book stimulated discussions on various emerging trends, innovations, practices, and applications in the field of soft computing, ranging from image processing, health care, medicine, supply chain management, and cryptanalysis. This book that we wish to bring forth with great pleasure is an encapsulation of research papers, presented during the three-day international conference on the fifth International Conference on Soft Computing: Theories and Applications (SoCTA 2020) organized in virtual format in association with STEM Research Society. We hope that the effort will be found informative and interesting to those who are keen to learn on technologies that address to the challenges of the exponentially growing information in the core and allied fields of soft computing. We are thankful to the authors of the research papers for their valuable contribution to the conference and for bringing forth significant research and literature across the field of soft computing. Offering valuable insights into soft computing for teachers and researchers alike, the book will inspire further research in this dynamic field.

We express special thanks to Springer and its team for their valuable support in the publication of the proceedings. With great fervor, we wish to bring together researchers and practitioners in the field of soft computing year after year to explore new avenues in the field.

Gangoh, India — Tarun K. Sharma
Gwangju, Korea (Republic of) — Chang Wook Ahn
Jalandhar, India — Om Prakash Verma
New Delhi, India — Bijaya Ketan Panigrahi

# About SoCTA Series

**SoCTA (Soft Computing: Theories and Applications)** is now a four-year young international conference. SoCTA was coined in the year 2016 in technical collaboration with Machine Intelligence Research (MIR) Labs, USA, with an aim of, to highlight the latest advances, problems, and challenges and to present the latest research results in the field of soft computing with a link to scientific research and its practical implementation. SoCTA especially encourages the young researchers at the beginning of their career to participate in this conference and invite them to present their work on this platform.

The objective of **SoCTA** is to provide a common platform to researchers, academicians, scientists, and industrialists working in the area of soft computing to share and exchange their views and ideas on the theory and application of soft computing techniques in multi-disciplinary areas.

Previous **SoCTA** series were successfully organized at the following venues:

- **SoCTA-2016**: Amity University Jaipur, Rajasthan, India.
- **SoCTA-2017**: Bundelkhand University, Jhansi, Uttar Pradesh, India.
- **SoCTA-2018**: Dr. B. R. Ambedkar National Institute of Technology, Jalandhar, Punjab, India.
- **SoCTA-2019**: National Institute of Technology Patna, Bihar, India.

Due to the pandemic and keeping in mind the health issue of the research fraternity, the organizing committee has decided to host the SoCTA2020 conference in a virtual format. Also SoCTA2020 is dedicated to the corona warriors, especially in the field of academics and research.

The tagline for this year is *Virtual Meet—Real Connections*

SoCTA2020 is organized in the technical support of Dr. B. R. Ambedkar National Institute of Technology, Jalandhar, India, and in association with a recently introduced Science, Technology, Engineering and Management (STEM)—Research Society.

The proceedings of all the previous years of SoCTA series were published in Advances in Intelligent Systems and Computing (AISC), Series of Springer, Indexed in SCOPUS.

The credit of the success of the SoCTA series goes to our mentors, keynote and invited speakers, chief guests, guest of honor(s), members of the advisory board (national and international), program committee members, Springer as a publishing partner, all the author(s), participants, and the reviewer's board. We sincerely appreciate your continued support, encouragement, and trust in us. We look forward to have this wonderful support in the coming SoCTA series as well.

We are glad to inform you that the next in the SoCTA series, i.e., SoCTA2021 is scheduled at the Indian Institute of Information Technology (IIIT) Kota, Rajasthan (MNIT Jaipur Campus).

Looking forward to have your significant contribution in SoCTA series…

# About STEM-Research Society

Est: 2020

The STEM-Research Society, a foundation, is registered in the year 2020 to support and promote research in the multi-disciplinary domain under the able guidance of renowned academicians and researchers from India and abroad.

The objective of the foundation is scientific, technical, research, and educational in nature. The foundation strives to advance the theory, practice, and application of science, technology, engineering, and management and maintains a high professional standing among its members. The basic purpose of the STEM-RS is to bring together researchers, academicians, industrialists, and experts from different parts of the country and abroad to exchange knowledge and ideas at a common platform by organizing national and international events such as conferences, seminars, and workshops that unite the science, technology, engineering, and management and topics which are not mentioned here for the empowerment of research and development.

## Vision

The STEM-RS foundation will build a dynamic, interactive, global community of researchers, academicians, and industrialists to advance excellence in science, technology, engineering, and management.

## Mission

The STEM-RS is a foundation of interested peoples worldwide that promotes research for the advancement of society in various spheres and the quality of life.

## Values

Being visionary, dynamic, interdisciplinary, inclusive, egalitarian and promoting research in all spheres of human life.

## Diversity Statement

Diversity drives innovation. STEM-RS engage all demographic teams worldwide in advancing science, technology, engineering, and management to improve the quality of life.

*All the author(s) would be given a free membership for one year.*

Please keep visiting the Web site for recent updates: www.stemrs.in/

## Message from Conveners

It is our great pleasure to welcome you to the International Conference on Soft Computing: Theories and Applications (SoCTA2020) in a virtual format. Soft computing methods are increasingly applied to solve problems in diverse domains. Hence, SoCTA is appropriately conceived to offer a forum to bring all such applied researchers together under one umbrella.

SoCTA is now almost five years old, which means more diligent handling and sense of responsibility are required for continuous improvement and growth.

There is no SoCTA series and SoCTA2020 without the quality contributions made by the authors. In addition, SoCTA2020 is very fortunate to have so many top-quality panel, keynote speakers in this tough time of pandemic COVID-19. We sincerely thank them all.

We are particularly looking forward to the invited talks. We are delighted to have such a strong and varied series of plenary talks at the conference. The underlying philosophy motivating this conference, which has become a flagship forum in the area of mathematics and computer science in general and in the area of soft computing in particular, has been to bring together researchers who apply, besides conventional traditional computing techniques, soft and other novel computing paradigms to problems and situations that have hitherto been intractable, complex, highly nonlinear, and difficult to solve. Soft computing is a cutting-edge field of research in which one of the main inspirations for problem solving is based on, for example, natural or biological systems that tend to be decentralized, are adaptive, and are known to be environmentally aware, and as a result, they have survivability, scalability, and flexibility properties. In addition to work on traditional serial computers, these researchers also exploit methods of efficiency with parallel computing techniques and tools to achieve high-performance computing capabilities in their work.

There are two further key features of this conference series that make this a unique event; i.e., these events are "go-green" environmentally friendly conferences where the emphasis is on the quality of academic endeavor rather than spin and gloss; these events see participation from a large number of young researchers and particularly women scientists which is an important aspect if we are to increase female participation in science, technology, engineering, and mathematics (STEM) areas. Conferences like these are only possible thanks to the hard work of a great many people, and the successful organization of SoCTA2020 has required the talents, dedication, and time of many volunteers and strong support from sponsors.

Chairs of each event contributed exceptionally by attracting contributions, getting them reviewed, making accept and reject recommendations, developing the programs, and so on. We also thank the national and international advisory committee. Publication of SoCTA2020 proceedings is not a simple task. Committee has contributed immensely. We are as ever grateful to Springer Plc. for their dedication and professionalism in helping us produce what is an excellent and high-quality proceedings.

We also give our sincere thanks to all our colleagues on the organizing committee for their sincere work and support throughout the year. We are very grateful to the technical sponsors who have supported the conference despite the continuing difficult pandemic conditions. It only remains for us to thank all of you for participating in the conference and helping to make it a success.

We hope that all of you will benefit from the extensive technical program (*in online mode*) and establish long-lasting interactions with fellow delegates at SoCTA2020.

# Contents

# About the Editors

**Dr. Tarun K. Sharma** holds Ph.D. in Soft Computing in 2013 from IIT Roorkee. Since June 2020, he is associated with Shobhit University, Gangoh, Saharanpur as a Professor and Head of CSE and Dean—School of Engineering and Technology. Earlier he worked with Amity University Rajasthan as an Associate Professor and Head—Department of Computer Science and Engineering/IT as well as Alternate Director—Outcome. He has supervised two Ph.D., seven M.Tech. Dissertations, several MCA and B.Tech. Projects. He has over 80 research publications in his credit. He has been to Amity Institute of Higher Education Mauritius on deputation. He has availed grants from Microsoft Research India, CSIR, New Delhi and DST New Delhi to visit Australia; Singapore and Malaysia respectively. He is a founding member of International Conference on Soft Computing: Theories and Applications (SoCTA Series) and Congress on Advances in Materials Science and Engineering (CAMSE). He has edited 5 volumes of Conference Proceedings published by AISC series of Springer (SCOPUS) Publication and two edited books with Asset Analytics, Springer.

**Prof. Chang Wook Ahn** received the Ph.D. degree from the Department of Information and Communications, GIST, in 2005. From 2005 to 2007, he was with the Samsung Advanced Institute of Technology, South Korea. From 2007 to 2008, he was a Research Professor at GIST. From 2008 to 2016, he was an Assistant/Associate Professor with the Department of Computer Engineering, Sungkyunkwan University (SKKU), South Korea. He is currently a Director of MEMI (Meta-Evolutionary Machine Intelligence) Lab and a Professor with the School of Electrical Engineering and Computer Science, Gwangju Institute of Science and Technology (GIST), South Korea. His research interests include genetic algorithms/programming, multi-objective optimization, evolutionary neural networks, and quantum machine learning. He has been Guest Edited various thematic issues in Journal of repute.

**Dr. Om Prakash Verma** is presently associated with Dr. B. R. Ambedkar National Institute of Technology Jalandhar, Punjab, India since January 2018 as an Assistant Professor in the Department of Instrumentation and Control Engineering. He has almost 11 years of teaching experience. He did his Ph.D. from IIT Roorkee,

M.Tech. from Dr. B. R. Ambedkar NIT Jalandhar and B.E. from Dr. B. R. Ambedkar University Agra. He is presently working on ISRO Sponsored Project as a PI. He has edited a book on Soft Computing: Theories and Applications and has been a potential reviewer of several International Journals of high repute. He has published more than 30 research papers in SCI/Scopus/ESI indexed Journals. He has published recently published a paper in *Renewable and Sustainable Energy Reviews*, (IF: 12.110). He has guided four M.Tech. Students and supervising six Ph.D. Students.

**Prof. Bijaya Ketan Panigrahi** is a Professor in the Department of Electrical Engineering, Indian Institute of Technology Delhi. His research interests include the security of cyber physical systems, digital signal processing, and soft computing applications to power systems. Professor Panigrahi received a Ph.D. in electrical engineering. He is an associate editor for IEEE Systems Journal and a Senior Member of the IEEE. Professor Panigrahi was elected Fellow of INAE in 2015. He is in teaching and research profession since 1990. He has published more than 500 Journals Articles in Journals of high repute. He has edited several conference proceedings with publishers of repute like Springer. He has file more than six patents. As per the Google Scholar his H-Index is 55 with more than 12650 Citations and CrossRef is above 7550. His research interest includes Power System Planning, Operation and Control, Machine Intelligence and Evolutionary Computing.

# A Study on the Effect of Optimal Control Strategies: An SIR Model with Delayed Logistic Growth

**Harendra Pal Singh, Sumit Kaur Bhatia, Riya Jain, and Yashika Bahri**

**Abstract** With the increased number of new infectious diseases coming up, the need to get deeper insights into these diseases is also increasing. Mathematical models developed using compartmental modeling approach have helped in understanding the dynamics of infectious diseases. In this paper, an $S$-$I$-$R$ epidemic model with delay in logistic growth and two control parameters is introduced. We discussed the boundedness of the system followed by the existence of the optimal control variables. Mathematical concepts pertaining to optimality are studied using Pontryagin's maximum principle. Also, the impact of vaccination and treatment as two control strategies to combat epidemic diseases is analyzed. We graphically illustrate the impact of the presence of control strategies in the system.

**Keywords** SIR model · Delay in logistic growth · Optimal control

## 1 Introduction

Mathematical models are significant instruments in epidemiology, for understanding the behavior of epidemics and foreseeing the outcomes of various strategies to be undertaken to control these diseases for general well being. As the time advances and the curiosity of humankind to explore new dimensions increases, humans are exposed to new, destructive and unavoidable diseases, with complex nature and structures. To cope with this, not only have humans advanced in medical research but researchers from various fields other than medicine have also dedicated themselves to discover the ideal methods to control a disease, if its complete eradication cannot be accomplished. Over the years, various researchers have used mathematical models to represent infectious diseases and study their dynamics, while also employing different techniques for reducing their spread and finding mathematical solutions

H. P. Singh
Cluster Innovation Centre, University of Delhi, Delhi, India

S. K. Bhatia (✉) · R. Jain · Y. Bahri
Amity University, Noida, Uttar Pradesh, India

 
T. K. Sharma et al. (eds.), *Soft Computing: Theories and Applications*,
Advances in Intelligent Systems and Computing 1381,
https://doi.org/10.1007/978-981-16-1696-9_1

that can be applied in real life. Suggestions based on analysis and simulations of mathematical models, therefore, become a powerful tool in the hands of specialists. The *S-I-R* model with one control strategy was proposed by Zaman et al. [1], but it is observed that in order to control an epidemic, a single control strategy is not enough. Various aspects of control theory and its applications have been studied by many researchers while incorporating two or more control strategies in the system, along with time delay parameters and optimality [2–9]. While introducing various control strategies in the system, we aim to constrain the spread of infection in the population, which is achieved using optimal control methodology. Since, treatment and vaccination are significant and successful strategies utilized for controlling the spread of infectious diseases, this paper aims to study the outcome when we allow for the two control strategies (vaccination and improved treatment) in our system, that is, when we vaccinate the susceptible class and give treatment to infectious class. Many researchers have also studied the dynamics of the SIR model with delay [10, 11].

Further, a lot of recent work [12–17] can be found in the domain where the studies involve SIR models which have focused on disease dynamics through mathematical modeling. Growth rate plays an important role in population dynamics and there are a few studies with delayed logistic growth [18]. In this paper, we develop a new model, significance of which lies in the incorporation of time delayed logistic growth along with the use of two control strategies (vaccination and treatment) to cope with an epidemic. The use of delayed growth adds a realistic touch to the model because human reproduction has time constraints, and thus, there will be delay in the growth of susceptible class in the system. Section 1 gives a brief introduction to the paper, and the formulation of the mathematical framework describing our system is given in Sect. 2. Boundedness of our model is studied in Sects. 3 and 4 discusses the optimality of the system using objective function and Pontryagin's maximum principle. In Sect. 5, the numerical simulations are performed, and in Sect. 6, we have given the conclusion.

## 2 Mathematical Model

We use susceptible-infected-recovered epidemic model with delay in logistic growth and nonlinear incidence rate. We have assumed that the recovered individuals are permanently immune to the disease for life, that is, they cannot catch the infection again. Here, $r$ and $K$, respectively, are the intrinsic growth rate and the carrying capacity of susceptible population. Further, $\mu$, $d$ and $\eta$ denotes the rate of natural deaths, disease-induced death and natural recovery, respectively. The term $\frac{\chi SI}{1+\delta I}$ is the incidence rate where $\chi$ denotes maximal disease transmission rate, $\chi SI$ denotes the force of infection and $\frac{1}{1+\delta I}$ corresponds to the inhibitory effect when the number of infectives is large. From a biological point of view, human births are constrained with time, resulting in a delayed growth of susceptible class, which is taken care of by the incorporation of delay parameter ($\tau$) [18, 19]. Based on the above discussion, we have formulated a system of three differential equations listed below:

$$\begin{aligned}\frac{dS}{dt} &= rS\left(1-\frac{S(t-\tau)}{K}\right)-\frac{\chi SI}{1+\delta I}\\ \frac{dI}{dt} &= \frac{\chi SI}{1+\delta I}-(d+\eta)I\\ \frac{dR}{dt} &= \eta I-\mu R\end{aligned} \tag{1}$$

## 3 Boundedness of System

We can prove the boundedness of our model with help of provided lemma:

**Lemma 1** *For the region,*

$$B=\left\{(S,I,R)\in R_+^3,\ 0\le S\le K,\ 0\le S+I+R\le\frac{rK}{\theta}\right\}$$

*entire solutions will lie in this region as* $t\to\infty$ *and* $\theta=\min(r,\mu,d)$. *The proof is similar as proved in [20].*

## 4 Optimal Control for Delayed System

In this section, we want to control the extent to which infectious and communicable diseases spread in the population. We want to achieve that a number of infected people are less when compared to susceptibles and more people undergo recovery post infection. Consider the system (1) with two control variables $\varrho_1(t)$ and $\varrho_2(t)$ where $\varrho(t)=(\varrho_1(t),\varrho_2(t))\in G$ relative to three state variables, where

$$G=\{(\varrho_1(t),\varrho_2(t))|\varrho_i(t)\text{ is measurable, }0\le\varrho_i(t)\le 1<\infty,\ t\in[0,t_{\text{end}}]\} \tag{2}$$

for $i=1,2$ be an admissible control set. We assume some restrictions on control parameters: $\varrho_1(t)+\varrho_2(t)\in[0,K)$, where $\varrho_1(t)$ and $\varrho_2(t)$ are non-negative for all $t\in[0,t_{\text{end}}]$. Biologically, this means that when we apply the control parameters, less number of suscpetibles and infectives will be there in the community. Here, $\varrho_1(t)$ denotes the optimal control vaccine given to susceptible class before they come in contact with infection, and $\varrho_2(t)$ denotes optimal control treatment that needs to be given to those infected with disease. To maximize the recovered class and minimize the infective with possible minimal control parameters, we use quadratic terms for control variables and hence the minimizing the objective functional is the optimal strategy. Firstly, we will describe that solutions exists for our control system. We have to estimate control parameters $\varrho_1(t)$ and $\varrho_2(t)$ and individuals $S(t), I(t), R(t)$ for objective functional to be minimized,

$$Z(\varrho) = \int_0^T \left[ I(t) + \frac{\epsilon_1 {\varrho_1}^2(t)}{2} + \frac{\epsilon_2 {\varrho_2}^2(t)}{2} \right] dt \tag{3}$$

subject to

$$\begin{cases} \frac{dS}{dt} = rS\left(1 - \frac{S(t-\tau)}{K}\right) - \frac{\chi SI}{1+\delta I} - \varrho_1 S \\ \frac{dI}{dt} = \frac{\chi SI}{1+\delta I} - (d+\eta)I - \varrho_2 I \\ \frac{dR}{dt} = \eta I - \mu R + \bar{\varrho} \end{cases} \tag{4}$$

with initial conditions $S(0) = S_0$, $I(0) = I_0$ and $R(0) = R_0$.
Here, $\bar{\varrho} = \varrho_1 S + \varrho_2 I$. Weight factors are $\epsilon_1$ and $\epsilon_2$ which are positive constants describing the individual's level of accepting the vaccination and treatment respectively. Here, our goal is that the objective functional is minimized so as to reduce probability of spread of infection through infected population and maximize the recovered population with help of possible minimal control variables $\varrho_1(t)$ and $\varrho_2(t)$.

## 4.1 Existence of Optimal Control Pair

**Theorem 1** *There exists an optimal control pair* $({\varrho_1}^*, {\varrho_2}^*)$ *which satisfies*

$$Z({\varrho_1}^*, {\varrho_2}^*) = \min\{Z(\varrho_1, \varrho_2) : (\varrho_1, \varrho_2) \in G\}$$

*with respect to control system (4).*

***Proof*** For a finite time interval, the boundedness of solution of system (4) has been utilized to show the existence of optimal control. We need to verify the following properties in order apply an existence result in [19].

1. The control parameters $\{\varrho_1, \varrho_2\}$ along with state variables are nonempty.
2. The set $G = \{\varrho_1, \varrho_2\}$ is closed and convex.
3. The state control parameters in system (4) are linear functions.

4. Objective functional integrand is convex on $G$.
5. The objective functional integrand satisfies,

$$I(t) + \frac{1}{2}(\epsilon_1 \varrho_1{}^2(t) + \epsilon_2 \varrho_2{}^2(t)) \geq c_1(|\varrho_1|^2 + |\varrho_2|^2)^{\frac{q}{2}} + c_2$$

where there exist constants $c_i > 0$, $(i = 1, 2)$ and $q > 1$.

For system (4) with bounded coefficients, in order to prove the existence of solution, we can use an existence result mentioned in Lukes [21], which proves statement 1. Statement 2 is proved by the definition of control set itself. The right-hand side of the state system (4) satisfies statement 3 due to the linearity of the state system in $\varrho$ along with the system being bounded. The positivity of the second derivative of the integrand in the Eq. (3) meets the criteria of convexity in statement 4. Also, there exist constants $c_i > 0$, $(i = 1, 2)$ and $q > 1$ satisfying

$$I(t) + \frac{1}{2}(\epsilon_1 \varrho_1{}^2(t) + \epsilon_2 \varrho_2{}^2(t)) \geq c_1(|\varrho_1|^2 + |\varrho_2|^2)^{\frac{q}{2}} + c_2$$

which gives the statement 5. Hence, our theorem is proved.

## 4.2 Characterization of Optimal Control

The Hamiltonian $H(t) = H(x, x_\tau, \varrho_1, \varrho_2, \psi)(t)$ of our control problem is defined as:

$$H(x, x_\tau, \varrho_1, \varrho_2, \psi)(t) = L(I, \varrho_1, \varrho_2) + \psi_1\left(rS\left(1 - \frac{S(t-\tau)}{K}\right) - \frac{\chi SI}{1+\delta I} - \varrho_1 S\right)$$
$$+\psi_2\left(\frac{\chi SI}{1+\delta I} - (d+\eta)I - \varrho_2 I\right) + \psi_3\left(\eta I - \mu R + \bar{\varrho}(t)\right)$$

We have used the Pontryagin's maximum principle to find the optimal solution as follows:

If in an optimal control problem, $(x^*(t), \varrho^*(t))$ is an optimal solution, then a non-trivial vector function $\psi(t) = (\psi_1(t), \ldots, \psi_n(t))$ exists which satisfies the following conditions,

1. State Equation

$$\frac{dx}{dt} = \frac{\partial H(t, x^*(t), \varrho^*(t), x_\tau(t), \psi(t))}{\partial \psi} \tag{5}$$

2. The optimality condition

$$0 = \frac{\partial H(t, x^*(t), \varrho^*(t), x_\tau(t), \psi(t))}{\partial \varrho} \tag{6}$$

3. The adjoint equation

$$-\psi'(t) = \frac{\partial H(t, x^*(t), \varrho^*(t), x_\tau(t), \psi(t))}{\partial x} + \psi(t+\tau)\frac{\partial H(t, x^*(t), \varrho^*(t), \psi(t))}{\partial x} \tag{7}$$

**Theorem 2** *If $S^*(t), I^*(t)$ and $R^*(t)$ are the optimal state solutions along with optimal control variable ${\varrho_1}^*(t)$ and ${\varrho_2}^*(t)$, then there exists adjoint variables $\psi_1(t)$, $\psi_2(t)$ and $\psi_3(t)$ satisfying*

$$\frac{d\psi_1}{dt} = -r\psi_1 + \frac{\chi I^*\psi_1}{1+\delta I^*} - \frac{\chi I^*\psi_2}{1+\delta I^*} + \varrho_1(\psi_1 - \psi_3) + \psi_1(t+\tau)\psi_1\frac{rS^*}{K}$$

$$\frac{d\psi_2}{\partial t} = -1 + \frac{\chi S^*\psi_1}{(1+\delta I^*)^2} - \frac{\chi S^*\psi_2}{(1+\delta I^*)^2} + (d+\eta) - \psi_3\eta + \varrho_2(\psi_2 - \psi_3)$$

$$\frac{d\psi_3}{\partial t} = \mu\psi_3$$

*with boundary or transversality conditions*

$$\psi_i(t_{end}) = 0, i = 1, 2, 3 \tag{8}$$

*Next, optimal control pairs are provided below:*

$${\varrho_1}^*(t) = max\left\{min\left\{\frac{S(\psi_1 - \psi_3)}{\epsilon_1}, 1\right\}, 0\right\} \tag{9}$$

$${\varrho_2}^*(t) = \max\left\{\min\left\{\frac{I(\psi_2 - \psi_3)}{\epsilon_2}, 1\right\}, 0\right\} \tag{10}$$

***Proof*** The adjoint and transversality equations can be determined when the Hamiltonian $H$ is differentiated with respect to $x(t)$ and $x_\tau(t)$. Using Eqs. (5) and (7), we have the following equations:

$$\frac{\partial H}{\partial \psi_1} = rS\left(1 - \frac{S(t-\tau)}{K}\right) - \frac{\chi SI}{1+\delta I} - \varrho_1 S = \frac{dS}{dt}$$

$$\frac{\partial H}{\partial \psi_2} = \frac{\chi SI}{1+\delta I} - (d+\eta)I - \varrho_2 I = \frac{dI}{dt}$$

$$\frac{\partial H}{\partial \psi_3} = \eta I - \mu R + \bar{\varrho}(t) = \frac{dR}{dt}$$

$$\frac{\partial H}{\partial S} = r\psi_1 - \frac{\chi I\psi_1}{1+\delta I} + \frac{\chi I\psi_2}{1+\delta I} + \varrho_1(\psi_3 - \psi_1) - \psi_1\frac{rS(t-\tau)}{K}$$

$$\frac{\partial H}{\partial I} = 1 - \frac{\chi S\psi_1}{(1+\delta I)^2} + \frac{\chi S\psi_2}{(1+\delta I)^2} - (d+\eta)\psi_2 + \psi_3\eta + \varrho_2(\psi_3 - \psi_2)$$

$$\frac{\partial H}{\partial R} = -\mu\psi_3$$

$$\frac{d\psi_1}{dt} = -r\psi_1 + \frac{\chi I\psi_1}{1+\delta I} - \frac{\chi I\psi_2}{1+\delta I} + \varrho_1(\psi_1 - \psi_3) + \psi_1(t+\tau)\psi_1\frac{rS}{K}$$

$$\frac{d\psi_2}{\partial t} = -1 + \frac{\chi S \psi_1}{(1+\delta I)^2} - \frac{\chi S \psi_2}{(1+\delta I)^2} + (d+\eta)\psi_2 - \psi_3 \eta + \varrho_2(\psi_2 - \psi_3)$$
$$\frac{d\psi_3}{\partial t} = \mu \psi_3$$

Using the condition of optimality (6), we obtain

$$\frac{\partial H}{\partial \varrho_1} = \epsilon_1 \varrho_1 + S(\psi_3 - \psi_1)$$
$$\frac{\partial H}{\partial \varrho_2} = \epsilon_2 \varrho_2 + I(\psi_3 - \psi_2)$$

Now, set $S(t) = S^*(t), I(t) = I^*(t)$ and $R(t) = R^*(t)$ to obtain the adjoint equations stated in the preceding theorem.
Next, when $\varrho_i(t) = \varrho_i^*(t),\ i = 1, 2$

$$\varrho_1{}^*(t) = \max\left\{\min\left\{\frac{S(\psi_1 - \psi_3)}{\epsilon_1}, 1\right\}, 0\right\} \tag{11}$$

and

$$\varrho_2{}^*(t) = \max\left\{min\left\{\frac{I(\psi_2 - \psi_3)}{\epsilon_2}, 1\right\}, 0\right\} \tag{12}$$

Next applying the property of control set in Eq. (2) to obtain Eqs. (9) and (10). The formulas provided by Eqs. (9) and (10) for $\varrho_i^*(t),\ i = 1, 2$ are known as the characterization of optimal control. We have found the optimal control and state equations by evaluating the optimality system that includes state system, adjoint equations, boundary conditions and characterization of optimal control. Also, it is observed that when the Lagrangian is differentiated twice with respect to each of the variables $\varrho_i(t),\ i = 1, 2$, it turns out to be positive showing that the optimal control problem is minimum at control $\varrho_i{}^*(t),\ i = 1, 2$. Finally, we obtain the control system provided below:

$$\frac{dS^*}{dt} = rS^*\left(1 - \frac{S^*(t-\tau)}{K}\right) - \frac{\chi S^* I^*}{1+\delta I^*} - \max\left\{\min\left\{\frac{S^*(\psi_1 - \psi_3)}{\epsilon_1}, 1\right\}, 0\right\} S^*$$
$$\frac{dI^*}{dt} = \frac{\chi S^* I^*}{1+\delta I^*} - (d+\eta)I^* - \max\left\{\min\left\{\frac{I^*(\psi_2 - \psi_3)}{\epsilon_2}, 1\right\}, 0\right\} I^*$$
$$\frac{dR^*}{dt} = \eta I^* - \mu R^* + \max\left\{\min\left\{\frac{S^*(\psi_1 - \psi_3)}{\epsilon_1}, 1\right\}, 0\right\} S^*$$
$$+ \max\left\{\min\left\{\frac{I^*(\psi_2 - \psi_3)}{\epsilon_2}, 1\right\}, 0\right\} I^*$$

and the Hamiltonian $H^*$ at $(t, S^*, I^*, R^*, \varrho_1{}^*, \varrho_2{}^*, \psi_1, \psi_2, \psi_3)$ is denoted by

$$H^* = I^*(t) + \psi_1(t)\frac{dS^*}{dt} + \psi_2(t)\frac{dI^*}{dt} + \psi_3(t)\frac{dR^*}{dt}$$
$$+ \frac{1}{2}\left[\epsilon_1\left(\max\left\{\min\left\{\frac{S^*(\psi_1-\psi_3)}{\epsilon_1}, 1\right\}, 0\right\}\right)^2\right.$$
$$\left.+\epsilon_2\left(\max\left\{\min\left\{\frac{I^*(\psi_2-\psi_3)}{\epsilon_2}, 1\right\}, 0\right\}\right)^2\right]$$

## 5 Numerical Section

Here, we will present the simulations for optimal control system. The algorithm of forward and backward Euler method is used to solve the optimality graphically. For hypothetical and biologically feasible set of parameters $r = 0.15$, $K = 1000$, $\delta = 0.51$, $d = 0.1$, $\eta = 0.039$, $\mu = 0.1$ and $\chi = 0.09$, the impact of vaccination ($\varrho_1$) and treatment ($\varrho_2$) is studied in the absence of delay in our model. We have obtained the simulationfvs for a time period of 300 d. From Fig. 1a, it is observed that there is a sharp decline in count of susceptibles when non-zero controllers are applied. Thus, this ensures that the optimal strategy of vaccination given to susceptible class is effective. Behavior of infected population, both with and without controls, is depicted in Fig. 1b. It is observed that the count of infectives $I = 10.428$ (in the absence of control strategy) while $I^*$ wipes out completely in presence of controls, thus, depicting that the treatment given to infected class is effective and the count of infected individuals decreases with time. From Fig. 1c, it is observed that in initial days, a sudden increase in the recovered individuals is observed when control strategies are applied; thus, both vaccination and treatment are effective and best suitable optimal controls for our model. Figure 1d depicts the controllers $\varrho_1$ (vaccination, blue solid curve) and $\varrho_2$ (treatment, red dashed curve) as optimal control strategies given to susceptible and infected class, respectively. It is observed that a combination of both plays an important role in an SIR model and has a significant effect in maximizing the recovered and minimizing the infected class.

## 6 Conclusion

In the preceding sections, we developed the $S$-$I$-$R$ model with delay in logistic growth and nonlinear incidence rate. We have discussed the boundedness of the system and the existence of the optimal control pair. Next, we have found an optimal solution using Hamiltonian of the control problem and Pontryagin's maximum principle. The uniqueness of the model is with the incorporation of time delayed logistic growth

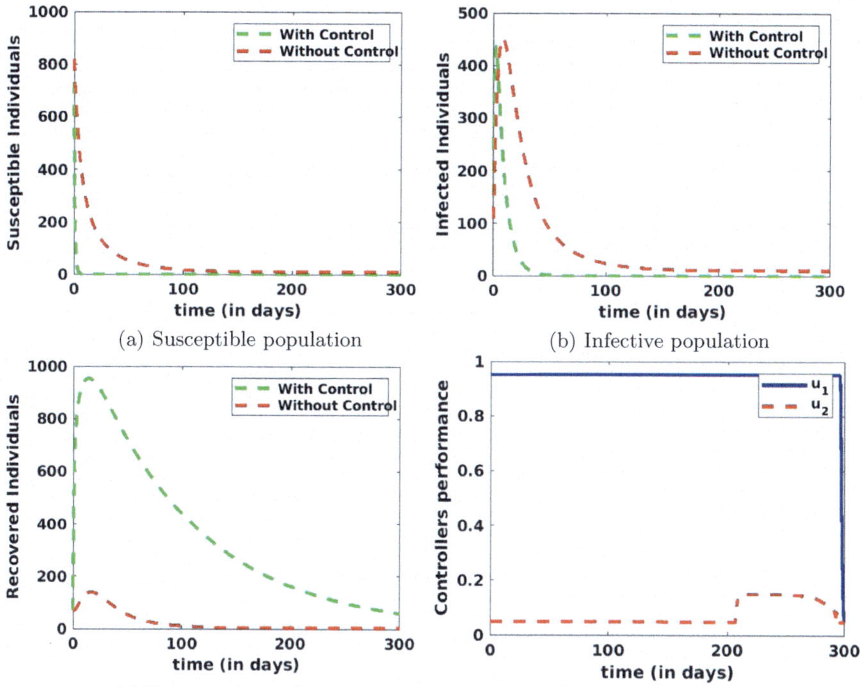


(a) Susceptible population

(b) Infective population

(c) Recovered population

(d) Controllers performance

**Fig. 1** Behavior of the system in presence of controllers

along with the use of two control strategies (vaccination and treatment) to cope with an epidemic. After discussing the analytical results, we have done numerical simulations in order to graphically see the impact of control parameters on the spread of the disease. We have observed that when controllers are applied the susceptible count decrease over time, infected individuals ultimately wipe out of the system and the count of recovered individuals increase. This highlights the importance of control strategies in coping with infectious diseases and proves how control strategies help in the faster elimination of infection from the system. It is also worthwhile to note that this model is limited in the sense that it lacks certain useful compartments (like exposed, symptomatic, asymptomatic and hospitalized), along with neglecting the use of potential delay parameters (like incubation delay), and the unpredictable emigrations and immigration that the human population is accustomed to, which could have given this model a more realistic front.

## References


1. Zaman, G., Kang, Y.H., Jung, I.H.: Stability analysis and optimal vaccination of an SIR epidemic model. Biosystems **93**, 240–249 (2008). https://doi.org/10.1016/j.biosystems.2008.05.004
2. Bashier, E.B.M., Patidar, K.C.: Optimal control of an epidemiological model with multiple time delays. Appl. Math. Comput. **292**, 47–56 (2017). https://doi.org/10.1016/j.amc.2016.07.009
3. Zaman, G., Kang, Y.H., Jung, I.H.: Optimal strategy of vaccination and treatment in an SIR epidemic model. Math. Comput. Simul. **136**, 63–77 (2017). https://doi.org/10.1016/j.matcom.2016.11.010
4. Huo, H.F., Huang, S.R., Wang. X.Y., Xiang, H.: Optimal control of a social epidemic model with media coverage. J. Biol. Dyn. **11**, 226–243 (2017). https://doi.org/10.1080/17513758.2017.1321792
5. Barro, M., Guiro, A., Ouedraogo, D., Optimal control of a SIR epidemic model with general incidence function and a time delays. CUBO A Math. J. **20**, 53–66 (2018). http://dx.doi.org/10.4067/S0719-06462018000200053
6. Ghosh, J.K., Ghosh, U., Biswas, M.H.A., Sarkar, S.: Qualitative analysis and optimal control strategy of an SIR model with saturated incidence and treatment (2018). https://arxiv.org/abs/1807.05954
7. Laarabi, H., Labriji, E., Rachik, M., Kaddar.: A, Optimal control of an epidemic model with a saturated incidence rate. Nonlinear Anal Model Control **17**, 48–459 (2012). https://doi.org/10.15388/NA.17.4.14050
8. Grigorieva, E.V., Khailov, E.N., Optimal vaccination, treatment, and preventive campaigns in regard to the SIR epidemic model. Math. Model. Nat. Phenom., **9**, 105–121 (2014). https://doi.org/10.1051/mmnp/20149407
9. Liu, C., Lu, Na, Zhang, Q., Dynamical analysis in a hybrid bioeconomic system with multiple time delays and strong Allee effect. Math. Comput. Simul. **136**, 104–131 (2017). https://doi.org/10.1016/j.matcom.2016.12.005
10. Kumar., A., Nilam.: Stability of a time delayed SIR epidemic model along with nonlinear i1850055-1-1850055-17 (2018).https://doi.org/10.1142/S021987621850055X
11. Hattaf, K., Lashari, A.A., Louartassi, Y., Yousfi, N.: A delayed SIR epidemic model with general incidence rate. Electron J. Qual. Theory Differ. Equ. **3**, 1–9 (2013). https://doi.org/10.14232/EJQTDE.2013.1.3
12. Liu, C., Heffernan, J., Stochastic dynamics in a delayed epidemic system with Markovian switching and media coverage. Adv. Differ. Equ. article no. 439 (2020). https://doi.org/10.1186/s13662-020-02894-5
13. Zhou, W., Xiao, Y., Heffernan, J.M.: Optimal media reporting intensity on mitigating spread of an emerging infectious disease. PLoS ONE **14**(3), e0213898 (2019). https://doi.org/10.1371/journal.pone.0213898
14. Kumar, A., Goel, K., Nilam.: A deterministic time-delayed SIR epidemic model: mathematical modeling and analysis. Theory Biosci. **139**, 67–76 (2020). https://doi.org/10.1007/s12064-019-00300-7
15. Khatun, M.S., Biswas, M.H.A.: Mathematical analysis and optimal control applied to the treatment of leukemia. J. Appl. Math. Comput. **64**, 331–353 (2020). https://doi.org/10.1007/s12190-020-01357-0
16. Khan, M.A., Iqbal, N., Khan, Y., Alzahrani, E.: A biological mathematical model of vector-host disease with saturated treatment function and optimal control strategies. Math. Biosci. Eng. **17**, 3972–3997 (2020). https://doi.org/10.3934/mbe.2020220
17. Song, C., Xu, R., Bai, N., Tian, X.H., Lin, J.Z.: Global dynamics and optimal control of a cholera transmission model with vaccination strategy and multiple pathways. Math. Biosci. Eng. **17**, 4210–4224 (2020). https://doi.org/10.3934/mbe.2020233. PMID: 32987576

18. Shi, X., Zhou, X., Song, X.: Dynamical behavior for an eco-epidemiological model with discrete and distributed delay. J. Appl. Math. Comput. **33**, 305–325 (2010). https://doi.org/10.1007/s12190-009-0288-8
19. Fleming, W.H., Rishel, R.W.: Deterministic and Stochastic Optimal Control. Springer, NewYork (1975)
20. Elhla, M., Rachik, M., Benlahmar, E.: Optimal Control of an SIR Model with Delay in State and Control variables. Hindawi Publishing Corporations, pp. 1–7 (2013). https://doi.org/10.1155/2013/403549
21. Lukes, D.L.: Differential Equations: Classical to Controlled, Mathematics in Science and Engineering. Academic Press, New York (1982)

# Emperor Penguin Optimized Clustering for Improved Multilevel Hierarchical Routing in Wireless Sensor Networks

**Satyajit Pattnaik and Pradip Kumar Sahu**

**Abstract** In recent years, wireless sensor network (WSN) is a very effective in adaptive application for research community. The WSN nodes have limited amount of energy that needs that the energy should be used in an efficient way. The energy depletion factor should be reduced by combining the nodes for data collection and routing. This paper proposed an energy-efficient-based emperor penguin optimized (EPO) clustering for improved multilevel hierarchical routing (IMHR) in wireless sensor networks. The whole network region is divided by EPO algorithm into number of clusters and regions where every cluster has one cluster head (CH). According to energy condition, the WSN nodes are combined which perform routing in hierarchical level. The sensor nodes (SN) exchange different information such as distance and energy to perform clustering and routing process. The communicated messages are conveyed by IMHR for cluster association. In our work, multilevel hierarchical routing (MHR) is improved by optimizing MHR with social ski-driver (SSD) algorithm. MATLAB platform is used to implement the work and take the graphical analysis by means of throughput, delay, PDR, etc., by comparing with exiting techniques. The performance of proposed throughput is 99.3%, packet delivery ratio is 97%, and the delay is 6.1 s.

**Keywords** Multilevel hierarchical routing · Energy-efficient clustering · Cluster head · Emperor penguin optimization · Social ski-driver

## 1 Introduction

The WSN is a dynamic network which has self-organization and large-scale properties. WSN is utilized in many applications like military applications, environmental monitoring, volcanic earthquake detection and temperature acquisition [1–3]. A large amount of small-scale sensor nodes (SNs) is composed of WSN which has the ability of wireless communication, calculation and perception [4]. It consists of SNs with

S. Pattnaik (✉) · P. K. Sahu
Veer Surendra Sai University of Technology, Burla, Sambalpur, Odisha, India

 
T. K. Sharma et al. (eds.), *Soft Computing: Theories and Applications*,
Advances in Intelligent Systems and Computing 1381,
https://doi.org/10.1007/978-981-16-1696-9_2

limited amount of energy. In a particular region, the WSNs are randomly applied to acquire different types of environmental parameters, and the information is transmitted to the base station (BS) for detecting and monitoring applications [5]. The SNs rechargeable batteries are very difficult due to the deployment of WSNs in hazardous environments.

The SNs are able to sense the signals and support the transmission of data among various SNs like BS [6]. The energy consumption of SNs is generally for communicating with other nodes, processing the information and sensing data of physical phenomena. For data communication, more energy is required than any other process [7, 8]. Therefore, to reduce the energy consumption is the objective so as to extend the lifetime of entire network. The communication is depending upon clustering, and the CH nodes are responsible for the data exchange among the BS and node, so the communication cost of cluster members is reduced efficiently [9–11].

In such a way, the CH selection is performed to attain higher throughput depending upon nodes energy which presented in the cluster [12]. CHs will prompt the overall nature of the structure which is similarly circled and afterwards the network lifetime will be calculated at the end [13, 14]. Another approach to manage CH determination relies upon limiting the separation to the cluster nodes which decrease the utilization of energy [15]. Thus, we want to improve the energy efficiency by using the meta-heuristic optimization algorithm. Contribution of the work is organized as follows.

- Design the WSN by using the nodes.
- Energy-efficient clustering by EPO.
- To select the optimal path by IMHR.

The remaining section will be organized as follows.

The next section gives the some recent analysis of related works. Section 3 gives the system model, energy model, clustering by EPO and IMHR, Sect. 4 gives the performance of results with graphical analysis, and final section includes the conclusion.

## 2 Related Work

In WSN, an energy-efficient cluster-based multilevel hierarchical routing (EECMHR) method was presented by Kumaresan et al. [16] for optimal data collection. The whole network region was divided by using their method into number of clusters and regions. According to energy condition, the WSN nodes are combined which perform routing in hierarchical level.

A low-energy adaptive clustering hierarchy (LEACH) was implemented by Liu et al. [17]. The energy consumption of the network was increased by using LEACH method without considering the CHs distributions. A modified routing protocol was proposed to improve the energy efficiency of WSN. Moreover, they consider that the nodes closer to the BS were not participated in the cluster formation for getting the less energy consumptions. Furthermore, in data transmission instead of single communication mode, they employed single, multiple and hybrid communication.

Based on hierarchical clustering, a new routing algorithm was presented by Yarinezhad et al. [18]. The appropriate nodes are selected as CHs in each hierarchical clustering. For routing, a multilevel clustering algorithm was presented in WSN. An optimal routing tree was formed in the network to form the cluster and select the cluster heads. With the help of virtual layer structure, an optimal routing tree was created.

A routing protocol for heterogeneous WSN (HWSN)-based modified grey wolf optimizer (GWO) called as (HMGWO) was presented by Zhao et al. [19]. Initially, based on different fitness function, the initial clusters were selected for heterogeneous energy nodes. In GWO, fitness values from the nodes are treated and designed as initial weights. Simultaneously, according to the distance among wolves and their pray, the weight values are dynamically updated.

For efficient routing, Pattnaik et al. [20] proposed fuzzy method for clustering and elephant herding optimization (EHO) with greedy algorithm (GA) for routing. To decrease the energy utilization, separate sink nodes are considered. By utilizing the enhanced expectation–maximization algorithm, the numeral of nodes is designed into multiple clusters. For efficient data transmission, the hybrid EHO-GA routing protocol was utilized.

Swami et al. [21] suggested an enhanced approach to revise the position of solution in local leader phase. This method is very well adjusted for exploration and exploitation in most of the cases. The process of data collection may be enhanced with this approach. Mainly, it can solve the unconstrained optimization problems.

In wireless sensor network, finding an optimal route is the key issue to lessen the energy consumption. Bhasin et al. [22] presented the applicability of genetic algorithm in longest path problem. The algorithm fixes the issue of longest path problem. The outcome showed the better result as compared with several benchmarks.

BOA is most advanced function optimization algorithm proposed by Sharma et al. [23] which eliminates the weak exploitation ability of SOS. This algorithm is evaluated with twenty-five benchmarks based on its simplicity, efficiency and robustness and proved improved results.

To increase the lifetime of WSN by minimizing the energy consumption is now a demanding and challenging task. Many researchers have adopted various techniques to lessen the energy utilization. Kalaivaani et al. [24] presented the new approach by extracting the faulty information by using PFDIAES and PFDIF algorithm. The network performance is analysed in critical condition.

Energy draining issue in sensor nodes of WSN must be minimized. Now, it becomes paramount. Sensors inference during sensing, transmitting and receiving the data may seveourly affect the whole network. So, in order to resolve the issue, Naik et al. [25] suggested interference minimization technique by optimal positioning of sensors.

# 3 Proposed Methodology

In this paper proposed an emperor penguin optimized clustering for improved multilevel hierarchical routing in WSN. The whole network region is splitted by EPO algorithm into number of clusters and regions where every cluster has one CH. According to energy condition, the WSN nodes are combined which perform routing in hierarchical level. The SN exchanges different information such as distance and energy to perform clustering and routing process. The communicated messages are conveyed by IMHR for cluster association. In our work, multilevel hierarchical routing (MHR) is improved by optimizing MHR with social ski-driver (SSD) algorithm.

## 3.1 System Model

The WSN consists of N number of ordinary nodes and one BS which are randomly deployed in the monitoring areas. Identification (ID) number is given for every node which is advantageous for recognizing various nodes during the process of routing. Furthermore, we study WSN with the accompanying properties. Figure 1 shows the system model.

- In monitoring area, a large amount of SNs must be applied to guarantee the existence of redundant nodes.
- All the SNs are fixed because WSN is a static network.
- SNs are not location-aware which means based on transmission data packets the sink node will assess the whole network distribution.
- Every nodes have the ability to choose the CH.

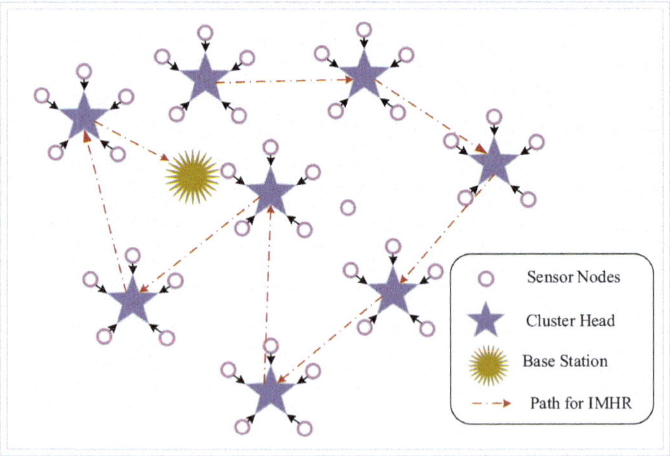


**Fig. 1** System model

- The process of clustering process is completely distributed, and every nodes have equal rights to take the decision independently.

## 3.2 Energy Model

A free space or multi-path fading channel models are utilized based on distance among the target and source nodes. The energy consumed by the SN radio to transmit an $l$-bit data over a distance $d$ is expressed as below

$$E_{Tx}(l,d) = \begin{cases} lE_{\text{elec}} + l\varepsilon_{\text{fs}}d^2 & \text{if } d < d_o \\ lE_{\text{elec}} + l\varepsilon_{\text{mp}}d^4 & \text{if } d \geq d_o \end{cases} \tag{1}$$

where $d$ is represented as distance, $E_{\text{elec}}$ is represented as energy depletion of the electronic circuit, $\varepsilon_{\text{mp}}$ is represented as energy consumed at free space, and $\varepsilon_{\text{fs}}$ is represented as multi-path fading channel. Moreover, the energy consumed in the $l$-bit generation of control message is expressed by

$$E_{\text{count}}(l,d) = lE_{\text{elec}} \tag{2}$$

$E_{\text{elec}}$ based on different factors like signal spreading, filtering, modulation and digital encoding.

## 3.3 Clustering by EPO

Clustering is used in the WSN to prolong the network lifetime. It includes group of SNs into cluster and selecting CHs for all clusters. Here, EPO technique is used for cluster forming and choosing the optimal CH.

The scientific name of EPO [26] is *Aptenodytes forsteri*, which is heaviest and tallest in all penguin species. The EPO huddling behaviour is decomposed into four stages. The first stage of generate and determine the EPO huddle boundary is determined as generate the boundary for clustering. After that, calculate the temperature is determined as energy calculation, determine the distance of EPO is calculated as distance among sensor nodes, and finally, relocate the effective mover is determined as optimal solution.

**Generate and Determine the Boundary for Clustering**. The SN is determined to set the boundary for forming the cluster. The Euclidean distance is determined to find the cluster boundary. The Euclidean distance is minimum than the boundary condition.

$$\Psi = \Delta\Phi \tag{3}$$

where $\Psi$ denotes the cluster boundary, and $\Phi$ denotes the Euclidean distance.

$$\Phi = \left\| x_i^j - c_j \right\|^2 \tag{4}$$

where $x$ denotes the number of cases, and $c$ represents the centroid for cluster.

$$\text{Fitness Function} = \text{CH} = \sum_{\max} E + \sum_{\min} D \tag{5}$$

**Energy Calculation around Boundary**. The energy for sensor node between the boundary locations is calculated. To mathematically model, we assume the energy $E' = 1$ when the power $R < 1$ and $E' = 0$ when the power $R > 1$. The energy is responsible for sensor nodes with various locations. The energy calculation around boundary is calculated as follows:

$$E' = \left( T - \frac{\text{Max}_{\text{iter}}}{x - \text{Max}_{\text{iter}}} \right) \tag{6}$$

$$E = \begin{cases} 0, & \text{if } P > 1 \\ 1, & \text{if } P < 1 \end{cases} \tag{7}$$

where maximum amount of iteration is denoted as $Max_{iter}$, current iteration is defined as $x$, time is denoted as $T$, and power is denoted as $P$.

**Distance between Sensor Nodes**. After the generation of boundary, the distance among the sensor nodes and best optimal solution is calculated. According to the current best optimal solution, the other search agent will update their position. The distance between the sensor nodes is as follows

$$\vec{D}_{\text{ep}} = \text{Abs}\left( S\left( \vec{A} \right) \cdot P(\vec{x}) - \vec{C} \cdot P_{\text{ep}}(x) \right) \tag{8}$$

where distance between the SNs is denoted as $\vec{D}_{\text{ep}}$, the current iteration is indicated as $x$, $\vec{A}$ and $\vec{C}$ are utilized to avoid the collision, the position of sensor node is denoted as $\vec{P}_{\text{ep}}$.

**Optimal Solution**. Based on SN distance and energy, the CH is selected. According to the best optimal solution the cluster heads are updated. This is responsible for selecting the CH for every cluster in a given space. The below equation is to update the CH for every cluster in a SN.

$$\vec{P}_{\text{ep}}(x + 1) = \vec{P}(x) - \vec{A} \cdot \vec{D}_{\text{ep}} \tag{9}$$

where updated CH is represented as $\vec{P}_{ep}(x+1)$. The updated equation is compared with the fitness Eq. (5), if it satisfied the fitness function, the CH is selected with optimal solution or else the condition will be continued.

| |
|---|
| **Step 1**: Initialize the sensor nodes $x$, where $x = 1, 2, 3, 4..., n$ |
| **Step 2**: Select the initial parameters such as $E', \vec{A}, \vec{C}, S$, &$\text{Max}_{\text{iter}}$ |
| **Step 3**: Generate and determine the boundary for clustering using Eqs. (3) and (4) |
| **Step 4**: Calculate the fitness function Eq. (5) of CH |
| **Step 5**: Calculate energy around boundary using Eq. (6) |
| **Step 6**: Calculate the distance among the sensor nodes using Eq. (8) |
| **Step 7**: Update the CH of each cluster using Eq. (9) |
| **Step 8**: Check whether the CH is more energy and less distance beyond the boundary SN |
| **Step 9**: Compute the updated fitness value and update the previously obtained CH optimum solution |
| **Step 10**: If the fitness function Eq. (5) is satisfied, then it will be stop or else it will be continued |
| **Step 11**: After stopping criteria, return the best optimal solution |

**Improved Multilevel Hierarchical Routing**. Multi-hierarchical-based routing is a feasible solution in WSNs for reducing the energy consumption because of the redundant data transmission reduction. The multilevel hierarchical aim is to ensure the connectivity between the SNs, extending the network lifetime and saving the each SN residual energy. Here, we used IMHR which is the combination of multilevel hierarchical routing optimized by SSD. Based on time, congestion, delay and network lifetime, the multilevel hierarchical routing selects the best path.

The packet which takes time to transmit or receive with respect to defined rate can be explained as below:

$$\text{Time } (T) = \frac{\text{Packet Size}}{D_{\text{rate}}} \tag{10}$$

where $D_{\text{rate}}$ is denoted as defined rate.

Transmitting a packet from source to the BS time is known as delay. The transmission delay of network formula is shown in equation.

$$\overline{D} = \frac{\sum_{i=1}^{N} D(i)}{N} = \frac{\sum_{i=1}^{N} [\text{RT}(i) - \text{ST}(i)]}{N} \tag{11}$$

where transmission time of $i$th packet is represented as $\text{ST}(i)$, reception time of $i$th packet is denoted as $\text{RT}(i)$, number of nodes is represented as $N$, and transmission delay of $i$th data packet is represented as $D(i)$.

The network congestion of any node is calculated by using the formula.

$$C = \frac{k}{n^2} \tag{12}$$

where the congestion factor is represented as $c$, and adjustment factor is represented as $n$ and $k$.

The complete definition for network lifetime is time to network partition, and it occurs if there is a cut-set in the network.

$$\text{Network Lifetime } (NL) = E - (\underline{U} + \sigma) \tag{13}$$

where $\underline{U} = \frac{\Sigma U_i}{N}$ and $\sigma^2 = \left(\frac{(U_i - \underline{U})^2}{N}\right)$.

Where initial energy at each node is defined as $E$, average used energy is represented as $U_i$, and number of nodes is defined as $N$.

SSD is used to optimize the multilevel hierarchical routing for finding the best path. SSD is one of the meta-heuristic algorithm [27] whose behaviour was inspired from various evolutionary algorithms. It mimics the path taken by the ski-drivers downhill.

$$\text{Fitness Function} = \text{PH} = \sum_{\min}[T + \overline{D} + C] \times \sum_{\max} \text{NL} \tag{14}$$

**Position of the node** $X_i \in R^n$: To calculate the best path at particular location, the position of the nodes is used. Where node dimension is denoted as $n$.

**Previous Best Path Position** $P_i$: By using the fitness function Eq. (14), the fitness value for all path position is calculated. The best path position is stored after the comparison of fitness function of current path solution.

**Mean Global Solution** $M_i$: The nodes move towards the global point that means the mean value of best three paths and the equation of global solution is as follows:

$$M_i^t = \frac{\text{PH}_\alpha + \text{PH}_\beta + \text{PH}_\gamma}{3} \tag{15}$$

where the best three paths are represented as $X_\alpha$, $X_\beta$ and $X_\gamma$.

Again, the fitness value is calculated for the current best three paths. From that, we have choosen the best path according to the fitness function equation. If the fitness function equation satisfied, then choose the best path or else the process will continue until it gets the best optimal path.

## 4 Results and Discussion

The whole work is implemented in MATLAB 2018a with system configuration of 4 GB RAM under Windows XP operating system. The performance of throughput, average hops, packet delivery ratio and delay ratio is compared with existing energy-efficient cluster routing (EECR) and EECMHR techniques [16].

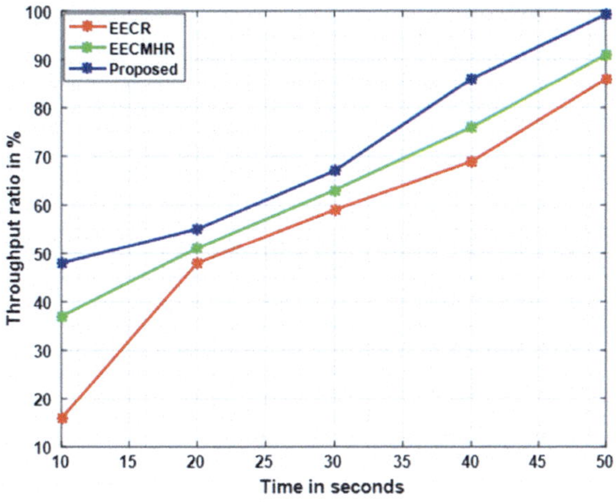


**Fig. 2** Performance of throughput

**Table 1** Comparison of throughput

| Time (s) | Proposed | EECMHR (%) | EECR (%) |
|---|---|---|---|
| 10 | 48 | 37 | 16 |
| 20 | 55 | 51 | 48 |
| 30 | 67 | 63 | 59 |
| 40 | 86 | 76 | 69 |
| 50 | 99.3 | 91 | 86 |

Figure 2 displayed the throughput performance with existing and proposed methods. The throughput calculation is done by varying the time. At 50 s, the throughput of proposed method is 99.3%, the existing method of EECMHR is 91%, and EECR is 86%. Our proposed method attains high throughput when compared than existing methods. Table 1 explains the throughput of both proposed and existing method comparison.

Figure 3 displays the average number of hops performance with proposed and existing techniques. The average number of hops performance is taken by varying the amount of nodes. At 50th node, the average hops performance of proposed method is 3.5, the existing method of EEC-MHR is 4.1, and EECR is 5.0. Our proposed method attains less average number of hops when compared than existing methods. The average hops of both proposed and existing method comparison are shown in Table 2.

The proposed and existing method of PDR performance are shown in Fig. 4. The PDR performance is taken by varying the time. At 50 s, the PDR of proposed method is 97%, the existing method of EEC-MHR is 92%, and EECR is 89%. Our proposed method attains high PDR than existing methods. The packet delivery ratio of both proposed and existing method comparison is displayed in Table 3.

Figure 5 shows the performance of delay by comparing existing methods. The delay calculation is done by varying the rate of sending packets. At 64 packet/s, the

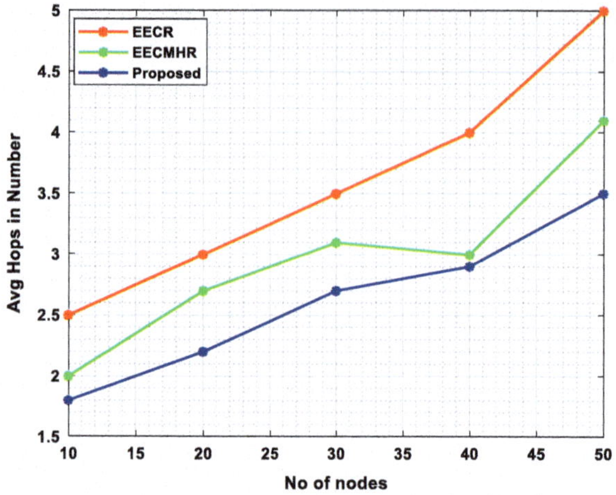


**Fig. 3** Performance of average hops

**Table 2** Comparison of average hops

| Number of nodes | Proposed | EECMHR | EECR |
|---|---|---|---|
| 10 | 1.8 | 2.0 | 2.5 |
| 20 | 2.2 | 2.7 | 3.0 |
| 30 | 2.7 | 3.1 | 3.5 |
| 40 | 2.9 | 3.0 | 4.0 |
| 50 | 3.5 | 4.1 | 5.0 |

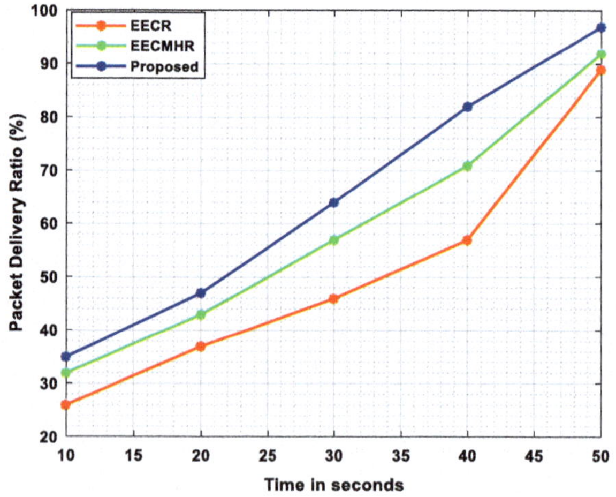


**Fig. 4** Performance of PDR

**Table 3** Comparison of PDR

| Time (s) | Proposed (%) | EECMHR (%) | EECR (%) |
|---|---|---|---|
| 10 | 35 | 32 | 26 |
| 20 | 47 | 43 | 37 |
| 30 | 64 | 57 | 46 |
| 40 | 82 | 71 | 57 |
| 50 | 97 | 92 | 89 |

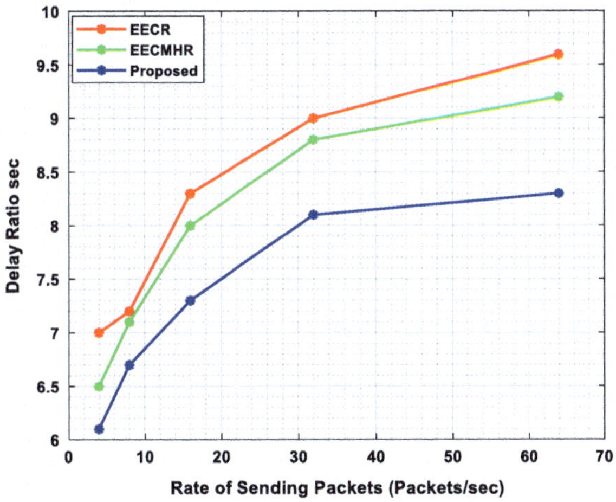


**Fig. 5** Performance of delay

delay of proposed method is 8.3 s, the existing method of EEC-MHR is 9.2 s, and EECR is 9.6 s. Our proposed method attains less delay when compared than existing methods. The delay of both proposed and existing method comparison is displayed in Table 4.

**Table 4** Comparison of delay

| Rate of sending packets (Packet/s) | Proposed (s) | EECMHR (s) | EECR (s) |
|---|---|---|---|
| 4 | 6.1 | 6.5 | 7 |
| 8 | 6.7 | 7.1 | 7.2 |
| 16 | 7.3 | 8 | 8.3 |
| 32 | 8.1 | 8.8 | 9 |
| 64 | 8.3 | 9.2 | 9.6 |

## 5 Conclusion

In this paper, an energy-efficient-based EPO clustering for IMHR in WSNs was proposed. The SN exchanges different information such as distance and energy to perform clustering and routing process. The communicated messages are conveyed by IMHR for cluster association. Here, MHR is improved by optimizing MHR with SSD algorithm. MATLAB platform is used to implement the work and take the graphical analysis by means of throughput, delay, PDR, etc., by comparing with EEC-MHR and EECR. The performance of proposed throughput is 99.3%, packet delivery ratio is 97%, average hop is 3.5, and the delay is 6.1 s.

## References

1. Muthukumaran, K., Chitraand, K., Selvakumar, C.: An energy efficient clustering scheme using multilevel routing for wireless sensor network. Comput. Electr. Eng. **69**, 642–652 (2018)
2. Kareem, M.M., Ismail, M., Altahrawi, M.A., Arsad, N., Mansorand, M.F., Ali, A.H.: Grid based clustering technique in wireless sensor network using hierarchical routing protocol. In: 4th International Symposium on Telecommunication Technologies (ISTT), pp. 1–5. IEEE (2018)
3. Dener, M.: A new energy efficient hierarchical routing protocol for wireless sensor networks. Wireless Pers. Commun. **101**(1), 269–286 (2018)
4. Bhatia, T., Kansal, S., Goeland, S., Verma, A.K.: A genetic algorithm based distance-aware routing protocol for wireless sensor networks. Comput. Electr. Eng. **56**, 441–455 (2016)
5. Popovićand, G., Dukanović, G.: Cluster formation techniques in hierarchical routing protocols for Wireless Sensor Networks. Jita-J. Inf. Technol. Appl. **11**(1) (2016)
6. Fei, X., Wang, Y., Liuand, A., Cao, N.: Research on low power hierarchical routing protocol in wireless sensor networks. In: International Conference on Computational Science and Engineering (CSE) and International Conference on Embedded and Ubiquitous Computing (EUC)2, pp. 376–378. IEEE (2017)
7. Khabiriand, M., Ghaffari, A.: Energy-aware clustering-based routing in wireless sensor networks using cuckoo optimization algorithm. Wireless Pers. Commun. **98**(3), 2473–2495 (2018)
8. Huang, J., Hong, Y., Zhao, Z., Yuan, Y.: An energy-efficient multi-hop routing protocol based on grid clustering for wireless sensor networks. Clust. Comput. **20**(4), 3071–3083 (2017)
9. Hassan, A.A.H., Shah, W.M., Iskandarand, M.F., Mohammed, A.A.J.: Clustering methods for cluster-based routing protocols in wireless sensor networks: comparative study. Int. J. Appl. Eng. Res. **12**(21), 11350–11360 (2017)
10. Wang, T., Zhang, G., Yangand, X., Vajdi, A.: Genetic algorithm for energy-efficient clustering and routing in wireless sensor networks. J. Syst. Softw. **146**, 196–214 (2018)
11. Zhou, Y., Wangand, N., Xiang, W.: Clustering hierarchy protocol in wireless sensor networks using an improved PSO algorithm. IEEE Access **5**, 2241–2253 (2016)
12. Pan, J.S., Kong, L.P., Sung, T.W., Chuand, S.C., Roddick, J.F.: Hierarchical routing strategy for wireless sensor network. J. Inf. Hiding Multimed. Signal Process. **9**(1), 256–264 (2018)
13. Batraand, P.K., Kant, K.: LEACH-MAC: a new cluster head selection algorithm for wireless sensor networks. Wireless Netw. **22**(1), 49–60 (2016)
14. Wei, Z., Liu, F., Ding, X., Feng, L., Lyu, Z., Shiand, L., Ji, J.: A clustering hierarchical routing algorithm for wireless rechargeable sensor networks. IEEE Access **7**, 81859–81874 (2018)
15. Oudani, S.K.H., Kabrane, M., Bandaoud, K., Elaskri, M., Karimi, K., Elboustyand, H., Elmaimouni, L.: Energy efficient in wireless sensor networks using cluster-based approach routing. Int. J. Sens. Sens. Netw. **5**(5–1), 6–12 (2017)

16. Kumaresanand, K., Kalyani, S.N.: Energy efficient cluster based multilevel hierarchical routing for multi-hop wireless sensor network. J. Ambient Intel. Human. Comput. 1–10 (2020)
17. Liu, Y., Wu, Q., Zhao, T., Tie, Y., Baiand, F., Jin, M.: An improved energy-efficient routing protocol for wireless sensor networks. Sensors **19**(20), 4579 (2019)
18. Yarinezhadand, R., Sarabi, A.: MLCA: a multi-level clustering algorithm for routing in wireless sensor networks. J. Commun. Eng. **8**(2), 249–265 (2019)
19. Zhao, X., Ren, S., Quanand, H., Gao, Q.: Routing protocol for heterogeneous wireless sensor networks based on a modified grey wolf optimizer. Sensors **20**(3), 820 (2020)
20. Pattnaik, S., Sahu, P.K.: Assimilation of fuzzy clustering approach and EHO-Greedy algorithm for efficient routing in WSN. Int. J. Commun. Syst. **33**(8), 4354 (2020)
21. Swami, V., Kumar, S., Jain, S. : An Improved Spider Monkey Optimization Algorithm. In: Soft Computing: Theories and Applications, pp. 73–81.Springer (2018).
22. Bhasin, H., Gupta, N.: Critical path problem for scheduling using genetic algorithm. In: Soft Computing: Theories and Applications, pp. 15–24. Springer (2017)
23. Sharma, S., Saha, A.K., Nama, S.: An enhanced butterfly optimization algorithm for function optimization. In: Soft Computing: Theories and Applications, pp. 593–603. Springer (2020)
24. Kalaivaani, P.T., Krishnamoorthy, R.: Critical an improved energy-efficient faulty information extraction scheme using PIFDIAES and PFDIF algorithm. In: Soft Computing: Theories and Applications, pp. 883–893. Springer (2020)
25. Naik, C., Shetty, D.P.: Intelligent inference minimization algorithm for optimal placement of sensors using BBO. In: Soft Computing: Theories and Applications, pp. 955–969. Springer (2020)
26. Dhimanand, G., Kumar, V.: Emperor penguin optimizer: a bio-inspired algorithm for engineering problems. Knowl. Based Syst. **159**, 20–50 (2018)
27. Tharwatand, A., Gabel, T.: Parameters optimization of support vector machines for imbalanced data using social ski driver algorithm. Neural Comput. Appl. 1–14 (2019)

# A Collaborative Filtering-Based Recommendation System for Preliminary Detection of COVID-19

**Gourav Jain, Tripti Mahara, and S. C. Sharma**

**Abstract** COVID-19, an infectious virus caused by the severe acute respiratory syndrome coronavirus 2 (SARS-CoV-2), made its worldwide appearance in the last few months. This virus does not show consistent symptoms in all and spreads rapidly through contact with an infected person, so it is not easy to detect it without testing. Due to the limited resources and increased patient count, a technique needs to be deployed to help the common man in the decision-making process. The use of computer technology in the medical field saves both time and money, and in the present situation, it can be one of the possible ways for preliminary detection of this disease. In this paper, a CF-based recommendation system is presented to detect the possibility of COVID-19 in a suspected person at an early stage. For this, the symptoms of a suspected person is compared with symptoms of COVID-19 infected patients and over all similarity is computed between the symptoms. According to the calcuated similarity, a suspected person can be classified into either of the three risk categories: high, medium, and low. However, this is not a confirmatory test but only a mechanism to alert an individual. A sample data set has been created with the help of healthcare professionals and secondary data sources to depict the working of the model.

**Keywords** Recommender systems · COVID-19 · Coronavirus · Collaborative filtering · Similarity measure · Prediction

## 1 Introduction

At the end of 2019, a novel virus originated in Wuhan, China, called "2019-novel coronavirus (2019-nCoV)" or "severe acute respiratory syndrome coronavirus 2 (SARS-CoV-2)". Since then, it has spread worldwide and become a major threat

G. Jain (✉) · S. C. Sharma
Indian Institute of Technology Roorkee, Roorkee, India

T. Mahara
Christ University Bengaluru, Bengaluru, India

 
T. K. Sharma et al. (eds.), *Soft Computing: Theories and Applications*,
Advances in Intelligent Systems and Computing 1381,
https://doi.org/10.1007/978-981-16-1696-9_3

to global economies. The disease caused by this virus is called COVID-19. It is an infectious air borne disease transmitted from one person to another through droplets produced during sneezing or coughing. It was declared a "global pandemic" by the World Health Organization (WHO) on 11 March 2020. Till 25 November 2020, a total of 59,816,510 positive cases have been registered, with 1,410,378 deaths, and still, the effect of pandemic continues [1]. India recorded the first COVID-19 positive case on 30 January 2020. Despite India's dense population, the spread of virus was under control due to lockdown implementation in the initial phase. But, due to the relaxation of lockdown in India [2] number of cases is increasing rapidly. Even though the vaccine is available now, but as not all have been vaccinated and mutated virus becoming more dangerous, it is necessary to detect COVID-19 at an early stage so that the spread can be curbed. One of the effective ways for the early detection of this virus is to perform mass testing but, due to the vast population and limited medical facility, it is not a feasible solution. Another way to control the spread is the usage of technology so that an individual can be alerted on time according to their symptoms and appropriate suggestions can be provided. The Recommender system (RS) is one such application of data mining that can be a cost-effective solution for early detection of this disease. The popularity of RS is growing and is being used in various domains like music, movies, news, jokes, article recommendations, health care, etc. [10–13]. In the healthcare domain, RS analyses a large amount of medical data and collects valuable data to identify diseases so that appropriate treatment may be provided. In the healthcare domain, RS analysis large amounts of medical data and extracts valuable data from it so that disease can be identified quickly and appropriate treatment provided accordingly. An RS also helps in selecting the best physician [14] and suitable therapy in a particular area [15]. This system is wise in estimating a patient's health status by analyszing his their lifestyle, health history, social activities, etc. Due to its ability to make appropriate decisions in the healthcare sector, the RS has become an imperative tool in this domain.

Content-based (CB) [3] and collaborative filtering (CF) [4, 5] are the two much known methods used to build a RS [6]. Content-based filtering is not very effective as, the recommendation is based on the profile of users/items that requires more information. On the other hand, collaborative filtering [7] is a successful, mature, and widely used neighbourhood-based technique that considers the user's ratings on items without exploring the information about users and items. It predicts the active users' rating based on the similarity between users/items. Therefore, selecting an appropriate similarity measure [8] is vital to calculate the similarity efficiently. CF's [9] key benefit is that it does not require much information about users/items to create profiles and is more accurate than content-based. Furthermore, it can find undetected patterns easily, which is difficult in other techniques.

Hence this paper aims to develop a collaborative filtering-based RS system that can aid the user to detect COVID-19 disease based on the initial symptoms and recommend a suitable course of action. The CF-based RS considers each patient's symptoms, whose severity may differ from others, but overall the symptoms remain the same. An appropriate similarity measure is used in CF to find patients with similar symptoms and recommend specific treatments to them. For this, an improvised Gower's coefficient that computes the similarity among users of the sample data set taken. The reason to use Gower's is that it can work effectively with any type of data. Initially, it calculates the similarity between users of the sample data set and corona patient (CP) and then classifies the users into different risk categories according to the calculated similarity value. After predicting the risk category of a patient, specific recommendations have been provided to the user.

## 2 Related Work

In [16], to tackle many aspects of the COVID-19 crisis at different scales, including molecular, clinical, and societal applications, Bullock et al. presented an overview of recent studies using artificial intelligence. Rao et al. [17] proposed a mobile phone-based web survey using machine learning algorithms to improve the possible case detection of COVID-19. Maghdid et al. [18] developed an artificial intelligence (AI) powered framework that reads the smartphone sensors' signal measurements to predict pneumonia's intensity and predict the disease's outcome. In [19], MRAN et al. proposed a preliminary diagnosis for COVID-19 from cough samples using artificial intelligence. A methodology to identify cough types is discussed in [20], where correlation and wrapper-based selection algorithms are used to reduce the sample's dimensionality. The smartphone-based human fatigue detection technique is proposed by Karvekar et al. [21]. In this, the author detects human fatigue in different environments via human gait analysis. The importance of artificial intelligence and machine learning to fight COVID-19 is reported by Allimadi [22]. In [23], Rahmatizadeh proposed the AI-based model to improve the critical care of COVID-19 patients. Arti MK et al. [24] examined the impact of social distancing and lockdown in coronavirus dissemination to predict the new patients using a tree-based model. Salman et al. [25] constructed a deep learning model to detect COVID-19 pneumonia after analyzing X-ray images. Anuradha et al. [26] employed short-term memory and curve-fitting strategies to estimate COVID-19 spread. They also explore the impact of preventive measures by the Government of India. In [27], an RS is used to enhance personal health records (PHR) solutions. Abhaya et al. [28] suggested the smart HRS using restricted Boltzmann machine (RBM)-convolutional neural network (CNN) deep learning system. In this, how big data analytics can introduce an efficient healthcare recommender engine and demonstrate an opportunity for the healthcare industry to shift from a conventional scenario to a more customized model in a telehealth environment is discussed.

## 3 Methodology

The CF approach is one of the most popular methods to design a recommendation system. In the health domain, the primary purpose of a RS is to either detect the disease based on the symptoms or provide valuable suggestions to users according to their symptoms [29]. In this research, a system architecture for is proposed the preliminary diagnosis of COVID-19 in an individual based on the symptoms.

In general, four symptoms are mainly found in individuals infected with COVID-19: temperature, cough, breathing problem, and fatigue. A patient may exhibit these symptoms at different intensity levels. For example, some may have more fatigue, while in others, it may be moderate. Depending on the symptoms' intensity, a suspected person is classified into symptomatic, pre-symptomatic, atypical symptomatic, or asymptomatic categories. An individual who exhibits all the four symptoms is called symptomatic. On the other hand, if a suspected patient (SP) feels better today and exhibits symptoms two days later, he/she is classified as pre-symptomatic. A person who has mild symptoms, but after some time, these symptoms disappear is called atypical symptomatic. Those who do not show any symptoms are called asymptomatic [30]. The proposed recommender system considers the four symptoms at varying levels and predict outcomes in terms of the risk of having COVID-19 along with the suggestions for the action to be taken. Figure 1 depicts the model and each step of this model is discussed ahead in detail.

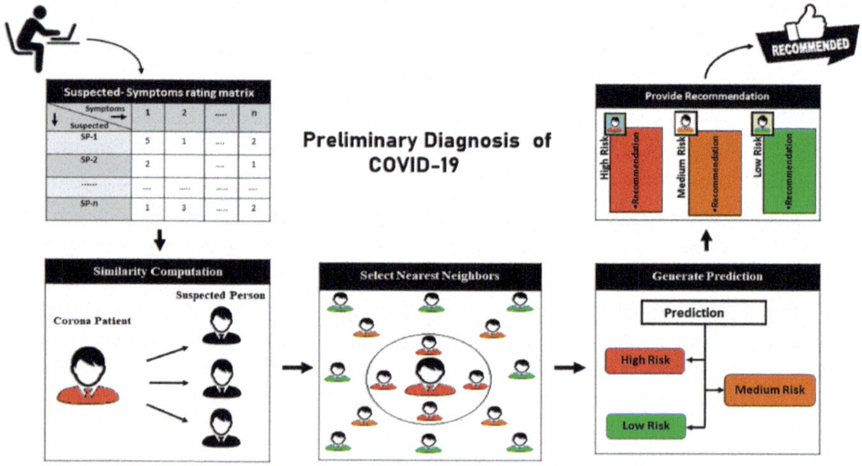


**Fig. 1** Proposed model for preliminary detection of COVID-19

## 3.1 Creating User Item Rating Matrix

In traditional CF-based RS, information is gathered from the browsing or purchasing activity of each user. This information is converted into a matrix form, where every row is a user-evaluated rating vector, and each column has the ratings provided by a user on a specific item. The user item rating matrix can be expressed as $R = [r_{ij}]\ m * n$ where $m$ and $n$ are expressed as the number of users and items, respectively. In the matrix, $U = \{u_1, u_2, \ldots, u_m\}$ is set of users and item $I = \{I_1, I_2, \ldots, I_n\}$ is a set of items, and $r_{ij}$ represents the rating given by user $i$ on item $j$.

A sample of user item rating matrix is depicted in Table 1 consisting of ten users (rows) and four items (columns). Each row represents the symptoms of an individual. Here, an individual is a person whose symptoms are entered into the system to detect whether he/she might be infected with COVID-19. For this, four most common symptoms are considered after consultation with doctors.

These symptoms include temperature, cough, fatigue, and breathing problem, having various levels, as depicted in Table 2. This has been created after taking

**Table 1** Suspected person (SP)—symptoms rating matrix

| | Symptom 1 (temperature) | Symptom 2 (cough) | Symptom 3 (fatigue) | Symptom 4 (breathing problem) |
|---|---|---|---|---|
| SP-1 | 5 | 4 | 3 | 1 |
| SP-2 | 1 | 2 | 1 | 0 |
| SP-3 | 2 | 2 | 2 | 1 |
| SP-4 | 5 | 3 | 2 | 0 |
| SP-5 | 4 | 5 | 1 | 1 |
| SP-6 | 1 | 1 | 1 | 0 |
| SP-7 | 1 | 3 | 3 | 0 |
| SP-8 | 2 | 4 | 2 | 1 |
| SP-9 | 5 | 1 | 3 | 1 |
| SP-10 | 2 | 5 | 2 | 0 |

**Table 2** Symptoms categorization

| Temperature ($T$) °F | | | Cough (C) | | Fatigue (F) | | Breathing problem (B) | |
|---|---|---|---|---|---|---|---|---|
| $99 > T$ | No fever | 1 | Whooping cough | 1 | Low | 1 | No | 0 |
| $99 \leq T < 100$ | Mild | 2 | Chest cough | 2 | Medium | 2 | Yes | 1 |
| $100 \leq T < 101$ | Moderate | 3 | Post viral cough | 3 | High | 3 | | |
| $100 \leq T < 102$ | High | 4 | Bronchitis | 4 | | | | |
| $102 \leq T$ | Very severe | 5 | Dry cough | 5 | | | | |

relevant input from the doctors. This categorization will lead to better accuracy of recommendation. Other than these, patients may suffer from aches and pains, inflammation of the nose, runny nose, sore throat, or diarrhoea, but these symptoms are mild, occur slowly, and are not present in all patients of COVID-19. Hence, in the preliminary test of COVID-19 detection, they are ignored.

Several studies [31] have shown that the symptoms of COVID-19 are very similar to the influenza virus as both are transmitted by contact, droplets, and fomites. Despite this, the transmission speed of these two viruses is also important. Influenza has a shorter median incubation period (time from diagnosis to symptom appearance) and a shorter serial interval (time between consecutive cases) than the COVID-19 virus. The serial interval for the COVID-19 virus is estimated at 5–6 days, whereas it is three days for the influenza virus.

The user rating matrix is populated when an individual enters his symptoms according to the classification mentioned in Table 2. For instance, if the first suspected individual (SP-1) has a very severe fever, bronchitis cough, high fatigue, and breathing problem, the user will enter (5, 4, 3, 1) in the sample matrix depicted in Table 1.

## *3.2 Similarity Computation*

Similarity computation between the users or items is one of the most important steps to determine the system's accuracy. In our proposed work, the similarity between the suspected person (given in Table 1) and the corona patient (symptoms mentioned in Table 3) is explored. Several studies have shown that a symptomatic person infected with COVID-19 has a very severe temperature (T), dry cough (cough), shortness of breath, and high fatigue symptoms. Table 3 depicts the symptoms of a confirmed symptomatic corona patient (CP). A similarity measure is used to calculate the similarity. It compares the values of SP and CP symptoms and displays the result in between 0 and 1. The similarity value zero indicates that none of the symptoms match, while a value one means that all symptoms match. Thus, the similarity value helps us in predicting whether an individual might be infected with COVID-19 or not.

The system performance primarily depends on the similarity computation; henceforth, selecting an appropriate measure is very important. There are many traditional and widely used similarity measures like Pearson correlation coefficient (PCC), cosine correlation coefficient (COS), Jaccard, and MSD, but no one can work with

**Table 3** COVID-19 patient symptoms

| | Symptom 1 (temperature) | Symptom 2 (cough) | Symptom 3 (fatigue) | Symptom 4 (breathing problem) |
|---|---|---|---|---|
| Corona patient (CP) | 5 | 5 | 3 | 1 |

both qualitative and quantitative data. Considering this, we proposed the improvised Gower (iG) similarity measure based on the Gower's coefficient [32]. Also it has an additional benefit that it works well even when some entries in the data matrix are absent. In 1971, J. C. Gower proposed the basic Gower's coefficient. In 1999, Podani [33] presented the extended version of Gower's coefficient, where the similarity between ordinal characters has been determined. Fontecha et al. [34] introduce the novel mobile service infrastructure to improve the accuracy of frailty diagnosis in the elder population using Gower's similarity coefficient. The similarity using Gower's coefficient is calculated as follows:

$$\text{sim}(u_a, u_b)^{\text{Gower's}} = \frac{\sum_{k=1}^{n} D_{u_a u_b k} \cdot W_{u_a u_b k}}{\sum_{k=1}^{n} W_{u_a u_b k}} \tag{1}$$

where $D_{u_a u_b k}$ calculates the rating difference between two user $u_a$ and $u_b$ for the $k$th item, and $W_{u_a u_b k}$ is the binary weight. Its value will be 1 when the user's rating on an item is known; otherwise, it is set to zero. For the different categories of data variables, the values of $D_{u_a u_b k}$ and $W_{u_a u_b k}$ are calculated differently. In this paper, we focussed on the ordinal variable. The value of $D_{u_a u_b k}$ is computed by Eq. 2.

$$D_{u_a u_b k} = 1 - \frac{\left| r_{u_{a,k}} - r_{u_{b,k}} \right|}{R_k} \tag{2}$$

where $r_{u_{a,k}}$ and $r_{u_{b,k}}$ are the ratings given by user $u_a$ and $u_b$, respectively, and $R_k$ display the difference between maximum and minimum rating of each item.

The Gower's coefficient is used to compute the similarity between the suspected person (SP's of Table 1) and corona patient (CP of Table 3). Each SP's symptom is compared with CP's symptoms and results are displayed in the form of similarity value. In Eqs. 1 and 2, $\text{u}_\text{a}$ means a suspected user, $\text{u}_\text{b}$ means a corona patient, and $k$ represents the symptoms. The Gower's coefficient computes the value of $D_{u_a u_b k}$ using Eq. 2 for each $k$ and places the result into Eq. 1 to compute the effective similarity.

After calculating similarity, we notice that Gower's coefficient computes false similarity in some instances. For example, it computes zero similarity between SP-6 and CP, but from Table 1, it is evident that some similarities should exist between them as SP-6 depicts some symptoms. It is the case when the values of $\left| r_{u_{a,k}} - r_{u_{b,k}} \right|$ and $R_k$ become equal. To overcome this, we proposed the improvised Gower's (iG) coefficient, where the correction factor is introduced while computing similarity. The similarity using improvised Gower's coefficient is calculated as follows:

$$\text{sim}(u_a, u_b)^{\text{improvised - Gower's}} = \frac{\sum_{k=1}^{n} D_{u_a u_b k}^{\text{improvised}} \cdot W_{u_a u_b k}}{\sum_{k=1}^{n} W_{u_a u_b k}} \tag{3}$$

where

$$D^{\text{improvised}}_{u_a u_b k} = 1 - \left\{ \frac{\left| r_{u_{a,k}} - r_{u_{b,k}} \right|^{\text{improvised}}}{R_k} \right\} \tag{4}$$

In Eq. 4, If

$$\left| r_{u_{a,k}} - r_{u_{b,k}} \right|^{\text{improvised}} = R_k$$

then,

$$\left| r_{u_{a,k}} - r_{u_{b,k}} \right|^{\text{improvised}} = \left| r_{u_{a,k}} - r_{u_{b,k}} \right| - \text{random.uniform}(0, 0.5),$$

Here, uniform () is a method specified in a random library in Python. This will improve the recommendation accuracy. Table 8 depicts the similarity between a suspected person (Each SP of Table 1) and a corona patient (CP of Table 3) calculated using improvised Gower's coefficient based on the symptoms mentioned in Table 2.

## 3.3 Selecting Nearest Neighbours

The result of this step entirely depends on the previous step. After calculating similarity, the neighbours for each SP are found out. This step aims to identify similar users so that we can classify them into a group and provide appropriate treatment based on the group's belongingness.

## 3.4 Generating Prediction

This section categorizes a SP into a particular risk level as depicted in Table 4 based on the similarity value obtained in the previous step. A study noted that most patients infected with COVID-19 have a very severe temperature (T), dry cough (cough), shortness of breath, and high fatigue [35, 36]. Taking this as the benchmark for confirmation of COVID-19, we take the respective values from Table 2. It is the case when all the symptoms are prominent and visible.

There are suspected patients who do not show severity for every symptom but still can be classified as high-risk patients and requiring immediate attention. For

**Table 4** Classification of risk level

| Similarity | Severity |
|---|---|
| Similarity $\geq$ 0.75 | High risk |
| 0.5 $\leq$ Similarity $<$ 0.75 | Medium risk |
| Similarity $<$ 0.5 | Low risk |

**Table 5** Sample CP symptoms

| | Symptom 1 (temperature) | Symptom 2 (cough) | Symptom 3 (fatigue) | Symptom 4 (breathing problem) |
|---|---|---|---|---|
| CP1 | 2 | 5 | 2 | 1 |
| CP 2 | 3 | 4 | 3 | 1 |
| CP-3 | 2 | 4 | 3 | 1 |

instance, Table 5 shows the sample data of corona patients. CP1 has a mild fever, dry cough, mild fatigue, and has a breathing problem. Considering all the possible combinations of CP symptoms, minimum values are used to determine patients' threshold to be classified in the Risk 1 category. Expert knowledge after rigorous iterations is needed to find this threshold value. Here, (2, 4, 3, 1) is the minimum value arrived after discussion with doctors. Taking this, the 0.75 similarity value is calculated. Similarly, a threshold is set for each risk category, as shown in Table 4. After generating the prediction, we provide a specific set of recommendations in the subsequent section depending on the risk.

## *3.5 Providing Recommendations*

In the final phase, a set of recommendations is provided to the users based on the risk category they fall.

**High-Risk Category**: This group will constitute the suspected patient who have high similarity with symptoms of CP and is depicted with red colour in Fig. 1. The patients placed in this group required immediate action in terms of medical attention. The following recommendation is provided to users of this group:

- Immediate visit the nearest test centre for further investigations.

**Medium-Risk Category**: One of this paper's objectives is to focus on the suspected patients who fall in the medium-risk category, and they are shown by orange colour in Fig. 1. The individuals in this group have neither very high nor very low similarity with CP. It indicates that a partial match exists between SP and CP, which is not decisive enough to put them in a high-risk category. SP symptoms are further investigated in this category with a few more personal questions as shown in Table 6 to decide the recommendations. The data is utilized to provide personalized recommendations.

The answers obtained for the questions given in Table 6 will help to define the recommendations. After analyzing a published data set [37], it is evident that males are more affected by COVID-19 than women in India. Also, analysis reveals that people in the age bracket of 20–40 and 40–60 are more susceptible to COVID-19. The graphical representation of both the outcomes is shown in Fig. 2. Studies have shown that a person suffering from underlying chronic medical problems like

**Table 6** Questionnaire

| Questionnaire | Answers | | | | |
|---|---|---|---|---|---|
| What is your gender? | ☐ Male | | ☐ Female | | |
| Which age group you belong to? | ☐ 0–20 | ☐ 20–40 | ☐ 40–60 | ☐ 60–80 | ☐ 80+ |
| Do you have any underlying chronic disease? | ☐ Diabetes | | ☐ Hypertension | | ☐ Obesity |
| | ☐ Lung disease | | ☐ Heart disease | | ☐ Others |
| Do you have any travel history for the last 30 days? (If Yes: Mention that Place) | ☐ No | | ☐ Yes<br>Place: | | |

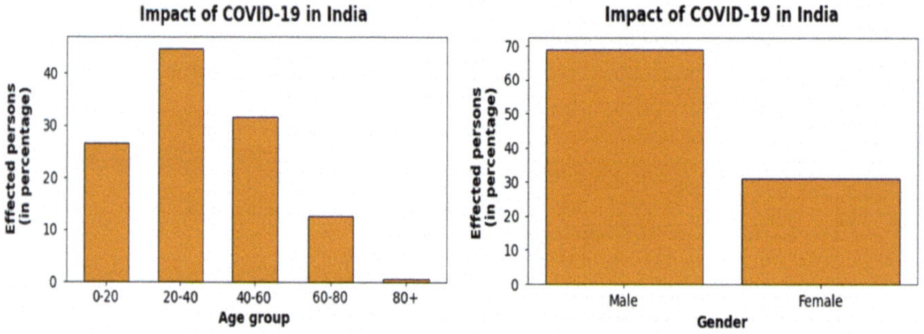


**Fig. 2** Impact of COVID-19 on persons of different age group and genders in India

hypertension, diabetes, and heart problems has more chances of getting affected by the disease [38], hence underlying chronic ailment is also considered for providing personalized recommendation to SP.

Analyzing the response of the questions provided by SP and matching it against Table 7 attributes, a decision is taken about the risk category of the SP. All possible combinations must be considered to categorize SP into a particular risk category.

In general, for a medium-risk category patient, the following recommendations are provided.

**Table 7** Risk categories on basis of primary data analysis

| | Category 1 | Category 2 | Category 3 | Category 4 |
|---|---|---|---|---|
| Gender | Male/Female | Male/Female | Male | Female/Male |
| Age group | 20–60 | All ages | 60–80 | All ages |
| Chronic disease | Yes | No | Yes | No |
| Travel history | Yes | No | No | Yes |
| **Risk category** | **High risk** | **Low risk** | **Medium risk** | **High risk** |

**Table 8** Risk prediction

| Suspected patient (SP) | Similarity between SP and CP | Risk category |
|---|---|---|
| SP-1 | 0.938 | High |
| SP-2 | 0.208 | Low |
| SP-3 | 0.500 | Medium |
| SP-4 | 0.667 | Medium |
| SP-5 | 0.750 | High |
| SP-6 | 0.167 | Low |
| SP-7 | 0.542 | Medium |
| SP-8 | 0.625 | Medium |
| SP-9 | 0.812 | High |
| SP-10 | 0.583 | Medium |

- Enter your details the next day and check for the recommendations.
- If symptoms remain consistent for 3 days, go to the doctor or test center.
- Isolate yourself at home till symptoms start reducing or you visit the doctor.

**Low-Risk Category**

In Fig. 1, we represent such users by green colour. The symptoms of such individuals are very mild to negligible. The following recommendation is provided:

- If there are any changes in the symptoms, please recheck your symptoms and see the recommendations.
- If the symptoms are persistent for more than 3 days, go to the doctor.

## 4 Results and Discussion

The primary focus of the work is to assist an individual by providing recommendations based on their COVID-19 symptoms. So far, no data set that contains information about users and their symptoms is available in the public domain; hence, a sample dummy data set as given in Table 1 was created. The data set is populated when an individual enters his/her symptoms as per the specifics mentioned in Table 2. Table 3 shows the symptoms of the confirmed COVID-19 patients. The SPs symptoms presented in Table 1 are compared with symptoms of CP in Table 3, and similarity is calculated using improvised Gower's coefficient (iG). The similarity results, along with their risk prediction, are presented in Table 8.

From Table 8, it is evident that symptoms of SP-1 and SP-9 with CP are very similar, resulting in a high similarity value. These results are an indication that the disease might infect them. Hence, they are recommended to proceed to the nearest test center for further diagnosis immediately. Similarly, we classify the other users according to Table 4 and provide relevant recommendations. The primary benefit of this work is that the common man who is not aware much about the disease's technicalities and has apprehension to go to a doctor can get a particular direction to proceed based on preliminary symptoms.

## 5 Conclusion and Future Work

A recommender system in the health domain can predicts the chance of an individual being contracted by the disease based on the severity of the symptoms. In this research, prediction is done for COVID-19 disease. As of now, no data is available in the public domain about the patient and their symptoms; thereby, a sample data set has been taken. To futher validate the conceptual model real-time data set is needed. The future work encompasses creating an application so that a common person can use it and take decisions accordingly. It becomes more relevant when patients with mild symptoms are self-quarantined at home. This work can be extended further to capture the suspected patients' geographical position for further action.

## References

1. "Coronavirus disease 2019." [Online]. Available: https://www.who.int/emergencies/diseases/novel-coronavirus-2019. Accessed: 25-Nov-2020
2. "India: WHO Coronavirus Disease (COVID-19) Dashboard." [Online]. Available: https://covid19.who.int/region/searo/country/in. Accessed: 25-Nov-2020
3. Wang, D., Liang, Y., Xu, D., Feng, X., Guan, R.: A content-based recommender system for computer science publications. Knowl. Based Syst. **157**, 1–9 (2018)
4. Jain, G., Mishra, N., Sharma, S.: CRLRM: category based recommendation using linear regression model. In: Proceedings of—2013 3rd International Conference on Advance Computing on Communication. ICACC 2013, pp. 17–20 (2013)
5. Jain, G., Mahara, T., Tripathi, K.N.: A survey of similarity measures for collaborative filtering-based recommender system. In: Soft Computing: Theories and Applications, pp. 343–352. Springer, Singapore (2020)
6. Tan, Z., He, L.: An efficient similarity measure for user-based collaborative filtering recommender systems inspired by the physical resonance principle. IEEE Access **5**, 27211–27228 (2017)
7. Jain, G., Mahara, T.: An efficient similarity measure to alleviate the cold-start problem (2019)
8. Kant, S., Mahara, T., Jain, V.K., Jain, D.K., Sangaiah, A.K.: LeaderRank based k-means clustering initialization method for collaborative filtering. Comput. Electr. Eng. **69**, 598–609 (2018)
9. Kant, S., Mahara, T.: Nearest biclusters collaborative filtering framework with fusion. J. Comput. Sci. **25**, 204–212 (2018)
10. Kant, S., Mahara, T.: K-means Atkinson clustering approach for collaborative filtering-based recommendation system. Int. J. Intel. Syst. Des. Comput. **1**(3–4), 272–285 (2017)
11. Tripathi, K.N., Sharma, S.C., Jain, G.: A new reputation-based algorithm (RBA) to detect malicious nodes in vehicular Ad Hoc networks (VANETs). In: Soft Computing: Theories and Applications, pp. 395–404. Springer, Singapore (2020)

12. Bhardwaj, T., Sharma, S.C.: An autonomic resource provisioning framework for efficient data collection in cloudlet-enabled wireless body area networks: a fuzzy-based proactive approach. Soft Comput. **23**, 10361–10383 (2019)
13. Naim, I., Mahara, T., Aqib Khan, M.: Deep neural network model for monthly natural gas prediction. In: Pant M., Sharma T., Verma O., Singla R., Sikander, A. (eds.) Soft Computing: Theories and Applications. Advances in Intelligent Systems and Computing, vol. 1053. Springer, Singapore (2020)
14. Fernandez-Llatas, C., García-Gómez, J.M.: Data mining in clinical medicine. Data Min. Clin. Med. **1246**, 1–267 (2014)
15. Gräßer, F., Beckert, S., Küster, D., Schmitt, J., Abraham, S., Malberg, H., Zaunseder, S.: Therapy decision support based on recommender system methods. J. Healthc. Eng. (2017)
16. Bullock, J., Luccioni, A., Pham, K.H., Lam, C.S.N., Luengo-Oroz, M.: Mapping the Landscape of Artificial Intelligence Applications against COVID-19, pp. 1–32 (2020)
17. Rao, A.S.R.S., Vazquez, J.A.: Identification of COVID-19 can be quicker through artificial intelligence framework using a mobile phone-based survey in the populations when cities/towns are under quarantine. Infect. Control Hosp. Epidemiol. **1400** (2020)
18. Maghdid, H.S., Ghafoor, K.Z., Sadiq, A.S., Curran, K., Rabie, K.: A Novel AI-Enabled Framework to Diagnose Coronavirus COVID 19 Using Smartphone Embedded Sensors: Design Study, pp. 1–5 (2020)
19. Imran, A., et al.: AI4COVID-19: AI Enabled Preliminary Diagnosis for COVID-19 from Cough Samples via an App, pp. 1–12 (2020)
20. Nemati, E., Rahman, M.M., Nathan, V., Vatanparvar, K., Kuang, J.: Poster abstract: a comprehensive approach for cough type detection. In: Proceedings—4th IEEE/ACM Conference on Connected Health: Applications, Systems and Engineering Technologies. CHASE 2019, pp. 15–16 (2019)
21. Babasaheb Karvekar, S.: Smartphone-based Human Fatigue Detection in an Industrial Smartphone-based Human Fatigue Detection in an Industrial Environment Using Gait Analysis Environment Using Gait Analysis (2019)
22. Alimadadi, A., Aryal, S., Manandhar, I., Munroe, P.B., Joe, B., Cheng, X.: Artificial intelligence and machine learning to fight covid-19. Physiol. Genom. **52**(4), 200–202 (2020)
23. Rahmatizadeh, S., Valizadeh-Haghi, S., Dabbagh, A.: The role of artificial intelligence in management of critical COVID-19 patients. J. Cell. Mol. Anesth. **5**(1), 16–22 (2020)
24. Arti, M.K.: Modeling and Predictions for COVID 19 Spread in India (2020)
25. Salman, F.M., Abu-Naser, S.S., Alajrami, E., Abu-Nasser, B.S., Ashqar, B.A.M.: COVID-19 detection using artificial intelligence. Int. J. Acad. Eng. Res. 4(3), 18–25 (2020)
26. Tomar, A., Gupta, N.: Prediction for the spread of COVID-19 in India and effectiveness of preventive measures. Sci. Total Environ. **728**, 138762 (2020)
27. Wiesner, M., Pfeifer, D.: Health recommender systems: concepts, requirements, technical basics and challenges. Int. J. Environ. Res. Public Health **11**(3), 2580–2607 (2014)
28. Sahoo, A.K., Pradhan, C., Barik, R.K., Dubey, H.: DeepReco: deep learning based health recommender system using collaborative filtering. Computation **7**(2) (2019)
29. Nasiri, M., Minaei, B., Kiani, A.: Dynamic recommendation: disease prediction and prevention using recommender system. Int. J. basic Sci. Med. **1**(1), 13–17 (2016)
30. "Here Are 3 Classifications of Asymptomatic COVID-19 Patients—engteco_life_style Tempo.co." [Online]. Available: https://en.tempo.co/read/1339559/here-are-3-classifications-of-asymptomatic-covid-19-patients. Accessed: 24-Nov-2020
31. "Q&A: Influenza and COVID-19—similarities and differences." [Online]. Available: https://www.who.int/emergencies/diseases/novel-coronavirus-2019/question-and-answers-hub/q-a-detail/q-a-similarities-and-differences-covid-19-and-influenza. Accessed: 24-Nov-2020
32. Gower, J.C.: A general coefficient of similarity and some of its properties. Society **27**(4), 857–871. https://www.jstor.org/stable/2528823 (2011)
33. Podani, J.: Extending Gower's general coefficient of similarity to ordinal characters. Taxon **48**(2), 331–340 (1999)

34. Fontecha, J., Hervás, R., Bravo, J.: Mobile services infrastructure for frailty diagnosis support based on Gower's similarity coefficient and treemaps. Mob. Inf. Syst. **10**(1), 127–146 (2014)
35. "Corona virus main symptoms—Google Search." [Online]. Available: https://www.google.com/search?q=corona+viirsu+main+symptoms&oq=corona+viirsu+main+symptoms&aqs=chrome..69i57j0.7506j0j7&sourceid=chrome&ie=UTF-8. Accessed: 24-Nov-2020
36. "Coronavirus disease (COVID-19)" [Online]. Available: https://www.who.int/emergencies/diseases/novel-coronavirus-2019/question-and-answers-hub/q-a-detail/coronavirus-disease-covid-19#:~:text=symptoms. Accessed: 24-Nov-2020
37. "COVID-19 in India | Kaggle." [Online]. Available: https://www.kaggle.com/sudalairajkumar/covid19-in-india. Accessed: 24-Sep-2020
38. "Nearly All Patients Hospitalized With Covid-19 Had Chronic Health Issues, Study Finds—The New York Times." [Online]. Available: https://www.nytimes.com/2020/04/23/health/coronavirus-patients-risk.html. Accessed: 24-Sep-2020

# Frequencies of Nonuniform Triangular Plate with Two-Dimensional Parabolic Temperature

**Snehlata, Naveen Mani, Kamal Kumar, Amit Sharma, Reeta Bhardwaj, and Pravesh Kumar**

**Abstract** Frequencies of nonuniform triangular plate (isosceles, right-angled, and scalene triangle) on clamped edges under temperature field are computed using Rayleigh–Ritz method. For non-uniformity, two-dimensional linear thickness variation on the plate is considered. The two-dimensional parabolic temperature is considered on the plate. First three modes of vibration are computed on different variation of plate parameters and presented with the help of tables. The objective of the study is to reduce the frequency of the plates. A comparative study of the frequencies with other available published results well presents the objective of the study.

**Keywords** Temperature effect · Triangle plate · Frequency · Vibration

## 1 Introduction

The vibration of plates is a special case of the more general problem of mechanical vibration. A lot of research work [1–3] and studies have been done. The effect of material temperature dependency on frequencies is discussed. The importance of study of vibration of plates is due to its wide application in engineering. Effect of temperature on vibration of plates is necessary due to its application in manufacturing of ships, aircraft, submarines, etc. The following literature study shows how plate parameters effect the vibration of plates.

To analysis the vibration of triangular plates, ray-tracing method [4] has been applied specifically to a class of triangular plates with certain aspect ratios. Free vibration frequencies as well as nodal patterns has been analyzed for completely free triangular plates [5]. Natural vibration of nonhomogeneous tapered square plate

Snehlata · K. Kumar · A. Sharma · R. Bhardwaj (✉)
Amity University Haryana, Gurugram, Haryana, India

N. Mani
Chandigarh University, Mohali, Punjab, India

P. Kumar
Rajkiya Engineering College, Bijnor, India

 
T. K. Sharma et al. (eds.), *Soft Computing: Theories and Applications*,
Advances in Intelligent Systems and Computing 1381,
https://doi.org/10.1007/978-981-16-1696-9_4

[6, 7] has been computed under temperature variation on the plate. Vibration of tapered isotropic rectangular plate [8] under thermal conditions has been analyzed and presented by a mathematical model. Frequency for two modes of vibration for different types of clamped triangular plates [9] has been computed by using Rayleigh–Ritz method on clamped edges. Parabolic thermal vibrations of circular plates [10] have been studied with the help of PSS method: first four frequencies of vibrations have been computed. The time period of frequency modes of skew plate [11] with circular density and linear thickness on *CCCC* and *CCSS* edge condition has been computed under temperature field. Finite element technique has been used to solve the problem of the natural frequencies and mode shape of cantilevered triangular plates [12] with variable thickness and arbitrary planform. An analysis has been performed for triangular plate [13] with free clamped-free boundary condition for different materials by using FEM analysis and experimental modal analysis. Rayleigh–Ritz technique has been used to compute the frequency modes of skew plate [14] with circular density and linear thickness and rectangle plate [15] with linear thickness and circular Poisson's ratio under temperature field. The effect of circular density and circular Poisson's ratio on time period of vibration of nonuniform skew plate [16] under temperature environment has been investigated using Rayleigh–Ritz method. Damped vibrations of an orthotropic rectangular plate [17] with thermal gradient and variable thickness have been analyzed by using the quintic spline technique. Vibration characteristics of functionally graded rectangular [18] plates have been investigated by developing a theoretical method with different thermal conditions. Rayleigh–Ritz method has been used to study circular thickness and circular Poisson's ratio impact on frequency of skew plate [19] on various edge conditions. Using the Rayleigh–Ritz technique, the effect of parabolic temperature variation on frequency of non-uniform and non-homogeneous square plate [20] and skew plate [21] has been obtained. On the basis of Mindlin theory, the vibration of isosceles triangle plates [22] with different edge conditions using pb-2 Rayleigh–Ritz method has been investigated.

The aim of the present study is to investigate two-dimensional thermal effect on the vibration of triangular plates which is shown with the help of tables. The second objective of the study is to show that, by choosing an appropriate variation in plate parameters frequencies can be minimized.

## 2 Analysis

Consider a viscoelastic triangle plate having aspect ratio $\theta = b/c$ and $\mu = c/a$ , two-dimensional thickness $l$ as shown in Fig. 1. Now transform the given triangle into right-angled triangle using the transformation $x = a\zeta + b\psi$ and $y = c\psi$ as shown in Fig. 2.

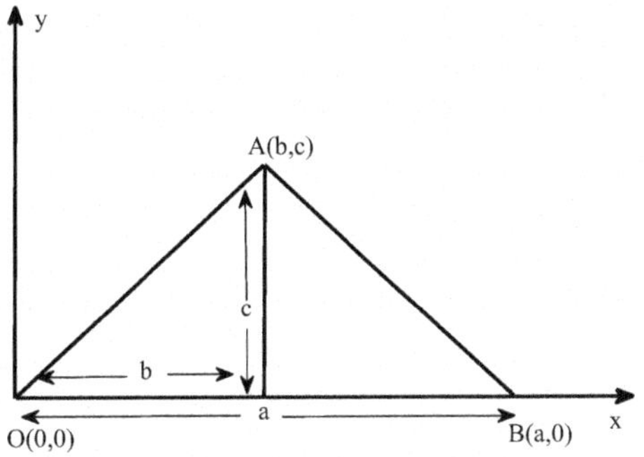


**Fig. 1** Triangle plate

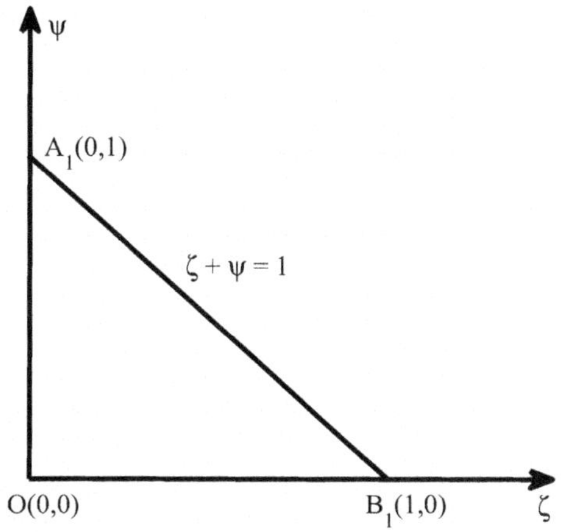


**Fig. 2** Transformed triangle plate

The kinetic energy and strain energy [23] for vibration of triangle plate are

$$T_s = \frac{1}{2}\rho\omega^2 \int_0^1 \int_0^{1-\zeta} l\Phi^2 ac\,\mathrm{d}\psi\,\mathrm{d}\zeta \tag{1}$$

$$\begin{aligned} V_s = \frac{1}{2}\int_0^1 \int_0^{1-\zeta} D_1 \Bigg[ & \frac{1}{a^4}\left(\frac{\partial^2\Phi}{\partial\zeta^2}\right)^2 + \left(\frac{b^2}{a^2c^2}\frac{\partial^2\Phi}{\partial\zeta^2} + \frac{1}{c^2}\frac{\partial^2\Phi}{\partial\psi^2} - \frac{2b}{ac^2}\frac{\partial^2\Phi}{\partial\zeta\partial\psi}\right)^2 \\ & + 2\nu\left(\frac{1}{a^2}\frac{\partial^2\Phi}{\partial\zeta^2}\right)\left(\frac{b^2}{a^2c^2}\frac{\partial^2\Phi}{\partial\zeta^2} + \frac{1}{c^2}\frac{\partial^2\Phi}{\partial\psi^2} - \frac{2b}{ac^2}\frac{\partial^2\Phi}{\partial\zeta\partial\psi}\right) \\ & + 2(1-\nu)\left(-\frac{b}{a^2c}\frac{\partial^2\Phi}{\partial\zeta^2} + \frac{1}{ac}\frac{\partial^2\Phi}{\partial\zeta\partial\psi}\right)^2\Bigg](ac)\,\mathrm{d}\psi\,\mathrm{d}\zeta \end{aligned} \tag{2}$$

where $\Phi$ is deflection function, $\omega$ is natural frequency, and $D_1 = El^3/12\left(1-\nu^2\right)$ is flexural rigidity, here $E$and $\nu$ is Young's modulus and Poisson's ratio of the plate.

Rayleigh–Ritz method requires

$$L = \delta(V_s - T_s) = 0 \tag{3}$$

Using Eqs. (1), (2), we have

$$\begin{aligned} L = \frac{1}{2}\int_0^1\int_0^{1-\zeta} D_1\Bigg[\frac{1}{a^4}\left(\frac{\partial^2\Phi}{\partial\zeta^2}\right)^2 + \left(\frac{b^2}{a^2c^2}\frac{\partial^2\Phi}{\partial\zeta^2} + \frac{1}{c^2}\frac{\partial^2\Phi}{\partial\psi^2} - \frac{2b}{ac^2}\frac{\partial^2\Phi}{\partial\zeta\partial\psi}\right)^2 \\ + 2\nu\left(\frac{1}{a^2}\frac{\partial^2\Phi}{\partial\zeta^2}\right)\left(\frac{b^2}{a^2c^2}\frac{\partial^2\Phi}{\partial\zeta^2} + \frac{1}{c^2}\frac{\partial^2\Phi}{\partial\psi^2} - \frac{2b}{ac^2}\frac{\partial^2\Phi}{\partial\zeta\partial\psi}\right) \\ + 2(1-\nu)\left(-\frac{b}{a^2c}\frac{\partial^2\Phi}{\partial\zeta^2} + \frac{1}{ac}\frac{\partial^2\Phi}{\partial\zeta\partial\psi}\right)^2\Bigg](ac)\mathrm{d}\psi\mathrm{d}\zeta \\ - \frac{1}{2}\rho\omega^2\int_0^1\int_0^{1-\zeta} l\Phi^2(ac)\mathrm{d}\psi\mathrm{d}\zeta. \end{aligned} \tag{4}$$

Introducing two-dimensional [9] linear thickness as

$$l = l_0\left(1-\beta_1\zeta\right)\left(1-\beta_2\psi\right), \tag{5}$$

where $l_0$ are the thickness at origin. Also $\beta_1,\ \beta_2$ are tapering parameters.

Two-dimensional parabolic temperature on the plate is assumed as

$$\tau = \tau_0\left(1-\zeta^2\right)\left(1-\psi^2\right), \tag{6}$$

where $\tau$ and $\tau_0$ denote the temperature on and at the origin, respectively. The modulus of elasticity is given by

$$E = E_0\left(1-\gamma\tau\right) \tag{7}$$

where $E_0$ is Young's modulus at $\tau = 0$ and $\gamma$ is called slope of variation. Using Eqs. (6), (7) becomes

$$E = E_0\left(1-\alpha\left(1-\zeta^2\right)\left(1-\psi^2\right)\right) \tag{8}$$

where $\alpha = \gamma\tau_0$, $(0 \leq \alpha < 1)$ is called thermal gradient. Using Eqs. (5) and (8), the functional in Eq. (4) becomes

$$
\begin{aligned}
L = \frac{D_0}{2}\int_0^1\int_0^{1-\zeta} & \left[\left(1-\alpha\left(1-\zeta^2\right)\left(1-\psi^2\right)\right)(1-\beta_1\zeta)^3(1-\beta_2\psi)^3\right] \\
& \left[(1+\theta^2)^2\left(\frac{\partial^2\Phi}{\partial\zeta^2}\right)^2 + \left(\frac{\partial^2\Phi}{\partial\psi^2}\right)^2 + \frac{2\left(2\theta^2+1-\nu\right)}{\mu^2}\left(\frac{\partial^2\Phi}{\partial\zeta\partial\psi}\right)^2\right. \\
& + \frac{2\left(\nu+\theta^2\right)}{\mu^2}\left(\frac{\partial^2\Phi}{\partial\zeta^2}\right)\left(\frac{\partial^2\Phi}{\partial\psi^2}\right) - \frac{4\theta\left(1+\theta^2\right)}{\mu}\left(\frac{\partial^2\Phi}{\partial\zeta^2}\right)\left(\frac{\partial^2\Phi}{\partial\zeta\partial\psi}\right) \\
& \left. - \frac{4\theta}{\mu^3}\left(\frac{\partial^2\Phi}{\partial\psi^2}\right)\left(\frac{\partial^2\Phi}{\partial\zeta\partial\psi}\right)\right](ac)\mathrm{d}\psi\mathrm{d}\zeta \\
& - \frac{1}{2}\rho\omega^2\int_0^1\int_0^{1-\zeta}(1-\beta_1\zeta)(1-\beta_2\psi)\,\Phi^2 ac\mathrm{d}\psi\mathrm{d}\zeta
\end{aligned}
\tag{9}
$$

where

$$D_0 = E_0 l_0^3/12\,\left(1-\nu^2\right)$$

and

$$\lambda^2 = \rho\omega^2 l_0 a^2/D_0\,.$$

Deflection function is taken as

$$\Phi(\zeta,\psi) = \left[(\zeta)^e(\psi)^f(1-\zeta-\psi)^g\right]\left[\sum_{i=0}^{n}\Psi_i\{(\zeta)(\psi)(1-\zeta-\psi)\}^i\right] \tag{10}$$

where $\Psi_i$, $i = 0, 1, 2 \ldots n$ are $n$ unknowns and the value of $e, f, g$ can be 0, 1, and 2, corresponding to given edge condition (free, simply supported, clamped respectively). To minimize Eq. (10), we have

$$\frac{\partial L}{\partial \Psi_i} = 0, \quad i = 0, 1, \ldots n. \tag{11}$$

Solving Eq. (11), we have frequency equation as

$$\left|P - \lambda^2 Q\right| = 0, \tag{12}$$

where $P = \left[p_{ij}\right]_{i,j=0,1,\ldots,n}$ and $Q = \left[q_{ij}\right]_{i,j=0,1,\ldots,n}$ are square matrix of order $(n+1)$.

## 3 Results and Discussions

To analyze the impact of two-dimensional parabolic temperature on various triangle plate with two-dimensional linear thickness, first three modes of frequency are computed and presented in tabular form.

Table 1 presents first three modes of frequencies for right-angled isosceles triangle corresponding to tapering parameters $\beta_1, \beta_2$ for fixed aspect ratios $\theta = 0.0$, $\mu = 1.0$, and for variable value of thermal gradient $\alpha$, i.e., $\alpha = 0.2, 0.4, 0.6$. From Table 1, we analyzed that frequency modes decreases with the increasing value of both tapering parameters $\beta_1, \beta_2$. Also increase in temperature gradient on the plate results the decrease in frequency modes. The rate of decrement in frequency modes corresponding to tapering parameters $\beta_1, \beta_2$ increases as thermal gradient $\alpha$ varies from 0.2 to 0.6 on the plate.

**Table 1** Frequencies of right-angled isosceles triangle plate corresponding to tapering parameters $\beta_1, \beta_2$

| $\theta = 0, \mu = 1$ | | | | | | | | | |
|---|---|---|---|---|---|---|---|---|---|
| $\alpha = 0.2$ | | | | | | | | | |
| $\beta_1$ | $\beta_2 = 0.0$ | | | $\beta_2 = 0.4$ | | | $\beta_2 = 0.8$ | | |
| | $\lambda_1$ | $\lambda_2$ | $\lambda_3$ | $\lambda_1$ | $\lambda_2$ | $\lambda_3$ | $\lambda_1$ | $\lambda_2$ | $\lambda_3$ |
| **0.0** | 99.52 | 363.90 | 826.14 | 84.91 | 309.93 | 701.27 | 71.94 | 262.58 | 593.18 |
| **0.2** | 91.78 | 336.29 | 761.82 | 77.49 | 285.50 | 644.18 | 65.52 | 241.03 | 543.69 |
| **0.4** | 84.51 | 309.93 | 701.33 | 70.88 | 262.17 | 591.69 | 59.45 | 220.41 | 496.13 |
| **0.6** | 77.85 | 285.22 | 644.54 | 64.82 | 240.32 | 541.43 | 53.85 | 201.00 | 452.36 |
| **0.8** | 71.94 | 262.58 | 593.25 | 59.45 | 220.39 | 496.90 | 48.88 | 183.32 | 412.97 |
| **1.0** | 66.92 | 242.43 | 548.35 | 54.04 | 202.79 | 457.60 | 44.70 | 167.79 | 376.87 |
| $\alpha = 0.4$ | | | | | | | | | |
| **0.0** | 91.52 | 332.65 | 757.00 | 77.41 | 282.80 | 641.01 | 65.70 | 238.99 | 540.80 |
| **0.2** | 84.17 | 307.15 | 697.38 | 70.89 | 260.25 | 588.05 | 59.75 | 219.09 | 494.78 |
| **0.4** | 77.44 | 282.81 | 640.92 | 64.75 | 238.69 | 539.41 | 54.10 | 200.02 | 450.68 |
| **0.6** | 71.21 | 259.96 | 588.15 | 59.11 | 218.47 | 492.66 | 48.88 | 182.08 | 409.84 |
| **0.8** | 65.70 | 238.9 | 540.72 | 54.10 | 200.02 | 451.16 | 44.24 | 165.68 | 373.44 |
| **1.0** | 61.02 | 220.32 | 499.08 | 49.89 | 183.72 | 414.68 | 40.34 | 151.32 | 341.86 |
| $\alpha = 0.6$ | | | | | | | | | |
| **0.0** | 82.34 | 297.99 | 680.72 | 69.57 | 252.67 | 574.57 | 58.79 | 212.75 | 482.32 |
| **0.2** | 75.76 | 274.84 | 626.20 | 63.60 | 232.18 | 526.01 | 53.34 | 194.65 | 440.27 |
| **0.4** | 69.57 | 252.68 | 574.38 | 57.96 | 212.56 | 481.30 | 48.15 | 177.30 | 399.87 |
| **0.6** | 63.87 | 231.86 | 526.09 | 52.77 | 194.15 | 438.31 | 43.35 | 160.13 | 362.41 |
| **0.8** | 58.79 | 212.75 | 482.34 | 48.15 | 177.29 | 404.52 | 39.06 | 145.95 | 328.95 |
| **1.0** | 54.47 | 195.69 | 444.22 | 44.26 | 162.40 | 366.80 | 35.45 | 132.80 | 299.86 |

**Table 2** Frequencies of right-angled scalene triangle plate corresponding to tapering parameters $\beta_1, \beta_2$

$\theta = 0, \mu = 1.5$

| $\beta_1$ | $\beta_2 = 0.0$ | | | $\beta_2 = 0.4$ | | | $\beta_2 = 0.8$ | | |
|---|---|---|---|---|---|---|---|---|---|
| | $\lambda_1$ | $\lambda_2$ | $\lambda_3$ | $\lambda_1$ | $\lambda_2$ | $\lambda_3$ | $\lambda_1$ | $\lambda_2$ | $\lambda_3$ |
| $\alpha = 0.2$ | | | | | | | | | |
| **0.0** | 89.77 | 328.22 | 744.69 | 76.64 | 281.13 | 636.76 | 65.77 | 240.19 | 544.70 |
| **0.2** | 82.98 | 304.07 | 689.17 | 70.41 | 259.51 | 586.79 | 60.00 | 220.86 | 500.27 |
| **0.4** | 76.64 | 281.12 | 636.80 | 64.57 | 238.98 | 540.26 | 54.54 | 202.40 | 458.07 |
| **0.6** | 70.86 | 259.70 | 588.26 | 59.24 | 219.13 | 496.63 | 49.52 | 185.12 | 418.65 |
| **0.8** | 65.77 | 240.20 | 544.56 | 54.54 | 202.41 | 457.83 | 45.07 | 169.38 | 383.78 |
| **1.0** | 61.48 | 222.89 | 507.07 | 50.61 | 187.10 | 424.57 | 41.34 | 155.56 | 353.74 |
| $\alpha = 0.4$ | | | | | | | | | |
| **0.0** | 82.34 | 299.72 | 681.35 | 70.14 | 256.21 | 581.01 | 60.01 | 218.36 | 495.65 |
| **0.2** | 76.03 | 277.42 | 629.89 | 64.35 | 236.28 | 534.77 | 54.65 | 200.53 | 454.42 |
| **0.4** | 70.14 | 256.22 | 581.20 | 58.92 | 217.56 | 491.82 | 49.98 | 183.47 | 415.38 |
| **0.6** | 64.76 | 236.42 | 536.13 | 53.96 | 199.60 | 451.23 | 44.91 | 167.50 | 378.84 |
| **0.8** | 60.01 | 218.36 | 495.62 | 49.58 | 183.48 | 415.20 | 40.77 | 152.94 | 346.55 |
| **1.0** | 56.01 | 202.34 | 460.72 | 45.92 | 169.32 | 384.45 | 37.29 | 140.18 | 318.61 |
| $\alpha = 0.6$ | | | | | | | | | |
| **0.0** | 74.12 | 268.07 | 611.63 | 62.94 | 228.53 | 519.80 | 53.62 | 194.04 | 441.26 |
| **0.2** | 68.35 | 247.81 | 564.41 | 57.54 | 210.42 | 477.33 | 48.72 | 177.84 | 403.68 |
| **0.4** | 62.94 | 228.52 | 519.97 | 52.66 | 193.16 | 437.89 | 44.07 | 162.35 | 367.67 |
| **0.6** | 58.00 | 210.50 | 478.33 | 48.10 | 177.03 | 400.82 | 39.77 | 147.78 | 334.42 |
| **0.8** | 53.62 | 194.03 | 441.30 | 44.07 | 162.35 | 367.61 | 35.95 | 134.91 | 304.66 |
| **1.0** | 49.92 | 179.43 | 409.18 | 40.69 | 149.43 | 339.48 | 32.75 | 122.88 | 279.14 |

First three modes of frequency of right-angled scalene triangles corresponding to both tapering parameters $\beta_1, \beta_2$ for variable value of thermal gradient $\alpha$, i.e., $\alpha = 0.2, 0.4, 0.6$ is presented in both Table 2 (for fixed value of aspect ratios $\theta = 0.0$, $\mu = 1.5$) and Table 3 (for fixed value of aspect ratios $\theta = 0.0, \mu = 2.0$). Both Tables 2 and 3 enlighten the fact that behavior of frequency modes for the variation in tapering parameters $\beta_1, \beta_2$ and thermal gradient $\alpha$ is same as the behavior of frequency modes reported in Table 1, i.e., the impact of variation in tapering parameters $\beta_1, \beta_2$ and thermal gradient $\alpha$ on the behavior of frequency modes are same for both right-angled isosceles triangle and right-angled scalene triangle. But the frequency modes tabulated for right-angled scalene triangle are much less in comparison with the frequency modes of right-angled isosceles triangle. Also, the frequency modes provided in Table 3 is less when compared with the frequency modes provided in Table 2.

Table 4 accommodates first three modes of frequencies for scalene triangle plate corresponding to thermal gradient $\alpha$ for fixed value of aspect ratios $\theta =$

**Table 3** Frequencies of right-angled scalene triangle plate corresponding to tapering parameters $\beta_1, \beta_2$

| $\theta = 0, \mu = 2$ | | | | | | | | | |
|---|---|---|---|---|---|---|---|---|---|
| $\alpha = 0.2$ | | | | | | | | | |
| $\beta_1$ | $\beta_2 = 0.0$ | | | $\beta_2 = 0.4$ | | | $\beta_2 = 0.8$ | | |
| | $\lambda_1$ | $\lambda_2$ | $\lambda_3$ | $\lambda_1$ | $\lambda_2$ | $\lambda_3$ | $\lambda_1$ | $\lambda_2$ | $\lambda_3$ |
| **0.0** | 86.10 | 314.81 | 713.71 | 73.69 | 270.89 | 612.84 | 63.47 | 231.82 | 526.56 |
| **0.2** | 79.67 | 291.91 | 661.76 | 67.75 | 249.77 | 565.28 | 57.93 | 213.33 | 484.07 |
| **0.4** | 73.69 | 270.31 | 612.65 | 62.20 | 230.29 | 521.33 | 52.70 | 195.69 | 443.66 |
| **0.6** | 68.25 | 250.15 | 567.16 | 57.15 | 212.15 | 480.25 | 47.90 | 179.18 | 406.24 |
| **0.8** | 63.47 | 231.82 | 526.55 | 52.70 | 195.69 | 443.57 | 43.64 | 164.16 | 373.05 |
| **1.0** | 59.39 | 215.59 | 491.77 | 48.99 | 181.23 | 412.57 | 40.07 | 150.98 | 344.54 |
| $\alpha = 0.4$ | | | | | | | | | |
| **0.0** | 78.94 | 287.31 | 652.84 | 67.40 | 246.22 | 558.93 | 57.88 | 210.63 | 479.00 |
| **0.2** | 72.97 | 266.22 | 604.52 | 61.90 | 227.29 | 514.79 | 52.75 | 193.59 | 439.53 |
| **0.4** | 67.40 | 246.22 | 558.86 | 56.74 | 209.29 | 474.15 | 47.89 | 177.31 | 402.2 |
| **0.6** | 62.34 | 227.59 | 516.65 | 52.03 | 192.52 | 436.01 | 43.43 | 162.05 | 367.43 |
| **0.8** | 57.88 | 210.63 | 479.03 | 47.89 | 177.31 | 401.94 | 39.47 | 148.17 | 336.65 |
| **1.0** | 54.13 | 195.63 | 446.59 | 44.44 | 163.93 | 373.32 | 36.14 | 136.01 | 310.15 |
| $\alpha = 0.6$ | | | | | | | | | |
| **0.0** | 71.02 | 256.78 | 585.51 | 65.56 | 237.63 | 541.10 | 51.69 | 187.04 | 425.91 |
| **0.2** | 65.56 | 237.63 | 541.15 | 55.41 | 202.24 | 459.04 | 46.99 | 171.55 | 390.03 |
| **0.4** | 60.45 | 219.44 | 499.33 | 50.67 | 185.88 | 421.74 | 42.54 | 156.76 | 355.74 |
| **0.6** | 55.69 | 202.48 | 460.69 | 46.35 | 170.62 | 386.76 | 38.43 | 142.88 | 323.93 |
| **0.8** | 51.69 | 187.04 | 425.93 | 42.54 | 156.76 | 355.58 | 34.79 | 130.23 | 295.70 |
| **1.0** | 48.23 | 173.35 | 396.31 | 39.36 | 144.38 | 329.30 | 31.73 | 119.17 | 271.53 |

**Table 4** Frequencies of scalene triangle plate corresponding to thermal gradient $\alpha$

| $\theta = 1/\sqrt{3}, \mu = \sqrt{3}/2$ | | | | | | | | | |
|---|---|---|---|---|---|---|---|---|---|
| $\alpha$ | $\beta_1 = \beta_2 = 0.0$ | | | $\beta_1 = \beta_2 = 0.4$ | | | $\beta_1 = \beta_2 = 0.8$ | | |
| | $\lambda_1$ | $\lambda_2$ | $\lambda_3$ | $\lambda_1$ | $\lambda_2$ | $\lambda_3$ | $\lambda_1$ | $\lambda_2$ | $\lambda_3$ |
| **0.0** | 84.94 | 311.56 | 705.52 | 65.11 | 237.40 | 538.56 | 48.82 | 175.90 | 401.29 |
| **0.2** | 78.51 | 287.57 | 652.25 | 59.96 | 222.40 | 494.44 | 44.74 | 161.36 | 368.19 |
| **0.4** | 71.50 | 261.38 | 593.87 | 54.33 | 197.90 | 450.11 | 40.25 | 145.37 | 331.97 |
| **0.6** | 63.72 | 232.24 | 528.97 | 48.04 | 174.84 | 398.29 | 35.20 | 127.38 | 290.94 |
| **0.8** | 54.80 | 198.73 | 455.04 | 40.79 | 148.18 | 338.89 | 29.29 | 106.40 | 243.25 |

$1/\sqrt{3}$ , $\mu = \sqrt{3}/2$ and for variable value of tapering parameters $\beta_1 = \beta_2$ i.e., $\beta_1 = \beta_2 = 0.0, 0.4, 0.8$. Table 4 shows with the increasing value of thermal gradient $\alpha$ as well as tapering parameters $\beta_1, \beta_2$ decrease the frequency modes. The rate of decrement in frequency modes is higher in case of variation in tapering parameters $\beta_1, \beta_2$ in comparison with rate of decrement in frequencies in case of variation in thermal gradient $\alpha$.

# 4 Results Comparison

In order to validated the finding of the present study, a frequency comparison of scalene triangle plate with [9] is also given.

A comparison of frequency modes (first two modes) of scalene triangle plate with the frequency modes obtained in [9] corresponding to thermal gradient $\alpha$ for fixed value of aspect ratios $\theta = 1/\sqrt{3}$ , $\mu = \sqrt{3}/2$ and for variable values of tapering parameters $\beta_1 = \beta_2$, i.e., $\beta_1 = \beta_2 = 0.0, 0.2, 0.4, 0.6, 0.8$ is presented in Table 5.

Table 5 concludes that frequencies obtained in present analysis is much lesser than the frequencies obtained in [9] for all the variable value of tapering parameters $\beta_1, \beta_2$ corresponding to thermal gradient $\alpha$.

# 5 Conclusions

The behavior of vibrational frequency modes (on variation of tapering parameters and temperature) of different triangle plate on clamped edges is computed and presented in tabular form. Based on computational results and comparison, authors would like to drag the attention of readers on the following facts:

**Table 5** Comparison of frequencies of scalene triangle plate with [9] corresponding to thermal gradient $\alpha$

$\theta = 1/\sqrt{3}, \mu = \sqrt{3}/2$

| $\alpha$ | $\beta_1 = \beta_2 = 0.0$ | | $\beta_1 = \beta_2 = 0.2$ | | $\beta_1 = \beta_2 = 0.4$ | | $\beta_1 = \beta_2 = 0.6$ | | $\beta_1 = \beta_2 = 0.8$ | |
|---|---|---|---|---|---|---|---|---|---|---|
| | $\lambda_1$ | $\lambda_2$ | $\lambda_1$ | $\lambda_2$ | $\lambda_1$ | $\lambda_2$ | $\lambda_1$ | $\lambda_2$ | $\lambda_1$ | $\lambda_2$ |
| 0.0 | 84.96 | 346.11 | 74.61 | 304.12 | 65.13 | 265.85 | 56.53 | 231.43 | 48.85 | 201.12 |
| | **100.95** | **411.24** | **88.07** | **358.94** | **76.35** | **311.73** | **65.82** | **269.74** | **56.51** | **233.22** |
| 0.2 | 78.54 | 320.65 | 68.86 | 281.19 | 59.99 | 245.24 | 51.94 | 212.90 | 44.77 | 184.44 |
| | **93.98** | **382.84** | **81.87** | **333.58** | **70.86** | **289.13** | **60.97** | **249.62** | **52.25** | **215.28** |
| 0.4 | 71.53 | 292.98 | 62.57 | 256.22 | 54.36 | 222.72 | 46.91 | 192.60 | 40.28 | 166.09 |
| | **86.45** | **352.17** | **75.16** | **306.12** | **64.91** | **264.61** | **55.71** | **227.73** | **47.60** | **195.71** |
| 0.6 | 63.75 | 262.41 | 55.57 | 228.54 | 48.07 | 197.66 | 41.27 | 169.89 | 35.22 | 145.44 |
| | **78.19** | **318.54** | **67.79** | **275.94** | **58.35** | **237.57** | **49.89** | **203.51** | **42.45** | **173.96** |
| 0.8 | 54.85 | 227.79 | 47.53 | 197.01 | 40.82 | 168.93 | 34.73 | 143.63 | 29.31 | 121.34 |
| | **68.96** | **280.93** | **59.52** | **242.03** | **50.96** | **207.03** | **43.29** | **175.98** | **36.55** | **149.06** |

Bold font values are from [9]

- Two-dimensional parabolic temperature variation (present study) impacts more on frequency modes of plate in comparison with one-dimensional linear temperature variation [9], i.e., two-dimensional parabolic temperature variation provides less frequency modes in comparison with one-dimensional linear temperature variation.
- Two-dimensional parabolic thermal gradient provides less variation in frequencies when compared with the variation in frequencies in case of two-dimensional linear tapering.
- Increase in temperature on the plate results that rate of decrements in frequency modes increases corresponding to variation in plate parameters.
- Increase in both tapering parameters and thermal gradient results decreases in frequencies of the plate.

The above facts clearly show that frequency modes as well as variation in frequency modes can be controlled by taking an appropriate variation in plate parameters and desired frequency can be obtained.

## References


1. Leissa, A.W.: Vibration of Plates. NASA, Washington DC (1969)
2. Leissa, A.W.: Recent research in plate vibrations 1973–1976: complicating effects. Shock Vibr. Dig. **10**(12), 21–35 (1978)
3. Leissa, A.W.: Plate vibration research, 1976–1980: classical theory. Shock Vibr. Dig. **13**(9), 11–22 (1981)
4. Vijayakar, S.M., Singh, R., Gunda, R.: Vibration analysis of triangular plates using the Ray Tracing technique. J. Acoust. Soc. Am. **97**, 33–82 (1995)
5. Leissa, A.W., Jaber, N.A.: Vibrations of completely free triangular plates. Int. J. Mech. Sci. **34**(8), 605–616 (1992)
6. Sharma, A.: Free vibration of square plate with temperature effect. J. Measure. Eng. **5**(4), 222–228 (2017)
7. Sharma, A., Kumar, P.: Natural vibration of square plate with circular variation in thickness. Soft Comput. Theories Appl. Adv. Intell. Syst. Comput. **742**, 311–319 (2019)
8. Khanna, A., Singhal, A.: Effect of plate's parameters on vibration of isotropic tapered rectangular plate with different boundary conditions. J. Low Freq. Noise Vibr. Active Control **35**(2), 139–151 (2016)
9. Kaur, N.: Vibrational behavior of tapered triangular plate with clamped ends under thermal condition. J. Inst. Eng. (India): Series C 1-9 (2020)
10. Sarswat, N.K., Bansal, V., Singh, A.: Effect of temperature on vibrations of elastic circular plate under the influence of shear effect. Int. J. Pure Appl. Math. **120**(6), 10577–10589 (2018)
11. Bhardwaj, R., Mani, N., Sharma, A.: Time period of transverse vibration of skew plate with parabolic temperature variation. J. Vibr. Control (2020). https://doi.org/10.1177/1077546320926887
12. Bijlani, M., Mirza, S.: Vibration of triangular plates of variable thickness. Comput. Struct. **21**(6), 1129–1135 (1985)
13. Chaudhary, R.R., Falak, Y.R.: Vibration analysis of laminated triangular plate by experimental and finite element analysis. Int. J. Eng. Res. Gen. Sci. **3**(2), 786–791 (2015)
14. Sharma, A.: Natural vibration of parallelogram plate with circular variation in density. Acta Tech. **63**(6), 763–774 (2019)

15. Sharma, A.: Vibration frequency of a rectangle plate with linear variation in thickness and circular variation in Poisson's ratio. J. Theor. Appl. Mech. **57**(3), 605–615 (2019)
16. Bhardwaj, R., Mani, N.: Modelling on vibration of skew plate with thickness and temperature variation. Vibroengineering Procedia **22**, 6–12 (2019)
17. Rana, U.S.: Robin: effect of damping and thermal gradient on vibration of orthotropic rectangular plate of variable thickness. Appl. Appl. Math. Int. J. (AAM) **12**(1), 201–216 (2017)
18. Kim, Y.W.: Temperature dependent vibration analysis of functionally graded rectangular plates. J. Sound Vib. **284**(3–5), 531–549 (2005)
19. Lather, N., Sharma, A.: Natural vibration of skew plate on different set of boundary conditions with temperature gradient. Vibroengineering Procedia **22**, 74–80 (2019)
20. Sharma, A.: Vibration of square plate with parabolic temperature variation. Romanian J. Acoust. Vibr. **14**, 107–114 (2017)
21. Sharma, A.: Vibrational frequencies of parallelogram plate with circular variations in thickness. Soft Comput. Theores Appl. Adv. Intell. Syst. Comput. **583**, 317–326 (2017)
22. Kitipornchai, S., Liew, K.M., Xiang, Y., Wang, C.M.: Free vibration of isosceles triangular Mindlin plates. Int. J. Mech. Sci. **35**, 89–102 (1993)
23. Chakraverty, S.: Vibration of Plates. CRC Press, Taylor and Francis Group, Boca Raton, London New York (2009)

# Machine Learning in Finance: Towards Online Prediction of Loan Defaults Using Sequential Data with LSTMs

**V. A. Kandappan and A. G. Rekha**

**Abstract** In recent times, machine learning is finding many important use cases in the financial services industry. Financial datasets usually pose significant statistical challenges, and hence, traditional econometric methods fail in many practical applications. Though several approaches are suggested and being used, to predict whether a borrower would default on the loan availed, it is still an open challenge. In this work, two approaches are proposed based on LSTM (Long Short Term Memory) along with a hybrid neural network architecture to understand the context between financial transactions and loan defaults. The novelty of the proposed methods is in how they handle structured data and the associated temporal data. Further, the performance of the proposed approaches using debit card transactions and loan application information of customers to predict default is demonstrated. The experiments provided promising accuracies. Bidirectional LSTMs with a hybrid architecture achieve an accuracy of 94%. Hybrid neural network architectures used in this work provide an appropriate direction for making an early warning system through online loan default prediction.

**Keywords** Loan default prediction · Recurrent neural networks · LSTM · Ensemble learning · Machine learning in finance

## 1 Introduction

Credit products form an indispensable part of the banking business as well as to the economy of a country. The lending process entails risks. Financial institutions usually try to minimize the risk involved in lending loans at its approval stage. Banks nowadays operate with their standards to approve a loan. Credit scores to approve a loan are a conventional system to help the loan approval process. Predicting

V. A. Kandappan (✉)
Indian Institute of Technology Madras, Chennai, Tamil Nadu 600036, India

A. G. Rekha
Research wing-SBIL, State Bank of India, Kolkata 700160, India

 
T. K. Sharma et al. (eds.), *Soft Computing: Theories and Applications*,
Advances in Intelligent Systems and Computing 1381,
https://doi.org/10.1007/978-981-16-1696-9_5

whether a borrower will default on a loan is of significant concern to the banks. Recently, machine learning is finding many important use cases in the financial services industry, especially in the risk management area [1]. Statistical and machine learning tools can be used for credit scoring as well as to get a fair judgment on loan applicants, which helps determine the collateral, loan amount, the period for repayment, and the interest rate [2–4]. Though several approaches are suggested and being used to predict whether a borrower would default on the loan availed, it is still an open challenge.

Banking institutions take much care in identifying the creditworthiness of a customer and mitigate the risk involved at the stage of loan approval [5]. To monitor the loan account post-approval, factors that impact the loan being default is in need. One such quantity is the time-stamped credit–debit information of a customer's savings account information [6]. Prediction or forecasting on time-series data employs recurrent neural networks based on LSTM (Long Short Term Memory) due to their effectiveness [7]. Most of the natural language processing applications apply LSTMs to understand the perspective of sentences and words involved. This paper attempts to predict loan defaults using LSTMs on cross-account information, which includes loan application information and credit–debit information from savings account. Two methods proposed in this paper get the context of transaction data with the help of LSTMs.

This paper is organized through five sections, starting with Introduction. Section 2 deals with related work done to predict loan defaults and provides some preliminaries required in building the methods developed. Section 3 discusses the methods employed to predict loan defaults and discusses the nature of the data set. Section 4 examines the results, followed by a conclusion.

## 2 Preliminaries and Related Work

### *2.1 Loan Default Prediction Problem*

Ozbayoglu et al. discuss various financial problems for which different machine learning techniques employed so far [8]. One such problem discussed is the prediction of loan default problem. The loan default prediction problem formulated in the literature uses the information available on loan availed. The output expected is either a binary or a score that provides creditworthiness. The use of credit scores based on the customer's financial state is the most commonly used measure in this problem. Individual agencies prepare credit scores using the customer's financial data as the prerequisite for the loan application process [6]. The popular loan default prediction approaches depend on loan applicants' details and apply classification based on historical data.

## 2.2 Classifiers for Loan Default Prediction

The application of classifiers can be found in fields ranging from engineering to medicine. Binary classification models are abundant in the literature [9, 10]. Though for a binary problem considered, statistical techniques such as linear regression, probit analysis, and discriminant analysis are the starting point in approaching such problem. Learning algorithms such as decision tree methods, logistic regression, and artificial neural networks proved to work better than statistical methods. Machine learning algorithms extract information from previously labeled data better than statistical methods. An initial classifier choice to predict loan defaults is the artificial neural network, preceded by tree-based classifiers.

Decision tree methods are far superior in binary classification problems [11]. Decision tree methods create a classification tree by forming a decision node and partitions the data recursively based on decision nodes variables. This method helps find the significance of features in a dataset, which is evident from the value used for decision variables. A boosted decision tree method named XGBoost outperforms other methods when it comes to classification [12].

Several classifiers use DNN (Deep Neural Network) for loan default prediction. The input to such DNN-based classifiers is either the customer's information provided during loan application or the information obtained through pre-tested forms for behavior identification [11, 13–15]. Chongren Wang et al. proposed LSTM-based credit scoring to predict default of peer-to-peer loan lending process [16].

## 2.3 Ensemble Learning

Each learning algorithm handles and interprets the available data differently, which reflects in their predicted output. Ensemble learning methods combine outputs from multiple learning algorithms (named as inducers) to predict a judicious result [17]. This method resembles the real-life voting scheme adopted to achieve a conclusion. The simplest version of an ensemble method is to take the average value of predicted values from different learning algorithms. Recent ensemble methods use bagging, boosting, stacking, and random decision trees for their prediction. The ensemble learning method helps mitigate classifiers' common issues, such as class imbalance and overfitting the training data. The exclusivity in training involved inducers makes the technique more efficient [18].

## 2.4 Recurrent Neural Network (RNN)

RNNs are a particular case of an artificial neural network that deals with time-dependent data. However, there are more ways to handle time-series data; for example, neural networks are used to forecast rainfall [19]. Neurons in the same layer of RNNs transfer information among them. Each neuron in the input layer takes information corresponding to a time step. In RNN, the neuron's output depends on the previous layer and neurons from the same layer.

There are different types of RNN cells, such as gated recurrent unit (GRU), long short-term memory (LSTM). For datasets with long time steps, RNNs encounter an issue of exploding gradients. Long short-term memory (LSTM) eliminates this issue with a different variable named cell state and by introducing forget gates. LSTM-based RNN captures the long-term dependencies in time-series data. Bidirectional LSTM is two copies of the LSTM network that consider input sequence in both directions (forward and backward in time); hence, it helps capture information/trends in both time-series directions. Mathematical aspects of the LSTMs described in the paper by Hochreiter et al. [20].

## 3 Methodology

Cross-account information is valuable in mitigating the risks involved in lending loans to customers. Cross-account information is as simple as a savings account through which a loan gets processed. The dynamic information available through the savings account linked with the loan helps to predict loan default. Agarwal et al. used a large dataset to establish that cross-account details are valuable in predicting loan defaults [21]. The premise revolves around the lender and borrower's relationship, which strengthens through the information available on the borrower's transaction accounts. Our methods use cross-account information from a customers' loan account and transactions from savings account related to it.

### *3.1 Data Preparation and Cleaning*

The proposed methods to predict loan defaults were tested using the dataset from a large public sector bank in India. The dataset contains features with categorical, quantitative, and time series in nature. For a customer considered, some of the features corresponding to categorical nature are gender, geographical location, type of loan availed, marital status. Some of the quantitative features are the outstanding amount on the availed loan, age of the customer, the number of loan repayments made, the number of active loan accounts.

Along with structured data mentioned before, a univariate time-series data on daily transactions made through the savings account tagged with the loan. The dataset is a combination of structural and univariate time-series data. One-hot encoding is used for features of categorical type. Quantitative features are scaled using a min-max scaler, and the time-series data available with individual timestamps are summed up monthly for two years, which then scaled using a min-max scaler. The dataset has approximately 46,994 customers with 51,000 loans. Each customer data has a label on whether the loan account is defaulted or not. Data cleaning and preparation are done with the help of scikit-learn [22].

## 3.2 Method 1-RNN-Based Ensemble

An ensemble-like approach is used in this method to predict whether a loan defaults or not. This method uses the recurrent neural network (RNN) architecture to extract the information from the data's time-varying component, and a classifier such as deep neural network (DNN), decision tree-based classifiers, naive Bayes, SVM for the structured part. As a start, structured data is pre-classified using a classifier. Instead of the final binary output, the probability output is provided to the next stage. A stacked RNN architecture with bidirectional LSTM and deep neural network layers was used in this method. Figure 1 shows the network architecture of this method. The network's output layer takes the concatenated vector of the probability from the classifier and the output from RNN-DNN.

The structured part of the data ($x_s$) and the time-varying part ($x_t$) was trained separately as in ensemble methods. This method's advantage lies in training each classifier exclusivity from others, and the modifications are done independently. Like most ensemble methods, the decision output depends on two different classifiers. A single-layer neural network is used to predict the final output with an encoded vector from the time-dependent part of data and the standalone classifier's probability. This method makes online prediction easier, taking advantage of the ensemble learning approach.

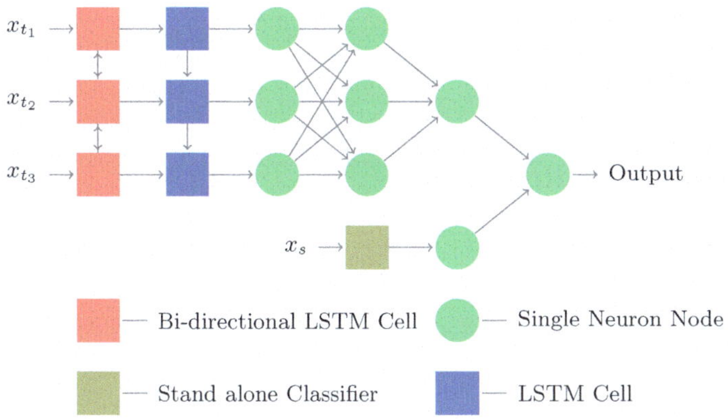


**Fig. 1** Network architecture for Method 1

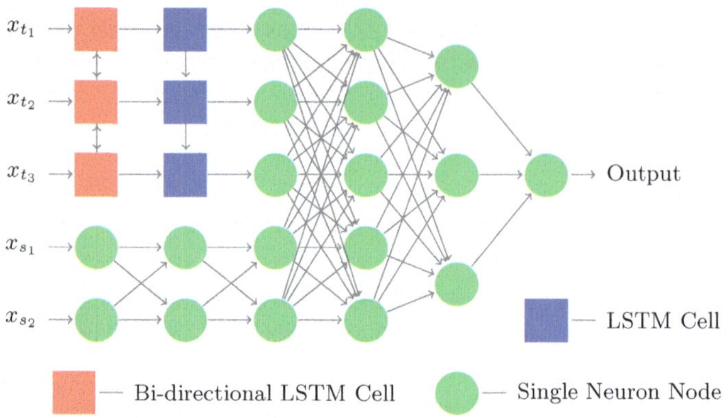


**Fig. 2** Network architecture for Method 2

### *3.3 Method 2—Hybrid DNN-RNN*

This method uses a hybrid network architecture with recurrent layers and deep neural network layers. Similar to the case above, this method uses RNN using LSTMs for the data's time-varying part and deep neural network layer for the structured data. The variation occurs in placing the concatenation of two output layers of RNN and DNN, which resembles as shown in Fig. 2. This method stacks deep neural network layers for the concatenated output. This method sees the cross-account information as a single entity.

Using deep neural network layers for the concatenated output of intermediate layer bests LSTM-based classifier for the sequential transaction data used in Method 1. Change in periodicity of the time-dependent data enforces the reconfiguration of the network's complete structure, whereas, in Method 1, the reconfiguration is done only to LSTM-based classifier.

## 4 Evaluation and Results

The loan savings dataset is split 80% for training and 20% for testing (validation set) randomly. The stopping criteria for training the model is a breakpoint in the graph between the epoch and loss function, after which validation loss increases, whereas training set loss decreases. Performance metrics employed in evaluating the model are the accuracy of prediction and the AUC score. Accuracy and loss after each step of training helped us tune the hyperparameters of the network.

As discussed before, Method 1 uses standalone classifiers such as XGBoost, linear discriminant analysis, and DNN for the structured part of data. For the time-series part of the data, a RNN-based classifier is used. A separate neural network with two

**Table 1** Comparison of Method 1's performance based on the standalone classifier

| Stand-alone | Accuracy | | AUC Score | |
|---|---|---|---|---|
| Classifier | Training set | Validation set | Training set | Validation set |
| Linear discriminant analysis | 0.9192 | 0.9195 | 0.6825 | 0.6758 |
| XGBoost | 0.9860 | 0.9599 | 0.9248 | 0.8089 |
| DNN | 0.9426 | 0.9418 | 0.9078 | 0.8955 |

**Table 2** Comparison of Method 2's performance based on RNN cell type used

| RNN | Accuracy | | AUC Score | |
|---|---|---|---|---|
| Cell type | Training set | Validation set | Training set | Validation set |
| Simple RNN | 0.9360 | 0.9342 | 0.8994 | 0.8698 |
| GRU | 0.9393 | 0.9311 | 0.9099 | 0.8671 |
| LSTM | 0.9412 | 0.9321 | 0.9124 | 0.9091 |
| Bi-directional LSTM | 0.9455 | 0.9427 | 0.9289 | 0.9256 |

input neurons takes the output of two classifiers. Table 1 shows the performance of the ensemble like method (Method 1).

Method 1 with XGBoost performs better than other settings considered. The high accuracy in fitting the training data's structured part gets reflected in the overall prediction. The model is biased towards the loan account information than the savings account transaction data. Method 2 uses a hybrid neural network with few neurons recurrent in nature. Table 2 lists the training and test accuracy for different RNN cell types and their AUC score.

Hybrid RNN-DNN architecture (Method 2) provides significant importance to a customer's static and temporal data. Four different types of RNN cells are applied to the temporal part of the data. Among the types used, RNN cell with bidirectional LSTMs proves to be effective in improving a classifier's performance, which is evident through the AUC score. Figure 3 shows the variation of training and testing accuracy at each epoch for both methods. Training the model posts the accuracy listed in the table resulted in the model overfitting the test dataset. TensorFlow is used for the modeling and evaluation of the networks [23].

Time and space complexity for training and testing the networks presented above depend on the features and labeled dataset available, while that for the output prediction depends on the network's parameters and the features considered. Standalone classifiers used in method 1 are far better in terms of time complexity; still, the need for a classifier based on sequential data outweighs its complexity advantage. Among the different network architectures discussed, bidirectional LSTMs take a longer duration for training, testing, and prediction. The advent of high-performance archi-

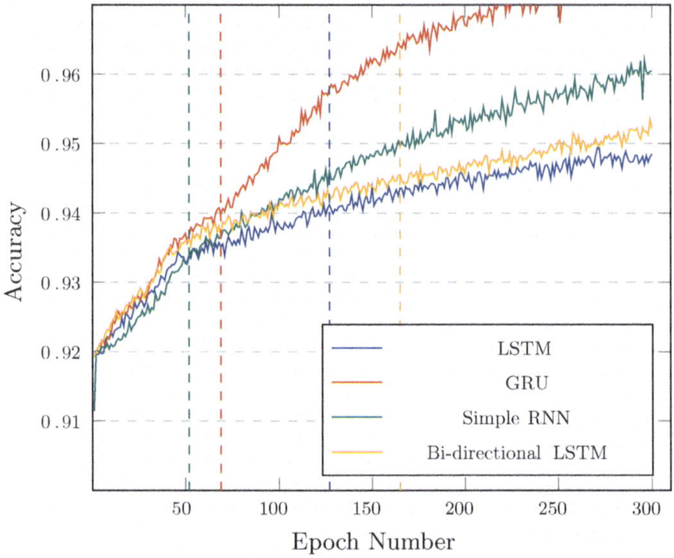


**Fig. 3** Variation of accuracy at the end of an epoch with different RNN cell types

tectures and codes that use it helps reduce the time required drastically for training and testing recurrent neural networks.

The proposed methods depend on close ties between loan account information and the transaction data from the savings account. Customers involved in, may have many savings/deposit accounts at different banks. Method 2's prediction capability decreases significantly in such cases, whereas Method 1's prediction leans towards the standalone classifier's output.

## 5 Conclusion

Online loan default prediction using temporal changes in loan applicants' financial terms aids banking institutions to handle the risks involved. Such systems assist in being an early warner. This paper proposed two methods that drive us towards online loan default prediction using loan application data and time-varying credit debit information from savings account. LSTMs applied to the credit–debit sequence helped in understanding the context towards default. Among the proposed methods, the hybrid DNN-RNN approach with bidirectional LSTMs gives better results on previously unseen data making it more generalizable.

The utilization of the proposed methods to credit card defaults is straightforward through transactions done. Though our approach aims towards online prediction and an early warning system, these methods will help us in the loan lending decision process by analyzing an applicant's past financial status. Thereby the discussed techniques can also serve as an alternative to available credit scoring methods. As stressed earlier, the significance of the customer's savings account bound with the loan account is reflected in the methods' performance. The next significant factor in improving/decreasing the accuracy of methods is the duration under consideration for the sequential data. Thus, it is essential to establish the relationship that prevails between duration considered and its importance towards default.

Thus, this work's future scope will be using a multivariate time series that provides context to the seasonal variations through openly available economic indicators. Further, the availability of information such as repayment status, remaining repayable amount at every time step will improve the credibility of loan default prediction methods discussed above across different data distributions. With the potential multivariate time-series data described, anomaly detection techniques can improve the loan default prediction.

## References

1. Dumitrescu, E.I., Hué, S., Hurlin, C., Tokpavi, S.: Machine learning or econometrics for credit scoring: let's get the best of both worlds. SSRN Electron. J. (2020)
2. Munkhdalai, L., Munkhdalai, T., Namsrai, O.E., Lee, J., Ryu, K.: An empirical comparison of machine-learning methods on bank client credit assessments. Sustainability **11**, 699 (2019)
3. Pang, S.-L., Wang, Y.-M., Bai, Y.-H.: Credit scoring model based on neural network. In: Proceedings. International Conference on Machine Learning and Cybernetics, vol 4, pp. 1742–1746 (2002)
4. Van Liebergen, B.: Machine learning: a revolution in risk management and compliance. J. Finan. Transform. **45**, 60–67 (2017)
5. Zurada, J., Zurada, M.: How secure are good loans: validating loan-granting decisions and predicting default rates on consumer loans. RBIS **6**(3), 65–84 (2002)
6. Treigueiros, D., Berry, R.: The application of neural network based methods to the extraction of knowledge from accounting reports. In: Proceedings of the Twenty-Fourth Annual Hawaii International Conference on System Sciences, vol 4, pp. 136–146 (1991)
7. Rather, A.M., Agarwal, A., Sastry, V.: Recurrent neural network and a hybrid model for prediction of stock returns. Expert Syst. Appl. **42**(6), 3234–3241 (2015)
8. Ozbayoglu, A.M., Gudelek, M.U., Sezer, O.B.: Deep learning for financial applications: a survey. Appl. Soft Comput. **93** (2020)
9. Rekha, A., Abdulla, M.S., Asharaf, S.: Lightly trained support vector data description for novelty detection. Expert Syst. Appl. **85**, 25–32 (2017)
10. Sharma, P., Wahlang, I., Sanyal, S., Maji, A. In: Classification of Brain MRI Using Deep Learning Techniques, pp. 559–569. Springer (2020)
11. Chou, T.: An explainable hybrid model for bankruptcy prediction based on the decision tree and deep neural network. In: 2019 IEEE 2nd International Conference on Knowledge Innovation and Invention (ICKII), pp. 122–125 (2019)
12. Chen, T., Guestrin, C.: XGBoost.: A scalable tree boosting system. In: Proceedings of the 22nd ACM SIGKDD International Conference on Knowledge Discovery and Data Mining, pp. 785–794. KDD '16, ACM (2016)
13. Atiya, A.F.: Bankruptcy prediction for credit risk using neural networks: a survey and new results. IEEE Trans. Neural Netw. **12**(4), 929–935 (2001)
14. Kruppa, J., Schwarz, A., Arminger, G., Ziegler, A.: Consumer credit risk: individual probability estimates using machine learning. Expert Syst. Appl. **40**(13), 5125–5131 (2013)

15. Naidu, G.P., Govinda, K.: Bankruptcy prediction using neural networks. In: 2018 2nd International Conference on Inventive Systems and Control (ICISC), pp. 248–251 (2018)
16. Wang, C., Han, D., Liu, Q., Luo, S.: A deep learning approach for credit scoring of peer-to-peer lending using attention mechanism LSTM. IEEE Access **7**, 2161–2168 (2019)
17. Dong, X., Yu, Z., Cao, W., Shi, Y., Ma, Q.: A survey on ensemble learning. Front. Comput. Sci. **14** (2019)
18. Marqués, A.I., García, V.: Exploring the behaviour of base classifiers in credit scoring ensembles. Expert Syst. Appl. **39**, 12 (2012)
19. Mahajan, A., Rastogi, A., Sharma, N. In: Annual Rainfall Prediction Using Time Series Forecasting, pp. 69–79. Springer (2020)
20. Hochreiter, S., Schmidhuber, J.: Long short-term memory. Neural Comput. **9**, 1735–80 (1997)
21. Agarwal, S., Chomsisengphet, S., Liu, C., Song, C., Souleles, N.: Benefits of relationship banking: evidence from consumer credit markets. J. Monet. Econ. **96** (2018)
22. Pedregosa, F., Varoquaux, G., Gramfort, A., Michel, V., Thirion, B., Grisel, O., Blondel, M., Prettenhofer, P., Weiss, R., Dubourg, V., Vanderplas, J., Passos, A., Cournapeau, D., Brucher, M., Perrot, M., Duchesnay, E., Louppe, G.: Scikit-learn: machine learning in python. J. Mach. Learn. Res. **12** (2012)
23. Abadi, M., Agarwal, A., Barham, P., Brevdo, E., Chen, Z., Citro, C., Corrado, G.S., Davis, A., Dean, J., Devin, M., Ghemawat, S., Goodfellow, I., Harp, A., Irving, G., Isard, M., Jia, Y., Jozefowicz, R., Kaiser, L., Kudlur, M., Levenberg, J., Mané, D., Monga, R., Moore, S., Murray, D., Olah, C., Schuster, M., Shlens, J., Steiner, B., Sutskever, I., Talwar, K., Tucker, P., Vanhoucke, V., Vasudevan, V., Viégas, F., Vinyals, O., Warden, P., Wattenberg, M., Wicke, M., Yu, Y., Zheng, X.: TensorFlow: large-scale machine learning on heterogeneous systems (2015) Software available from tensorflow.org

# Use of Marker-Controlled Watershed Segmentation to Classify Cumulonimbus Cloud with Pre-trained CNN

**Sitikantha Chattopadhyay, Subhra Prokash Dutta, Sanjukta Mishra, and Suprativ Saha**

**Abstract** Cumulonimbus cloud is the main ingredient of thunderstorm over a specific area. Although a lot of other weather parameters like temperature, air pressure, humidity, etc., are needed to correctly predict the probability of thunderstorm, presence of cumulonimbus cloud is definitely accelerate the decision taking process. In this paper, we have taken around 545 of cloud images which are divided into three classes namely cumulonimbus, cirrus, and stratus cloud. From those images, only cloud part is segmented using watershed transformation and finally these three thousand cloud portions are used to train a pre-trained convolutional neural network (CNN), namely AlexNet. After training, it is correctly predicting the presence of cumulonimbus cloud from a cloud image. We have tested this method in MATLAB and it has given a promising result of 84% accuracy.

**Keywords** Cumulonimbus cloud · AlexNet · Watershed segmentation

## 1 Introduction

Our environment is chaotic in nature means the behavior of it cannot be fully predicted. Lot of whether parameters like air pressure, temperature, humidity, cloud types, changing of cloud types, etc., can make a drastic change in the atmosphere over a specific region. Among many natural disasters, thunderstorm is the most common. Generally the chance of thunderstorm is predicted by analyzing the changes over time of these weather parameters over a specific area. In this paper, we have proposed a method to find out the cumulonimbus cloud from directly the cloud images since this type of cloud is the main ingredient of thunderstorm.

S. Chattopadhyay · S. P. Dutta · S. Mishra · S. Saha (✉)
Department of Computer Science and Engineering, Brainware University, Barasat, Kolkata 700125, India



T. K. Sharma et al. (eds.), *Soft Computing: Theories and Applications*, Advances in Intelligent Systems and Computing 1381,
https://doi.org/10.1007/978-981-16-1696-9_6

In digital image processing, image segmentation [1–3] is mainly used to segment a specific region of interest (ROI) from a digital image. Different methods of image segmentation are available. Among them morphological image segmentation is a popular one. Here, we have used marker-controlled watershed segmentation [4] to segment the cloud part from those images.

Machine learning and deep learning are the burning research topics nowadays. Those are the techniques by which a computer learns from the experience to do a job which it has not done previously. In other words, by these techniques a computer will learn from the experiences without being explicitly programmed. There are three types of machine learning techniques namely supervised, unsupervised and reinforcement learning [5].

Here, we have taken three thousand cloud images and segmented only the cloud portions from those images using marker-controlled watershed segmentation technique. Then we have used these images to train a pre-trained convolutional neural network (CNN), namely AlexNet [6]. After this training, the network can classify three types of clouds from a cloud image on which AlexNet was not previously trained on.

We have organized this paper in following modules. Followed by introduction in Sects. 1, 2, 3 and 4 will consist of Different cloud types, marker-controlled watershed segmentation, and pre-trained CNN-AlexNet, respectively. In Sect. 5, proposed method is given followed by implementation in Sect. 6. Section 7 is the experimental results followed by conclusion in Sect. 8.

## 2 Different Cloud Types

There are lots of cloud types present in our environment. But our focus is in three primary types of clouds namely cumulonimbus, cirrus, and stratus cloud.

1. Cumulonimbus cloud is a puffy or cotton-like cloud has high density and comparatively vertical shaped. It formed from water vapor and is the main ingredient of several severe weather like tornadoes and thunderstorm. In most of the situations, its top portion is of "anvil shaped". Actually cumulous cloud gives pleasant weather but it can turn into cumulonimbus cloud which is dangerous in nature [7]. In this work, our main objective is to identify this type of cloud to predict thunderstorm over a specific region.
2. Cirrus clouds are look like curling lock of hair and considered as one of the high-level cloud. They are made of ice crystals and usually thin. This type of cloud gives pleasant weather, so they do not produce rain. Most of the time its color is white.

3. Stratus cloud is considered as a low-level cloud, generally lies below 6500 ft. They are layers like shaped and appear rainy, snowy, and dense. They are also visible with cottony white clumps and spread over a region. They are horizontally layering cloud with uniform base. Color of stratus cloud varies from dark gray to white.

Sometimes cloud forms in a combination of these three types of basic clouds. Some examples are cirrocumulus, cirrostratus, Altocumulus, nimbostratus, stratocumulus cloud, etc.

## 3 Marker-Controlled Watershed Segmentation

In case of a noisy image, gradient mask of a high-pass spatial filter performs poorly. The better alternative is to take a controlled mask that can operate with a noisy image. Watershed transformation is a widely accepted technique to identify the touching objects present in an image. It works well with an image with no noise. But in case of noisy image, it produces "Over-Segmentation" problem. This can be controlled by the use of controlled marker which is the connected components in an image. This algorithm treats the image as a surface where light pixels have high value and dark pixels have low value to find out cachement basins and watershed ridge lines. In this method, two types of markers are used namely internal markers and external markers. Internal markers are mainly attached with objects of interest which we want to segment (cloud part in an image) and external markers are attached with background (sky and other objects). The use of controlled marker like stochastic gradient along with watershed transformation is a powerful tool to handle over segmentation problem.

If foreground and background objects can be identified, watershed transformation works better. There are basically five steps are involved in this technique

1. Calculation of segmentation function.
2. Identification of foreground marker.
3. Identification of background marker.
4. Modification of segmentation function to contain only minima at both the markers.
5. Calculation of watershed transformation with this modified segmentation function.

Figure 1 represents the use of marker-controlled watershed segmentation in a normal image. In the left, the original image has been taken, and in the right, the segmented parts are shown.

**Fig. 1** An input image (Left) and corresponding segmented image (Right) by applying marker-controlled watershed segmentation

## 4 Pre-trained CNN (ALEXNET)

To extract features from different cloud images, we have to use convolutional neural network which is a multi-layer artificial neural network. All the hidden layers are designed to extract some feature from a digital image. To perform a machine learning task, we can adopt any one of two approaches. Either we have to design a fully functional CNN and train that with lot of digital images of same kind or use a pre-trained CNN. In the second approach, only few hundreds of digital images are sufficient to train that CNN to perform the machine learning task. There are many pre-trained CNN exist to use in the domain of digital image segmentation. Among them AlexNet is a popular one. Some other these kinds of networks are GoogLeNet [8], VGG-16 [9], VGG-19 [10] etc.

AlexNet [11, 12] is a fast and efficient pre-trained CNN which is already trained with a subset of ImageNet database. This database is actually used in ImageNet Large-Scale Visual Recognition Challenge (ILSVRC). This dataset is a collection of over 15 million high-resolution digital images which are subdivided into 22,000 categories. AlexNet is trained on almost 1.2 million digital images from the ImageNet dataset and can identify 1000 different image categories. Some of the examples of these categories are mouse, keyboard, different animals, etc. Since it is already trained with a very large numbers of digital images, feature extraction from an image is more effective which in turns improves the performance for new set of images.

Figure 2 represents the total structure of AlexNet [13]. It consists of total 25 layers among which 1 is input layer, 1 is output layer, and 23 are hidden layers. Among these hidden layers, there are eight layers with learnable weights, five numbers of convolutional layers, and three numbers of fully connected layers. For our new set of cloud images, we have used transfer learning approach to use this pre-trained network. All the images in AlexNet are of $227 \times 227 \times 3$ resolution [14].

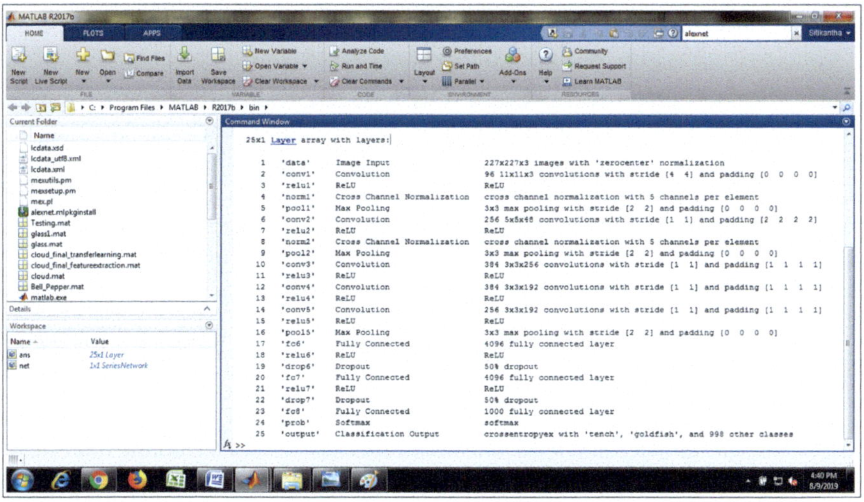


**Fig. 2** Structure of AlexNet

## 5 Proposed Method

AlexNet is already trained with millions of digital images and can categorize 1000 different types of objects. Some of these object categories are different fish like tiger shark, great white shark, etc., different birds like hen, cock, etc., and there are others this kind of categories. we have used the same network to identify three different types of clouds namely cumulonimbus, cirrus, and stratus cloud. Since we cannot change the internal architecture of the network, we have removed three existing categories and included these three cloud types as three different categories along with 997 existing one. This technique is known as transfer learning.

In transfer learning, compared with a fresh and new CNN, small numbers of images are required from each new category for training purpose. Therefore, training with pre-trained network is much more time effective than designing a new network. So, we have taken some digital images of cumulonimbus, cirrus, and stratus cloud. There are 172 numbers of cirrus cloud images, 169 numbers of cumulonimbus cloud images, and 204 numbers of stratus cloud images. Then we have trained AlexNet with this new set of cloud images. Before that all these images have been resized as $227 \times 227 \times 3$ resolution because this network accepts images of only this size. When training is complete, then it can successfully categorize these three types of clouds.

## 6 Implementation

To implement this proposed work, at first we have taken 545 different cloud images and made three groups as cumulonimbus, cirrus, and stratus cloud. Then all the images are reduced to the size of 227 × 227 × 3 resolution. This is the resolution AlexNet accepts for any digital images. Then we have loaded this image collection into the training domain. Now we have divided this image collection into two groups namely training set and test set. We have taken 80% of these images as training set and remaining as test set or validation set. Since AlexNet can identify only 1000 different image categories, we have modified the second last fully connected layer "fc7" to identify these three new categories of cloud images in place of last three existing categories.

Now we have trained AlexNet with this new cloud images with some additional attributes like mini-batch size = 10, maximum epochs = 4, initial learning rate = le-4, etc. After the training, this network is effectively categorizing these three types of clouds at an accuracy of 82%.

There is another way to implement the proposed approach. Instead of transfer learning, we can also use feature extraction from a CNN and use these features to predict the correct category among these new set of images. But this approach is giving the accuracy level of 79% instead of 82% in transfer learning approach. Since in this approach actual CNN is not trained with this new image set, the accuracy level can only be achieved by the mean of training levels and validation levels.

## 7 Experimental Results

We have simulated the experiment by MATLAB R2017B. We have used Image Processing Toolbox and Machine Learning Toolbox to implement the experiment.

Figure 3 represents the simulation process of transfer learning method using the pre-trained network and Fig. 4 represents the corresponding feature extraction method. It has clearly seen that Although the progress is similar, transfer learning is giving even better result compared with feature extraction. Figure 5 represents some validation result corresponding to the transfer learning technique.

## 8 Conclusion

In this method, we have taken only 545 different cloud images and trained with a pre-trained CNN. We have got the accuracy level of correct prediction as 84%. This can be improved in several ways. Firstly, if the numbers of cloud images is increased, then training will be better and corresponding result will be definitely improved. Secondly, these images are not pre-segmented images. All these images

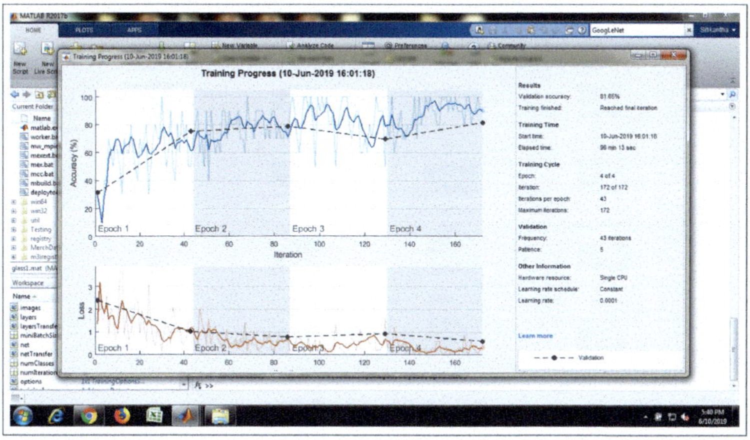

Fig. 3 Transfer learning method progress

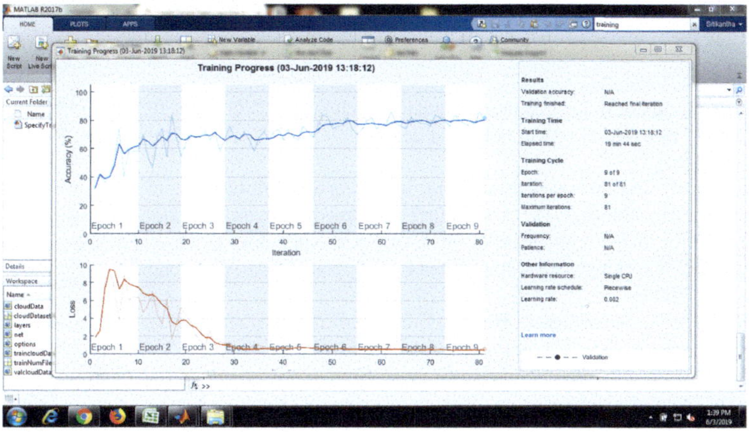

Fig. 4 Feature extraction method progress

have cloud portion along with other parts such as trees, birds, rocks, and some captions. If all these images will be pre-segmented and only the cloud portions are used for training purpose, then result will be improved. Furthermore, we have other pre-trained CNN as options like GoogLeNet, vgg-16, etc. May be these network will give even better result compared with AlexNet. We have used marker-controlled watershed segmentation to segment the cloud portions from those images inside the network. If we use different structuring element for segmentation purpose, then also result can be improved.

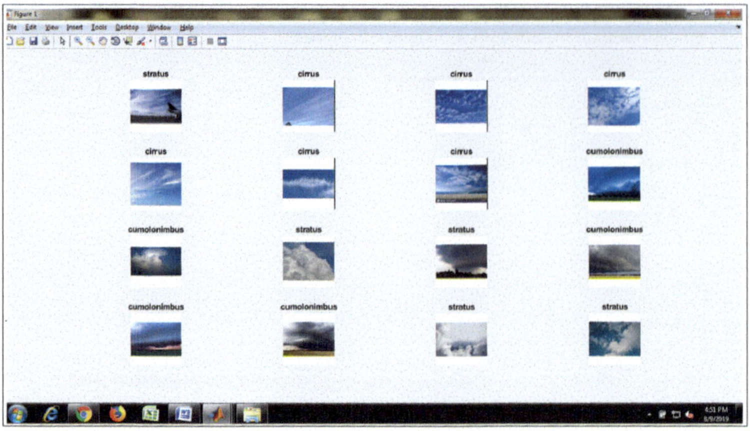


**Fig. 5** Sample validation images of different cloud types

# References


1. Harlick, R., Shapiro, L.: Image segmentation technique. CVGIP **29**, 100–137 (1985)
2. Meyer, F., Beucher, S.: Morphological segmentation. J. Vis. Commun. Image Represent. **1**, 21–46 (1990)
3. Gonzalez, R.: Digital Image Processing, 3rd'. Pearson Hall (2008)
4. Acharjya, P.P., Ghoshal, D.: A modified watershed segmentation algorithm using distances transform for image segmentation. Int. J. Comput. Appl. **52**(12), 47–50 (2012)
5. Bengio, Y.: "Learning deep architectures for AI", foundations and trends in machine. Learning **2**, 1–127 (2009)
6. Krizhevsky, A., Sutskever, I., Hinton, G.E.: ImageNet classification with deep convolutional neural networks. Adv. Neural Inf. Process. Syst. (2012)
7. Liou, K.N.: An Introduction of Atmospheric Radiation, 2nd edn. Academic Press, Chapter 8 (2002)
8. BVLC GoogLeNet Model
9. Very Deep Convolutional Networks for Large-Scale Visual Recognition
10. Russakovsky, O., Deng, J., Su, H.: ImageNet large scale visual recognition challenge. J. Vis. Commun. Image Represent. **1**, 21–46 (1990)
11. Sudha, K.K., Sujatha, P.: A qualitative analysis of Googlenet and Alexnet for fabric defect detection. Int. J. Recent Technol. Eng. **8**(1), 86–92 (2019)
12. Muhammad, N.A., Nasir, A.A., Ibrahim, Z., Sabri, N.: Evaluation of CNN, Alexnet and Googlenet for fruit recognition. Indo. J. Electr. Eng. Comput. Sci. **12**(2), 468–475 (2018)
13. Bhalerao, S., Ansari, I.A., Kumar, A.: Performance comparison of SVM and ANN for reversible ECG data hiding. Soft Comput. Theories Appl. Adv. Intell. Syst. Comput. **1154**, 197–207 (2019)
14. Kishore, V.V., Kalpana, V.: Effect of noise on segmentation evaluation parameters. Soft Comput. Theories Appl. Adv. Intell. Syst. Comput. **1154**, 443–453 (2019)

# Optimizing Drug Schedule for Cell-Cycle Specific Cancer Chemotherapy

**Bharti Panjwani, Vijander Singh, Asha Rani, and Vijay Mohan**

**Abstract** This article presents optimization of drug schedule for cycle-specific chemotherapy by multi-objective evolutionary algorithm NSGA-II. The possible response of tumor and normal cells under the influence of cycle-specific drugs is formulated as an optimal control problem. The objective is to minimize the tumor burden, with minimum harm to the patient's health. Elitist non-dominated sorting genetic algorithm-II optimizes the drug schedule under constrained toxicity and drug concentration. Results provide multiple solutions to oncologists to choose the drug regimen according to patient's physiology. Comparative analysis of optimized drug schedules with previous work for cycle-specific chemotherapy is also presented. NSGA-II optimized drug schedule shows promising results in effectively solving the conflicting multi-objective constrained problem of chemotherapy drug scheduling.

**Keywords** Drug scheduling · Cell cycle-specific chemotherapy · Non-dominated sorting genetic algorithm-II (NSGA-II)

## 1 Introduction

Cancer is caused by abnormal cell growth in body that leads to death in the absence of treatment. In arsenal of treatments for cancer, chemotherapy is an essential remedy proven to be highly effective [1]. Cytotoxic drugs are infused in the patient's body in a specific pattern, i.e., repeated in cycles during chemotherapeutic treatment. These drugs kill the fast-growing cancer cells in a planned manner but they also affect the normal healthy cells in patient body [2]. The main aim of chemotherapy is to reduce the tumor burden in few treatment cycles but with minimum overall side effects. Therefore, the drug schedule needs to be carefully designed according to patient's physiology. It is crucial to optimize the drug schedule in order to achieve a balance

B. Panjwani (✉) · V. Singh · A. Rani
Netaji Subhas Institute of Technology, University of Delhi, Delhi, India

V. Mohan
School of Automation, Banasthali Vidyapith, Tonk, Rajasthan, India

 
T. K. Sharma et al. (eds.), *Soft Computing: Theories and Applications*,
Advances in Intelligent Systems and Computing 1381,
https://doi.org/10.1007/978-981-16-1696-9_7

between cancer cell killing and its side effects. Optimization of the drug schedule may result in better health condition of the patient while saving money and time.

The mathematical models describe the relationship between cancer cells, normal cells and effect of chemotherapeutic drugs depending upon its pharmacokinetics and pharmacodynamics. Various models [3–5] are available in literature to predict tumor growth under effect of chemotherapy. These formulated models provide cost-effective platform for trial and optimization of new drugs schedules and combination therapies. Swan in [6] presented a review on the development of various cancer chemotherapy models based on different mechanisms of tumor growth. Recently, Moore [7] presented the importance of application of optimal control in finding drug regimens in silico by taking examples of patients infected with human immunodeficiency virus (HIV). Authors in [8, 9] utilized Gompertz cell growth model for cell cycle non-specific drugs and optimized the schedule by utilizing memetic algorithm and adaptive elitist population-based genetic algorithm (AEGA), respectively. Two different models of chemotherapeutic treatment are presented in [10] for drug schedule optimization. An OCP is designed, which is solved by utilizing mixed integer dynamic optimization for minimization of tumor size.

In recent studies [11], closed-loop control based upon traditional PID controllers, fuzzy logic controllers [12] and advanced hybrid controllers [13–16] are employed to determine the optimal drug schedules for cancer chemotherapy. Authors in [17] used a combination of fractional calculus and IMC to find a continuous chemotherapeutic drug dose to the patient on daily basis. Alam et al. [18] designed an integral–proportional–derivative (I-PD) controller to regulate the drug infused to the patient's body for cell cycle-specific cancer tumor models and also used MOGA to find acceptable drug concentration at tumor site and optimize parameters of the controller. Closed-loop control of drug infusion requires fixed reference drug input and continuous monitoring of various clinical output parameters which is not practical possible. Also, continuous drug administration may lead to higher side effects causing harm to patients and is therefore infeasible technique for drug scheduling. Therefore, open-loop drug schedule optimization by NSGA-II is proposed in this paper.

In this work, multi-objective elitist non-dominated sorting genetic algorithm-II is proposed to determine optimal drug schedule for cell cycle-specific chemotherapy. A constrained optimization problem is formulated to minimize the tumor size in fixed duration of treatment. NSGA-II determines effective drug dosages to maintain drug concentration and required toxicity in patient body. Multiple optimized schedules obtained by NSGA-II are analyzed, and comparison with previous work in this field is also presented. This paper is organized as follows. Section 2 describes the drug administration dynamic model for cell cycle-specific chemotherapy. The problem formulated for optimization is presented in Sect. 3. Implementation of optimization algorithm NSGA-II is explained in Sects. 4 and 5 and presents the results and analysis of drug schedules. Conclusions are drawn in Sect. 6.

## 2 Mathematical Formulation of Cell Cycle-Specific Chemotherapy

A mathematical model defines the relationship between growth of tumor cells, normal cells, drug dosages and toxicity caused by infused drugs [19]. Based on drug mechanism, two compartment model by [20] are proposed for cell cycle-specific chemotherapy. As these drugs act specifically on the cells in division phase, the total tumor population is classified as proliferating cells and quiescent cells, forming two compartments. The proliferating compartment consists of actively dividing cells in cycling state, whereas quiescent population consists of cells in resting state which can proliferate later in life cycle [21]. Figure 1 shows the dynamics between these compartments along with drug action of cycle-specific chemical agents. As per clinical studies, tumor burden present at the time of diagnosis is $10^{12}$ cells. Moreover, only 20% of total cell population are in proliferating compartment, while 80% of total cell population is in quiescent state [10].

Drugs delivered are cytotoxic in nature which kills both fast-growing cancer as well as normal cells. Therefore, healthy cells are also considered in this work with initial population of $10^9$ cells. The differential equations describe the variation in tumor and normal cell population under effect of cycle-specific chemotherapy.

$$\mathrm{d}P/\mathrm{d}t = (a - b - c)P(t) + dQ(t) - g(t)P(t) \tag{1}$$

$$\mathrm{d}Q/\mathrm{d}t = cP(t) - (d + e)Q(t) \tag{2}$$

$$\mathrm{d}M/\mathrm{d}t = \alpha M(t)(1 - M(t)/\beta) - g(t)M(t) \tag{3}$$

Here, $P(t)$, $Q(t)$ and $M(t)$ represent proliferating cells, quiescent cells and normal cells, respectively. Equation (1) describes the rate of change of proliferating cells at tumor site during the treatment period. Equation (2) depicts the rate of change in quiescent cells, while change in normal cell population due to effect of chemotherapy is given by logistic Eq. (3). The drug effect on the cell proliferating and normal cell population is produced by parameter $g(t)$. Moreover, the immigration of cells between the compartments depends upon constant parameters $a$, $b$, $c$, $d$ and $e$ that are defined with values in Table 1. Also, $\alpha$ and $\beta$ represent the growth rate and carrying capacity

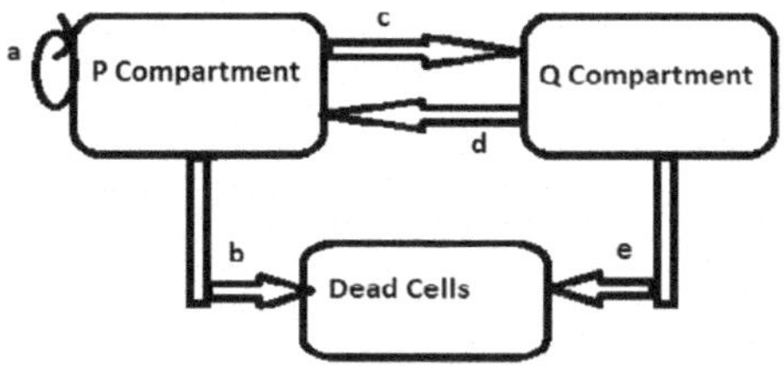


**Fig. 1** Two compartment model of cancer chemotherapy

**Table 1** Parameters for cycle-specific compartmental model [10]

| Parameter | Description | Value |
|---|---|---|
| $a$ | $P$ cells growth rate | 0.5 cells per day |
| $b$ | $P$ cells natural death rate | 0.477 cells per day |
| $c$ | Rate of immigration from $P$ to $Q$ cells | 0.218 cells per day |
| $d$ | Rate of immigration from $Q$ to $P$ cells | 0.05 cells per day |
| $e$ | $Q$ cells natural death rate | 0.0173 cells per day |
| $\alpha$ | Growth rate for normal cells | 0.1 per day |
| $\beta$ | Carrying capacity of normal cells | $10^9$ cells |
| $\gamma$ | Drug decay rate | 0.27 per day |
| $\mu$ | Toxicity elimination rate | 0.4 per day |
| $\delta$ | Cell kill effect | 0.0084 |

of normal cells. Pharmacokinetics of drug determines its effect on tumor as well as healthy normal cell population [22]. Equations (4) and (5) show the rate of change of drug concentration and toxicity in patient body for complete duration of treatment, where $u(t)$, $D(t)$ and $T(t)$ specify drug dose to be infused, drug concentration and toxicity in patient body, respectively. Equation (6) defines the killing effect produced by drug depending upon amount of drug concentration and its metabolism.

$$\mathrm{d}D/\mathrm{d}t = u(t) - \gamma D(t) \tag{4}$$

$$\mathrm{d}T/\mathrm{d}t = D(t) - \mu T(t) \tag{5}$$

$$g(t) = \delta D(t) \tag{6}$$

Here, $\gamma$ is the drug decay rate, $\mu$ is parameter related to elimination of toxicity, and $\delta$ is parameter related to cell kill. Values of these parameters are also recorded in Table 1

Dynamic mathematical model of tumor under the effect of single cell cycle-specific drug is simulated in MATLAB/Simulink using the described equations. Optimal drug schedule is determined for the simulated patient model.

## 3 Problem Formulation

The objective is to determine optimal chemotherapy regime according to the tumor state [23] and patient's condition. The dosage should be such that the minimization of cancer cells is achieved with minimum harm to healthy cells. If the drugs delivered are low, cancer cell may turn resistant to drugs, and if the drugs delivered are higher, toxicity may increase which puts patient in danger with increased side effects. Therefore, an optimal problem formulation is required to determine efficient drug schedule. Equations (7) and (8) described below define the objectives to be minimized.

$$\text{Objective 1: } Z1 = P(t) + Q(t) \tag{7}$$

$$\text{Objective 2: } Z2 = \int_{0}^{86} \left\{10^9 - M(t)\right\} \mathrm{d}(t) \tag{8}$$

In this work, final day of treatment is 86 days, on which monitoring of the tumor burden is carried out. In order to ensure patient safety, constraints are applied on drug concentration and toxicity levels. Constraint is also placed on minimum number of normal cells remaining during treatment to limit the harm on healthy cells.

$$\text{Constraint 1: } 0 \leq D(t) \leq 50 \tag{9}$$

$$\text{Constraint 2: } T(t) \leq 100 \tag{10}$$

$$\text{Constraint 3: } M(t) > 10^8 \text{ cells} \tag{11}$$

Hence, the problem to determine drug dosage $u(t)$ for duration of treatment which minimizes both the objectives while satisfying each constraint is formulated. In clinical practice, standard regime and schedules based on trials and empirical data are utilized. Therefore, in this work benchmark schedule of administering weekly dosage is considered and optimized. It provides additional advantage of improving patient's physiological condition with decreased toxicity in the gap period between consecutive drug deliveries. Furthermore, evolutionary NSGA-II is proposed to solve the multi-objective constrained problem of cell cycle-specific chemotherapy scheduling.

## 4 Non-dominated Sorting Genetic Algorithm-II (NSGA-II)

NSGA-II is a multi-objective optimization technique that provides multiple good solutions instead of a single best solution for any given problem [24]. NSGA-II is advanced and improved in comparison with basic genetic algorithm [25] in terms of less computational complexity, preserving good solutions and better constraint handling strategy. It gives a choice to designer for selection of suitable solution based upon the requirement by trading off between various objectives. The algorithm [26, 27] provides all possible good solutions for problems with conflicting objectives. It optimizes the solution by improving the adaptive fitness function of population set, to form a Pareto front of optimal solution on the basis of non-dominated sorting and crowding distance [28, 29]. The basic design steps for NSGA-II are recorded below:

1. Generate initial population of size $P$ randomly.
2. Evaluate the objective functions for first generation.
3. Sort the population on basis of non-domination and provide rank to each solution.
4. Select parents for producing next generation by binary tournament selection by comparing their non-dominance ranks.
5. Generate child population by applying genetic operators such as crossover and mutation on selected parents.
6. Evaluate the objective functions for child population generated.
7. Combine parent and child population to get total population of size $2P$ and rank them according to non-dominance level.
8. Select best rank solutions to obtain population size $P$ and discard other solutions with lower ranks.
9. Check stopping criterion of algorithm or maximum number of generations, if satisfied stop further generation and plot the Pareto front of optimized solutions.
10. Repeat steps 4–9, if criterion not satisfied.

## 5 Simulations and Results

This work focuses on optimal drug scheduling by elitist NSGA-II algorithm for two compartment cell cycle-specific chemotherapy. The dynamics of tumor growth under influence of cycle-specific drugs is simulated in MATLAB with sampling time of 1 day. Optimization problem is then formulated by considering objectives and constraints specified in Eqs. (7)–(12), which is solved by employing non-dominated sorting genetic algorithm-II. The population size of 50 and maximum generations is 100 for the NSGA-II optimization. Polynomial mutation and simulated binary crossover methods are utilized. The drug schedule provides optimized input drug dosages from day 1 to 86 with dosage limited to the range (10–20). End of treatment cycle is considered on 86th day on which the tumor burden is monitored. Multiple optimized drug regimens are plotted in objective space, thus forming Pareto front as shown in Fig. 2.

All the optimized solutions are categorized as aggressive, moderate and mild therapy. In aggressive chemotherapy, the highest number of cancer cells is killed at cost of higher number of healthy cells as well. In moderate therapy, the dosage is such that satisfactory reduction of tumor cells is achieved at tolerable loss of healthy

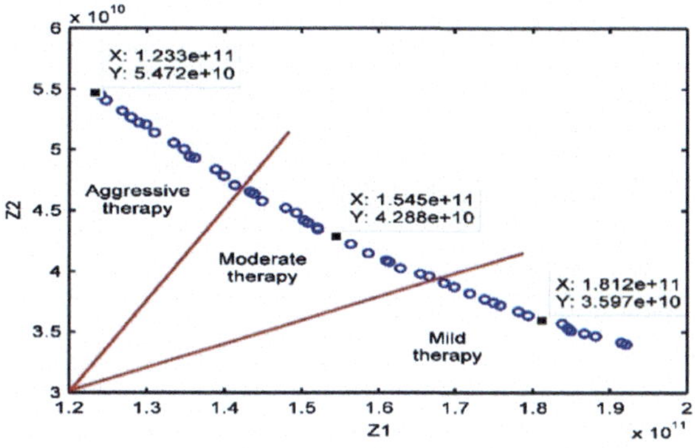


**Fig. 2** Pareto front of optimized solutions

**Table 2** Performance measures for different chemotherapy schedules by NSGA-II

| Therapy | Proliferating cells | Quiescent cells | Normal cells | % reduction in *P* cells | % reduction in *Q* CELLS |
|---|---|---|---|---|---|
| Aggressive | $1.95 \times 10^{10}$ | $1.04 \times 10^{11}$ | $1.88 \times 10^{8}$ | 90.25 | 87.00 |
| Moderate | $2.46 \times 10^{10}$ | $1.29 \times 10^{11}$ | $2.32 \times 10^{8}$ | 87.70 | 83.86 |
| Mild | $2.88 \times 10^{10}$ | $1.52 \times 10^{11}$ | $3.13 \times 10^{8}$ | 85.60 | 81.00 |

cells. Mild therapy is for patients, who cannot deal with side effects of therapy, and therefore, maximum healthy cells are preserved. One schedule from each section is considered and highlighted in the Pareto front. The performance of the selected schedules is analyzed at the end of treatment and recorded in Table 2.

The main is to obtain the optimally designed schedule in order to reduce maximum tumor burden. It is observed that all the optimized schedules provide the dosage $u(t)$ such that toxicity is within limits, and the minimum level of normal cells is maintained for complete duration of treatment. The selected drug schedule from aggressive chemotherapy is shown in Fig. 3a, and corresponding drug concentration in patient body after administering this schedule is shown in Fig. 3b.

As the start of treatment, initial tumor cell population is $10^{12}$ cells, out of which $2 \times 10^{11}$ cells are in proliferating phase and $8 \times 10^{11}$ cells are in quiescent phase. Optimized schedule from aggressive chemotherapy schedule is then delivered to patient model and tumor burden at the end of treatment is monitored. Figure 4a shows the reduction in proliferating cells, while Fig. 4b depicts reduction in quiescent cells with progression of treatment. At the end of treatment, the number of cancer cells remaining is $1.235 \times 10^{11}$ cells which is 87.65% reduction in total tumor burden. Figure 5a depicts the overall toxicity caused by infused drugs, and it is observed that toxicity remains within specified bounds for complete treatment duration. Moreover, effect of toxic drugs on normal cells can be seen from Fig. 5b. Total normal cells surviving at the end of treatment by aggressive chemotherapy are $1.88 \times 10^{11}$ cells.

Cell cycle-specific chemotherapy schedule is optimized by various authors [10, 17, 18, 30]. A comparative analysis on the basis of performance measures of cell

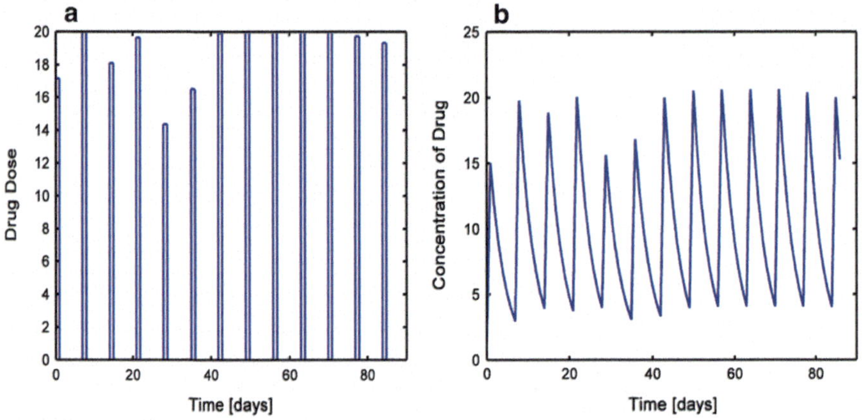


**Fig. 3** Aggressive chemotherapy **a** optimal drug dose, **b** drug concentration

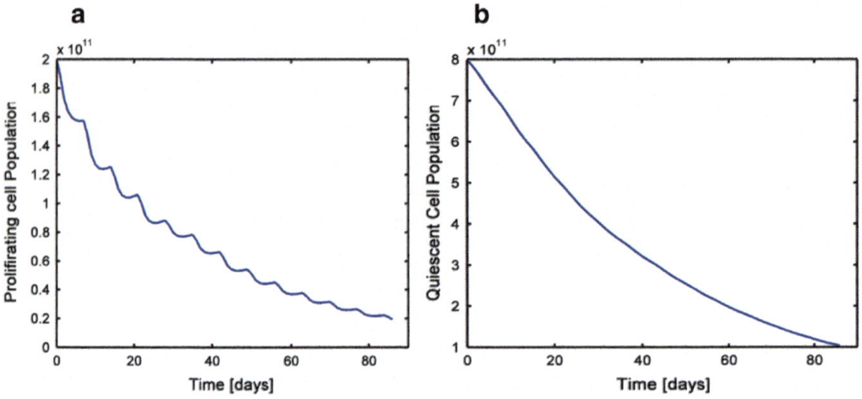


**Fig. 4** Aggressive chemotherapy effect on **a** proliferating cells, **b** quiescent cells

population profile is presented in Table 3. The tumor growth model and drug characteristics in each reported work are cycle specific, and therefore, fair comparison could be done. It is revealed from Table 3 that proposed schedule achieves maximum reduction in total tumor population with minimum harm to healthy cells. It can be seen that proposed NSGA-II optimized drug schedule outperforms other schedules presented in literature. Moreover, the schedules reported in previous works [10, 17, 18, 30] delivered continuous drug dosages from day 1 to 86 which could be harmful for patient. Therefore, in this article, the schedule gives ample interval between consecutive drug dosages which helps patient to recover from side effects of harmful drugs.

Hence, it is observed from table that NSGA-II optimized schedule performs better in terms of reduction of tumor burden while saving maximum healthy cells. The

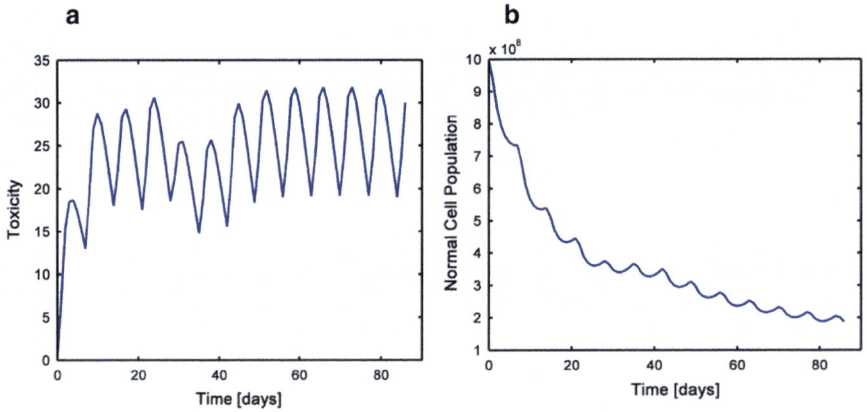


**Fig. 5** Aggressive chemotherapy **a** toxicity profile, **b** effect on normal cells

**Table 3** Comparative analysis with respect to cell population profile

| Cell population profile | Proposed schedule | Dua et al. [10] | Pachauri et al. [15] | Algoul et al. [21] | Alam et al. [16] |
|---|---|---|---|---|---|
| % reduction in *P* cells | 90.25 | 70 | 72.4 | 72.5 | 72.2 |
| % reduction in *Q* cells | 87.00 | 50 | 60.5 | 61 | 60.4 |
| Number of normal cells | $1.8 \times 10^8$ | $1.05 \times 10^8$ | $1.009 \times 10^8$ | $1.03 \times 10^8$ | $1.0002 \times 10^8$ |

schedules obtained by multi-objective optimization in this work also provide choice to oncologist to select appropriate schedule.

## 6 Conclusion

In this article, multi-objective evolutionary algorithm NSGA-II is proposed for optimal drug scheduling of cell cycle-specific chemotherapy. Drug scheduling for chemotherapy is a complex problem with conflicting objectives and various constraints on toxicity and drug concentration. The aim is to reduce tumor burden at the end of treatment while keeping normal cell population above safe levels. Optimization with NSGA-II provides many good schedules, thus giving multiple options for oncologist to decide drug dosages depending upon patient's condition. The NSGA-II optimized schedules achieve maximum tumor cell killing with minimum harm to normal cells keeping toxicity and drug concentration in specified bounds. Hence, it is concluded that NSGA-II provides efficient drug schedules to address complex constrained chemotherapy schedule optimization. The mathematical model could be further accurately described by considering the effect of drug resistance and multiple drugs with immunotherapy which could be further studied.

## References


1. Corrie, P.G.: Cytotoxic chemotherapy: clinical aspects. Medicine **36**, 24–28 (2008)
2. Ku-Carrillo, R.A., Delgadillo-Aleman, S.E., Chen-Charpentier, B.M.: Effects of the obesity on optimal control schedules of chemotherapy on a cancerous tumor. J. Comput. Appl. Math. **309**, 603–610 (2017)
3. Martin, R., Teo, K.L.: Optimal Control of Drug Administration in Cancer Chemotherapy, chap. 9, pp. 112–141. World Scientific, Singapore (1994)
4. Panetta, J.C., Fister, K.R.: Optimal control applied to competing chemotherapeutic cell-kill strategies. SIAM J. Appl. Math. **63**, 1954–1971 (2003)
5. de Pillis, L.G., Radunskaya, A.E.: A mathematical tumor model with immune resistance and drug therapy: an optimal control approach. Theor. Med. **3**, 79–100 (2001)
6. Swan, G.W.: Role of optimal control theory in cancer chemotherapy. Math. Biosci. **101**, 237–284 (1990)
7. Moore, H.: How to mathematically optimize drug regimens using optimal control. J. Pharmacokinet Pharmacodyn. **45**(1), 127–137 (2018)
8. Tse, S.-M., Liang, Y., Leung, K.-S., Lee, K.-H., Mok, S.-K.T.: Multi-drug cancer chemotherapy scheduling by a new memetic optimization algorithm. In: 2005 IEEE Congress on Evolutionary Computation, pp. 699–706. IEEE (2005)
9. Liang, Y., Leung, K.-S., Mok, T.S.-K.: A novel evolutionary drug scheduling model in cancer chemotherapy. IEEE Trans. Inf. Technol. Biomed. **10**(2), 237–245 (2006)
10. Dua, P., Dua, V., Pistikopoulos, E.N.: Optimal delivery of chemotherapeutic agents in cancer. Comput. Chem. Eng. **32**, 99–107 (2008)
11. Panjwani, B., Mohan, V., Rani, A., Singh, V.: Optimal drug scheduling for cancer chemotherapy using two degree of freedom fractional order PID scheme. J. Intel. Fuzzy Syst. **36**(3), 2273–2284 (2019)
12. Patel, H.R., Shah, V.A.: Integrated design of model-based passive fault-tolerant control for nonlinear systems based on PID and fuzzy control. In: Pant, M., Sharma, T., Verma, O., Singla, R., Sikander, A. (eds.) Soft Computing: Theories and Applications. Advances in Intelligent Systems and Computing, vol. 1053. Springer, Singapore. https://doi.org/10.1007/978-981-15-0751-9_15 (2020)
13. Mohan, V., Chhabra, H., Rani, A., Singh, V.: Robust self-tuning fractional order PID controller dedicated to non-linear dynamic system. J. Intel. Fuzzy Syst. **34**, 1467–1478 (2018)
14. Mohan, V., Rani, A., Singh, V.: Robust adaptive fuzzy controller applied to double inverted pendulum. J. Intel. Fuzzy Syst. **32**, 3669–3687 (2017)
15. Verma, O.P., Manik, G., Jain, V.K.: Simulation and control of a complex nonlinear dynamic behavior of multi-stage evaporator using PID and Fuzzy-PID controllers. J. Comput. Sci. **25**, 238–251 (2018)
16. Sahoo, S.K., Prasad, S., Srivastava, V., Chhabra, H., Yadav, J.: Design of FOPID controller for regulating anesthesia. In: Pant, M., Sharma, T., Verma, O., Singla, R., Sikander, A. (eds.) Soft Computing: Theories and Applications. Advances in Intelligent Systems and Computing, vol. 1053. Springer, Singapore. https://doi.org/10.1007/978-981-15-0751-9_97 (2020)
17. Pachauri, N., Yadav, J., Rani, A., Singh, V.: Modified fractional order IMC design based drug scheduling for cancer treatment. Comput. Biol. Med. **109**, 121–137 (2019)
18. Alam, M.S., Hossain, M.A., Algoul, S., Majumader, M., Al-Mamun, M., Sexton, G., et al.: Multi-objective multi-drug scheduling schemes for cell cycle specific cancer treatment. Comput. Chem. Eng. **58**, 14–32 (2013)

19. Shindi, O., Kanesan, J., Kendall, G., Ramanathan, A.: The combined effect of optimal control and swarm intelligence on optimization of cancer chemotherapy. Comput. Methods Prog. Biomed. (2020). https://doi.org/10.1016/j.cmpb.2020.105327
20. Panetta, J.C., Adam, J.: A mathematical model of cycle-specific chemotherapy. Math. Comput. Model. **22**(2), 67–82 (1995)
21. Kandubothula, V., Uppada, R., Nandan, D.: A review on detection of breast cancer cells by using various techniques. In: Soft Computing: Theories and Applications. Advances in Intelligent Systems and Computing, vol. 1154 (2020)
22. Martin, R.B.: Optimal control drug scheduling of cancer chemotherapy. Automatica **28**(6), 1113–1123 (1992)
23. Pareek, M., Jha, C.K., Mukherjee, S.: Brain tumor classification from MRI images and calculation of tumor area. In: Pant, M., Sharma, T., Verma, O., Singla, R., Sikander, A. (eds.) Soft Computing: Theories and Applications. Advances in Intelligent Systems and Computing, vol. 1053. Springer, Singapore. https://doi.org/10.1007/978-981-15-0751-9_7 (2020)
24. Pati, S., Yadav, D., Verma, O.P.: Synergetic fusion of energy optimization and waste heat reutilization using nature-inspired algorithms: a case study of Kraft recovery process. Neural Comput. Appl. (2020). https://doi.org/10.1007/s00521-020-04828-4
25. Verma, O.P., Manik, G., Jain, V.K., Jain, D.K., Wang, H.: Minimization of energy consumption in multiple stage evaporator using genetic algorithm. Sustain. Comput. Inf. Syst. **20**, 130–140 (2018)
26. Deb, K., Pratap, A., Agarwal, S., Meyarivan, T.: A fast and elitist multi objective genetic algorithm: NSGA-II. IEEE Trans. Evol. Comput. **6**(2), 182–197 (2002)
27. Mohan, V., Chhabra, H., Rani, A., Singh, V.: An expert 2DOF fractional order fuzzy PID controller for nonlinear systems. Neural Comput. Appl. 1–18
28. Verma, O.P., Mohammed, T.H., Mangal, S., Manik, G.: Minimization of energy consumption in multi-stage evaporator system of Kraft recovery process using interior-point method. Energy **129**, 148–157 (2017)
29. Priyadarshini, K.A.: A comparative performance of sorting algorithms: statistical investigation. In: Soft Computing: Theories and Applications. Advances in Intelligent Systems and Computing, vol. 1154 (2020)
30. Algoul, S., Alam, M.S., Hossain, M.A., Majumder, M.: Feedback control of chemotherapy drug scheduling for phase specific cancer treatment. In: 2010 IEEE Fifth International Conference on Bio-Inspired Computing: Theories and Applications (BIC-TA), pp. 1443–1450. IEEE (2010)

# A Review of Feature Extraction Techniques for EEG-Based Emotion Recognition System

**Rupali Gill and Jaiteg Singh**

**Abstract** Emotional recognition based on electroencephalogram (EEG) signals is an important area of research in the field of compatible computing. There exist many feature extraction methods to extract EEG features, but it is dependent on the extensive knowledge of the EEG domain. The existing research has studied various feature extractions used for studying EEG for the understanding of emotions for decision-making. The major limitation is that there exists less research to provide details of EEG features in a single research. The major studies are selected based on the search string. With the help of a search string, we are reviewing only 50 studies. Therefore, the current study aims to review the mechanisms for the removal of the sensory perception element from the EEG and suggests the use of multiple strategies as part of EEG-based emotional recognition.

**Keywords** Electroencephalogram (EEG) · Feature extraction · Emotion recognition · Time domain · Frequency domain · Time–frequency domain

## 1 Introduction

Emotions play a vital role in understanding human behaviour and the human brain. Emotion is a way for communication through expressions. Emotion recognition has a broad application area in education and research, clinical science, human–computer interaction (HCI), marketing, and commerce [30, 50]. The interest in emotion recognition research can be done through different modalities like facial expression, brain action, speech, and walking trails [32]. With the advancement of research work in emotion recognition, EEG-based studies have gained much importance [12] with its early research done in 1997 [34]. EEG-based sensory recognition is based on placing electrodes at the head of the study according to a 10–20 international system. The assumption of feature extraction and selection of electrodes is based on neuroscience study [5, 17]. It is based on understanding the spectral power in various

R. Gill (✉) · J. Singh
Chitkara University Institute of Engineering and Technology, Chitkara University, Punjab, India
e-mail: rupali.gill@chitkara.edu.in



T. K. Sharma et al. (eds.), *Soft Computing: Theories and Applications*,
Advances in Intelligent Systems and Computing 1381,
https://doi.org/10.1007/978-981-16-1696-9_8

frequency bands, coherence, and electrode [20]. Also, frontal asymmetry in spectral band power plays a major role in understanding emotion states [7]. Besides understanding of neuroscience assumptions, the study needs the advanced learning of signal processing techniques for feature extraction and dimensionality recognition and reduction [43]. Computational methods used in machine learning and deep learning for emotion recognition are to be applied for accurate feature extraction and channel selection [20]. Various frequency variants and their frequency bands were explored by Hadjidimitriou [13] on a small data set. The comparison of features was studied by Petrantonakis [37] on small data sets [33]. Few studies have worked on EEG feature selection [46]. Applied feature selection technique mRMR is to find electrode positioning on a self-prepared small data set. Single-subject testing was done by Ansari-Asl [4] for EEG-based channel selection. The main purpose of this paper is to provide research on various techniques for the removal of the EEG-based sensory perception feature. The subsequent sections detail the various feature extraction techniques and by outcome and conclusion.

## 2 Search Process

The selected studies are based on the search string. To make a search string, we used the major keywords which are found in selected studies. The search string is

***((“Feature extraction technique” OR “EEG Feature extraction technique”) AND (“EEG Emotion Recognition” OR “Electroencephalogram Emotion Recognition”)).***

## 3 EEG-Based Emotion Recognition System

EEG-based emotion recognition system is a five-step process as shown in Fig. 1. The steps start with the acquisition of the EEG signal from an EEG headset. The headset can be from any of the manufacturers like Emotiv Epoc, Neurosky, etc. After the acquisition of the signal, it involves cleansing of data, i.e., pre-processing of EEG data [28, 44]. It includes reducing noise, spatial, and temporal filtering. Followed by this is the classification of extracted features to predict the emotions. The classification task involves studying various algorithms like support vector machine, k-nearest neighbour, random forest, decision trees, regression, linear discriminant analysis,

**Fig. 1** EEG-based emotion recognition system

**Table 1** EEG signal processing

| Frequency band (Hz) | Brain region | Affective state |
|---|---|---|
| Delta (<4) | Central region | – |
| Theta (4–7) | Central region | Gaze focus |
| Alpha (8–15) | Posterior brain | Positive and negative emotional valences |
| Beta low range (12.5–28) | Central region | Alterations during affective arousal |
| Gamma (25–100) | Highly localized | Arousal effects |

principal component analysis, and various deep learning techniques. The study of these algorithms is not covered as part of this study, but previous work done for classification has been listed in Table 3.

The EEG electrical activity is distinguished by dividing the frequencies into classes, called alpha, delta, theta, and gamma. The human brain wave consists of the five most common bands called delta (1–3 Hz), theta (4–7 Hz), alpha (8–13 Hz), beta (14–30 Hz), and gamma (31–50 Hz) [13, 18]. These tasks are logged utilizing electrodes placed over the cortex right on the scalp Table 1.

## 4 EEG-Based Feature Extraction Methods

This section presents the study of various EEG-relevant features for emotion recognition which has been studied, and a summary of the same is presented in Table 3. Features are usually divided into three types—time domain, frequency, and time domain.

### *4.1 Time Statistical Domain Features*

The time domain features do not play a major role in EEG feature extraction, but they are being used to detect emotions. Time domain features are classified into six major types as detailed below:

1. *Event-related potential (ERP):* The most commonly used ERPs are P100, N100, N200, P200, and P300 [10, 45], although it is difficult to measure ERP-based features.
2. *Hjorth features:* Hjorth features are extracted based on three features—activity, mobility, and complexity. These features help in indicating the statistical properties of the signal [35]

Activity is represented in Eq. (1)

$$\text{Activity} = v(f(t)) \tag{1}$$

*The mobility parameter* is defined as the square root of the measure of the difference between the initial signal findings and that of the actual signal. This parameter has a standard deviation of the normal power deviation. Travel must be in Eq. (2)

$$\text{Mobility} = \surd(v(f'(t)/\surd(v(f(t))) \tag{2}$$

*The complex parameter* shows an indication of signal similarity to a pure sine wave. The complexity value is equal to 1 as the signal form is similar to a pure sine wave. Anxiety must be in Eq. (3)

$$\text{Complexity} = \text{Mobility}\frac{f'(t)}{\text{Mobility}(f(t))} \tag{3}$$

3. *Non-stationary index (NSI)*: It helps to analyse the variance of the local scale over time, and this will represent the degree of difficulty. The signal is divided into smaller parts, and the ratio of each component is calculated as NSI the normal deviation of those intermediates [8, 17, 21]
4. *Fractal dimension (FD)*: It is a complex measure that can be calculated by logarithm layout. The structure incorporates multiple volumes to cover the signal. Higuchi algorithm is the most popular way to make FD. Higuchi FD algorithm details [25] are given as follows in Eq. (4):

Let $X(1)$, $X(2)$, …, $X(N)$ be a finite set of time series samples. Then, the newly constructed time series is

$$X_m^t = \text{X(m)}, X(m+t), \ldots, X\left(m + \left[\frac{N-m}{t}\right].t\right) \tag{4}$$

where $m$ =1,2, …, $t$ is the start time, and $t$ are the interval.

5. *Higher-order crossings (HOC)*: HOC captures the oscillatory patterns of EEG using high-pass filters. In this process, a sequence of filters is applied. Different HOCs can be obtained by varying filter design [37].
6. *Statistical methods*: There are various statistical measures used to extract the EEG feature [33, 46]. The statistical features used ($X(n)$, $n = 1,\ldots, N$, and the raw $N$-sample EEG signal can be measured using mean, power, standard deviation, first-order difference, normalized first difference, second-order difference, and the normalized second difference [37].

However, time domain features are less effective for EEG-based feature extraction due to the complexity of the EEG signals [39].

## 4.2 Frequency Domain

The frequency domain is the most popular domain for emotion recognition from EEG. It can be classified into two types. The EEG signals are recorded in bands.

1. Band power: It is a widely used method based on the division of frequency bands. EEG-based frequency bands are divided into independent frequency bands [48], such as delta $\delta$ (0.5–4 Hz), theta $\theta$ (4–8 Hz), alpha $\alpha$ (8–12 Hz), beta $\beta$ ( 12–30 Hz), and the gamma $\gamma$ (30–100 Hz). EEG signal decay into frequency components is obtained by Fourier conversion. Another method is power spectral density which is based on the squared magnitude of FFT [2, 3, 8, 25, 27].
2. Higher-order spectrum (HOS) is a step-by-step extension of FFT order. It is a second-order spectrum. Quadratic phase coupling (QPC) detection is used to distinguish between two frequency factors [15]. They study third-order statistics for EEG features known as bispectrum. This bispectrum of third-order Fourier transform is given by the following equation (Eq. 5)

$$\mathrm{Bs}(f_1, f_2) = E\left[X(f_1) \cdot X(f_2) \cdot X^*(f_1 + f_2)\right] \tag{5}$$

where Bs is bispectrum, * is complex conjugate, $X(f)$ is Fourier transform of the signal $x(nT)$, and $E[\cdot]$ is expectation operation. The normalized value of bispectrum is bicoherence Bc through Eq. (6).

$$\mathrm{Bc}(f_1, f_2) = \frac{\mathrm{Bs}(f_1, f_2)}{\sqrt{X(f_1)} \cdot X(f_2) \cdot X^*(f_1 + f_2)} \tag{6}$$

## 4.3 Time–Frequency Domain

The domain elements of the time–frequency are used to extract the dynamic features. Among all the time domain features, wavelet transform is the most used time domain feature extraction technique.

1. Hilbert–Huang transform (HHT): HHT is an algorithm that requires two steps for data analysis. In the first step, data pre-processing is done by the empirical mode decomposition (EMD) [26]. The next step is based on the corruption of the signals followed to create internal mode functions (IMF). The IMF function represents the magnitude of the signal amplitude and frequency in the average time and provides appropriate results for illegal and nonlinear features [17].

Let $X(t)$ be a time series; the EMD method decomposes $X(t)$, by a selected process, in MIMFs named as a filtering algorithm. A repeat filtering algorithm provides different types of extraction $h(t)$ based on local maxima and local minima. The result of this action is the IMF. The algorithm configuration method is the standard deviation between successive IMFs shown in Eq. (7).

$$\sigma = \frac{\sum_{t=0}^{T} |h_{1n}(t) - h_{1(n-1)}(t)|^2}{\sum_{t=0}^{T} |h_{1(n-1)}^2(t)}, <\in (0.2 <\in< 0.3) \quad (7)$$

Significant $M$ IMF protection protects the original signal $X(t)$ is shown in Eq. (8).

$$X(t) = \sum_{k=1}^{M} \mathrm{IMF}_k(t) + r_{\mathrm{M}}(t) \quad (8)$$

where $r_{\mathrm{M}}(t)$, that is, means the remainder of the EMD process and is a continuous process [13].

2. *Wavelet transform (WT)*: It is the most common form of the waveform used for neuroscience research [17]. These can be classified as discrete WT and continuous WT. It is based on a multi-resolution analysis of WT. The combined time and frequency extracted through this method enable it to extract emotion feature from an EEG signal. The signal approximations are obtained through FFT and STFT [33]. The decompositions of EEG signals into bands are done based on the sampling frequency ranging from 0 to 512 Hz (128, 256, 512 Hz).

Absolute and relative power, and other entropy-based method and correlation dimension coefficient are not discussed as part of this study. Table 2 presents a brief of the left-over EEG feature extraction techniques [24, 40, 41].

**Table 2** Miscellaneous EEG features

| EEG feature | Description |
|---|---|
| Amplitude | Calculates maxima and minima of magnitudes |
| Eigenvalues | Calculates signals' frequency and power for the dominant signals |
| Mean | Identifies the central value from the raw signal |
| Entropy and its variants | Quantifies the regularity of the signal and measures its complexity |
| Skewness | Measures asymmetry of the signal |
| Kurtosis | Measures fourth-order central moment of a signal distribution |
| Energy | Measures strength of the EEG signal |
| Average power | Central power is a square sign |
| Root mean square (RMS) | Square root of central power |
| Hurst exponent | Shows the familiarity of the time series |
| Autoregressive | Time series model parameters are included in the DWT equation sequence |

**Table 3** Summary of feature extraction techniques

| Reference | Year | Emotion classification | Classifier | Feature extraction | Type of feature extraction | Stimuli |
|---|---|---|---|---|---|---|
| Mikels [31] | 2005 | Dimensional | Linear ridge regression and support vector regression (SVR) | Wavelet and Gaussian mixture model | Time–frequency | DEAP |
| Petrantonakis [38] | 2012 | Dimensional | Quadratic discriminant analysis (QDA), Mahalanobis distance (MD), k-nearest neighbour (kNN), support vector machine (SVM) | Adaptive way (AsI-based algorithms) | Mix | IAPS |
| Huang [16] | 2012 | Discrete and dimensional | kNN, Naive Bayes (NB), and support vector machine (SVM) | Asymmetric features and filter bank common spatial pattern (FBCSP) as a benchmark and proposing recursive Fisher linear discriminant (RFLD) | Mix | 20 min video for four participants |
| Liu [25] | 2014 | Dimensional | Kernel Fisher's discriminant kernel Eigen emotion pattern (KEEP) | Wavelet | Mix | Ten participants |
| Lin [23] | 2014 | Discrete | Gaussian Naïve Bayes (GNB) | STFT differential laterality (DLAT) and differential causality (DCAU) | Time domain and mix | Audio music, with 24 trails each |

(continued)

**Table 3** (continued)

| Reference | Year | Emotion classification | Classifier | Feature extraction | Type of feature extraction | Stimuli |
|---|---|---|---|---|---|---|
| Wichakam [47] | 2014 | Dimensional | SVM | Band power and PSD by wavelet transform | Frequency and time domain | DEAP |
| Gao [11] | 2015 | Discrete | Three layers of restricted Boltzmann machines (RBMs) | Deep learning | Advanced | 21 participants |
| Lin [22] | 2015 | Discrete | Fisher linear discriminant (FLD) | Approximate entropy (ApEn) | Mix | 25 participants playing gambling |
| Handayani 14 | 2015 | Discrete | Mel-frequency cepstral coefficient (MFCC) | Wavelet | Time–frequency | IAPS |
| Ackermann [1] | 2016 | Discrete | Random forests (RF) and SVM | STFT, mRMR | Time–frequency | DEAP |
| Samara [42] | 2016 | Dimensional | SVM | Wavelet, HOC | Time–frequency | DEAP |
| Patil [36] | 2016 | Discrete | Radial basis function neural network (RBFNN) and SVM | PCA, FT, STFT slow cortical potential (SCP) and wavelet packet transform (WPT), DWT | Time–frequency | 5 Subjects |
| Yano [49] | 2016 | Dimensional | ElasticNet, LDA, QDA, SVM | The spatial filter of the common spatial pattern (CSP) | Mix | Set of movie clips 23 participants |
| Mangalagowri [29] | 2016 | Discrete | Feed forward back propagation | Wavelet, DWT | Time–frequency | Ten participants |

(continued)

**Table 3** (continued)

| Reference | Year | Emotion classification | Classifier | Feature extraction | Type of feature extraction | Stimuli |
|---|---|---|---|---|---|---|
| Djamal [9] | 2017 | Dimensional | Support vector machine (SVM) and learning vector quantization (LVQ) | Wavelet | Time–frequency | Ten Subjects |
| Byun [6] | 2017 | Dimensional | Relief algorithm, Bhattacharyya distance | Statistical feature, Hjorth features, non-stationary index, HOC | Frequency domain | DEAP |
| Kim [19] | 2018 | Dimensional | Multilayer perceptron (MLP) multilayer convolutional neural network (CNN) | Deep physiological affect network (deep learning model) | Advanced | 1280 videos, as well as 64 body language combinations per video |

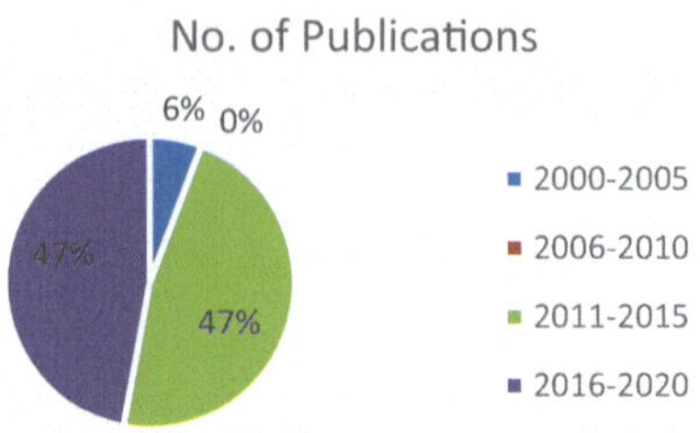


**Fig. 2** Publication count per 5 year

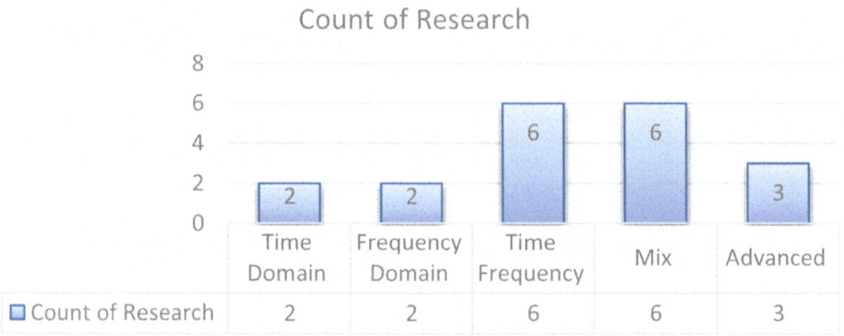


**Fig. 3** Publication count as per type of feature extraction

## 5 Discussions

The section provides the summary of various feature extraction techniques along with the classifier used. It shows that various researches have been carried out to extract EEG-based features from the study included from 2005 to 2019. The graph shows the number of publications per 5 year and number of research being carried out as per the type of feature extraction technique as shown in Figs. 2 and 3.

The review shows that towards moving in a new era, the researchers are moving towards mix, advanced and time–frequency domain methods. So, there is a scope of using deep learning-based approaches for feature extraction in combination with tradition feature extraction techniques.

## 6 Conclusion

The present study focuses on studying various techniques of feature extraction. Feature extraction strategies are divided into three domains namely time domain, frequency domain, and time–frequency domain. Subjects are selected based on the search string. Therefore, the search string is built with the help of keywords. Research shows that time-varying domain features are less commonly used in EEG-based study researchers and quantity and time-varying domain features are more commonly used in EEG-based sensory perceptions. As the future scope, the authors propose to explore various classification and deep learning techniques using these combined approaches.

## References


1. Ackermann, P.K.: EEG-based automatic emotion recognition: feature extraction, selection and classification methods. In: 2016 IEEE 18th International Conference on E-health Networking, pp. 1–6. IEEE (2016)
2. ALdayel, M.Y.-N.: Deep learning for EEG-based preference classification in neuromarketing. Appl. Sci. **10**(4), 1525–1548 (2020)
3. Al-Nafjan, A.H.-O.-W.: Review and classification of emotion recognition based on EEG brain-computer interface system research: a systematic review. Appl. Sci. **7**(12), 1239 (2017)
4. Ansari-Asl, K.C.: A channel selection method for EEG classification in emotion assessment based on synchronization likelihood. In: 2007 15th European Signal Processing Conference, pp. 1241–1245. IEEE (2007)
5. Berka, C.L.D.: {EEG} correlates of task engagement and mental workload in vigilance, learning, and memory tasks 78. Aviat. Space Environ. Med. **78**, B231–B244 (2007)
6. Byun, S.W.: Feature selection and comparison for the emotion recognition according to music listening. In: 2017 International Conference on Robotics and Automation Sciences (ICRAS), pp. 172–176. IEEE (2017)
7. Cacioppo, J.T.: Feelings and emotions: roles for electrophysiological markers. Bio. Psychol. **67**(1–2), 235–243 (2004)
8. Christensen, L.R.: EEG emotion detection review. In: 2018 IEEE Conference on Computational Intelligence in Bioinformatics and Computational Biology (CIBCB), pp. 1–7. IEEE (2018)
9. Djamal, E.C.: EEG based emotion monitoring using wavelet and learning vector quantization. In: 2017 4th International Conference on Electrical Engineering, Computer Science and Informatics (EECSI), pp. 1–6. IEEE (2017)
10. Frantzidis, C.A.: Toward emotion aware computing: an integrated approach using multichannel neurophysiological recordings and affective visual stimuli. IEEE Trans. Inf. Technol. Biomed. **14**(3), 589–597 (2010)
11. Gao, Y.L.: Deep learninig of EEG signals for emotion recognition. In: 2015 IEEE International Conference on Multimedia and Expo Workshops (ICMEW), pp. 1–5. IEEE (2015)
12. Gunes, H.P.: Automatic, dimensional and continuous emotion recognition. Int. J. Syn. Emotions (IJSE) **1**(1), 68–99 (2010)
13. Hadjidimitriou, S.K.: Toward an EEG-based recognition of music liking using time-frequency analysis. IEEE Trans. Biomed. Eng. **59**(12), 3498–3510 (2012)
14. Handayani, D.Y.: Statistical approach for a complex emotion recognition based on EEG features. In: 2015 4th International Conference on Advanced Computer Science Applications and Technologies (ACSAT), pp. 202–207. IEEE (2015)
15. Hosseini, S.A.-S.: Higher order spectra analysis of EEG signals in emotional stress states. In: 2010 Second International Conference on Information Technology and Computer Science, pp. 60–63. IEEE (2010)
16. Huang, D.G.: Asymmetric spatial pattern for EEG-based emotion detection. In: The 2012 International Joint Conference on Neural Networks (IJCNN), pp. 1–7. IEEE (2012)
17. Jenke, R.P.: Feature extraction and selection for emotion recognition from EEG. IEEE Trans. Affect. Comput. **5**(3), 327–339 (2014)
18. Kaur, R.G.: Cognitive emotion measures of brain. In: 2019 6th International Conference on Computing for Sustainable Global Development (INDIACom), pp. 290–294. IEEE (2019)
19. Kim, B.H.: Deep physiological affect network for the recognition of human emotions. IEEE Trans. Affect. Comput. (2018)

20. Kim, M.K.: A review on the computational methods for emotional state estimation from the human EEG. Comput. Math. Methods Med. (2013)
21. Kroupi, E.V.: Subject-independent odor pleasantness classification using brain and peripheral signals. IEEE Trans. Affect. Comput. **7**(4), 422–434 (2015)
22. Lin, O.L.: Neurophysiological markers of identifying regret by 64 channels EEG signal. In: 2015 12th International Computer Conference on Wavelet Active Media Technology and Information Processing (ICCWA), pp. 395–399. IEEE (2015)
23. Lin, Y.P.: Exploring day-to-day variability in EEG-based emotion classification. In: 2014 IEEE International Conference on Systems, Man, and Cybernetics (SMC), pp. 2226–2229. IEEE (2014)
24. Liu, Y.H.: EEG-based emotion recognition based on kernel Fisher's discriminant analysis and spectral powers. In: 2014 IEEE International Conference on Systems, Man, and Cybernetics (SMC), pp. 2221–2225. IEEE (2014)
25. Liu, Y.S.: EEG-based subject-dependent emotion recognition algorithm using fractal dimension. In: 2014 IEEE International Conference on Systems, Man, and Cybernetics (SMC), pp. 3166–3171. IEEE (2014)
26. Lo, M.T.: The nonlinear and nonstationary properties in EEG signals: probing the complex fluctuations by Hilbert-Huang transform. Adv. Adapt. Data Anal. **1**(03), 461–482 (2009)
27. Lotte, F.B.L.: A review of classification algorithms for EEG-based brain–computer interfaces: a 10 year update. J. Neural Eng. **15**, 031005 (2018)
28. Mahajan, R.: Emotion recognition via EEG using neural network classifier. In: Soft Computing: Theories and Applications, pp. 429–438. Springer, Singapore (2018)
29. Mangalagowri, S.G.: EEG feature extraction and classification using feed forward backpropogation algorithm for emotion detection. In: International Conference on Electrical, Electronics, Communication, Computer and Optimization Techniques (ICEECCOT), pp. 183–187. IEEE (2016)
30. Marrero-Fernández, P.M.-P.-I.-C.: Evaluating the research in automatic emotion recognition. IETE Tech. Rev. **31**(3), 220–232 (2014)
31. Mikels, J.A.-L.: Emotional category data on images from the international affective picture system. Behav. Res. Methods **37**(4), 626–630 (2005)
32. Mühl, C.B.: Modality-Specific Affective Responses and Their Implications for Affective BCI, pp. 120–123. Verlag der Technischen Universität, Graz, Austria (2011)
33. Murugappan, M.R.: EEG feature extraction for classifying emotions using FCM and FKM. Int. J. Comput. Commun. **1**(2), 21–25 (2007)
34. Musha, T.T.: Feature extraction from EEGs associated with emotions. Artif. Life Rob. **1**(1), 15–19 (1997)
35. Oh, S.H.: A novel EEG feature extraction method using Hjorth parameter. Int. J. Electron. Electr. Eng. 106–110 (2014)
36. Patil, A.D.: Feature extraction of EEG for emotion recognition using Hjorth features and higher order crossings. In: 2016 Conference on Advances in Signal Processing (CASP), pp. 429–434. IEEE (2016)
37. Petrantonakis, P.C.: Emotion recognition from brain signals using hybrid adaptive filtering and higher order crossings analysis. IEEE Trans. Affect. Comput. **1**(2), 81–97 (2010)
38. Petrantonakis, P.C.: Adaptive emotional information retrieval from EEG signals in the time-frequency domain. In: IEEE Transactions on Signal Processing, pp. 2604–2616. IEEE (2012)
39. Phadikar, S.S.: A survey on feature extraction methods for EEG based emotion recognition. In: International Conference on Innovation in Modern Science and Technology, pp. 31–45. Springer, Cham (2019)
40. Ramadan, R.A.: Basics of brain computer interface. In: Brain-Computer Interfaces, vol. 74, pp. 31–50. Springer, Switzerland (2015)
41. Ramadan, R.A.: Brain computer interface: control signals review. Neurocomputing **223**, 26–44 (2017)

42. Samara, A.M.: Feature extraction for emotion recognition and modelling using neurophysiological data. In: 2016 15th International Conference on Ubiquitous Computing and Communications and 2016 International Symposium on Cyberspace and Security (IUCC-CSS), pp. 138–144. IEEE (2016)
43. Sanei, S.C.: EEG Signal Processing. Wiley (2013)
44. Subasi, A.: EEG signal classification using wavelet feature extraction and a mixture of expert model. Expert Syst. Appl. **32**(4), 1084–1093 (2007)
45. Sur, S.: Event-related potential: an overview. Ind. Psychiatry J. **18**(1), 70–73 (2009)
46. Wang, X.W.: EEG-based emotion recognition using frequency domain features and support vector machines. In: International conference on neural information processing, pp. 734–743. Springer, Berlin, Heidelberg (2011)
47. Wichakam, I.V.: An evaluation of feature extraction in EEG-based emotion prediction with support vector machines. In: 2014 11th International Joint Conference on Computer Science and Software Engineering (JCSSE), pp. 106–110. IEEE (2014)
48. Yadav, B.K.: A robust digital image watermarking algorithm using DWT and SVD. In: Soft Computing: Theories and Applications, pp. 25–36. Springer, Singapore (2018)
49. Yano, K.S.: Fixed low-rank EEG spatial filter estimation for emotion recognition induced by movies. In: 2016 International Workshop on Pattern Recognition in Neuroimaging, pp. 1–4. IEEE (2016)
50. Zander, T.O.: Towards passive brain-computer interfaces: applying brain–computer interface technology to human–machine systems in general. J. Neural Eng. **8**(2), 025005 (2011)

# Task Scheduling and Load Balancing Techniques Using Genetic Algorithm in Cloud Computing

**Meenu Vijarania, Akshat Agrawal, and Mayank Mohan Sharma**

**Abstract** Task scheduling and load balancing are challenging problems in cloud computing, as the demand of cloud computing resources is escalating everywhere anytime. The load on cloud nodes is very high with the increase in client request for cloud services. This condition needs load balancing techniques to uniformly distribute the client request among existing resources so that underutilization and overutilization of resources can be minimized. An impoverished scheduling strategy may over burdensome virtual machines while remaining virtual machines are idle. In this paper, load balance-aware task scheduling policy is proposed using genetic algorithm. The performance of the proposed technique is compared with existing schemes. The simulation output represents that the proposed algorithm outperforms than the current algorithms.

**Keywords** Genetic algorithm · Scheduling · Load balancing · Cloud computing

## 1 Introduction

Cloud computing is one of the emerging field in today's era. It is a concept that physically organizes platforms, resources, and software application using Internet to satisfy client request [1]. With the massive growth in cloud computing, server's workload is rapidly increasing and becomes overloaded. In order to serve huge number of client requests, the resources must be utilized effectively and efficiently to maximize the productivity and reducing the response time [2, 3]. Load balancing and task scheduling are the vital issues in cloud computing. Task scheduling is assigning task to the available resources while reducing the completion time. Task scheduling significantly affects the performance of the system. The parameters taken into account for

M. Vijarania · A. Agrawal (✉)
Amity University Haryana, Manesar 1222413, India

M. M. Sharma
Principal Software Test Lead Zillow Inc, San Francisco, USA



T. K. Sharma et al. (eds.), *Soft Computing: Theories and Applications*,
Advances in Intelligent Systems and Computing 1381,
https://doi.org/10.1007/978-981-16-1696-9_9

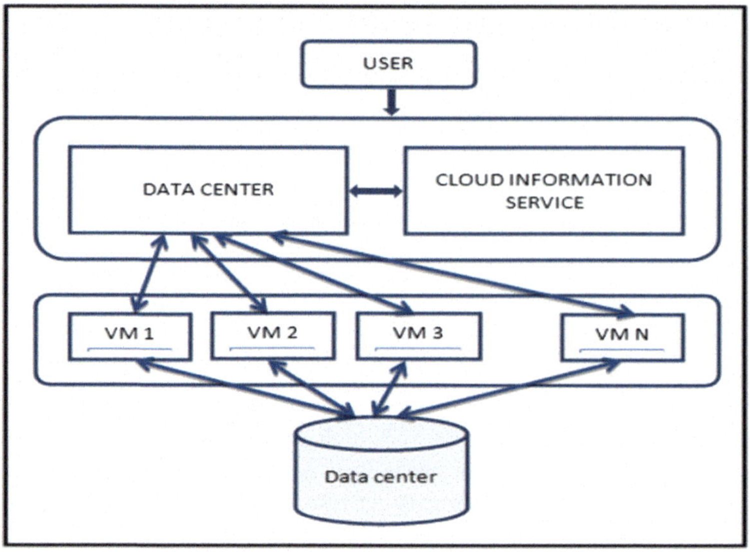


**Fig. 1** Cloud environment [8]

developing task scheduling algorithm are response time, task completion time, and cost [4].

The total processing power of each VM is constrained by the resources available. The computing resources available for each VM are constrained by total processing power [5]. In cloud environment, the pattern of task arrival is also random and virtual machine capability also varies from one another. Therefore, uniform load balancing is an important factor, lack of which may lead to degradation of system performance. Hence, load balancing becomes a critical task along with task scheduling [6, 7]. So, it becomes very essential to develop a technique to schedule task efficiently considering the load balancing factor among virtual machines. The intention of the article is to develop GA-based load balanced task scheduling mechanism which aims to reduce time of completion and execution cost of the task and maximizing the resource utilization (Fig. 1).

## 2 Related Work

Task scheduling and load balancing are the main challenges in distributed environment nowadays. There are numerous task scheduling mechanisms that are available in the literature, viz. min-min algorithm, max–min algorithm, and round-robin algorithm. These existing techniques have various drawbacks like load imbalance among VMs and response time and completion time are more. Junwei [3] proposed a GA-based algorithm which considered cost constraint, average task completion time, and total task completion time. Kaur et al. [9] presented improved GA-based scheme. A fitness function is generated using mean and grand mean values which can be used for both resource and task scheduling and thereby increased efficiency. Singh et al. [1] presented modified GA where initial population is generated using max–min in order to achieve better results and improved makespan of the tasks.

## *2.1 Task Scheduling Problem*

In task scheduling, n tasks are mapped onto m available resources; every task is independent of each other with expected completion time. Considering heterogeneity of different computing power of devices, the same task can have different completion time allocating with different resources [10]. The ETC represents the expected execution time of a task set on virtual machine which is computed by using ETC matrix. $\mathrm{ETC}_{i,j}$ defines the expected time for task $T_i$ to be completed on resource $R_j$ and $n$ is the task set and $m$ is the number of resources available.

The main purpose of scheduling is to reduce the task set overall completion time, which is represented as follows [11]:

$$\mathrm{ETC}_{i,j} = \begin{bmatrix} \mathrm{ETC}_{1,1} & \mathrm{ETC}_{1,2} & \cdots\cdots\cdots & \mathrm{ETC}_{1,m} \\ \mathrm{ETC}_{2,1} & \mathrm{ETC}_{2,2} & \cdots\cdots\cdots & \mathrm{ETC}_{2,m} \\ \vdots & \vdots & \vdots \;\; \vdots \;\; \vdots & \vdots \\ \mathrm{ETC}_{n,1} & \mathrm{ETC}_{n,2} & \cdots\cdots\cdots & \mathrm{ETC}_{n,m} \end{bmatrix} \tag{1}$$

$$\mathrm{ETC}_{i,j} = \frac{\text{Job Length}_i}{\text{VM Power}_j} \tag{2}$$

$$\text{makespan} = \max_{j \in m} \sum_{i=1}^{n} \mathrm{ETC}_{i,j} \tag{3}$$

where $\mathrm{ETC}_{i,j}$ represent the total time needed by the resource $R_j$ to process the assigned task.

## *2.2 Load Balancing*

Load balancing is a technique which fairly distributes the set of tasks to available resources in order to maximize the throughput, maximum resource utilization, and minimization of response time [12]. To attain these objectives, a load balancing approach is developed to uniformly divide the workload among servers as shown in Fig. 2. Therefore, workload status of every server must be continuously updated. Load

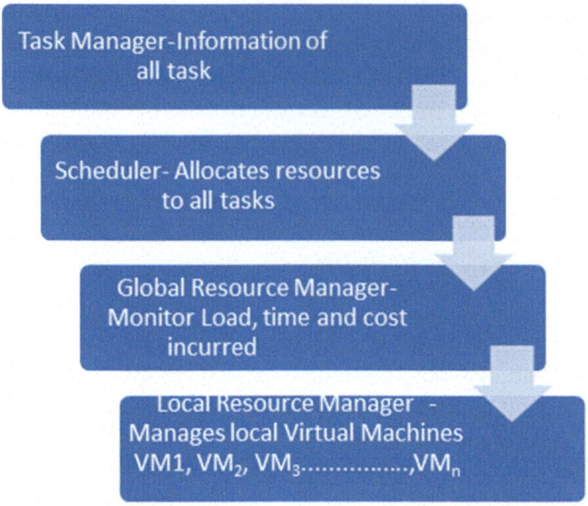


**Fig. 2** Framework of proposed model

balancing techniques have been grouped into various types—static and dynamic, decentralized and centralized, and non-periodic and periodic. Load balance module computes the load of each machine in the cloud. The load of a resource in the cloud can be computed as [13]:

$$L_j = \sum_{i=0}^{n} \mathrm{ETC}_{ij} * \mathrm{ESC}_{ij} \tag{4}$$

where $\mathrm{ETC}_{ij}$ represents task completion time on a particular resource. $\mathrm{ESC}_{ij}$ represents the schedule matrix of task sets and available resources, and value of $i$ and $j$ lies in between as follows- $0 \leq j \leq m$ and $0 \leq i \leq n$.

The standard load of every resources can be computed as follows [17]:

$$\mathrm{EL}_j = \frac{1}{m} \sum_{j=0}^{m} L_j \tag{5}$$

Difference of average load and load of a resource $R_j$ is represented as $L_j - \mathrm{EL}_j$ (Fig. 2).

## 3 Proposed Algorithm

Genetic algorithm as shown in Fig. 3 is a search heuristic imitating the process of natural genetics and natural selection. Darwin's evolution theory is used in GA where "survival of fittest" concept is used for selecting the best resource for a particular task to execute while considering the existing load of the machine.

**Chromosome Encoding**

For encoding of a chromosome, direct encoding method is used. The number of task defines the chromosome length, and gene in the chromosome represents the resource being used by the task as:

$$Z_i = [Z_{i1}, Z_{i2}, \ldots, Z_{in}] \tag{6}$$

where $n$ represents the number of task and $i$ represents the resource allocated to particular task.

**Fitness Function**

Fitness function measures the fitness of the chromosome. The fitness value of an individual represents its performance in the population. Here, fitness function for balancing load is defined by Eq. (7) and fitness function for response time is shown if Eq. 8. Final fitness function will be computed by adding Eqs. (7) and (8).

$$F_L = \frac{1}{L_j} \tag{7}$$

$$F_{CT} = \frac{1}{\mathrm{ETC}_{ij}} \tag{8}$$

$$F = \alpha F_L + (1 - \alpha) F_{CT} \tag{9}$$

where $\alpha$ is coefficient and its value lies between 0 and 1.

**Selection**

The selection operator evaluates the individual chromosome performance and determines whether the chromosome will move to the next generation or not based on its fitness value [14]. Here, proportion selection method is used to select the chromosome based on their fitness values.

**Crossover and Mutation Operators**

The crossover operator selects two chromosomes, and intersection point of the chromosomes is selected. Depending upon the intersection point two chromosomes are exchanged which produces two new offsprings [15, 16]. In this study, single-point crossover operator is used to generate new offsprings and intersection point is selected randomly (Fig. 3).

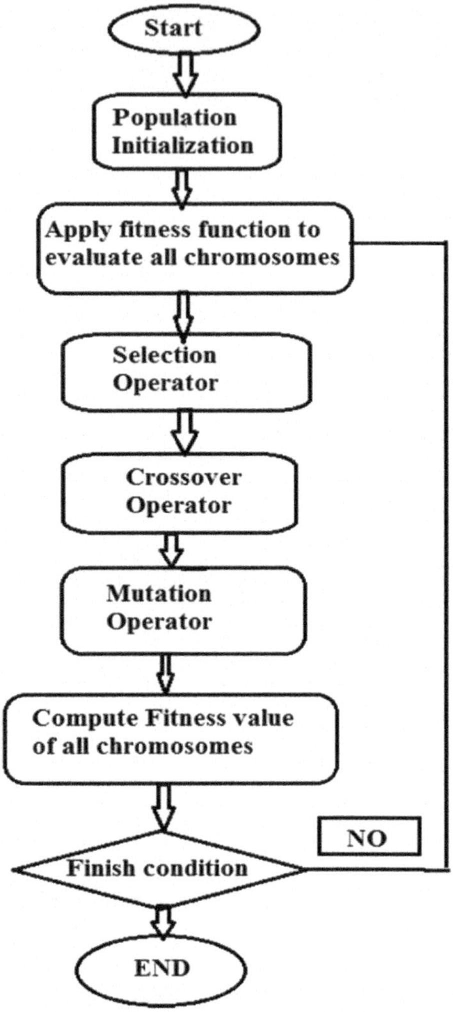


**Fig. 3** Working of GA

**Algorithm: Proposed Algorithm**

Step 1: Population initialization and chromosome encoding using Eq. (6)

Step 2: Compute the fitness value of all individual in the population using Eq. (9)

Step 3: Repeat the following steps until the optimal solution is achieved

(a) Apply selection operator and eliminate the chromosomes with lower fitness value.

(b) Apply single-point crossover operator to generate new offspring

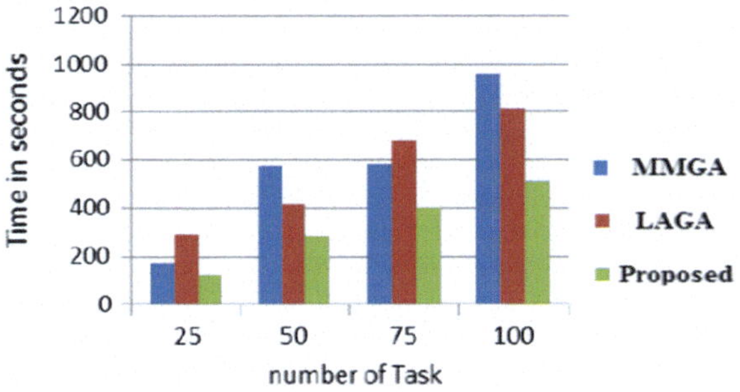


**Fig. 4** Completion time of MMGA, LAGA, and proposed algorithm

(c) Apply mutation operator with mutation probability 0.005 to generate new offspring
(d) Newly generated offspring is placed in the population for next round of iteration
(e) Exit condition is tested.

Step 4: End.

# 4 Performance Evaluation

To evaluate the performance of the presented technique, it is compared with the existing schemes LAGA and MMGA. MATLAB software is used to implement the proposed algorithm. Parameters taken into account are completion time and cost to evaluate the performance of the proposed scheme.

## *4.1 Completion Time*

It is the time VM takes to complete one task. During simulation, a number of task considered were 25, 50, 75 and 100 using five VMs as shown in Fig. 4.

## *4.2 Execution Cost*

Execution cost of all tasks on the available VMs is computed using Eq. (10)

$$\text{Total Cost} = \frac{\text{Task Length} * \text{Cost per seconds}}{\text{VM}} \tag{10}$$

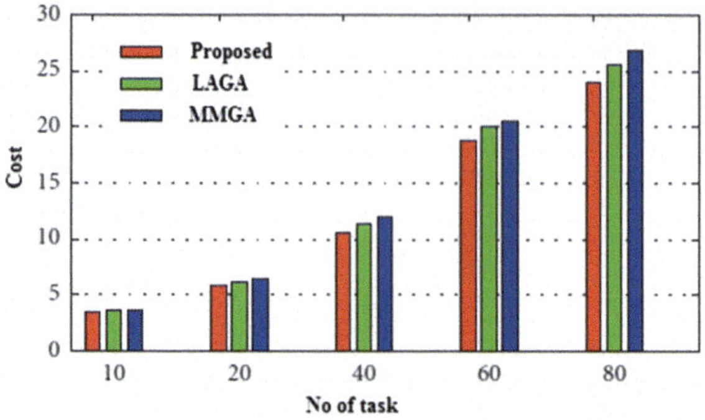


**Fig. 5** Execution cost of MMGA, LAGA, and proposed algorithm

As shown in Fig. 5, the execution cost of the proposed, LAGA, and MMGA schemes is under variable task sizes. The execution cost of the proposed algorithm is less than as compared to LAGA and MMGA, respectively.

## 5 Conclusion

This paper presented load balanced task scheduling mechanism for cloud computing environment. The presented mechanism achieved the improved results as compared to existing scheme LAGA and MMGA. A fitness function has been prepared based on load of individual VM and number of task set. The cost and completion time of presented mechanism are reduced significantly and achieved better results.

In future work, some other factors can be considered in addition to time span, ETC, and load balancing. A multiobjective model can be proposed for real-time applications.

## References

1. Hamad, S.A., Omara, F.A.: Genetic-based task scheduling algorithm in cloud computing environment. Int. J. Adv. Comput. Sci. Appl. **7**(4), 550–556 (2016)
2. Chitra Devi, D., Rhymend Uthariaraj, V.: Load balancing in cloud computing environment using improved weighted round Robin algorithm for non-preemptive dependent tasks. Hindawi Sci. World J. (2016)
3. Ge, J.W., Yuan, Y.S.: Research of cloud computing task scheduling algorithm based on improved genetic algorithm. Appl. Mech. Mater., 2426–2429 (2013)
4. Zhu, Z., Du, Z.: Improved GA-based task scheduling algorithm in cloud computing. Comput. Eng. Appl. **49**(5), 77–80 (2013)
5. Attiya, I., Abd Elaziz, M., Xiong, S.: Job scheduling in cloud computing using a modified Harris Hawks optimization and simulated annealing algorithm. Hindawi Comput. Intell. Neurosci. **2020**

6. Zheng, Z., Wang, R., Zhong, H., Zhang, X.: An approach for cloud resource scheduling based on Parallel Genetic Algorithm. In: 2011 3rd International Conference on Computer Research and Development (ICCRD), pp. 444–447 (2011)
7. Wang, T., Liu, Z., Chen, Y., Xu, Y., Dai, X.: Load balancing task scheduling based on genetic algorithm in cloud computing. In: IEEE International Conference on Dependable, Autonomic and Secure Computing, vol. 12, pp. 146–152 (2014)
8. Zhan, Z.-H., Zhang, G.-Y., Gong, Y.-J., Zhan, J.: Load balance aware genetic algorithm for task scheduling in cloud computing. In: Asia-Pacific Conference on Simulated Evolution and Learning, pp. 644–655. Springer International Publishing, Switzerland (2014)
9. Kaur, R., Kinger, S.: Enhanced genetic algorithm based task scheduling in cloud computing. Int. J. Comput. Appl. **101** (2014)
10. Rajak, R., Shukla, D., Alim, A.: Modified Critical Path and Top-Level attributes (MCPTL)-based task scheduling algorithm in parallel computing. In: Proceedings of soft computing: theories and applications. Advances in Intelligent Systems and Computing, vol. 583 (2017)
11. Teraiya, J., Shah, A.: Analysis of dynamic and static scheduling algorithms in soft real-time system with its implementation. In: Proceedings of Soft Computing: Theories and Applications. Advances in Intelligent Systems and Computing, vol. 1053, pp. 757–768. Springer, Singapore (2020)
12. Wang, B., Li, J.: Load balancing task scheduling based on multi-population genetic algorithm in cloud computing. In: Proceedings of the 35th Chinese Control Conference, 27–29 July 2016, Chengdu, China (2016)
13. Jana, B., Chakraborty, M., Mandal, T., Proceedings of the soft computing: theories and applications. Advances in Intelligent Systems and Computing, vol. 742, pp. 525–536. Springer, Singapore (2018)
14. Saadat, A., Masehian, E.: Load balancing in cloud computing using genetic algorithm and fuzzy logic. In: International Conference on Computational Science and Computational Intelligence (CSCI) (2020)
15. Gupta, A., Bhadauria, H.S., Singh, A.: Load balancing based hyper heuristic algorithm for cloud task scheduling. J. Ambient Intell. Humanized Comput. (2020)
16. Gundu, S.R., Anuradha, T.: Improved hybrid algorithm approach based load balancing technique in cloud computing. Global J. Comput. Sci. Technol. B Cloud Distrib. **19**(2) (2019)

# Riemann Problem in Generalized Chaplygin Gas

**Mustafa Rahman, Mohd Junaid Siddiqui, Azhar Iqbal, and Amjad A. Pasha**

**Abstract** Riemann problems are beneficial in computational fluid dynamics (CFD). Almost all computational techniques in CFD require testing the code for the Riemann problem. The aim is to capture the time evaluation of an initial discontinuity. Recently, researchers have found the solution of the Riemann problem in Dam break where the water of different heights is separated through a wall. There is a vast literature on solutions of the Riemann problem by numerical or analytical techniques in gas dynamics. However, very few studies have appeared on the Riemann problem in Chaplygin gas. To find the analytical solutions to these problems is quite cumbersome. We have applied the Harten-Lax-van Leer (HLL) numerical schemes to find the solution of the Riemann problem in generalized Chaplygin gas. The figures are plotted with different initial data to notice the behavior of delta-shock, and its interaction with rarefaction waves. The result shows that delta-shock can strengthen with time. The obtained results have been compared with the available solution of Wang (J Math Anal Appl 403:434–450 [1]).

**Keywords** Approximate Riemann solver · HLL · Chaplygin gas

M. Rahman
Department of Mechanical and Mechatronics Engineering, University of Waterloo, Waterloo, ON N2L 3G1, Canada

M. Junaid Siddiqui (✉) · A. Iqbal
Department of Mathematics and Natural Sciences, Prince Mohammad Bin Fahd University, Al Khobar 31952, Saudi Arabia

A. A. Pasha
Department of Aeronautical Engineering, King Abdulaziz University, Jeddah 21589, Saudi Arabia



T. K. Sharma et al. (eds.), *Soft Computing: Theories and Applications*, Advances in Intelligent Systems and Computing 1381,
https://doi.org/10.1007/978-981-16-1696-9_10

## 1 Introduction

Many problems in aerodynamics [2–5], combustion [6], astrophysics, and inertial confinement fusion can be modeled by Euler's conservation equations or magneto-hydrodynamics (MHD) equations. Finding the solution to this system of equations is very important to capture the dynamics of the system. In the context, Riemann problem is a problem where an initial discontinuity between two uniform steady states evolves in time. The structure that would be developed after infinitesimal little time depends on initial conditions. This problem is useful in understanding the Euler's/MHD equations. Almost every researcher in computational fluid dynamics and MHD use the problem for simulation to test their codes. Very recently, researchers in Switzerland [7] have simulated Riemann problem for analyzing the safety measures of small dams.

The exact solution of the Riemann problem by analytical methods is quite expensive and time-consuming. Therefore, approximate Riemann solvers have been developed by the researchers. Among the various available Riemann solvers, the Riemann solvers by Dai and Woodward [8], Brio and Wu [9], Roe and Balsara [10], Rahman and Shevare [11], and Balsara [12] are worth mentioning.

Steger and Warming [13] have applied the flux vector splitting scheme to the inviscid gas-dynamic equations with application to finite difference methods. In gas dynamics, the Riemann problem for the one-dimensional system of conservations laws with zero pressure has been solved by Cheng [14]. He obtained two kinds of solutions, one which is a vacuum and another one is a delta-shock solution by using characteristic analysis method. Raja Sekhar and Sharma [15] solved the Riemann problem for a quasilinear hyperbolic system of equations governing the one-dimensional unsteady simple wave flow of an isentropic fluid in magneto-gas dynamics. They also have shown the existence and uniqueness of the solution to the Riemann problem for arbitrary initial data. Kuila et al. [16] presented analytical and computational techniques for solving the Riemann problem of the drift-flux two-phase flow model. On computational techniques, we would like to mention the work by Sharma and Pant, Swami and Sandeep, Yadav et al. and Harsh and Nandeesh [17–20] and Rahman et al. [21–23] as well as Faridi and Seema [24]. Wang [1] solved the Riemann problem for one-dimensional generalized Chaplygin gas dynamics. Li [25] used approximate Riemann solvers HLL, for solving the Euler equations of magneto-hydrodynamics (MHD).

In the paper, we have considered the Riemann problem for one-dimensional generalized Chaplygin gas dynamics. We applied the HLL scheme to find the solution of the Riemann problem. Figures depict the results and are compared with the published result of Wang [1]. The detailed mathematical formulation of the model and results are presented in the following manner: In Sect. 2, we have introduced the Euler equations describing the conservation of mass and momentum for a compressible fluid in generalized Chaplygin gas dynamics. In Sect. 3, we have developed HLL numerical scheme for solving the Riemann problem. In Sect. 4, we have found the solution of the Riemann problem by HLL together with some initial and boundary condition. It is also devoted to the comparison between our obtained results by Wang [1] through the figures.

## 2 Governing Equations

The Euler equations describing the conservation of mass and momentum for a compressible fluid in generalized Chaplygin gas dynamics can be written in conservation form as

**Continuity equation**:

$$\frac{\partial}{\partial t}\int \rho \mathrm{d}x + (\rho_1 u_1 - \rho_2 u_2) = 0,$$

Or

$$\frac{\partial}{\partial t}\int \rho \mathrm{d}x + \frac{\partial}{\partial x}\int \rho u \mathrm{d}x = 0,$$

when $\Delta x \to 0$ in above equation.

$$\int \left(\frac{\partial \rho}{\partial t} + \nabla \cdot (\rho u)\right)\mathrm{d}x = 0,$$

Or

$$\frac{\partial \rho}{\partial t} + \frac{\partial}{\partial x}(\rho u) = 0, \tag{1}$$

**Momentum equation**:

$$\frac{\partial(\rho u)}{\partial t} + \frac{\partial\left(\rho u^2\right)}{\partial x} - \frac{\partial\left(s\left(\rho^{-\alpha}\right)\right)}{\partial x} = 0, \tag{2}$$

where the variable $u$ denotes the velocity and $\rho$ the mass density.

The equation of state for generalized Chaplygin gas is

$$P = -s\rho^{\alpha} \tag{3}$$

where $s > 0$ and $0 < \alpha \leq 1$.

The above system (1)–(2) is a nonlinear system which can be written in the compact form as follows:

$$\frac{\partial U}{\partial t} + \frac{\partial F(U)}{\partial t} = 0, t \in R^{+}, x \in R, \tag{4}$$

where $U = \begin{bmatrix} \rho \\ \rho u \end{bmatrix}$ and $F(U) = \begin{bmatrix} \rho u \\ \rho u^2 - s\rho^{-\alpha} \end{bmatrix}$, doing Jacobian matrix analysis [26] gives

$$\begin{bmatrix} 1 & 0 \\ 0 & 1 \end{bmatrix} \begin{bmatrix} \rho \\ \rho u \end{bmatrix}_t + \begin{bmatrix} 0 & 1 \\ \frac{\alpha s}{\rho}\rho^{-\alpha} - u^2 & 2u \end{bmatrix} \begin{bmatrix} \rho \\ \rho u \end{bmatrix}_x = 0 \tag{5}$$

This leads to the following eigenvalues:

$$\lambda_1 = u + \frac{\rho^{-\alpha}}{\rho}\sqrt{\alpha\rho^{\alpha+1}s} \text{and} \lambda_2 = u - \frac{\rho^{-\alpha}}{\rho}\sqrt{\alpha\rho^{\alpha+1}s}$$

Right eigenvector analysis

$$k = \begin{bmatrix} 1 & 1 \\ \lambda_1 & \lambda_2 \end{bmatrix}, k^{-1} = \frac{1}{\lambda_2 - \lambda_1}\begin{bmatrix} \lambda_2 & -1 \\ -\lambda_1 & 1 \end{bmatrix} \tag{6}$$

$$A^+ = \begin{bmatrix} -\frac{\lambda_1\lambda_2}{\lambda_1-\lambda_2} & \frac{\lambda_1}{\lambda_1-\lambda_2} \\ -\frac{\lambda_1^2\lambda_2}{\lambda_1-\lambda_2} & \frac{\lambda_1^2}{\lambda_1-\lambda_2} \end{bmatrix} A^- = \begin{bmatrix} \frac{\lambda_1\lambda_2}{\lambda_1-\lambda_2} & -\frac{\lambda_2}{\lambda_1-\lambda_2} \\ \frac{\lambda_1\lambda_2^2}{\lambda_1-\lambda_2} & -\frac{\lambda_2^2}{\lambda_1-\lambda_2} \end{bmatrix} \tag{7}$$

$$F^+ = \begin{bmatrix} -\frac{\lambda_1\lambda_2\rho}{\lambda_1-\lambda_2} + \frac{\lambda_1 u\rho}{\lambda_1-\lambda_2} \\ -\frac{\lambda_1^2\lambda_2\rho}{\lambda_1-\lambda_2} + \frac{\lambda_1^2 u\rho}{\lambda_1-\lambda_2} \end{bmatrix} F^- = \begin{bmatrix} \frac{\lambda_1\lambda_2\rho}{\lambda_1-\lambda_2} - \frac{\lambda_2 u\rho}{\lambda_1-\lambda_2} \\ \frac{\lambda_1\lambda_2^2\rho}{\lambda_1-\lambda_2} - \frac{\lambda_2^2 u\rho}{\lambda_1-\lambda_2} \end{bmatrix} \tag{8}$$

## 3 Numerical Schemes

### 3.1 HLL Scheme

Harten et al. [26] suggested the HLL scheme, which is a powerful tool to solve the Riemann problem approximately. The Riemann problem can be solved approximately by finding an approximation to the numeral flux $F_{i+\frac{1}{2}}$, i.e., the flux through the interface. Here, we assume that all the wave speed estimates $S_L$, $S^*$and$S_R$ are obtainable. Thus, for an interface $i + 1/2$ of cells $i$ and $i + 1$, the HLL numerical flux is as follows

$$\int_0^T \int_{x_L}^{x_R} \left(\frac{\partial U}{\partial t} + \frac{\partial F}{\partial x}\right) \mathrm{d}x\, \mathrm{d}t = 0, \tag{9}$$

The intermediate state $U^*$ is given by $U^* = \frac{S_R U_R - S_L U_L + (F_L - F_R)}{S_R - S_L}$, which gives the intermediate flux as

$$F^*_{i+\frac{1}{2}} = \frac{S_R F_L - S_L F_R + S_L S_R (U_R - U_L)}{S_R - S_L} \tag{10}$$

For the evaluation of numerical flux in above equation, we need the two preknown approximate wave speed $S_L$ and $S_R$. These can be obtained by using the following choice of wave speeds suggested by HLL [26]:

$$\mathrm{SL} = \min\left[\mathrm{ui} - \frac{\rho_i^{-\alpha}}{\rho_i}\sqrt{\alpha\rho_i^{\alpha+1}s},\ \mathrm{ui}+1 - \frac{\rho_{i+1}^{-\alpha}}{\rho_{i+1}}\sqrt{\alpha\rho_{i+1}^{\alpha+1}s}\right];$$

$$\mathrm{SR} = \max\left[\mathrm{ui} + \frac{\rho_i^{-\alpha}}{\rho_i}\sqrt{\alpha\rho_i^{\alpha+1}s},\ \mathrm{ui}+1 + \frac{\rho_{i+1}^{-\alpha}}{\rho_{i+1}}\sqrt{\alpha\rho_{i+1}^{\alpha+1}s}\right]$$

Thus, for an interface $i + 1/2$, interface of cells $i$ and $i + 1$,

$$F^{\mathrm{HLL}}_{i+\frac{1}{2}} = \begin{cases} F_i,\ S_L \geq 0 \\ F^*_{i+\frac{1}{2}},\ S_L \leq 0 \leq S_R, \\ F_{i+1},\ S_R \leq 0 \end{cases} \tag{11}$$

$$F_i = \begin{bmatrix} \rho_i u_i \\ \rho_i u_i^2 - s\rho_i^{-\alpha} \end{bmatrix} \tag{12}$$

These approximations for density and velocity can be used directly into Eq. 11 from Eq. 10 to obtain wave speed estimates for HLL schemes. Now, we can determine the estimates $\rho^*$ and $u^*$ for density and particle velocity by using the above proposed scheme.

## 4 Results and Discussions

We initially simulate verification test cases and then analyze the nonlinear wave propagation at low speeds. Three cases are chosen to determine the simulation of the previously stated HLL scheme and compare it with the literature. In all the cases, the constants $\alpha$ and $s$ are fixed as $\alpha = 0.5$ and $s = 5$ with CFL (Courant–Friedrichs–Lewy) condition of 0.4.

## 4.1 Solution of Riemann Problem by HLL Scheme

In this section, we will discuss Riemann solution to the generalized Chaplygin gas Eqs. (1) and (2) along the $x$-direction, it has the following piece-wise constant Riemann initial data for the left and right states.

We have used HLL scheme for finding the solution of Riemann problem as described by Eq. (9) with the initial conditions. The results are depicted in the figures below and comparison has been made with the Riemann problem solved by Wang [1] with respect to the time evaluation.

The test case I is considered such that the domain size 5 with a range from $x = -1$ to $x = 4$. The number of mesh cells is 200 with the diaphragm (discontinuity) at the location of $x = 0$ and $x = 0.4$. The parameters of the initial condition to the left (below location $x < 0$) of the first discontinuity are the density of 0.5 and velocity of 5.5. The parameters of the initial condition in between the discontinuity of location ($0 < x < 0.4$) are the density of 0.25 and velocity of −4. The parameters of the initial condition to the right (above location $x > 0.4$) of the second discontinuity are the density of 0.2 and velocity of −2.9137, as described in Eq. 13. Figure 1a plots the velocity at times 0.02 and 0.4, respectively. Figure 1b plots the density at corresponding times, respectively.

$$(\rho, u)(x, 0) = \begin{cases} (0.5, 5.5), x \leq 0 \\ (0.25, -4), 0 \leq x \leq 0.4 \\ (0.2, -2.9137), 0.4 \leq x \end{cases} \tag{13}$$

Test case II is chosen of domain size 5 with a range from $x = -1$ to $x = 4$. The number of mesh cells is 250 with the diaphragm (discontinuity) at the location of $x = 0$ and $x = 0.4$. The parameters of the initial condition to the left (below location $x < 0$)

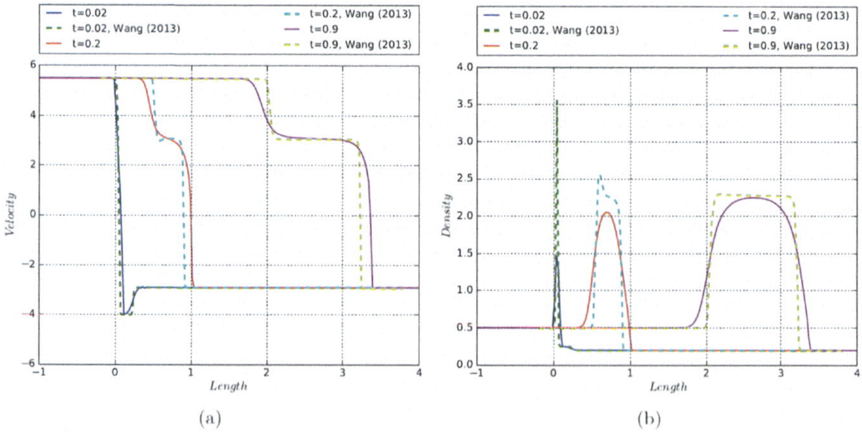


**Fig. 1** Plots of **a** velocity and **b** density at times 0.02 and 0.4, respectively

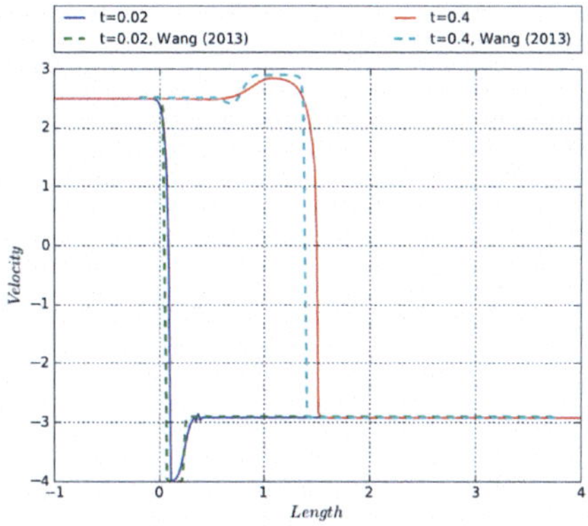


(a)

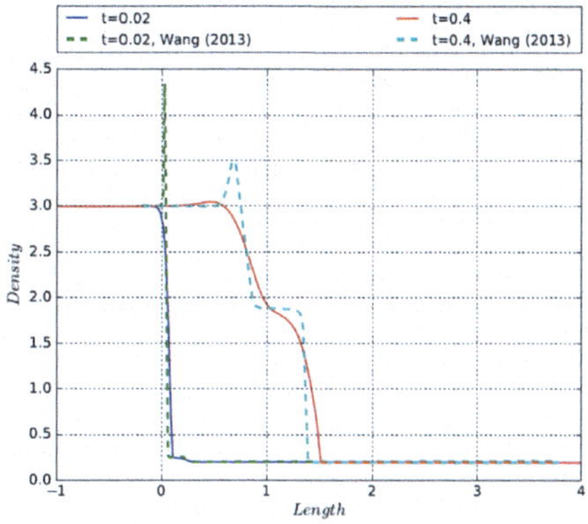


(b)

**Fig. 2** **a** Plots of velocity at times 0.02, 0.26 and 0.9, respectively, and **b** the plots of density at corresponding times, respectively

of the first discontinuity are the density of 3 and velocity of 2.5. The parameters of the initial condition in between the discontinuity of location ($0 < x < 0.4$) are the density of 0.25 and velocity of $-4$. The parameters of the initial condition to the right (above location $x > 0.4$) of the second discontinuity are the density of 0.2 and velocity of $-2.9137$, as described in Eq. 14. Figure 2a plots the velocity at times 0.02, 0.26, and 0.9, respectively. Figure 2b plots the density at corresponding times, respectively.

$$(\rho, u)(x, 0) = \begin{cases} (3, 2.5), x \leq 0 \\ (0.25, -4), 0 \leq x \leq 0.4 \\ (0.2, -2.9137), 0.4 \leq x \end{cases} \tag{14}$$

Test case III is chosen of domain size 4 with a range from $x = -2$ to $x = 2$. The number of mesh cells is 4000 with the diaphragm (discontinuity) at the location of $x = 0$. The parameters of the initial condition to the left (below location $x < 0$) of the discontinuity are the density of 3 and velocity of 4. The parameters of the initial condition to the right (above location $x > 0$) of the discontinuity are the density of 1 and velocity of $-4$, as described in Eq. 15. Figure 3a plots the velocity at times 0.05, 0.1, and 0.2, respectively. Figure 3b plots the density at corresponding times, respectively.

$$(\rho, u)(x, 0) = \begin{cases} (3, 4), x \leq 0 \\ (1, -4), 0 \leq x \end{cases} \tag{15}$$

The results are in good agreement. Some deviation at time $t = 0.02$ in the density profile Fig. 1b is due to the numerical scheme. Observing the solution of HLL scheme, the numerical results match well with the published results by Wang [1]. This study

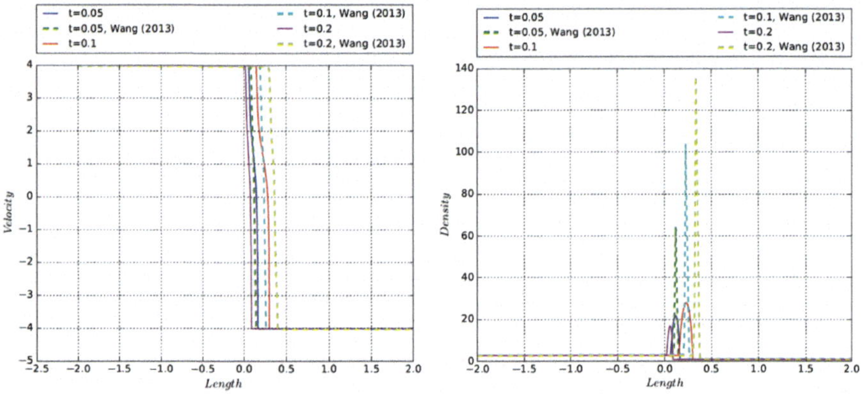


(a) (b)

**Fig. 3** **a** Plots of velocity at times 0.05, 0.1, and 0.2, respectively, and **b** the plots of density at corresponding times, respectively

also shows that the HLL scheme is quit powerful scheme to solve the Riemann problem compare to the other numerical scheme.

## *4.2 Nonlinear Wave Propagation*

System (2.4) leads to the corresponding Riemann problem as a special initial-value problem with the following Dirichlet condition: $U_{x,0} = \sin(-\pi x/2)$. The results are shown in Fig. 4 with respect to the time evaluation. From Fig. 4a, we can see that the velocity is symmetric in both side of the region. At time $t = 0.0$, the velocities are at peak and the magnitude decreases initially as the time increases. The velocity magnitude is tending to zero and at time near $t \approx 1.5$, the velocities magnitude are almost becoming zero. As the time is further increased, the velocity magnitude is increasing but the direction of the velocity is in opposite direction as an harmonic motion. The velocity magnitude at the center remains at the negligible magnitude. From Fig. 4b, it is clear that the density is also symmetric in both sides to the region of center. Initially, the density is uniform in both side to the region. As the time increases, the densities began to increase and reach the maximum at time $t \approx 1$, and then again tend to decrease as time increase and become almost uniform at time $t = 4.0$. From Fig. 4c, we can see that the pressure is also working in a proportional power manner.

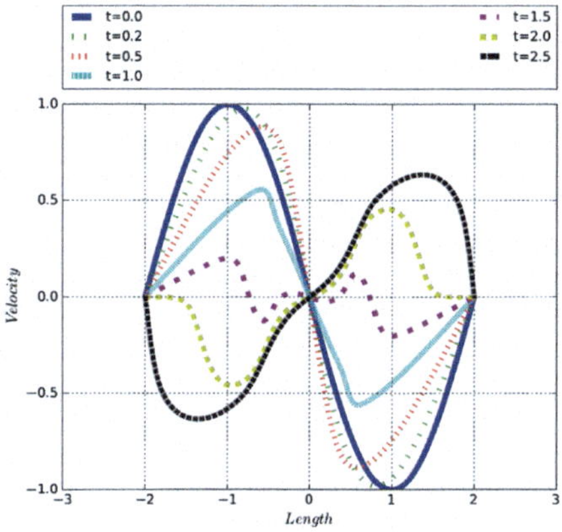

(a)

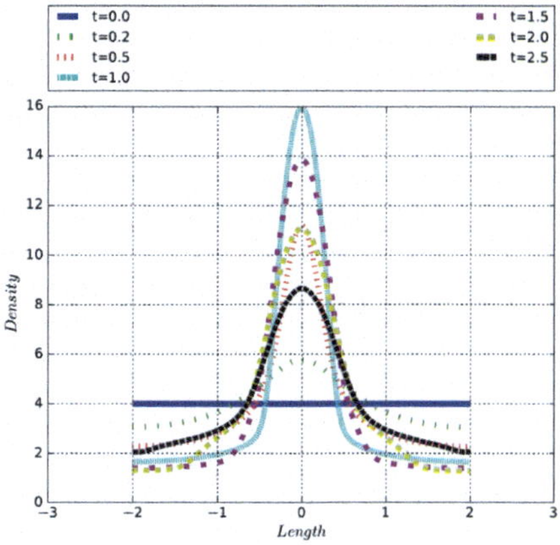

(b)

**Fig. 4** **a** Plots of velocity at times $0 \le t \le 0.4$, respectively, **b** the plots of density, and **c** the plots of pressure at corresponding times, respectively

# Appendix

## *Flux Vector Splitting*

$$F = AU = \left[K\left(\lambda^{+} + \lambda^{-}\right)K^{-1}\right]U = \left[A^{+} + A^{-}\right]U = F^{+} + F^{-} \tag{16}$$

Thus

$$F^{+} = A^{+}U = \left[K\left(\lambda^{+}\right)K^{-1}\right]U \tag{17}$$

For $F = \begin{bmatrix} \rho u \\ \rho u^2 - s\left(\rho^{-\alpha}\right) \end{bmatrix}$

$$\begin{gathered} F^{+} = F,\, F^{-} = 0,\, u > \frac{\rho^{-\alpha}}{\rho}\sqrt{\alpha\rho^{\alpha+1}s}; \\ F^{+} = \begin{bmatrix} \rho\lambda_1/2 \\ \rho\lambda_1^2/2 \end{bmatrix},\, F^{-} = \begin{bmatrix} \rho\lambda_2/2 \\ \rho\lambda_2^2/2 \end{bmatrix},\, -\frac{\rho^{-\alpha}}{\rho}\sqrt{\alpha\rho^{\alpha+1}s} < u < \frac{\rho^{-\alpha}}{\rho}\sqrt{\alpha\rho^{\alpha+1}s}; \\ F^{+} = 0,\, F^{-} = F,\, u < -\frac{\rho^{-\alpha}}{\rho}\sqrt{\alpha\rho^{\alpha+1}s}; \end{gathered} \tag{18}$$

## *Vanleer Scheme*

$$
\begin{aligned}
&F^{+} = F, F^{-} = 0, u > \frac{\rho^{-\alpha}}{\rho}\sqrt{\alpha\rho^{\alpha+1}s};\\
&F^{+} = \left[\frac{1}{4}\rho\lambda_1^2/\sqrt{k}\rho\lambda_1^2/2\right],\\
&F^{-} = \begin{bmatrix} -\frac{1}{4}\rho\lambda_2^2/\sqrt{k} \\ \rho\lambda_2^2/2 \end{bmatrix}, -\frac{\rho^{-\alpha}}{\rho}\sqrt{\alpha\rho^{\alpha+1}s} < u < \frac{\rho^{-\alpha}}{\rho}\sqrt{\alpha\rho^{\alpha+1}s}; \qquad (19)\\
&F^{+} = 0, F^{-} = F, u < -\frac{\rho^{-\alpha}}{\rho}\sqrt{\alpha\rho^{\alpha+1}s};
\end{aligned}
$$

# References

1. Wang, G.: The Riemann problem for one dimensional generalized Chaplygin gas dynamics. J. Math. Anal. Appl. **403**, 434–450 (2013)
2. Toro, E.F.: Shock-capturing methods for free-surface shallow flows. Wiley (2001)
3. Rahman, M.M., Samtaney, R.: Modeling and analysis of large-eddy simulations of particle-laden turbulent boundary layer flows. In 55th AIAA Aerospace Sciences Meeting, p. 981 (2017)
4. Rahman, M.M., Cheng, W., Samtaney, R., Urzay, J.: Large-eddy simulations of sandstorms as charged- particle suspensions in turbulent boundary layers. In: Proceedings of the Summer Program, 45 (2016)
5. Raheem, M.A., Edi, P., Pasha, A.A., Rahman, M.M., Juhany, K.A.: Numerical study of variable camber continuous trailing edge flap at off-design conditions. Energies **12**(16), 3185 (2019)
6. Reddy, V.M., Rahman, M.M., Gandi, A.N., Elbaz, A.M., Schrecengost, R.A., Roberts, W.L.: Cenosphere formation from heavy fuel oil: a numerical analysis accounting for the balance between porous shells and internal pressure. Combust. Theor. Model. **20**(1), 154–172 (2016)
7. Edom, E.: Numerical calculation of the Dam-Break Riemann problem with a detailed method and comparison with a simplified method. Dissertation Diploma Thesis of Leibniz University Hannover (2008)
8. Dai, W., Woodward, P.R.: An approximate Riemann solver for ideal magnetohydrodynamics. J. Comput. Phys. **111**(2), 354–372 (1994)
9. Brio, M., Wu, C.C.: An upwind differencing scheme for the equations of ideal magnetohydrodynamics. J. Comput. Phys. **75**(2), 400–422 (1988)
10. Roe, P.L., Balsara, D.S.: Notes on the Eigen system of magnetohydrodynamics. SIAM J. Appl. Math. **56**(1), 57–67 (1996)
11. Rahman, M.M., Shevare, G.R.: Analysis of conservation laws of non-linear springs. In: 13th AeSI CFD Symposium (2011)
12. Balsara, D.S.: A two-dimensional HLLC Riemann solver for conservation laws: application to Euler and magnetohydrodynamic flows. J. Comput. Phys. **231**(22), 7476–7503 (2012)
13. Steger, J.L., Warming, R.: Flux vector splitting of the inviscid gasdynamic equations with application to finite-difference methods. J. Comput. Phys. **40**(2), 263–293 (1981)
14. Cheng, H.: Riemann problem for one dimensional system of conservation laws of mass, momentum and energy in zero pressure gas dynamics. Diff. Equat. Appl. **4**(4), 653–664 (2012)
15. Sekhar, T.R., Sharma, V.D.: Riemann problem and elementary wave interactions in isentropic magnetogasdynamics. Nonlinear Anal. Real World Appl. **11**, 619–636 (2010)

16. Kuila, S., Sekhar, T.R., Zeidan, D.: On the Riemann problem simulation for the Drift-Flux equations of two-phase flows. Int. J. Comput. Meth. **13**(1), 1650009 (2016)
17. Sharma, T.K., Pant, M.: Opposition-based learning embedded shuffled Frog-Leaping algorithm. Soft Comput. Theor. Appl. **583**, 853–861 (2017)
18. Swami, S.V., Sandeep, K.: An improved spider monkey optimization algorithm. Soft Comput. Theor. Appl. **583**(1), 73–81 (2017)
19. Yadav, B., Kumar, A., Kumar, Y.: A robust digital image watermarking algorithm using DWT and SVD. Soft Comput. Theor. Appl. **583**(1), 25–36 (2017)
20. Harsh, B., Nandeesh, G.: Critical path problem for scheduling using genetic algorithm. Soft Comput. Theor. Appl. **583**(1), 15–24 (2017)
21. Rahman, M.M., Samtaney, R.: Generation and sustenance of electric fields in sandstorms. Phys. Rev. Res. **3**(1), L012008 (2021)
22. Rahman, M.M., Samtaney, R.: Particle concentration variation for inflow profiles in high Reynolds number turbulent boundary layer. In: Fluids Engineering Division Summer Meeting, vol. 83723, p. V002T04A022. American Society of Mechanical Engineers (2020)
23. Rahman, M.M.: An investigation of electric fields in sandstorms (2021)
24. Faridi, S.M., Seema, V.: Integration of GIS, spatial data mining, and Fuzzy logic for agricultural intelligence. Soft Comput. Theor. Appl. **583**(1), 171–183 (2017)
25. Shengtai, L.: An HLLC Riemann solver for magnetohydrodynamics. J. Comput. Phys. **203**, 344–357 (2005)
26. Harten, A., Lax, P.D., Leer, B.V.: On upstream differencing and Godunov-type schemes for hyperbolic conservation laws. SIAM Rev. **25**(1), 35–61 (1983)

# Teaching–Learning Perception Toward Blended E-learning Portals During Pandemic Lockdown

**Manish Dadhich, Kamal Kant Hiran, and Shalendra Singh Rao**

**Abstract** Education institutions across the country are discontinued because of a countrywide lockdown initiated by the administration to restrain the dissemination of COVID-19. To make sure the continuity of learning, many institutions (domestic, national, international) commenced a secure and healthy contactless environment for teachers and students. Therefore, it is obvious that teacher and student view is a paramount factor in the adoption and implementation of new-age virtual teaching pedagogy. It is relevant to analyze the perceptions of understanding the reasons behind the high rate of using such a virtual platform for content delivery. The students and faculties who have been using these online portals were selected and filled the questionnaire which comprises closed as well as open-ended questions. The numbers of online portals at the school and university levels have been rapidly gaining momentum for the past few months. Therefore, a hybrid method of exploratory study was conducted to probe teacher and learner discernment of the effectiveness of prevailing content delivery modes. A snowball sampling technique is applied for selecting students and faculties who enrolled and experienced Web-based learning from various Indian universities and colleges. The answers were analyzed with the help of t-test and outcomes revealed a statistically significant difference in teaching–learning perceptions among the faculty and students on three variables, i.e., lack of human interface, dummy presence of students, online teaching consume more resources.

**Keywords** E-learning · AICTE · Swayam · Remote centers · Technocratic pedagogy

---

M. Dadhich (✉) · K. K. Hiran
Sir Padampat Singhania University, Udaipur, India
e-mail: manish.dadhich@spsu.ac.in

S. S. Rao
Mohanlal Sukhadia University, Udaipur, India

 
T. K. Sharma et al. (eds.), *Soft Computing: Theories and Applications*,
Advances in Intelligent Systems and Computing 1381,
https://doi.org/10.1007/978-981-16-1696-9_11

## 1 Introduction

Advanced instruments of ICT have changed the entire scenario in every sphere of human endeavor [1]. Cloud-based technologies, interactive panels, remote centers, and Internet of things have coined the new digital learning platform [2, 3]. Blogs, content writing, virtual videos, and podcasting have created milestones in the teaching–learning environment [4]. The experts expect that the exponential growth of ICT will change the entire framework of delivering lectures, conducting meetings, webinars discussion, and sharing knowledge [5]. These unique states of art tools have proved their significance not only in the educational sector but also in every specialty in which people interact. Besides, they are equipped with enormous unique features to be adopted by inquisitors. For many, it is a platform for earning while for others sharing ideas, notions, and bringing awareness among learners. Society welcomed the technology-driven phenomena to promote the exciting journey of endless learning and it seems like adding icing on the cake [6].

Therefore, the usages of the latest technology open the door of virtual learning that is why leaner need not limit in a conformist boundary to be educated. The term e-learning refers to the delivery of valuable content through an electronic environment that composes learning easier [7].

Others outlook virtual learning as Internet-based learning that uses Web technology, knowledge sharing, LMS, and practical practice to add value to the people and organizations in which they work [8]. The potential for e-learning is high, and it is being gradually adopted by many organizations. Besides, some students are already interested in participating in e-learning to enhance their education, so students are taking advantage of e-learning opportunities offered by educational institutions. Over time, it is believed that more and more organizations are considering introducing e-learning worldwide as part of their new edge education [9]. India has a hefty pool of human resources, and the changing economic environment brought enormous learning opportunities for an active learner for higher education.

India is on the pathway to delivering qualitative education and development. The survey report of India brand equity foundation stated that the higher education sector in one of the largest chunks in the world, enrolling more than 70 million students. This segment is expected to grow at a stunning 18% to reach 34.12 billion $ in the next decade [10].

In the present pandemic situation, a drastic crisis has never been experienced by society. This challenging situation requires taking immediate action because desperate time calls for desperate measures. Education institutions athwart the country are discontinued because the countrywide lockdown forced by the administration to limit the spread of COVID-19. To make sure the continuity of learning, many institutions (domestic, national, international) commenced a secure and healthy contactless environment for teachers and students [11].

Education refoms at the international level brought UNICEF and Microsoft Corporation to expand the global learning platform for school students and youth who were affected by COVID-19 and assist them to continue their education on the virtual platform. Moreover, free Web learning tools, i.e., Big-Blue Button, Google Hangouts, Join-me, Meeting-Burner, Mikogo, Web-Huddle, Zoom, Google talk, Webex, etc., facilitate two-way interaction for the learning–teaching experience.

The paper is organized as follows: firstly, it describes the datasets of Indian education along with the review of literature; secondly, a methodology that comprises of teaching–learning perception of the respondents; the third section delineates two parameters, i.e., comparison of teaching–learning perceptions and perceptions of online technology and methods that comprises 20 variables; in last section, statistical outcomes followed by concluding remarks and implication of the related work.

### *1.1 Overview of Students Enrollment*

The statistics issued by the ministry of human resource development show a significant expansion in the higher education of India. Enrollment at all levels has augmented over the years. The CAGR has been 1.8% for the last 5 years, but interesting to see that the CAGR for integrated courses 11.2 and the rests other courses delineated the less than a digit as shown in Table 1.

## 2 Review of Literature

Huang and Hsiao [12] delineated factors influencing e-book reading, watching, and shopping through Web portals and discussed the impact of the pandemic lockdown on the perception of customers to purchase e-books and mobile applications by proposing a model/concept, application by conceptualizing cases such as situational influences. Moreover, it has also summarized some situational factors, the role of utilitarian and hedonic motives in predicting customer motive to buy e-books. Samsudeen and Mohamed [13] did confirmatory factor study and outlined that constructs of model/conceptual framework have a momentous impact on users and

**Table 1** Students enrolment and compound annual growth rate (in '000')

| Years | Ph.D. | M. Phil. | PG | UG | PGD | Dip. | Certi. | Integrated | G. total |
|---|---|---|---|---|---|---|---|---|---|
| 2014–15 | 117 | 33 | 3853 | 27425 | 215 | 2.5 | 170 | 141 | 34211 |
| 2015–16 | 126 | 42 | 3917 | 27420 | 229 | 2.54 | 144 | 155 | 34584 |
| 2016–17 | 141 | 43 | 4007 | 28348 | 213 | 2.61 | 166 | 173 | 35705 |
| 2017–18 | 161 | 34 | 4114 | 29016 | 235 | 2.7 | 177 | 195 | 36642 |
| 2018–19 | 169 | 30 | 4042 | 29829 | 224 | 2.69 | 162 | 241 | 37399 |
| CAGR | 7.6 | −1.71 | 1 | 1.9 | 0.9 | 1.5 | −0.9 | 11.2 | 1.8 |

indicated the imperative role of behavioral intent to utilize and using virtual learning system in the selective Government University of Sri Lanka.

Daniels et al. [5] outlined that the students from public establishments have higher expectations from e-learning than students from private organizations, and there exist is a noteworthy difference in expectation, feedback system, and problems encountered between male and female student among selected institutions.

Stack [1] focused on the limitations of the online delivery courses in the present context and suggested remedies for updating the system. It was found that the accomplishment and knowledge scale of online students is quite advanced than the score achieved in the classes taught in traditional classes. This study also highlighted the advantage of the online course that is controlling the options of cheating in examination. Substantial studies have contributed to the peer-reviewed literature by examining the impact of the present pandemic and the motivation of readers to buy and read books online. The Vietnam study offers market opportunities for retailers, including e-bookstores. Despite the current dominance of physical bookstores, book stocks in Vietnam are increasing. The largest online bookstores have tried to attract more customers to their sites because social gatherings could be harmful to the entire society [14].

The study is consistent with Dinh et al. [15] and outcomes delineated that there was expanding of book-buying motive through e-portal and the readers were provided perks and incentives by publishers for promoting online purchases.

Several studies [4, 9, 16–18] have compared F to F teaching to online learning in order to try to define the effectiveness of virtual portals and its implications, e.g., the optimum learning outcomes, craft the nearly all satisfied students with the high degree of course completion status.

Drape et al. [2] portrayed mainstay strategies and challenges included access to advanced education and lack of groundwork at both primary and secondary levels of teaching. Due to the advent in ICT, the school enrollments have also augmented but there was a need of training program to equip the faculty through mentoring programs.

ICT has become paramount important for the advanced technology, and medical education system in India and the virtual platform has gained momentum in the recent past [12]. Hiran and Henten [22] delineated users' perception toward ICT-based e-learning and the result of t-test revealed that respondents strongly agree to the variables, i.e., user-friendly, Web platform flexibility, facilitate students to take examinations and submit assignments, interactive portal, best material and content, download ease, post messages, and innovative concept. These variables are statistically significant because their $p$-value is less than 0.05 with the least mean difference.

E-learning is a comparatively new concept implying learning using digital media such as computers, blogs, and virtual meeting systems. Virtual classrooms consisting of tools such as text, graphics, podcasts, videos, three-dimensional objects, and animations can be used to broaden educational services [3, 21]. It is predicted that e-learning will facilitate student–teacher for sharing of resources and making learning environments conducive and promote comprehensive learning [11].

The global e-learning market would show stride growth over the subsequent few years [7]. Instead of fully online learning, the concept of hybrid learning has gained much attention in the recent past. Moreover, one-to-one meetings and Web-based interaction and conversation will change the entire learning pedagogy in the coming decade [13].

Dadhich [20] studied and compared faculty and students' perception to see e-learning as a new gateway for self-directed knowledge management platform and create impetus of self-learning. On the flip side, the students in online courses experience the sense of disconnectedness from the teacher. Besides, faculties tend to see the role of the college as extra indispensable to the success of online teaching than the students.

# 3 Objectives of the Study

The following research statement is proposed to investigate the learning–teaching perceptions toward blended e-learning modules:

- To find out the comprehensive perceptions of faculties and students toward online learning.

## *3.1 Research Methodology*

### 3.1.1 Collection of Data

The study investigates both primary and secondary sources of data, first-hand data collected through a structured questionnaire sent through Google form. The convenience sample technique was used to collect data from graduate and postgraduate students and faculties of major universities. The Google form was originally distributed by e-mail during March 2020 when the concept of taking online classes or work from home began due to COVID-19 pandemic.

### 3.1.2 Statistical Tool

The recorded data compiled, analyzed, and construed with the help of SPSS software, and independent sample t-test is administered on Likert five-point scales.

### 3.1.3 Sample Size

Having considered 140 respondents in the paper, the questionnaire consists of two sections. The first part comprises of four questions about the general profile of the students and faculties. The second part consists of two parameters, i.e., comparison of teaching–learning perceptions and methods that comprise of 10 variables. The period of the study was from March 2020 to May 2020.

## 4 Data Analysis and Interpretation

Table 2 outlines the demographic profile of students and faculty, of the 140 students out of which 90 respondents were males and 50 respondents were females, 60 respondents belonged to the IT, 70 from engineering, 10 from business studies, or others. The respondents' age groups less than 18 years were 10, 19–30 years 100, and above 30 years 30 respondents. 100 were graduate students and 40 were postgraduate. raduate students considered for this study.

Of the total, 152 faculty surveys submitted but 12 teachers were ultimately removed from the analysis because of a lack of momentous completion responses. Of the 140 faculty responses, 95 respondents were males and 45 were females, 80 respondents belonged to the IT, 50 from engineering, 10 from business studies whereas there were 20 adjunct faculty, 80, 20, and 20 were assistant professor, associate professor, and professor, respectively, (see Table 3). The following constructs or statements were considered for e-learning perception:

In Table 5, Levene's test for equality of variances depicts f-statistics and $p$-value of respective variables. Having calculated the descriptive statistics (see Table 4), inputs from faculty and student were compiled using t-test. These results outlined statistically considerable difference in teaching–learning perceptions among of the

**Table 2** Demographic profile of students

| Factors | Classification | Frequency | Percent |
|---|---|---|---|
| Gender | Male | 90 | 64.3 |
| | Female | 50 | 35.7 |
| | Total | 140 | 100 |
| Courses | (a) IT (IT, C.S. and Math) | 60 | 42.8 |
| | (b) Engineering | 70 | 50 |
| | (c) Business studies | 10 | 7.2 |
| | Total | 140 | 100 |
| Age | Less than 18 yrs | 10 | 7.2 |
| | 19–30 yrs | 100 | 71.4 |
| | Above 30 yrs | 30 | 21.4 |
| | Total | 140 | 100 |
| Education level | Graduate | 100 | 71.4 |
| | Postgraduate | 40 | 28.6 |
| | Total | 140 | 100 |

**Table 3** Demographic profile of faculties

| Factors | Classification | Frequency | Percent |
|---|---|---|---|
| Gender | Male | 95 | 67.8 |
| | Female | 45 | 32.2 |
| | Total | 140 | 100 |
| Courses | (a) IT (IT, C.S., and basic sciences) | 80 | 57.1 |
| | (b) Engineering | 50 | 35.7 |
| | (c) Business Studies | 10 | 7.2 |
| | Total | 140 | 100 |
| Academic rank | Adjunct | 20 | 14.3 |
| | Assistant professor | 80 | 57.1 |
| | Associate professor | 20 | 14.3 |
| | Professor | 20 | 14.3 |
| | Total | 140 | 100 |

respondents (faculty and students) on three variables such as lack of human interface (F-0.04, *p*-value 0.011), dummy presence of students (F-11.48, p-value 0.04), online teaching consume more resources (F-7.67, *p*-value 0.04). Rest other factors, i.e., easier online learning, incomplete learning on the virtual platform, facilitate two-way communications, learning online is time-consuming, best material and rich content, innovative concept of teaching–learning perceptions are not statistically noteworthy.

**Table 4** Group statistics of variables (Teaching–learning)

| Variables | *N* | Min. | Max. | Mean | Std. deviation |
|---|---|---|---|---|---|
| Easier online learning | 280 | 1 | 5 | 3.457 | 1.266 |
| Incomplete learning on virtual platform | 280 | 1 | 5 | 3.867 | 1.157 |
| Facilitate two-way communications | 280 | 2 | 5 | 3.05 | 1.137 |
| Learning online is time-consuming | 280 | 2 | 5 | 3.25 | 1.279 |
| Lack of human interface | 280 | 2 | 5 | 3.281 | 1.296 |
| Dummy presence of students | 280 | 1 | 5 | 3.25 | 1.366 |
| Consume more resources | 280 | 1 | 5 | 3.65 | 1.362 |
| Best material and rich Content | 280 | 1 | 5 | 3.457 | 1.322 |
| Future for comprehensive studies | 280 | 2 | 5 | 3.986 | 1.106 |
| Innovative concept | 280 | 2 | 5 | 3.764 | 1.069 |

**Table 5** Comparison of teaching–learning perceptions

| Variables | | Levene's test for equality of variances | | *t*-test for equality of means | | | |
|---|---|---|---|---|---|---|---|
| | | *F* | Sig. | *T* | *df* | Sig. (2-tailed) | Mean diff. |
| Easier online learning | Eq-vari assumed | 0.148 | 0.7 | −0.66 | 278 | 0.51 | −0.1 |
| | Eq-vari not assumed | | | −0.66 | 277.82 | 0.51 | −0.1 |
| Incomplete learning on virtual platform | Eq-vari. assumed | 2.278 | 0.132 | 0.98 | 278 | 0.328 | 0.135 |
| | Eq-vari not assumed | | | 0.98 | 277.49 | 0.328 | 0.135 |
| Facilitate two-way communications | Eq-vari. assumed | 0.017 | 0.895 | −0.524 | 278 | 0.6 | −0.071 |
| | Eq-vari not assumed | | | −0.524 | 277.99 | 0.6 | −0.071 |
| Learning online is time-consuming | Eq-vari. assumed | 0.108 | 0.742 | −0.467 | 278 | 0.641 | −0.071 |
| | Eq-vari not assumed | | | −0.467 | 277.96 | 0.614 | −0.071 |
| Lack of human interface | Eq-vari. assumed | 0.04 | 0.842 | 2.561 | 278 | 0.011 | 0.392 |
| | Eq-vari not assumed | | | 2.561 | 277.97 | 0.011 | 0.392 |
| Dummy presence of students | Eq-vari. assumed | 11.48 | 0.001 | −2.88 | 278 | 0.004 | −0.464 |
| | Eq-vari not assumed | | | −2.88 | 276.9 | 0.004 | −0.464 |
| Consume more resources | Eq-vari. assumed | 7.67 | 0.006 | −2.88 | 278 | 0.004 | −0.453 |
| | Eq-vari not assumed | | | −2.88 | 275.86 | 0.004 | −0.453 |
| Best material and rich Content | Eq-vari. assumed | 0.001 | 0.978 | 0.632 | 278 | 0.528 | 0.1 |

(continued)

**Table 5** (continued)

| Variables | | Levene's test for equality of variances | | *t*-test for equality of means | | | |
|---|---|---|---|---|---|---|---|
| | | *F* | Sig. | *T* | *df* | Sig. (2-tailed) | Mean diff. |
| | Eq-vari not assumed | | | 0.632 | 277.99 | 0.528 | 0.1 |
| Future for comprehensive studies | Eq-vari. assumed | 0.111 | 0.739 | 0.216 | 278 | 0.829 | 0.028 |
| | Eq-vari not assumed | | | 0.216 | 277.96 | 0.829 | 0.028 |
| Innovative concept | Eq-vari. assumed | 0.278 | 0.599 | −1.06 | 278 | 0.289 | −0.135 |
| | Eq-vari not assumed | | | −1.06 | 277.92 | 0.289 | −0.135 |

*Accepted at $p < 0.05$

## 5 Conclusion

Higher education in India has registered stride progress in the last few decades. India is nowadays one of the fastest creating international locations of the higher education system in the world with annual growth above 9% but still, a substantial section of the population abstained from education but now these online education portals can increase the penetration of leaning at every corner of life. The Indian education system has been facing many challenges regarding the quality teaching–learning experience for past decades [11]. Customary methods of teaching, academic learning, lack of practical disclosure, inconsistent use of the platform in different courses are the purview of this challenge in the higher education sector but recently pandemic COVID-19 has brought the phenomenal change in entire pedagogy. In the words of Stack [1], "momentous obstacle to the increased acceptance of online content delivery looks like to be the meager in resent scenario because e-learning can be supporting to teach but may not be a complete replacement." The result of t-test revealed a statistically significant difference in teaching–learning perceptions among the respondents (faculty and students) on three variables, i.e., lack of human interface, dummy presence of students, online teaching consumes more resources. Rest other factors do not differ statistically as the *p*-value is more than 0.05. This is consistent and relevant to the findings of [7, 13]. It can also be observed that the COVID-19 pandemic situation promotes the readers' and learners' intention toward not only to purchasing virtual learning modules but also to subscribe to prominent software for better access e-contents.

## References

1. Stack, S.: Learning outcomes in an online versus traditional course. Georg. Educ. Res. **9**(1), 1–20 (2015). https://doi.org/10.20429/ijsotl.2015.090105
2. Drape, T.A., Rudd, R., Lopez, M., Radford, D.: Challenges and solutions to higher education institutions in Africa. Int. J. Educ. **8**(1), 43 (2016). https://doi.org/10.5296/ije.v8i1.8742
3. Rich, A.J., Dereshiwsky, M.I.: Assessing the comparative effectiveness of teaching undergraduate intermediate accounting in the online classroom format. J. Coll. Teach. Learn. **8**(9), 19–28 (2011)
4. Chigeza, P., Halbert, K.: Navigating E-learning and blended learning for pre-service teachers: redesigning for engagement, access and efficiency. Aust. J. Teach. Educ. **39**(11), 133–146 (2014). https://doi.org/10.14221/ajte.2014v39n11.8
5. Daniels, M.M., Sarte, E., Dela Cruz, J.: Students' perception on e-learning: a basis for the development of e-learning framework in higher education institutions. In: IOP Conference Series Material Science and Engineering, vol. 482(1), pp. 2–8 (2019). https://doi.org/10.1088/1757-899X/482/1/012008
6. Kumar, A., Ambrish: Higher education: growth, challenges and opportunities. Int. J. Arts Humanit. Manage. Stud. **01**(2) (2015)
7. Engelbrecht, E.: Adapting to changing expectations: post-graduate students' experience of an e-learning tax program. Comput. Educ. **45**(2), 217–229 (2005). https://doi.org/10.1016/j.compedu.2004.08.001
8. Kelly, K.D., Bauer, T.: Managing intellectual resources via e-learning at Cisco. In: Handbook on Knowledge Management, vol. 2, pp. 511–532. INFOSYS (2004)
9. González-Gómez, D., Jeong, J.S., Airado Rodríguez D., Cañada-Cañada, F.: Performance and perception in the flipped learning model: an initial approach to evaluate the effectiveness of a new teaching methodology in a general science classroom. J. Sci. Educ. Technol. 25(3), 450–459 (2016). https://doi.org/10.1007/s10956-016-9605-9
10. Johnson, D.R.: Technological change and professional control in the professoriate. Sci. Technol. Hum. Values **38**(1), 126–149 (2013). https://doi.org/10.1177/0162243911430236
11. Soni, N.K., Patel, T.P.: Quality teaching & higher education system in India. Int. J. Sci. Res. Publ. 4(1), 2250–3153 (2014) [Online]. Available: www.ijsrp.org
12. Huang, E., Hsiao, X.: Synchronous and asynchronous communication in an online environment: faculty experiences and perceptions. Q. Rev. Distance Educ. 13(1), 15–30 (2012)
13. Samsudeen, S.N., Mohamed, R.: University students' intention to use e-learning systems: a study of higher educational institutions in Sri Lanka. Interact. Technol. Smart Educ. **16**(3), 219–238 (2019). https://doi.org/10.1108/ITSE-11-2018-0092
14. Nguyen, H.V., Tran, H.X., Van Huy, L., Nguyen, X.N., Do, M.T., Nguyen, N.: Online book shopping in Vietnam: the impact of the COVID-19 pandemic situation. Publ. Res. Q. **36**(3), 437–445 (2020). https://doi.org/10.1007/s12109-020-09732-2
15. Dinh, V.S., Nguyen, H.V., Nguyen, T.N.: Cash or cashless? Promoting consumers' adoption of mobile payments in an emerging economy. Strateg. Dir. **34**(1), 1–4 (2018). https://doi.org/10.1108/SD-08-2017-0126
16. Bernard, R.M., Borokhovski, E., Schmid, R.F., Tamim, R.M., Abrami, P.C.: A meta-analysis of blended learning and technology use in higher education: from the general to the applied. J. Comput. High. Educ. **26**(1), 87–122 (2014). https://doi.org/10.1007/s12528-013-9077-3
17. Otter, R.R., et al.: Comparing student and faculty perceptions of online and traditional courses. Internet High. Educ. **19**, 27–35 (2013). https://doi.org/10.1016/j.iheduc.2013.08.001
18. Southard, S., Meddaugh, J., France-Harris, A.: Can self-paced online course live long and prosper?Online J. Distance Learn. Adm. **18**(2), 8–20 (2015)
19. Ryan, S., Kaufman, J., Greenhouse, J., She, R., Shi, J.: The effectiveness of blended online learning courses at the community college level. Community Coll. J. Res. Pract. **40**(4), 285–298 (2016). https://doi.org/10.1080/10668926.2015.1044584
20. Dadhich, M.: An analysis of factors affecting on online shopping behavior of customers. ZENITH Int. J. Bus. Econ. Manag. Res. **7**(1), 20–30 (2017)

21. Hiran, K.K., Henten, A., Shrivas, M.K., Doshi, R.: Hybrid EduCloud model in higher education. In: IEEE 7th International Conference on Adaptive Science & Technology (ICAST), pp. 1–9 (2018). https://doi.org/10.1109/ICASTECH.2018.8507113
22. Hiran, K.K., Henten, A.: An integrated TOE-DoI framework for cloud computing adoption in higher education: the case of Sub-Saharan Africa, Ethiopia. In: Soft Computing: Theories and Applications, pp. 1281–1290. Springer, Singapore (2020)

# An Alternative Approach to Compress Text Files Using Fibonacci Sequence and Lucas Series

**Jatin Deep Singh, Mayur Agarwal, Manukansh Agrawal, and Sarvesh Kumar Swarnakar**

**Abstract** Data compression uses variable-length codes in which symbols that appear frequently are assigned with a specific code pattern (i.e. integer/alphanumeric). The ultimate goal behind assigning the characters with these code patterns is to facilitate easier compression and decompression. After concluding lots of research on different lossless text data compression techniques (i.e. Huffman coding, LZW coding, Shannon–Fano coding) along with the study of various concept of sequence and series (i.e. Fibonacci sequence (Bhattacharyya in complexity analysis of a lossless data compression algorithm using Fibonacci sequence. University of Calcutta, Kolkata, West Bengal (Bhattacharyya in Complexity analysis of a lossless data compression algorithm using Fibonacci sequence. Research Associate Department of Computer Science. University of Calcutta, Kolkata, West Bengal [3]) and Lucas Series), we have given an alternative approach through this research paper to improve the data compression ratio for the text file.

**Keywords** Data compression ratio · Huffman coding · LZW coding · Shannon–Fano coding · Fibonacci sequence · Lucas sequence

## 1 Literature Review

### *1.1 Introduction*

Data generally means information typed in a different format. Data cannot be randomly termed as information until it is analysed and is represented in a specific well-defined format. A software generally comprises data and the code associated with it. Data exists in various forms, and each formation leads us to extract the

J. D. Singh (✉) · M. Agarwal · M. Agrawal · S. K. Swarnakar
Information Technology, Galgotia College of Engineering and Technology, 1, Knowledge Park-III, Greater Noida, Uttar Pradesh, India

S. K. Swarnakar
e-mail: sarvesh.swarnkar@galgotiacollege.edu



T. K. Sharma et al. (eds.), *Soft Computing: Theories and Applications*,
Advances in Intelligent Systems and Computing 1381,
https://doi.org/10.1007/978-981-16-1696-9_12

hidden information that is related to the data. For example, alphanumeric for text, bits, and bytes for electronic mode of communication. A bit is the smallest unit that can be used to represent data in memory. A bit contains a value that is either 0 or 1. Hence, under storage format, data compression represents data in such a manner that the storage space occupied by target data is less than that occupied by the actual input data. Generally, for this purpose, we refer to two kinds of algorithms. The first is a compression algorithm that takes an input and generates its representation $X$ that consumes less space, followed by a reconstruction algorithm [1] that takes the compressed representation $X$ as its input to generate the reconstruction.

### *1.2 Lossless Data Compression*

Lossless compression techniques involve no loss of information [2]. If data is compressed in this manner, the input data can be obtained exactly in its original form from the processed data after a compress–expand cycle. Lossless compression is essential in files with non-uniform data distribution such as database records, spreadsheets, word-processing files, and even some kind of image and video information. For storing such information, we require certain lossless compression algorithms that function in such a manner that it compresses the text/file that we want and also decompresses the compressed text format to get the original text that was compressed.

## 2 Proposed Concepts

### *2.1 Fibonacci Sequence*

Fibonacci coding generates variable-length codes. It considers the basic method of replacing input characters with a specific code that is termed as code words. It uses the Fibonacci series [3] to implement this compression. In mathematical terms, the sequence $X(n)$ of Fibonacci numbers is represented by the under given recurrence relation.

#### 2.1.1 Recurrence Relation for Fibonacci Number

$X(N)$: It indicates the $N$th term of the Fibonacci series.

$$X(N) = X(N-1) + X(N-2)$$
$$\text{where } X(1) = 1 \,\&\, X(2) = 1$$
$$\text{OR}$$
$$X(0) = 0 \,\&\, X(1) = 1$$

By using brute force (recurrence tree) technique, the Fibonacci generation takes more time as well as it consumes greater space. Under this technique, the time complexity increases, as in each iteration, already computed Fibonacci terms are used for further computation which leads to a lot of memory utilization in the particular program. However, while using the dynamic programming approach, we can optimize the time. Under this technique, we evaluate the Fibonacci sequence for the new insertion of element and place it into a tabular structure and later fetch values from this table which leads to time optimization as the previously entered numbers are not evaluated. Section 2.1.2 shows the comparison of Fibonacci under brute force and that under the dynamic procedure.

### 2.1.2 Space and Time Complexity for Fibonacci Sequence

*M* represents the programming unit (i.e. number of loop execution to evaluate the time or space complexity to execute the logic and display the Fibonacci at *M*th location.)

1. Brute Force: (Recurrence Tree)

   Time Complexity (2^M)
   Space Complexity: O(M)

2. Dynamic Approach;

   Time Complexity: O(M)
   Space Complexity (M) (Fig. 1).

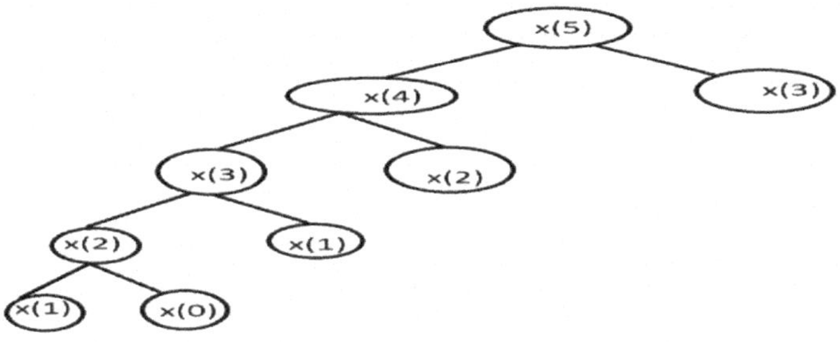


**Fig. 1** Recursion tree generated for computing 5th number of the Fibonacci sequence using the dynamic approach

## 2.2 *Lucas Number*

Edward Lucas studied the series: 2, 1, 3, 4, 7, 11, 18 … and named it as the Lucas numbers. It is related to the Fibonacci sequence in a manner that the extension of Fibonacci is the path that gave rise to a new mathematical concept to derive a new sequence called Lucas sequence.

The Fibonacci rule adds the recent two numbers of the series to generate the next input sequence. Similarly, a Lucas *n*th term can also be generated by Fibonacci and also by Lucas itself. The demonstration of this is given in Sect. 2.2.2. The recurrence relation for evaluating the member of Lucas sequence is given in Sect. 2.2.1 [where we write its *Z*th Lucas number as *P*(*Z*)].

### 2.2.1 Recurrence Relation [4] for Lucas Number

Listed below is the recurrence relation for evaluating the *Z*th Lucas number that appears in the Lucas series reoccurrence relation below:

$$P(Z) = \left\{ \begin{array}{ll} 2, & \text{for } Z = 0 \\ 1, & \text{for } Z = 1 \\ P(Z-1) + P(Z-2), & \text{for } Z > 0 \end{array} \right\}$$

### 2.2.2 Relationship Between Lucas and Fibonacci Sequence

The below explained example indicates the relationship between the Lucas and Fibonacci sequences.

$$\begin{aligned} &P(0) = 2,\ P(1) = 1,\ P(2) = P(1) + P(0) = 3 \\ &P(1) = X(0) + X(2) = 0 + 1 = 1 \\ &P(2) = X(1) + X(3) = 1 + 2 = 3 \end{aligned}$$

## 2.3 *Lucas Representation*

The conceptual thought for this paper has been developed and enriched from the various research papers that were read and various new concepts such as

### 2.3.1 Rank

In general, rank to a character is allotted concerning the frequency. The highest frequency character is allotted Rank 1, and similarly, in the case of a clash between two characters having the same frequency, the character that appears first in the text will be allotted a higher rank than the other.

### 2.3.2 Lucas Rank Representation

**Steps**:

In order to calculate the Lucas rank, we need to write the substitute:

1. If Rank == Lucas number, substitute 1 corresponding to the number.
2. If Rank! = Lucas number, then
3. Form combination in such a way that bits produce the given rank.
4. The combination is done in such a way that the value assigned will be from the max bit present in the previous rank formation.
5. The formation of bits combination will produce Lucas rank representation corresponding to a given rank allotted to a particular character.

### 2.3.3 Rank Formation

The updated Lucas rank is generated by appending 1 in the suffix of Lucas rank.

If Rank $= 10$

1. Lucas Rank
   (a) $10 = 2 + 1 + 7 = (11001)$
   (b) $10 = 3 + 7 = (00101)$.

Since the above-shown example describes Luca's representation of a number (10), we generally prefer the *b*th formation because such type of formation aids in the easier fetching of decoded string/text during the encoding phase.

2. Updated Lucas Rank
   (a) $10 = 2 + 1 + 7 = (110011)$
   (b) $10 = 3 + 7 = (001011)$.

The reason for opting the updated Lucas rank for 10 as (001011) and not (110011) is because on appending 1 to its suffix, it results to the formation of a unique pattern of string ending with 11, which leads to proper evidence of termination condition while extracting the encoded text to its original form.

Table 1 represents the few other ranks that will be allotted during the program execution. The highest frequency character in the text evaluation will be provided the highest rank, and the one with the lowest frequency will be provided the lowest

**Table 1** Illustrating the Lucas rank and updated Lucas rank

| Index | 0 | 1 | 2 | 3 | 4 | 5 | 6 | 7 | 8 | 9 | Lucas rank representation | Updated Lucas rank representation |
|---|---|---|---|---|---|---|---|---|---|---|---|---|
| Lucas number | 2 | 1 | 3 | 4 | 7 | 11 | 18 | 29 | 37 | 76 | | =Lucas Rank Repersentation + Suffix {1} |
| Rank | | | | | | | | | | | | |
| 1 | 0 | 1 | | | | | | | | | 01 | Oil |
| 2 | 1 | | | | | | | | | | 1 | 11 |
| 3 | 0 | 0 | 1 | | | | | | | | 001 | 001 |
| 4 | 0 | 0 | 0 | 1 | | | | | | | 0001 | 00011 |
| 5 | 0 | 1 | 0 | 1 | | | | | | | 0101 | 01011 |
| 6 | 1 | 0 | 0 | 1 | | | | | | | 1001 | 10011 |
| 7 | 0 | 0 | 0 | 0 | 1 | | | | | | 00001 | 000011 |
| 8 | 0 | 1 | 0 | 0 | 1 | | | | | | 01001 | 010011 |
| 9 | 1 | 0 | 0 | 0 | 1 | | | | | | 10001 | 100011 |
| 10 | 0 | 0 | 1 | 0 | 1 | | | | | | 00101 | 001011 |
| ⋮ | ⋮ | ⋮ | ⋮ | ⋮ | ⋮ | ⋮ | ⋮ | ⋮ | ⋮ | ⋮⋮ | ⋮ | ⋮ |
| 96 | 1 | 0 | 0 | 0 | 0 | 0 | 1 | 0 | 0 | 1 | 1000001001 | 10000010011 |
| 97 | 0 | 0 | 1 | 0 | 0 | 0 | 1 | 0 | 0 | 1 | 0010001001 | 00100010011 |
| 9S | 0 | 0 | 0 | **1** | 0 | 0 | **1** | 0 | 0 | **1** | 0001001001 | 00010010011 |
| 99 | 0 | 1 | 0 | **1** | 0 | 0 | **1** | 0 | 0 | **1** | 0101001001 | 01010010011 |
| 100 | 1 | 0 | 0 | **1** | 0 | 0 | **1** | 0 | 0 | **1** | 1001001001 | 10010010011 |
| ⋮ | ⋮ | ⋮ | ⋮ | ⋮ | ⋮ | ⋮ | ⋮ | ⋮ | ⋮ | ⋮⋮ | ⋮ | ⋮ |

rank. Later, applying the proposed concept and various terminologies, we generate Lucas rank and updated Lucas rank.

Update Lucas Rank = Lucas Rank + Suffix {1}.

# 3 Compression and Decompression

## *3.1 Compression Technique*

The compression technique uses the listed below pseudocode to generate the Lucas rank, and Fig 2 led to the formation of updated Lucas rank and helps in generating the encoded bits for the given input.

| CHAR | FREQUENCY | RANK | LUCAS REPRESENTATION |
|---|---|---|---|
|  | 6 | 1 | 10 |
| t | 4 | 2 | 1 |
| i | 4 | 3 | 100 |
| s | 3 | 4 | 1000 |
| r | 3 | 5 | 1010 |
| e | 3 | 6 | 1001 |
| y | 2 | 7 | 10000 |
| u | 2 | 8 | 10010 |
| m | 2 | 9 | 10001 |
| l | 2 | 10 | 10100 |
| a | 2 | 11 | 100000 |
| h | 1 | 12 | 100010 |
| c | 1 | 13 | 100001 |
| n | 1 | 14 | 100100 |
| o | 1 | 15 | 101000 |
| p | 1 | 16 | 101010 |
| P | 1 | 17 | 101001 |
| M | 1 | 18 | 1000000 |
| A | 1 | 19 | 1000010 |
| . | 1 | 20 | 1000001 |
| , | 1 | 21 | 1000100 |

**Fig. 2** Representation of Lucas rank for the example explained in 3.1

#### 3.1.1 Pseudocode

```
Let, n =total number of different characters.
      for i in n
        while (temp)
            if(fib[temp]==i+1) then
               while (j<=temp)
                 b=b+10;
                  j++;
                  end while.
                  arr[index]=b;
                   break;
            end if
            else if(fib[temp+1]>(i+1)) arr[index]=arr[k]+arr[index-k-1];
               break;
            end else if
       end while
      end for
```

#### 3.1.2 Compressed Bits

For Example:

"At this party usually Prime Minister, come."

0100001111011110100011001100011011010101100000110101111000011
0110100110001100011000001100101100101100001101110010110101100
11100011100110110000000110011001001100110001111100110101100100
011100001100010111000111001110000011

### 3.2 Decompression Technique

The decompression technique uses the listed below pseudocode to generate the original characters/strings for original text file from the encoded bits obtained in Sect. 3.1.2. by the original file.

#### 3.2.1 Pseudocode

```
Let n is the total number of encoded bits(cn)
   for i in n
       if (cn[i]==1 && cn[i+1] ==1)
           for (j in k)
                sum=sum*10+cn[j];
           end for;
           print m[ltr[sum]]
         k=i+2;
         i++;
     end if
   end for
```

#### 3.2.2 Decompressed Bits

For Example:

"At this party usually Prime Minister, come."

- At

  thi 0001101101010110000010101111000011011010011001101001100000
  1100101100111000011011100101101110111000111001101100000001101
  1001001100110001111001010110010001110000110001011100011110011
  10000011

- At this

  part000011011010011000110100110000011001011001011000011011110010
  1101011001110001110011011000000011001100100110011000111111001101
  0110010001110000110001011 1 0001110011100 00011

- At this party

  usua0010110010110000110111001011010110011100011100110110000000110
  0110010011001100011111001101011001000111000011000101110001110011
  10000011

- At this party usually

  Pr00111000111001101100000001100110010011001100011111001101010100100
  0111000011 0001011100011100111000001 1

- At this party usually Prime

  Mi0010011001100011111001101011001000111000011000 1011100011100111 0000011

- At this party usually Prime Minister 00100011100001100010111000111001110000011

- At this party usually Prime Minister, come.

## 3.3 Calculation

The example considered for explanation is "At this party usually Prime Minister, come." has the following data, associated with it. The text comprises of the total number of character's present in the text file is represented by total character ($t$). To calculate the total bytes, required originally to store a file into a given system is represented by (ofs) and is calculated by applying multiplication between ($t$*8), and the file compressed by applying the proposed algorithm is represented by (cfs)

In order to calculate % space saved, we apply the formula as [3]:

$$[\text{ofs} = t * 8\,\text{byte}$$
$$x = (\text{cfs}/\text{ofs}) * 100,$$
$$\%\,\text{space saved} = (1 - x) * 100]$$

For the assumed example, the calculation is

$$t{:}\,42$$
$$\text{ofs: } 42 * 8\,\text{byte} = 336\,\text{byte}$$
$$\text{cfs: } 220\,\text{byte}$$
$$\%\,\text{space saved: } 34.5238$$

The notation used in above calculation is listed below:

- $t$ = Total Character in a File
- ofs = Original File Size
- cfs = Compressed File Size
- $x$ = Compression Ratio.

## 4 Performance of Proposed Algorithm

The table listed below contains a few sample test cases of text file consisting of variable size, in which the uncompressed file size indicates the actual size of the file that has to be compressed. Applying the proposed algorithm compresses the file to a certain extent and is shown in the compressed file size column of Table 2. Later, by applying the formula listed above in Sect. 3.3, we get the amount of percentage space saved.

Figure 3 represents the graphical view depicting the changes that occur in % space saving, when comparing it will evaluate the different file sizes. The file size column in the graph contains both the compressed and uncompressed file size and is represented via different bar graphs, and the line graph depicts the percentage space saving of the given file when the given proposed algorithm is applied to it.

**Table 2** Sample test data for proposed algorithm

| Compressed file size | Uncompressed file size | %Space saving |
|---|---|---|
| 1275 | 1952 | 34.6824 |
| 3236 | 4928 | 34.3344 |
| 6300 | 9456 | 33.3756 |
| 14,726 | 21,880 | 32.0512 |
| 36,127 | 53,168 | 32.0512 |
| 52,965 | 77,464 | 31.6263 |
| 67,070 | 97,944 | 31.5221 |
| 85,567 | 131,040 | 34.7016 |
| 106,750 | 155,136 | 31.1894 |

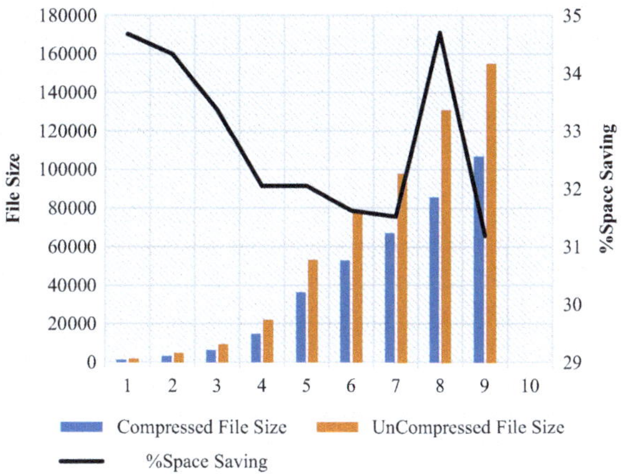


**Fig. 3** Analysis for the sample test data in Table 2

**Table. 3** Sample test for plotting graph between the best and the worst case

| Character size | Worst case<br>% space saving | Best case<br>% space saving |
|---|---|---|
| 5 | 52.50 | 68.05 |
| 6 | 52.77 | 68.75 |
| 7 | 50.96 | 68.29 |
| 8 | 50 | 68.33 |
| 9 | 49.26 | 68.38 |
| 10 | 48.02 | 68.42 |

**Fig. 4** Analysis for the sample test data in Table 3

## 5 Graphical Representation of Space Complexity

The listed data in Table 3 is the result of some assumed data (with known character size). This is done to know the behaviour of the program i.e. in what situation the logic implemented will give the best and the worst case. The assumptions are purely based on the different character sequences that might appear in the given text pattern scenario. The best case occurs when the text has a single repeated character only. Similarly, the worst case occurs when the characters in the text are completely different from each other. The average lies in between the worst and the best case.

Given Fig. 4 illustrates that the impact of various cases assumed in Table 3 along with its graphical representation to show the nature of the curve between both the worst and the best cases for unique file size.

## 6 Conclusion

The graphical representation in Fig. 3 depicts that the proposed algorithm is good to provide the best space saving in large-sized data, also, Fig. 4 depicts the nature of text file type under which the files give the best and the worst case, and the average lies under the mixed combination of both. The Lucas series is an extended version of the Fibonacci series, and it plays a vital role in deciding the public and private key in cryptography [5] and helps under encryption of various affine and Vigenere cipher. In this paper, we have compressed text type data using Lucas sequence, and our current work is focused on linking the similar concept decryption of such textual data using the above-proposed concept to maintain a higher order of secrecy.

## References


1. Bhadade, U.S., Trivedi, A.I.: Lossless text compression using dictionaries, Faculty of Technology & Engineering, University Vadodara (GJ), Shirsoli Road Jalgaon, India. Int. J. Comput. Appl. 13(8), 0975–8887 (2011)
2. https://cs.stanford.edu/people/eroberts/courses/soco/projects/datacompression/lossless/index.html
3. Bhattacharyya, S.: Complexity analysis of a lossless data compression algorithm using Fibonacci sequence. Research Associate Department of Computer Science. University of Calcutta, Kolkata, West Bengal (2017)
4. https://www.geeksforgeeks.org/lucas-numbers/
5. Pradana, S.A.: Lucas sequence and the application for cryptography. Informatics Engineering School of Electrical Engineering and Informatics (SEEI), Bandung Institute of Technology, ITB Bandung
6. https://patents.google.com/patent/US20080224905A1/en

# Smart Bin Management System with IoT-Enabled Technology

**Rishi Agrawal and Adamaya Sharma**

**Abstract** A hygienic environment is necessary for any nation in the world. Hygienic harmony leads to a healthy environment, when there is a healthy environment, there is the order in society, and where there is the order in society, there is the order in a nation, which will lead to a healthy and hygienic world. Like the population, the world garbage is increasing profusely day by day. We are having problems with environmental safety and pollution. It is getting much worse when it comes to society. Humans are destroying nature and leading it toward a polluted, unhealthy and unhygienic environment. So, in this paper, Web-based application is made for smart dustbins. A Web-based monitoring interface is also proposed to show the effective locations of smart dustbins and the percentage of garbage levels on a geographical map. Proposed system uses MQTT protocol for sending and receiving data between different devices. It gives the real-time update of garbage to an end user. Real-time update of the garbage level gives the facility to end user to find out the nearest empty dustbin so that garbage of home can be managed by the dustbin, protecting environment from pollution. The proposed system is divided into three different levels. In first level, data gathering is done by nodes. In the second level on the backend server, the analytics module analyzes data collected by the bin subsystem. In the third level, action of the event is performed according to conditions.

**Keywords** Smart bin management system · IoT · MQTT

## 1 Introduction

Pollution can be of different types, and one category of pollution is garbage overflow. It creates the unhygienic conditions for the people, and this leads to spreading of

R. Agrawal (✉) · A. Sharma
GLA University, Mathura, India
e-mail: rishi.agrawal@gla.ac.in

A. Sharma
e-mail: adamaya.sharma_iot18@gla.ac.in

 
T. K. Sharma et al. (eds.), *Soft Computing: Theories and Applications*,
Advances in Intelligent Systems and Computing 1381,
https://doi.org/10.1007/978-981-16-1696-9_13

some deadly diseases & human illness. To avoid all such situations, paper proposed smart bin management system using smart dustbin modules. Implementation of the proposed system is done with the help of IoT protocols. In this system, dustbins that are located throughout the city or the campus are provided a unique ID by a smart dustbin module which helps in tracking the level of the garbage bins and transmits the sensed data to the server. When the level of the dustbin reaches to the threshold limit, a garbage full message delivered by the server to a particular cleaning authority [1].

A Web-based application is made for smart dustbins monitoring Web interface that shows the effective locations of smart dustbins and the percentage of garbage levels on a geographical map. It gives the real-time update of garbage to an end user [2]. Real-time update of the garbage level gives the facility to end user to find out the nearest empty dustbin so that garbage of home can be managed by the dustbin, protecting environment from pollution.

## 2 Related Work

The dustbin plays an essential part in the clean environment, and it must be looked after well-organized way. The current system of collecting waste from the environment is very pathetic, laborious and practically inefficient, so the necessary step has to be executed before it collapses the whole idea of a hygienic and clean environment [3]. Adding some modifications in normal dustbins will solve most of the garbage problems, the smart bin has covered many aspects of a clean and smart environment, and some are the following points.

- Project architecture follows fog computing as well as cloud computing architecture. Because of the fog computing architecture, its processing speed and response time increase rapidly [4].
- It will reduce the problem of overflowing of dustbins along roadsides and society as bins are smartly managed in real time.
- It gives a unique ID to a smart dustbin so that we can easily track the garbage level of the particular dustbin.
- Module senses both the container levels of smart dustbin (plastic waste + organic waste)
- Any old dustbin can be converted into smart dustbin just simply attaching the module into the head of a dustbin.
- We are using the MQTT protocol which works with or without the Internet [5].
- Whenever a dustbin gets filled completely, the server sends a notification to the respective cleaning authority of that specific area to clean the smart dustbin.
- It will reduce the filling and cleaning time of bin, thus making empty and clean dustbins available to common people.

- Smart bin management system provides an interactive Web-based user interface that provides users to know about the smart dustbins near their respective areas and the level of garbage in smart dustbins [6].
- The system analysis the data that is coming from smart dustbin to check the status of the dustbin.
- It can find out the shortest distance using the prediction and route algorithm, thus, it helps to reduce the number of trucks required to clean, the workforce, the amount of fuel consumed by trucks, and thus, it saves large amounts of money [7].
- It is an eco-friendly system, and there is no harm to the health of nature, as it will reduce the fuel consumption which in turn reduces the pollution in the air.
- As technology goes parallel with the environment, it is important to protect the environment from pollution by the garbage. Paper proposed a system toward an echo-friendly technology.

## 3 Architecture of Smart Bin Management System

The proposed system is divided into three different levels. In first level, data gathering is done by nodes. In the second level on the backend server, the analytics module analyzes data collected by the bin subsystem. In the third level, action of the event is performed according to conditions.

### *3.1 Data Gathering via Nodes*

The data gathering modules consist of sensor nodes and gateway nodes. A sensor node module is installed in every smart dustbin. Modules are powered by a direct power supply or a battery. The level of smart dustbin capacity is measured using an ultrasound sensor. Upon receiving sensor reading, the data is transmitted to the gateway node using a low-power MQTT protocol. A gateway node is installed in every sensor module cluster. It serves the sensor module under its cluster. According to the density of smart dustbin and their transmission range, the size of the cluster is determined. The proposed model receives data from sensor with the help of sensor module nodes. The data received from sensor module nodes will be forwarded to the backend server. It may use any Internet connection (e.g., ethernet, Wi-Fi or 3G connection) for transmission of data [8].

### *3.2 Processing and Analytics*

At the backend server, the analytics module analyzes data collected by the bin subsystem. The bin sub-system collates and processes data and interfaces with external systems. The analytics module processes fullness readings compared against predefined rules and generates event upon exceeding threshold.

### *3.3 Projection and Action of Event*

The bin sub-system pushes processed data into the Web interface. The proposed model used a Web interface to show meaningful information to users. A Geographic Information System (GIS) is used to visualize the different dustbins with the help of icons. The app can also be used to display the level of fullness of smart dustbins (Fig. 1).

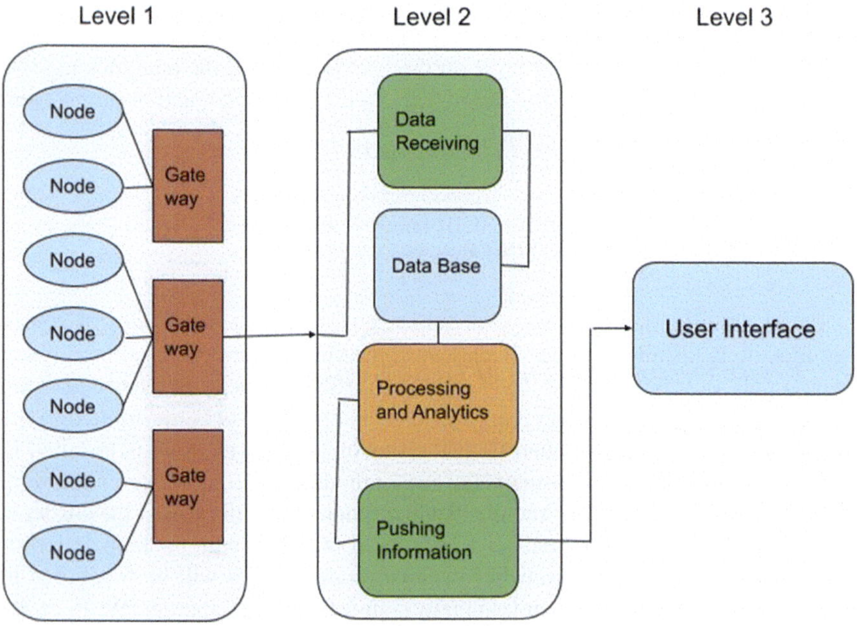


**Fig. 1** Architecture of smart bin management system

# 4 Architecture of Smart Bin Management System

The system can be divided into three different parts. First part contains hardware implementations (Table 1). Second part contains software and cloud framework. Third part of the system contains real-time data monitoring console.

## *4.1 Hardware Implementation*

The hardware consists of NodeMCU(esp8266) microcontroller, SIM900A GSM module, HC-SR04 ultrasonic sensor, microprocessor Raspberry Pi 3, IC7805 voltage regulator, resistor and capacitor. NodeMCU is an open-source Lua-based Firmware ESP8266 Wi-Fi SOC from Espressif having 17 GPIO pins. The ultrasonic sensor HC-SR04 has four pins, namely vcc, ground, echo and triggers. The microprocessor Raspberry Pi working as a mini server has a minimum of 1.4 GHz processing speed and based on arm architecture. GSM module which is used to send messages operates at RS232 logic. The circuit diagram is shown in Fig. 2.

**Table 1** Hardware requirements

| S. no. | Components | Specification |
|---|---|---|
| 1 | Microcontroller | NodeMCU(Esp8266) |
| 2 | Ultrasonic sensor | HC-SR04 |
| 3 | Microprocessor | Raspberry Pi 3 or above |
| 4 | GSM module | Sim 900A |
| 5 | Voltage regulator | IC7805 |
| 6 | Capacitors | |
| 7 | Resistor | |

**Fig. 2** Sensor module

The Ultrasonic sensor produces the sound waves. These sound waves are used in calculating the distance by using the formula:

$$\text{Distance} = (\text{duration}\, 2 * 0.034)/2$$

The distance between the object can be calculated by using the echo pin of Ultrasonic sensor. Microcontroller is used to perform the calculation and final distance will be transmitted to a local network. All the data that is coming to local network will be further processed by Raspberry Pi. The system is designed in such a way that Raspberry Pi will check the status of the dustbin and final status of dustbin will be stored on cloud. The system will also send as SMS to the corresponding person for the final status of dustbin.

### *4.2 Software and Cloud*

The forwarded data from Raspberry Pi to cloud service will be stored in the cloud database keeping all its attributes intact as time and date. Cloud database provides us to manage our dustbin data precisely and efficiently. The data is being stored in a relational database format and is processed by the different services of the cloud such as analytics to predict and analyze the data. If the value of the particular area remained above than threshold value, it means that the garbage of that area has not got cleaned for some days, then with the help of the cloud services, a reminder message is delivered to the municipal corporation via email and SMS. A history of collected data in months is very helpful for data analytics teams to analyze for prediction, deploy the new efficient algorithms on cloud and sophisticated reports making.

### *4.3 Real-Time Data Monitoring Console*

Data will be fetched by the real-time Web application to interpret it on a static user Web site which includes a real-time geo-location-based Map (Fig. 3). Map includes the locations of smart dustbins and the value of filled amount of garbage in that dustbin, and on clicking that dustbins, you will get the details sweeper and the cleaning society of that area. It gives users a transparent interface to monitor their environment. Web interface also provides the nearest dustbin location from the user location so that it reduces the chaos of finding the empty dustbin to throw the garbage.

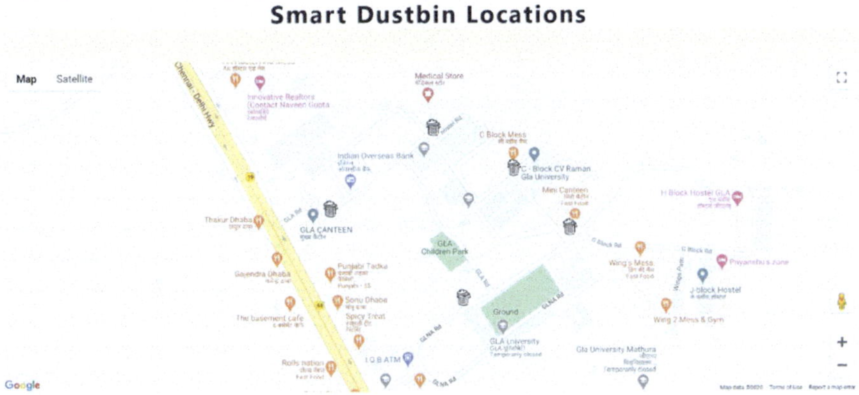


**Fig. 3** Web interface (geological map with smart dustbins)

## 5 Efficiency over Other Bin Management System

The sensor module is deployed with battery power. A low power consumption sensor node is selected due to the low power consumption battery. The sensor node had limited memory size to process sensed data. For this, a simple low powered efficient MQTT routing protocol is used.

By using MQTT protocol, the proposed system ensures that whether the sensor data is delivered to the gateway node and nodes are connected to the Internet or not. Thus, it gives the continual reliability of connectivity in rural areas (Fig. 4).

Processing the data on a server always took too much time and delay in output to reduce the delay, and we are using fog computing concept on local server which is Raspberry Pi in this case. Fog computing concept enhances the processing of data on local server instead of cloud server which decreases the transmission time of data to cloud server and acts immediately on local machine. But our system also uses the cloud services to process the large amount of data on a centralized system where all the data is processed and analyzed to make further predictions. Cloud computing provides different kind of services which are databases, computation, and many more which make our work too much easier and sophisticated.

## 6 Conclusion

As the population of the world is rapidly increasing and urbanization is rapid taking place, it is necessary to maintain the hygiene in cities as well as undeveloped area. Consumption of products is at his apex, but yet there is no proper solution to manage the garbage of consumed product. Because of lacking in management of garbage, nature is getting harm. Caring of our system is must to decrease the diseases and

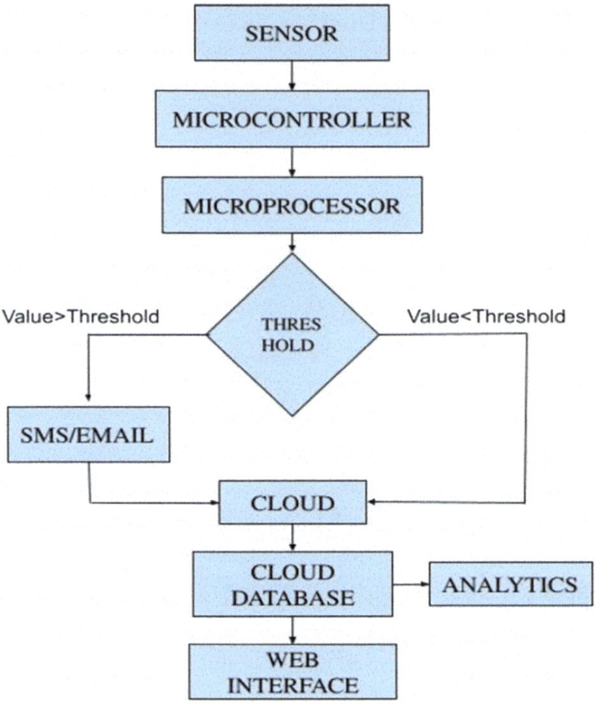


**Fig. 4** Work flow of system

to keep a healthy environment. Smart bin management system is a crucial project which is helping to maintain and monitor the environment efficiently, effectively and spreading awareness in people by its real-time data monitoring, robust connectivity and a transparent interactive user interface.

**Acknowledgements** Authors wish to thanks Anurag Sharma, Abhishek Vaish, Arpit Srivastava for their support to design the application.

## References

1. Chaudhari, S.S., Bhole, V.Y.: Solid waste collection as a service using IoT-solution for smart cities. In: 2018 International Conference on Smart City and Emerging Technology (ICSCET), Mumbai, pp. 1–5 (2018)
2. Tapashetti, A., Vegiraju, D., Ogunfunmi, T.: IoT-enabled air quality monitoring device: a low cost smart health solution. In: 2016 IEEE Global Humanitarian Technology Conference (GHTC), pp. 682–685 (2016)
3. Surapaneni, P., Maguluri, L., Syamala, P.M.: Solid waste management in smart cities using IoT. Int. J. Pure Appl Math. **118**(7), 635–640 (2018)

4. Kim, K.J.: Interacting socially with the Internet of Things (IoT): effects of source attribution and specialization in human–IoT interaction. J. Comput.-Mediat Commun **21**(6), 420–435 (2016)
5. Hannan, M.A., Abdulla Al Mamun, M., Hussain, A., Basri, H., Begum, R.A.: A review on technologies and their usage in solid waste monitoring and management systems: issues and challenges. Waste Manage. **43**, 509–523 (2015)
6. Hu, L., Ni, Q.: IoT-driven automated object detection algorithm for urban surveillance systems in smart cities. IEEE Internet Things J. **5**(2), 747–754 (2017)
7. Kannangara, M., Dua, R., Ahmadi, L., Bensebaa, F.: Modeling and prediction of regional municipal solid waste generation and diversion in Canada using machine learning approaches. Waste Manage. **74**, 3–15 (2018)
8. Anjomshoaa, A., Duarte, F., Rennings, D., Matarazzo, J., deSouza, T., Ratti, C.: City scanner: building and scheduling a mobile sensing platform for smart city services. IEEE Internet Things J. **5**(6), 4567–4579 (2018)

# Non-living Caretaker as a Medicine Box Based on Medical Equipment Using Internet of Things

**Ajitesh Kumar and Mona Kumari**

**Abstract** In today's world, as the population increases so do diseases and more and more people are becoming dependent on medicines for their living. With medicine being of such great importance, it is necessary for one to take his doses on time. But older people and virus-infected people find it difficult to remember their medicine schedule. To overcome this problem, this proposed work "Smart Medicine Dispenser" stores their daily medicine schedule and notifies them on the basis of the schedule about when and which medicine to take. Using Wi-Fi, it can also notify the caretaker of the patient about whether the patient has taken his medicine or not. It comprises a box containing several small sections in which we can store medicines according to their daily dose. It uses a real-time clock (RTC) module and an alarm system to make sure that the patient takes his dose on time. The experimental results show that the proposed model gives better results in comparison with existing work.

**Keywords** IoT · NodeMCU · Home automation

## 1 Introduction

In this emerging world of technologies, Internet of Things (IoT) plays a vital role in our day-to-day life by connecting us to the Internet-based automation systems. It connects all the devices inside a system internally. It plays a key role in our routine activities like people often forgotten and misplaces their valuable items such as keys, wallet, remote and many more. For such people, IoT takes a part in their life. Even the IoT devices can sense the nature in many procedures such as temperature, pressure, geo-positioning and many more. IoT helps us to connecting with other technologies and performs different functions. That's why the author has explained that how IoT

A. Kumar (✉) · M. Kumari
GLA University, Mathura, Uttar Pradesh, India
e-mail: ajitesh.kumar@gla.ac.in

M. Kumari
e-mail: mona.kumari@gla.ac.in

 
T. K. Sharma et al. (eds.), *Soft Computing: Theories and Applications*,
Advances in Intelligent Systems and Computing 1381,
https://doi.org/10.1007/978-981-16-1696-9_14

connects with other emerging technologies like Cloud Computing, Big Data and many other technologies. All these technologies which were discussed before helps in the form of designing in the IoT-based systems or projects where author stores data in cloud and other different information produced by sensors like GPS sensor, touch sensor and many more [1]. And big data helps to give smoothness to data storage efficiency. If there is an increase in the number of devices which are connected and more data consumption, then big data is being used with IoT to get the better of this problem. It overlays with IoT in manner to enhance data storage efficiency and provide data processing for a large amount of data. Without IoT, no one can imagine an automated system. The application of IoT is wide enough in almost each and every field like health and monitoring field, crime and defence field, education and industrial field and many more other fields. In this project, the author has shown that how IoT and its devices are useful in health and monitoring field in order to ease the medical system of a society or a country [2, 3].

Primarily, Internet of Things is a web of physical objects that consist of an embedded technology for sensing and interacting with the surroundings. The model below is about how IoT and its devices are being used in medical system. The proposed model could be such that there will be no compromise between performance, efficiency and security of the system, and the desired outcome must produce good precision and accuracy.

## *1.1 Objective*

The main objective behind this proposed work is to make people healthier and disease free by giving proper treatment on daily basis and make a system like this that every patient should be able to use this easily that's why this model is made with the help of technologies like IoT and cloud computing. Through this model, patient will get notification about to take medicines at the scheduled time or this model indicates LED and creates an alarm sound which helps patient to remember its medicines according to the scheduled time.

## *1.2 Scope*

The scope of this proposed work is widely distributed in whole world because it is used in the hospitals, old age home and even in the home itself to take care of the patient. It reminds the patient to take medicines according to the scheduled time and also reminds the caretaker when patient did not take medicines at the right time. This smart medicine dispenser becomes very useful in the coming world because in this world which is full of technology, people often forgets his necessary things then this box remind people to take medicines at the right time.

This paper contains five partitions. The authors write related work in Sect. 2.2; the model described in Sect. 3.1, where the author writes about the modules used in the model, the model layout, which shows the top view and the data transfer diagram of the model. The results illustrated in Sects. 4 and Sect. 5 contain the conclusion part of the system.

## 2 Related Work

### 2.1 Motivation

Everyone knows that health monitoring becomes necessary for everyone because in this world full of technology, no one remembers his necessary things which are used in daily life. So, how we can say a person remembers to take his prescribed medicines at the prescribed time. To solve this problem, the author makes this smart medicine dispenser to notify patients or users to take their medicines at the prescribed time. If patient forgets to take medicine then this box notifies him by smartphone notifications or an alarm sound or a LED indicator. This will help users to remind his medicines at the prescribed time and also helps caretaker of a patient to remind him that patient did not take medicine at the right time. This box is very useful for old patients because they often forget to take their medicines. This model was made by using IoT and big data like technology which is a future of technologies that's why this model will become very useful in the coming world.

### 2.2 Related Work

Nowadays, IoT is very useful and beneficial in the field of medical. The main reason to make the devices using IoT is to get the fast and accurate result. In this model, the authors use cloud to store time and data to get accurate results so that patient will take medicines at the prescribed time. These are the main things which have done in medical field with the help of IoT.

It will help the patient to remind to take medicines at the prescribed time. This project was made by using different technologies like cloud computing and big data. In this project, the author used GPS sensor to track medicine box which was explained in [4].

In order to improve system security and to access the database smoothly, author used Wi-Fi module which was explained in [4]. In this model, the author used RTC module to save current time means the time at which the patient have to take medicine which was explained in [5, 5]. The authors have used low-cost hardware to make this model to increase cost effectively. That's why they used ZigBee [6] protocol because it supports the low-power wireless networks.

The author made an android application to send SMS or email notifications to the patient according to the prescribed time to take medicines which were explained in [7, 8] and notify the caretaker when patient forgets to take the medicines at the prescribed time.

In existing systems, there were heartbeat sensors [9, 10] attached with medicine box to count the heart beats, but at present time, there is no need to attach heartbeat sensor because it need extra space in the box which complex the circuit and in this time, we count the heartbeats of a person by mobile app or smart watch that's why the author does not attach the heartbeat sensor in this box [11].

The author attached LED and buzzer [12] in the medical box with all the 21 sub-boxes to notify patient according to the prescribed time. According to the time saved in the cloud, these LEDs and buzzer indicate patient to take medicine.

# 3 Proposed Model

To overcome the defaults of the previous system, the proposed system has been made. This system target to reduce the work of care taker of older patients. Our solution is a Arduino Uno microcontroller-based smart medicine box controlled with the help of android app. In this proposed model, researchers deal with the time-efficient and effortless solution towards the daily medication of elder peoples. Arduino Uno along with RTC module is used to keep track of the current time, and when the medication time occurs, it will activate the notification module which in turn will notify the user in terms of visual and physical output. The Bluetooth module is used to establish connection between android app and Arduino Uno which will be useful to see upcoming medication time and also view and manage the schedule of medication. Our model comprises 21 small boxes arranged in a matrix of 7 rows and 3 columns. Each row is for one of 7 days, and column represents morning, noon and evening medicine. Each box is rigged with LED in order to notify the user, LEDs are controlled by Arduino Uno. The system also consists of a buzzer and a speaker for the audio notification. Our system is also capable of sending messages and emails to the caretakers of patient which will come handy in keeping track of patient medication. It will greatly help caretakers in saving their time which was used before to follow medication schedule, and it can also help in reducing the problem of overdosing.

## *3.1 Modules Description*

### Notification Module

In this module, we use RTC to keep track of the time, and with each scheduled time, LEDs and buzzer are being activated to notify the patient (a.k.a. usee).

### Schedule Update Module

In this module, we use Bluetooth module with Arduino to establish connectivity with smart phone via Bluetooth. User can update and view the current medicine schedule.

## 4 Simulation and Result

In this world, people are addicted to the various technologies and do anything to adopt them. And there is an IoT which left a mark in the world of technology or in the medical field. That's why our project Smart medicine dispenser is based on IoT. So, anyone can use this model easily. Our model consists of a box in which there are 21 identical sub-boxes having medicines according to three time in a day for a week. To make this model, the authors used resistances, diodes, jumper wires, 21 LED indicators and a buzzer to make a circuit and all are connected with Arduino Uno which helps to collect data. They also used GPS sensor to track medicine dispenser means if patient forgets where he put medical dispenser, then by enabling GPS in his mobile, he can find the dispenser. Our project is based on the concept of physics in which resistance is used to divide the current on a particular cell of the dispenser. The author used different resistors to divide the current and also used diodes to switch on or off the LEDs or buzzer. Here, Arduino Uno is operated at 5 V and used to send and retrieve data, and then, it sends the notification to the android application and then from android application, data is sent to the cloud (Fig. 1).

There are 21 LEDs in this model. So, this will create problem or complex the circuit. To overcome this problem, authors used Charlieplexing technique.

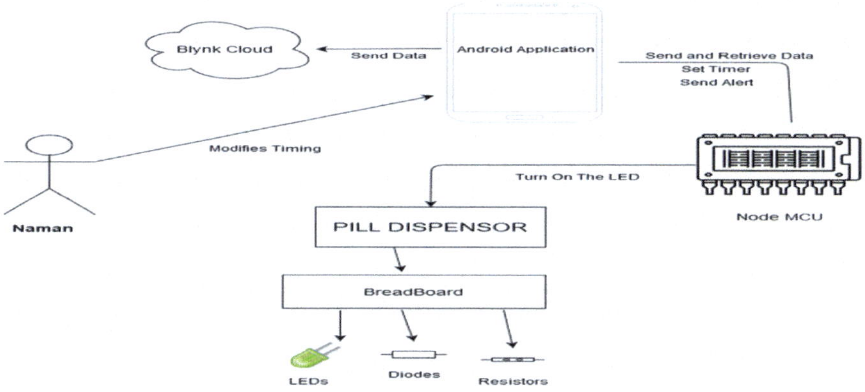


**Fig. 1** Layout of medicine dispenser

### Charlieplexing Technique

This technique is used to control many LEDs by using a few number of I/O pins. Charlieplexing is almost same as multiplexing, but it uses the tri-state logic to decrease the number of pins and to gain over multiplexing in terms of efficiency.

Charlieplexing is used to control $n * (n - 1)$ LEDs with n number of pins. We have to use 21 LEDs in our model, so we need 5 or 6 pins to control these LEDs according to the Charlieplexing technique $5 * (5 - 1) = 20$. If we suppose here LEDs are used as diodes, and in diodes, we know that current flows in one direction. So in this, we connect two LEDs in parallel to each other with opposite polarity so that only one LED switches on at a time.

Authors used this technique because if one use Arduino Uno in his project, he never have enough pins and if you have many LEDs in your project, then it creates problem because you also have sensors and other things to connect. In that tough situation, one can use the Charlieplexing technique to decrease the number of pins.

## 5 Conclusion and Future Direction

In this proposed work, Smart medicine box model is defined using IoT. Through this model, the patient never forgets to take their medicines in the given time. It is made like this that it is suitable for anyone who needs medicines daily. It indicates the patient about to take medicines at the prescribed time and notify the caretaker also when patient did not take medicines at the given time. It reduces the chances of human error, and it is made of low cost so, anyone can afford it easily.

In future, we are hoping that the medicine box will directly link to the medicine cart in which it will directly sends the message to the people of med-cart to refill the medicines. Firstly, we will do the modification in our system according to the people's need in future.

We can use the large LED, and we can convert our plastic box of medicine into the metal box; it will increase the strength of box and can sell this box to the hospital because this medicine box can be of great use for the nurses to take care of their patients as the nurses are also a human beings, they can also forget the timing of medicine so this box will remind them on time. We are hoping to make this box like to take care of many patients at a single time. we are hoping to put the large amount of medicine of many patients in a different cell of box.

## References

1. Bhati, S., Soni, H., Zala, V., Vyas, P., Sharma, Y.: Smart medicine reminder box. Int. J. Sci. Technol. Eng. **3**(10) (2017)

2. Motwani, K.R., Jitkar, B.D.: A model framework for enhancing document clustering through side information. In: Soft Computing: Theories and Applications, pp. 195–207. Springer, Singapore (2018)
3. Mahajan, R.: Emotion recognition via EEG using neural network classifier. In: Soft Computing: Theories and Applications, pp. 429–438. Springer, Singapore (2018)
4. Pang, Z., Tian, J., Chen, Q.: Intelligent packaging and intelligent medicine box for medication management towards the Internet-of-Things. In: 16th International Conference on Advanced Communication Technology, pp. 352–360. IEEE (2014)
5. Doshi, V., Dey, S., Mehta, N., Prasad, R.: An IoT based smart medicine box (2019)
6. Raga Lavima, P., Subhramanya Sarma, G.: An IoT based intelligent medicine box. Int. J. Comput. Sci. Mobile Comput. **4**(10) (2015)
7. Al-Shammary, R., Mousa, D., Esmaeili, S.E.: Design of a smart medicine box. In: Iranian Conference on Electrical Engineering, May (2018)
8. Singh, J., Singh, P., Malik, V.: Effect of intrinsic parameters on dynamics of STN model in Parkinson disease: a sensitivity-based study. In: Soft Computing: Theories and Applications, pp. 417–427. Springer, Singapore (2018)
9. Bedmuttha, P., Jain, N., Thigale, Y., Gargori, S.: A Health-IOT platform based on the biosensor and intelligent medicine box (2017)
10. Pattepu, S.: Performance analysis of two multi-antennas relays cooperation with hybrid relaying scheme in the absence of direct link for cooperative wireless networks. In: Soft Computing: Theories and Applications, pp. 147–160. Springer, Singapore (2018)
11. Yadav, S., Saroliya, A.: Patient-specific modeling for pattern identification in clinical events using data mining models and evolutionary computing. In: Soft Computing: Theories and Applications, pp. 209–216. Springer, Singapore (2018)
12. Nyapathi, N.U., Pendlimarri, B., Karishma, S.k., Kavya, Ch.: Smart medicine Box using ARM 7 Micro controller. Int. Res. J. Eng. Technol. (IRJET) **3**(05) (2016)

# Trends and Advancements in Genome Data Compression and Processing Algorithms

**Sanjeev Kumar**

**Abstract** The healthcare industry is moving toward drug development, which involves the use of human genetic information, so that medical treatment can be adjusted to the individuals specific characteristics. Modern healthcare study provides access to biological samples, including a patient's genome sequences and other medical documents. Personalized medicine also requires the capacity to store and interpret genome sequences and patients medical records. Genomics is the study of an organism by extracting its genome's secret (hidden) features. There are trillions of cells in the human body, which are the essential building frameworks of most living things. They provide the body with structure, take nutrients from the meal, turn certain nutrients into energy, and perform particular functions. With modern possibilities for new medical knowledge, the field of genomics is entering a new era. In order to incorporate heterogeneous information, such data requires highly distributed acquisitions, enormous storage requirements, broad distribution bandwidth, and highly involved analyses. To facilitate the storage and dissemination of genomic data and to facilitate its access and analysis, genome data compression and processing algorithms are used. The state-of-the-art algorithms to the compression and processing of genomic data are discussed in this paper. The ever-increasing production of genome data presents many storage, processing, and transmission challenges. The cost of storage, processing, and transmission is greater than the cost of the DNA sequence being produced. The main solutions that can solve these problems by minimizing the storage size and the cost and time of processing are compression and efficient processing.

**Keywords** Compression · Decompression · Processing · Genome · DNA

---

S. Kumar (✉)
Department of CEA, GLA University, Mathura, India
e-mail: sanjeev.kumar1@gla.ac.in



T. K. Sharma et al. (eds.), *Soft Computing: Theories and Applications*,
Advances in Intelligent Systems and Computing 1381,
https://doi.org/10.1007/978-981-16-1696-9_15

## 1 Introduction

With the development of high-throughput genome sequencing technology, vast quantities of genomic data are produced every day. Public genome sequence repositories now hold more genome data than any scientific/public genome repositories can accommodate [1]. In the near future, the situation is going to get worse, for several reasons:

- Faster and lower sequencing costs make it possible for larger and larger projects to produce vast genome data [2].
- With the knowledge of the relationship between genotype and phenotype (genotype-organisms complete inherited details, phenotype-actual characteristics observed by organisms), more and more individuals will also be interested in their genomes and genetic predispositions [3].

The ever-increasing growth of sequenced genomes makes it more important to examine the efficient process of storage and transmission. Present genome projects are now targeting several thousand human genomes for sequencing [4]. A simple method is to store and handle genome data using ASCII-encoded characters, i.e., one byte for each base, on computers. Although this representation in terms of memory is inefficient, it does not compress the genome [5]. It is still popular for many genome scientists to use this format as programming languages can easily access these data and human beings can also easily understand them. It has the downside that it is space-intensive. Suppose one wishes to sequence 10,000 human genomes in a project. It takes approximately 30 terabytes (TB) of memory to store this information (quality score information and identifier information are stripped) in ASCII [6]. Also in today's hardware architecture, storing such comprehensive genome data poses a serious issue. It will be an incredibly time-consuming process to send this information to a remote/cloud server for review [7]. When raw data(sequence) with relevant quality scores and meta-data data are also to be processed, the situation becomes worse. A factor of 20 or more, i.e., 600 TB, easily increases the amount of required space. Another factor is the expense of storing such a vast volume of information. The price of genome sequencing is decreasing much faster than the cost of storing genome data, as Stein [8] has pointed out. Storing genome data (i.e., purchasing hard drives) will become more expensive in the near future as a result. One choice is to delete the processed genome data after analysis in this case, and re-sequencing can be performed whenever possible. However, for a variety of reasons, such a fundamental method is still considered to be inapplicable by most researchers, including ethical scientific behavior that involves prolonged preservation of experimental data to provide re-assessment and replication of findings. Compression is another approach [9]. Compression is regarded as a key instrument for storing and handling genome data. There are different methods to accomplish this compression task, primarily based on bit manipulation techniques, i.e., they cram into one byte more bases (four bases). These techniques have been replaced by statistical and dictionary-based methods, providing substantially higher compression ratios over techniques of bit manipulation

[10]. Recently, referential methods for genome compression have come into being to compress datasets of related organisms. These techniques have a higher compression ratio than those of earlier techniques. In the last few years, the field of genome compression has been highly active, with hundreds of papers being created. The field is complex since projects vary widely, findings are published in numerous repositories of Medline databases, and techniques are typically specialized in particular problems (such as compression of bacterial genomes or compression of coding sequences only) [11]. With modern possibilities for new medical knowledge, the field of genomics is entering a new era. In order to incorporate heterogeneous information, such data requires highly distributed acquisition, enormous storage requirements, wide distribution bandwidth, and highly involved analyses [12].

## 2 Genome Data Compression: A Literature Survey

Naive bit encoding, dictionary-based encoding, statistical encoding, and referential encoding can be classified as genomic data compression techniques. For each symbol [13], a fixed length encoding is used in the Naive bit encoding. In essence, such algorithms are fast, but the compression ratio is not so strong. Without complete decompression, searching is not possible.

In dictionary-based algorithms, for repeated sequences, a dictionary is prepared and then encoding is done based on this dictionary. These algorithms' compression ratios are the same as those of Naive bit coding. The downside of such a method is that dictionary specifications are needed during decoding. A statistical model is prepared for input text in the statistical or entropy encoding methods, which forecasts the next symbol in the text [14]. The compression ratio of such methods completely depends on the reliability of the prediction model. One genome is taken as a reference genome in referential compression methods and all other genomes are coded by noting the discrepancies with the reference genome selected. These techniques have a high degree of compression ratio, but there are wide intervals of compression and decompression. Moreover, both encoding and decoding include reference genome and alignment algorithms. The main objective of the thesis is to suggest some new approaches for better storage of genome data along with quick and memory-efficient access [15]. Algorithms for genome data compression have been developed as reference-based as well as non-reference-based. Independent data files have non-reference-based techniques. It is ideal for multispecies databases. For datasets of the same species [16], reference-based techniques are feasible. One genome is selected here as a reference genome, and remaining genomes are represented w.r.t. to it.

# 3 Genome Data Processing: A Literature Survey

State-of-the-art techniques for rapid extraction of genomic data are outlined in this section. Two types of methods are used to promote quick access (retrieval): one is based on repeat data and another is based on index.

## *3.1 Repeat Information-Based Approaches for Genomic Data Processing*

DNA read sequence alignment is an expensive and time-intensive process. Public archives of genomes do not explicitly store genome data. Some statistical information such as repeat information (type of repeat, duration of repeat, etc.) [17], computer information sequencing, and type of organism are stored with these data. Knowledge on repetition is unique in each genome. To extract related genomes, this information is used. In computational biology and genome sequence analysis, the search for repeated sequences (maximal & super-maximal repeats) in DNA/Protein is a big issue [18]. Many inherited illnesses, forensic and genetic disorders, etc., are related to these repetitive sequences. In many other bioinformatics algorithms, time & space efficient approaches for determining such repeats are also beneficial. There are various algorithms and also automated tools in DNA/Protein genome sequences to locate both maximum and super maximum repeats. [23] suggested an algorithm to search for all maximum repeats using the data structure of the suffix tree. This algorithm's space consumption is very high. Kamel [19] suggested a space efficient algorithm using a novel data structure known as suffix array to search for maximum and super maximum repeats. Algorithms/tools based on this data structure are: "Vmatch" and "MUMmer". Lian et al. [20] suggested an algorithm called the "parent of the leaf" (POL) [21] for the search for super maximum repeats using an auxiliary index structure. POL is an improvement over the suffix tree, which uses less auxiliary data structures [21]. It just looks for the full super repeats. Uh, Wang et al. [22] suggested a method based on map reduction to detect all maximum repeats in both human & virus genomes in DNA/Protein sequences. This technique can only be extended to distributed-system environments, but not to stand-alone systems. Burrows-Wheeler transform (BWT) & wavelet tree are used by Oguzhan et al. [23] to locate all full repeats [24, 25]. It first uses BWT and wavelet tree to construct the LCP list, and then all the candidate-repeats (possible maximum-repeats) are calculated.

The maximum repetitions are derived from these candidate repetitions at the end. While it is a space-efficient approach, because it uses a variety of auxiliary data structures, the search time is high $O(n\log n)$ [26]. The drawback of this algorithm is that it only looks for maximum (not super-maximum) repeats. Most of the existing methods for determining for maximum repeats in the $n$-length sequence are $O(n^2)$. State-of-the-art methods rely on the suffix tree, suffix collection, BWT & wavelet tree to find maximum and super maximum repeats. Data structures that are used by these methods require enormous space. Suffix tree or suffix array-based techniques have good search time, but they need a large volume of memory (10–40 times the input genome sequence). While space efficiency relies on techniques based on BWT and wavelet tree, search time is high. There is therefore a possibility to create a stand-alone space-efficient method to scan for maximum and super maximum repeats in linear time [27].

## 3.2 *Index-Based Approaches for Genomic Data Processing*

As genome technology has improved, the size of the genomic database has been growing rapidly. One human genome has a size of about 3 billion genome characters, which needs about 3.2 GB to store in memory. The size of the database of genomes is therefore remarkably high. For studies in forensic science, genetic variation, and medical diagnosis, this database is used. There is a variety of conventional string matching approaches, such as Naive, Rabin-Karp, finite-state automation, Knuth-Morris-Pratt & Boyer-Moore, to retrieve information from the database [28]. These algorithms require $O(n)$ time ($n$-is the size of the genome repositories) to fit a sequence. The string matching approaches described above are not appropriate for human genome analysis because of their enormous scale. (i) The preprocessing (indexing, in compressed form) of the genome database needs (ii) an effective data structure to efficiently retrieve the information needed for better results. Numerous indexing methods exist to match short genome sequence reads with reference genome [29].

"Bowtie" [30] is relying on the FM-Index and BWT. The volume of the index created by Bowtie is high. Gapped alignment is not explicitly endorsed. The scale of the bowtie-based protein sequence index is 2–3 times greater than the genome data size. The "Bowtie-2" [31] variant of Bowtie is expanded. It encourages gaping alignment.

In Burrows-Wheeler Aligner (BWA) [32], both exact & approximate alignments of short sequence reads are carried out. "BWA" is relying on the FM-Index and BWT. It uses an FM-Index that is based on the method of backward search. The BWA-based techniques align a single reference genome with millions of short reads.

"BWBBLE" [33] is a compact (multigenome reference) genome representation that is also relying on the BWT & FM-Index. Indexing of numerous reference genomes is established. Via it, all kinds of alignments (exact & inexact) are possible. Subread [34] also uses the same methodology as BWA's expanded multiseed strategy in which the short reads are matched with the reference genome using seed and vote techniques. The seed and vote method is sequential in nature and requires a lot of time for long read alignment. Indexing based on hash was used by Sub-read. It utilizes more than 15 GB of memory to match 10 million reads with the human genome.

Both BWT and hashing are used in "Kart" [35] to index the reference genome. All kinds of reads are assisted—short reads as well as long reads. It uses divide and conquer tactics for long readings. The index size of the kart for both DNA and protein sequences is very high. "Minmap2" [36] is a long-read alignment instrument that reads DNA/mRNA aligned to a broad reference genome. Indexing based on hash has been used by Minmap2. To align the reads, it utilizes the seed and chain process. Minimap2 performs reliably for longer reads (about 100bp), but both alignment time and accuracy are not up to the mark for short reads (about 100bp). For both DNA and protein sequences, the index size of Minimap2 is very high. BWA [18], BWBBLE [35], Bowtie [26], and Kart [35] use BWT-based indexing, which takes up space equal to the reference genome (it is 3 billion characters long in the case of the human genome). The main downside of all the above BWT-based methods is that BWT does not provide compression on its own, but arranges the text in a compressible manner.

It uses several other external techniques for compression, such as switching to front coding (MTF), run length encoding, and variable length prefix coding that involves a lot of overhead computing and peak memory for the CPU. For protein sequence indexing, BWT-based indexes are also not suitable as the size of the BWT-based index is very high. Exact matching takes a few seconds with these techniques (the avg. rate is 220–550 s for one million short reads) to precisely match the reads while they take longer (avg. rate is 8500–11,000 s for one million short reads) as BWT is not proposed for approximate matching.

## 4 Research Gaps in Genome Data Compression and Processing

- A quicker and more cost-effective strategy for decoding genomic data has been encouraged by recent technical developments in genome sequencing. In the near future, examining the entire genome of every patient as regularly as we currently do for blood (blood testing) would be a standard practice. Due to the volume of data, the data structure, computer hardware limitations (memory, storage space, processing speed, I/O bandwidth, network bandwidth, etc.), transmission, storage, and processing of genome data have many technical challenges. While there are no set parameters for judging the compression and processing techniques of genome data, some points must still be taken into account.
- The majority of state-of-the-art indexing and reading alignment methods are based on the Burrows-Wheeler transform (BWT). In order to compact the reference series, a BWT-based algorithm utilizes many external methods, such as switching to front encoding, run length coding, and variable length coding. Another downside of the BWT-based algorithm is that the alignment time increases with the index size reduction.
- The suffix tree, suffix array, Burrows-Wheeler transform, and wavelet tree are based on current techniques for finding maximum and super maximum repeats. Data structures used by these methods need enormous space. Suffix tree or suffix

array-based techniques have decent search time, but require enormous memory (10–40 times text). Burrows-Wheeler transformation and wavelet tree-based techniques are memory-efficient, but search time is large.

## 5 Conclusion

In this paper, survey of the state-of-the-art techniques for compressing and processing of genomic data is presented. Ever-increasing growth of genome data poses several challenges to the storage, processing, and transmission. The costs related with the storage, processing, and data transmission are larger than the cost of generation of the DNA sequence. Compression and efficient processing are the key solutions that can overwhelm these challenges by decreasing the storage size and processing time.

## References


1. Adjeroh, D., Nan, F.: On Compressibility of Protein Sequences, pp. 422–434. DCC (2006)
2. Allison, L., et al.: Compression of Strings with Approximate Repeats, pp. 8–16. ISMB (1998)
3. Apostolico, A., Lonardi, S.: Compression of Biological Sequences by Greedy Off-Line Textual Substitution, pp. 143–152. DCC (2000)
4. Behzadi, B., Fessant, F.L.: DNA Compression Challenge Revisited: A Dynamic Programming Approach, pp. 190–200. CPM (2005)
5. Boulton, D.M., Wallace, C.S.: The information content of a multistate distribution. Theoret. Biol. **23**(2), 269–278 (1969)
6. Burrows, M., Wheeler, D.J.: A Block Sorting Lossless Data Compression Algorithm. Technical report, Digital Equipment Corporation, Palo Alto, CA (1994)
7. Chen, X., et al.: A Compression Algorithm for DNA Sequences and its Applications in Genome Comparison, p. 107. RECOMB (2000)
8. Chen, X., et al.: DNACompress: fast and effective DNA sequence compression. Bioinform. **18**(2), 1696–1698 (2002)
9. Cleary, J.G., Witten, I.H.: Data compression using adaptive coding and partial string matching. IEEE Trans. Comm. COM **32**(4), 396–402 (1984)
10. Cleary, J.G., Teahan, W.J.: Unbounded length contexts for PPM. Comput. J. **40**(2/3), 67–75 (1997)
11. Dix, T.I., et al.: Exploring long DNA sequences by information content. In: Probabilistic Modeling and Machine Learning in Structural and Systems Biology, Workshop Proc, pp. 97–102 (2006)
12. Dix, T.I., et al.: Comparative analysis of long DNA sequences by per element information content using different contexts. BMC Bioinformat. (2007)
13. Fenwick, P.: The Burrows-Wheeler Transform for block sorting text compression. Comput. J. **39**(9), 731–740 (1996)
14. Grumbach, S., Tahi, F.: Compression of DNA Sequences, pp. 340–350. DCC (1993)
15. Grumbach, S., Tahi, F.: A new challenge for compression algorithms: Genetic sequences. Inf. Process. Manage. **30**(6), 875–866 (1994)
16. Gupta, A., Agarwal, S.: A scheme that facilitates searching and partial decompression of textual documents. Intl. J. Adv. Comput. Eng. **1**(2) (2008)

17. Gupta, A., Agarwal, S.: A novel approach of data compression for dynamic data. In: Proceedings of IEEE third International Conference on System of Systems Engineering, California, USA (2008)
18. Gupta, A., Agarwal, S.: Transforming the natural language text for improving compression performance. Trends in Intelligent Systems and Computer Engineering (ISCE), Lecture Notes in Electrical Engineering Vol. 6, pp. 637–644. Spring, Heidelberg (2008)
19. Kamel, N.: Panel: data and knowledge bases for genome mapping: what lies ahead?. In Proceedings of International Very Large Databases (1991)
20. Korodi, G., Tabus, I.: An efficient normalized maximum likelihood algorithm for DNA sequence compression. ACM Trans. Inf. Syst. **23**(1), 3–34, 20058. Springer Verlag (1993)
21. Matsumoto, T., et al.: Biological sequence compression algorithms. Genome Inform. **11**, 43–52 (2000)
22. Nevill-Manning, C.G., Witten, I.H.: Protein is Incompressible, pp. 257–266. DCC (1999)
23. Kulekci, M.O., Vitter, J.S., Xu, B.: Efficient maximal repeat finding using the burrows-wheeler transform and wavelet tree. IEEE/ACM Trans. Computat. Biol. Bioinformat. **9**(2), 421–429 (2011)
24. Rivals, E., et al.: A Guaranteed Compression Scheme for Repetitive DNA Sequences, p. 453. DCC (1996)
25. Rubin, F.: Experiments in text file compression. Commun. ACM **19**(11), 617–623 (1976)
26. Stern, L., et al.: Discovering patterns in plasmodium falciparum genomic DNA. Molec. Biochem. Parasitol. **118**, 175–186 (2001)
27. Storer, J.A., Szymanski, T.G.: Data compression via textual substitution. J. ACM Assoc. Comput. Mach. **29**(4), 928–951 (1982)
28. Tabus, I., et al.: DNA Sequence Compression Using the Normalized Maximum Likelihood Model for Discrete Regression, p. 253. DCC (2003)
29. Willems, F.M.J., et al.: The context-tree weighting method: basic properties. IEEE Trans. Inf. Theory 653–664 (1995)
30. Witten, I.H.: Managing Gigabytes: Compressing and Indexing Documents and Images. Morgan Kaufman (1999)
31. Witten, I.H.: Arithmetic coding for data compression. Comm ACM **30**(6), 520–540 (1987)
32. Ziv, J., Lempel, A.: Compression of individual sequences via variable-rate coding. IEEE Trans. Inf. Syst. **24**(5), 530–536 (1978)
33. Ziv, J., Lempelm, A.: A universal algorithm for sequential data compression. IEEE Trans. Inf. Syst. **23**(3), 337–342 (1977)
34. Bose, T., Mohammad, M.H., Anirban, D., Sharmila, S.M.: BIND an algorithm for loss-less compression of nucleotide sequence data. J. Biosci. **37**, 785–789 (2012) ss
35. Haque, M.M., Dutta, A., Bose, T., Chadaram, S., Mande, S.S.: DELIMINATE-a fast and efficient method for loss-less compression of genomic sequences sequence analysis. Bioinformatics **28**, 2527–2529 (2012)
36. Sardaraj, M., Tahir, M., Ikram, A., Bajwa, H.: SeqCompress: an algorithm for biological sequence compression. Genomics **104**, 225–228 (2014)
37. Hosseni, M., Pratas, D., Pinho, J.: A survey on data compression methods for biological sequences. Informatics
38. ftp://ftp.ncbi.nih.gov/genomes/Bacteria/all.fna.tar.gz
39. ftp://ftp.ncbi.nih.gov/genomes/Bacteria/all.ffn.tar.gz
40. http://hgdownload.cse.ucsc.edu/goldenPath/hg18/Chromosomes
41. https://portal.camera.calit2.net
42. Sharma, T.K., Pant, M.: Opposition-based learning embedded shuffled frog-leaping algorithm. In: Soft Computing Theories Applications, 853–861. Springer, Singapores (2018)
43. Mahajan, R.: Emotion recognition via EEG using neural network classifier. In: Soft Computing: Theories and Applications, pp. 429–438. Springer, Singapore (2018)
44. Sayali, S., Iyer, B.: IoT-enabled early prediction system for epileptic seizure in human being. In: Soft Computing: Theories and Applications, pp. 37–46. Springer, Singapore (2020)

# Identification of Skin Diseases Using Convolutional Neural Network

**Pooja Pathak, Yash Punetha, and Kratika**

**Abstract** Dermatology is one of the most far reaching diseases in the world. Regardless of being common, its diagnosis is very difficult in view of its complexities of skin tone, shading, and nearness of hair. This paper gives a way to deal with use computer vision-based methods (deep learning) to consequently predict the different kind of skin diseases. The framework uses publicly available image recognition architecture with moderation for skin disease application and effectively predicts the skin disease. The architecture is published by image recognition giants for open use for different applications. The framework comprises of three stages: the feature extraction stage, the training stage, and the testing/approval stage. The framework utilizes deep learning technology to prepare itself with the different skin images. Transfer learning is used (used to build models in a timesaving way) to create the model, that means that we took a pre-trained network called MobileNet that is trained with ImageNet, a dataset of over 14 million images, and added some extra layers to it so it can classify skin diseases, tested with the HAM10000 dataset, the model obtains an accuracy of around 85%.

**Keywords** Skin disease · Dermatology · Diagnosis · Deep learning · Skin lesions

## 1 Introduction

In the previous 12-year time frame, from 2008 to 2020, the yearly scope of malignancy cases has extended with the guide of 53%, to a limited extent in light of broadened UV exposure [1, 2]. Despite the fact that melanoma is one of the most fatal sorts of pores and skin most malignant growths, a quick examination can bring about a totally high risk of endurance. The initial step inside the investigation of a

P. Pathak (✉) · Kratika
Department of Mathematics, IAH, GLA University, Mathura, India
e-mail: pooja.pathak@gla.ac.in

Y. Punetha
Department of CEA, GLA University, Mathura, India



T. K. Sharma et al. (eds.), *Soft Computing: Theories and Applications*,
Advances in Intelligent Systems and Computing 1381,
https://doi.org/10.1007/978-981-16-1696-9_16

threatening injury by a dermatologist is a visual test of the doubtful pores and skin territory. Without additional specialized assistance, dermatologists have a 65–80% exactness expense in melanoma analysis [3]. In doubtful cases, the visual review is enhanced with dermatoscopic photographs amped up for an extraordinary unnecessary choice and amplifying computerized camera. During the chronicle, the lighting installations are controlled and a channel out is utilized to reduce reflections on the skin, accordingly making further skin layers seen. With this specialized guide, the precision of skin sore visualization can be raised by means of an another 49% [4]. The blend of noticeable review and dermatoscopic previews at last outcome in a flat out malignant growth recognition accuracy of 75–84% with the guide of dermatologists [5, 6].

Using ML for analysis, we can dispense with preventable illnesses and make a future where never again need to stress over getting a precise finding. The ongoing development of AI and profound learning strategies for therapeutic picture examination has empowered the improvement of insightful restorative imaging-based finding frameworks that can help doctors in settling on better choices about a patient's well-being. In specific, skin imaging is where these new strategies can be applied with a high pace of victory.

## 2 Methodology

### *2.1 Data Collection*

The key step of the research on skin disease classification is the collection of datasets. The dermatoscopic images are collected from different populations having various diversities as shown in Fig. 1. The different cleaning and acquisition methods are applied to these collected dermatoscopic images.

The final dataset consists of 10,015 dermatoscopic images having seven different types of skin diseases. This exempted dataset can be further used in ML and for differentiation with the specialists. The HAM10000 dataset can be downloaded from the URL [7]. The images are stored in the JPEG format. The skin diseases (Fig. 2) found in the dataset are:

- Malignant Melanoma (mel)
- Melanocytic Nevi (nv)
- Basal Cell Carcinoma (bcc)
- Benign Keratosis (bkl)
- Dermatofibroma (df)
- Actinic Keratosis (akiec)
- Vascular Lesions (vasc).

**Melanoma (mel)**: It is a harmful neoplasm procured from melanocytes that may appear in various variations. It can be cured by surgical excision, whenever detected

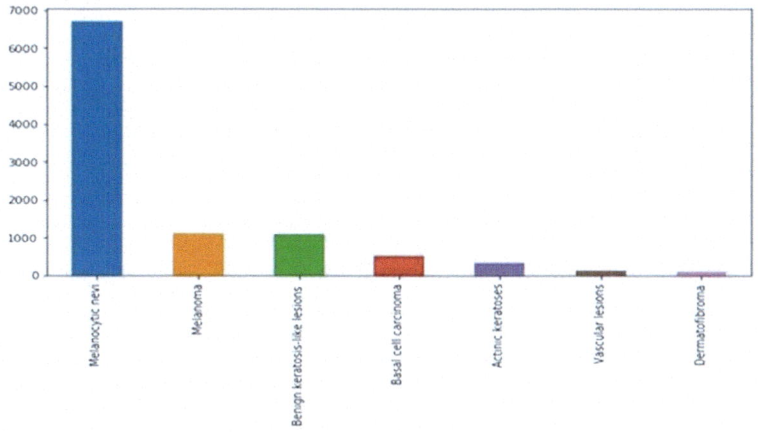


**Fig. 1** Dataset

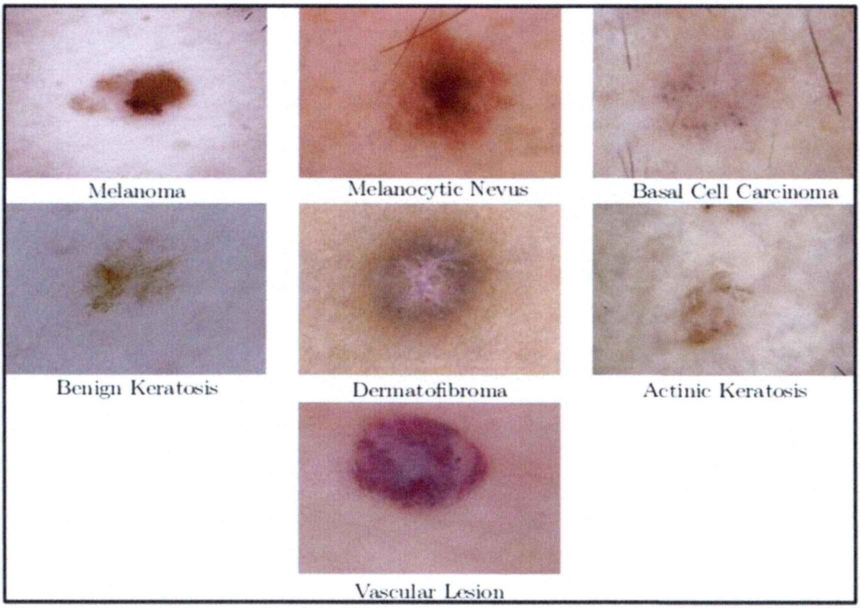


**Fig. 2** Skin disease type

in an early stage Melanomas can be intrusive or non-obtrusive. We incorporated all variations of melanoma involving melanoma for situ, yet excluded non-pigmented, subungual, visual or mucosal melanoma [1113 images].

**Melanocytic nevi (nv)**: They are usually non-cancerous disorder of pigment-producing skin cells which are commonly called birth marks or moles and show up in a horde of variations, which all are involved in our arrangement. The variations may vary essentially from a dermatoscopic point of view [6705 images].

**Basal cell carcinoma (bcc)**: It is a type of skin cancer that begins in the basal cells with variation of epithelial skin malignancy that rarely metastasizes however breed catastrophic if untreated. It shows up in various morphological variants (flat, nodular, pigmented, cystic, and so forth) which are included in this set [514 images].

**Benign keratosis (bkl)**: It is a non-cancerous skin condition that appears as a waxy brown, black, or tan growth. Seborrheic keratoses ("senile wart"), solar lentigo—which is one of the most common non-cancerous skin growths in older adults [1099 images].

**Dermatofibroma (df)**: It is a common benign skin sore viewed as either a benign escalation or an irritant response to minimal response also called cutaneous fibrous histiocytoma. It is brown, frequently, demonstrating a central zone of fibrosis dermatoscopically [115 images].

**Actinic Keratoses (akiec)**: These are regular non-obtrusive, variations of squamous cell carcinoma that can be dealt without medical procedure. These are rough, scaly patch on the skin caused by years of sun exposure. Actinic keratoses are usually occuring on the face and Bowen's disease on the other body parts [327 images].

**Vascular skin lesions (vcc)**: These are relatively common abnormalities of the skin and underlying tissues. The dataset varies from cherry angiomas to angiokeratomas and pyogenic granulomas. Hemorrhage is additionally involved in this classification [142 images].

## 2.2 *Selection of Deep Learning Models*

There are many deep learning models which are used like AlexNet [8, 9], Mobile Net, VGG, convolutional neural networks (CNNs). A deep learning system is used to differential diagnosis of skin conditions using images and clinical data [10], etc. CNN is considered as one of the most famous deep learning (DL) methods. The primary aim of CNNs is to recognize various image patterns and characteristics for the process of recognition and classification [11]. The input data of CNN can be images like human faces, various sites, and any other visual objects. CNN, being a supervised learning method, it is trained using labeled data with their respective classes. Initially, CNN learns the correlation among the input data and the labeled class and comprises of two parts: the hidden layer which helps in feature extraction and the fully connected layers that helps in classification. In contrast to neural networks, the hidden layers in CNN have a particular architecture. There are three main layers in CNN architecture (Fig. 3).

(1) Convolutional layer: It is used for extracting various key features from the input images.

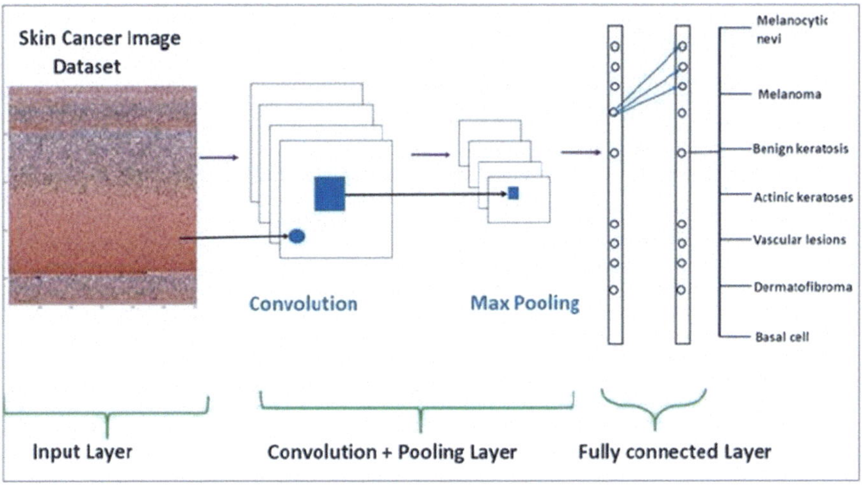


**Fig. 3** CNN architecture

(2) Hidden layer: This layer is used after the convolutional layer. It helps in the reduction of spatial dimensions of the input images.
(3) Fully Connected layer: In this layer, the nodes are fully connected to the preceding layer. This layer is considered as the second last step after the output layer.

## 2.3 *Performance Metrics for Classifiers*

An object to a class is assigned by a classifier. This task is, however, not an ideal one as an object may be assigned to a wrong class. The real class of the object must be known, to get the actual measure of a classifier [12]. To check the degree of classification, the classifiers output is compared with the true class. Thus, the objects are classified into four different sets:

1. True Positive (TP): the classifier/model accurately anticipates the positive category/class.
2. True Negative (TN): the classifier/model accurately anticipates the negative class/category.
3. False Positive (FP): the classifier/model falsely anticipates the positive class/category.
4. False Negative (FN): the classifier/model falsely anticipates the negative category/class.

Statistical quantities for the classifiers can be estimated by the cardinality of these subsets. A familiar and mostly used quantity is accuracy, which is a rational computation only if the different classes in the dataset are distributed equally [6].

**Accuracy** is computed by:

$$\frac{\text{TP} + \text{TN}}{\text{TP} + \text{FP} + \text{TN} + \text{FN}} \tag{1}$$

It tells the rate of objects that have been sorted accurately and correctly.

There are other metrics for classification. Even if the classes are not evenly distributed, these metrics can be used.

*Sensitivity/Recall* is how good the test is at detecting the positives. It is computed by:

$$\frac{\text{TP}}{\text{TP} + \text{FN}} \tag{2}$$

*Specificity* is how good a test is at avoiding false alarms. It is computed by:

$$\frac{\text{TN}}{\text{TN} + \text{FP}} \tag{3}$$

*Precision* figures out how many of the positively classified were relevant.

$$\frac{\text{TP}}{\text{TP} + \text{FP}} \tag{4}$$

*F1-Score* consolidates the precision and recall to give a single score. It is the harmonic mean of precision and recall. Essentially, it punishes extreme values and determined by

$$\frac{2 \times \text{Precision} \times \text{Recall}}{\text{Precision} - \text{Recall}} \tag{5}$$

*Support* is the number of samples of the true response that lie in that class.

The result of the binary classifier is rendered as a probability distribution over the classes. Generally, output of objects which are greater than 0.5 is classified as positive class/category in a binary classifier and object which have output value less than 0.5 are classified/sorted as negative class. There is other approach which is based on receiver operating characteristic (ROC). The threshold in this categorization lies between 0 and 1, and for each selected threshold, sensitivity and specificity are calculated. The ROC curve is drawn by plotting the sensitivity vs 1-specificity. It can also be used to examine the classifier. Further, the ROC curve drifts from the diagonal, and the finer is the classifier.

## 3 Implementation of CNN

In this paper, The set of images is obtained from the given dataset. Preprocessing and segmentation techniques are performed on each image [13]. After Image segmentation is done, one can determine whether the skin is infected with the disease or not. Once the disease is detected, the affected area will be highlighted indicating the exact location of the infected area.

Figure 3 diagram depicts the process of implementation of skin disease diagnosis. The system mainly consists of the following parts:

(1) Acquire Image: Firstly, the skin image is acquired for the model, for which the skin diseases has to be classified.
(2) Image Processing Unit: It is the key part of the process in the system as it is required to detect or identify the region of interest to undergo image processing.
(3) Feature Extraction: It is the process of reduction in which the original set of data is reduced into less number of useful data. It aims to summarize the useful information into less number of resources.
(4) Image Classification: It takes image as an input and produces the output for the classification of the disease. MobileNet is used for its lightweighted architecture.
(5) Generate Results: The outcome is generated which shows the skin disease is among the seven diseases. MobileNet is based on streamline architecture which uses depth-wise separable convolutions. Figure 4 represents the block diagram of proposed model. It performs a single convolution on each colour channel rather than combining all three and flattening it which is also called depth-wise

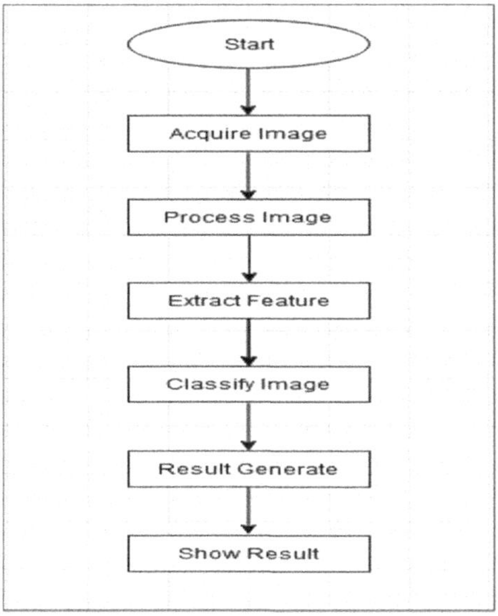


**Fig. 4** Block diagram of proposed model

separable convolutions. It also requires very low maintenance, thus performing quite well with high speed. Convolutional neural networks(CNN) differ from normal neural networks in a way that as they contain a special layer called convolutional layer that contains a filter which is able to recognize different types of arrangements in the image.

## 4 Result

In the present work in Spyder (a python IDE), ver. 3.3.6 is implemented. Authors have used a platform of Intel Core i5-6200U processor 2.30 GHz with 4 GB RAM [14]. The execution results are shown in Figs. 5 and 6. In skin disease identification, pre-training is performed by using MobileNet architecture. In spite of the non-medical visual domain, the learned features have enough quality for disease identification.

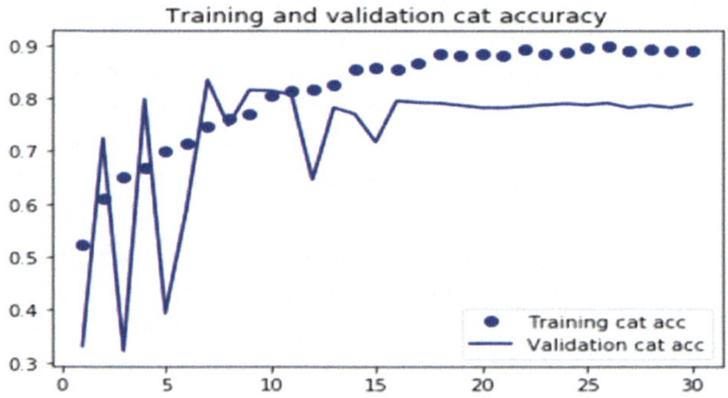


**Fig. 5** Training and validation categorical accuracy

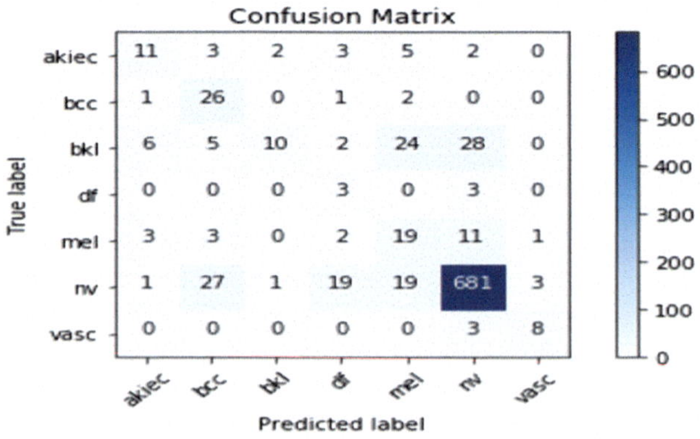


**Fig. 6** Confusion matrix

**Table 1** Performance matrix

| | Precision | Recall | F1-score | support |
|---|---|---|---|---|
| akiec | 0.50 | 0.42 | 0.46 | 26 |
| bcc | 0.41 | 0.87 | 0.46 | 30 |
| bkl | 0.77 | 0.13 | 0.23 | 75 |
| df | 0.10 | 0.50 | 0.17 | 6 |
| mel | 0.28 | 0.49 | 0.35 | 39 |
| Nv | 0.94 | 0.91 | 0.92 | 751 |
| vasc | 0.67 | 0.73 | 0.70 | 11 |
| Avg/Total | 0.86 | 0.81 | 0.81 | 938 |

**Categorical Accuracy**: It is used for calculating the predicted (yPred) values percentage that matches with the actual values, for one hot-labels; here we can provide yPred as logit or probability of predictions.

**Confusion Matrix**: It evaluates the accuracy of a classification. The number of correct and incorrect predictions are summarized with count values and broken down by each class. 20% images were used for validation and 80% images for training purposes. The system works well with approximately 85% accuracy rate. Table 1 shows different performance metrics accuracy for the seven different kinds of diseases.

## 5 Conclusion

There exist various deep learning architectures that can be trained for the solution of classification problems. Thus, this provides a technique called "transfer learning" due to which there is no need to create a new DL architecture from scratch. Sometimes, it is hard and may be impassable to differentiate the work of published classification results as most of the people use non-public datasets for training and testing. It is clearly seen from performance matrix the precision value for dermatofibroma (df) is 10%, whereas melanocytic nevi (nv) is 94%.

## References

1. Nami, N., Giannini, E., Burroni, M., Fimiani, M., Rubegni, P.: Teledermatology: state-of-the-art and future perspectives. Expert. Rev. Dermatol. **7**(1), 1–3 (2014)
2. Fabbrocini, G., Triassi, M., Mauriello, M.C., Torre, G., Annunziata, M.C., De Vita, V., et al.: Epidemiology of skin cancer: Role of some environmental factors. Cancers (Basel) **2**(4), 1980–1989 (2010)

3. Argenziano, G., Soyer, H.P.: Dermoscopy of pigmented skin lesions: a valuable tool for early diagnosis of melanoma. Lancet Oncol. **2**(7), 443–449 (2001)
4. Kittler, H., Pehamberger, H., Wolff, K., Binder, M.: Diagnostic accuracy of dermoscopy. Lancet Oncol. **3**(3), 159–165 (2002)
5. Ali, A.R.A., Deserno, T.M.: A systematic review of automated melanoma detection in dermatoscopic images and its ground truth data. Proc. SPIE Int. Soc. Opt. Eng. **8318**, 1–6 (2012)
6. Fabbrocini, G., De Vita, V., Pastore, F., D'Arco, V., Mazzella, C., Annunziata, M.C., et al.: Teledermatology: from prevention to diagnosis of nonmelanoma and melanoma skin cancer. Int. J. Telemed. Appl. **17**, 1–5 (2011)
7. Kaggle: https://www.kaggle.com/kmader/skin-cancer-mnist-ham10000. Last accessed 30 Jun 2020
8. Krizhevsky, A., Sutskever, I., Hinton, G.E.: ImageNet classification with deep convolutional neural networks. Adv. Neural. Inf. Process Syst. **1**, 1–23 (2012)
9. Simonyan, K., Zisserman, A.: Very deep convolutional networks for large-scale image recognition, 10 Apr 2014. https://arxiv.org/pdf/1409.1556. Accessed 06 Oct 2018
10. Liu, Y., Jain, A., Eng, C., et al.: A deep learning system for differential diagnosis of skin diseases. Nat. Med. **26**(1), 900–908 (2020)
11. Sharma, T.K., Pant, M.: Opposition-based learning embedded shuffled frog-leaping algorithm. Soft Comput. Theor. Appl. **583**(1), 853–861 (2018)
12. Swami, V., Kumar, S., Jain, S.: An improved spider monkey optimization algorithm. Soft Comput. Theor. Appl. **583**(1), 73–81 (2018)
13. Bhasin, H., Gupta, N.: Critical path problem for scheduling using genetic algorithm. Soft Comput. Theor. Appl. 15–24 (2018)
14. Pathak, P., Gangwar, H., Jalal, A.S.: Performance analysis of gradient descent methods for classification of oranges using deep neural network. In: 7th International Conference on Computing for Sustainable Global Development (INDIACom) 2020, IEEE Explore, pp. 68–72, IEEE (2020)

# Gene Sequence Classification Using K-mer Decomposition and Soft-Computing-Based Approach

**Sanjeev Kumar**

**Abstract** The healthcare industry is moving toward personalized medicine which requires the use of individual genetic information so that medical treatment can be customized to the specific properties of an individual. DNA sequence of a genome consists of several genes. These genes are the basic building blocks of an organism. A human genome consists of 20–30 thousand genes. Some of these genes are involved in the growth and development of the body and some are responsible for the production of critical diseases (influenza, ebola, dengue) remaining are the non-coding (junk) genes. Identification and classification of these genes into a few biological meaningful groups: coding, non-coding, and viral are useful for the treatment and diagnosis of an organism. In this paper, k-mer (substrings of length $k$) frequency decomposition and soft-computing-based approach is used to classify and identify the large set of unknown genes into some meaningful groups. It works by first taking a DNA sequence and computing a vector of the proportions of every possible k-mer. These vectors are used as feature vectors, and a well-known supervised classification algorithm (multi-mode Naive Bayes classifier) is trained on the vectors. Experiments show that the proposed approach achieves the highest accuracy ($\geq$90 %) along with the lowest running time in comparison with other state-of-the-art methods.

**Keywords** Classification · K-mer · Gene · Sequence · Soft-computing · Genome

## 1 Introduction

In genomics, analyzing DNA is an essential part of understanding about the organism. DNA sequence of a genome consists of several genes. These genes are the basic building blocks of an organism. A gene is a region of DNA that encodes the protein [1]. Classification of a large set of genes into a few biological meaningful groups facilitates the prediction of functions of genes. Mostly genes are classified into three

S. Kumar (✉)
Department of CEA, GLA University, Mathura, India
e-mail: sanjeev.kumar1@gla.ac.in



T. K. Sharma et al. (eds.), *Soft Computing: Theories and Applications*, Advances in Intelligent Systems and Computing 1381,
https://doi.org/10.1007/978-981-16-1696-9_17

groups: protein-coding genes, protein non-coding genes, and viral/bacterial genes. Protein-coding genes are code for protein, i.e., used for development and growth of the body of an organism. Non-coding DNA does not provide instructions for constructing proteins. Viral genes are responsible for various genes-related diseases [2].

Example of common human diseases caused by viral genes includes cold, influenza, chickenpox, and cold sores and many critical illnesses, such as rabies, Ebola, AIDS, and SARS. Understanding and identifying the unknown genes are very important in the diagnosis and treatment of genes based on diseases [3]. Identifying specific genes features and then classifying these genes into coding, non-coding, and viral groups with high accuracy is an essential step toward a deeper understanding of heredity over diseases. Computational approaches to genes classifications fall into two broad categories: alignment-based approach and alignment-free approach [4]. In alignment-based approach, gene sequence is identified by mapping of candidate genes to reference genes. Because of the time and space complexities of multiple gene sequence alignment, these approaches are inappropriate for the classification of large sequences. Several state-of-the-art tools are already developed including single-sequence alignment tools like BLAST and FASTA and multiple-sequence alignment tools like ClustalW and MUSCLE [5].

On the other hand, alignment-free tools do not compute character to character alignments which are very attractive for many applications as they are much more efficient in terms of accuracy and running times. Nguyen [2] was the first to use such methods, reporting success in constructing accurate phylo-genetic trees for mammal alpha and beta-globin genes and several other coding and non-coding DNA sequences. Accuracy of this tool is 80–85% but running time is not good. Jianmin et al. [6] proposed an approach for gene classification by adopting codon frequency as feature vector and SVM as a classifier to accurately classify the gene sequences. Running time and accuracy is better from state-of-the-art approaches but accuracy is less than 87%.

Massimo et al. [7] proposed another new approach for gene sequence classification which is based on probabilistic topic modeling. In probabilistic modeling first an corpus of recurrent features of genome and these corpus are used to classify the genes into some meaningful groups. RDP and SVM-based classifier is used for classification. Accuracy of this approach is about to 80–89% but running time is high [8].

In this paper, a new approach based on k-mer decomposition and soft-computing is proposed. Proposed method uses k-mer frequencies of a gene sequence as a feature vector and multinomial Naive Bayes classier for classification. Multinomial Naive Bayes classifier works better for data which can easily be turned into counts, such as subsequence counts in a text. From experiments it is confirmed that accuracy and running time achieved by propose method are outperformed in comparison with other state-of-the-art methods.

# 2 Proposed Method

## 2.1 Collection of Data and Preparation of Datasets

Gene sequence data are taken from NCBI and Pizza-Chili corpus. Each sequence is preprocessed by removing non-acgt characters and meta-data information.

## 2.2 K-mer Decompositions of Gene Sequence

The proposed approach uses k-mer frequency decomposition as a feature vectors and multinomial Naive Bayes classifier for classification. Here K-mer frequency proportion vectors are used as feature vectors. Running time and classification accuracy are measured for k-mers of length $k = 1 \ldots 10$, and experimentally, it is found that for $k = 6$ proposed approach provides the optimal balance of both accuracy and speed [9]. Proposed approach works in two phases. In first phase, all the gene sequences are transformed in uniform length vector. In second phase, these feature vectors are trained by suitable classifier method. Let $G$ be a genome sequence, and $k$ is the K-mer (subsequence) length. To create a uniform vectors, proposed approach first divides the genomes into k-mers each of size $k$. A dictionary of all possible k-mers of size k sorted in lexicographical order is also created temporarily. Proposed approach calculates the k-mer frequency distribution into the input genome by the use of above dictionary. These distributions of k-mers are used as feature vectors also re-scaled the feature vectors into the variance of 1. For example: Let genome sequence be $G$ = AGTCTGCCT and $k = 2$. The possible dictionary for $k = 2$ is [AA, AC, AG, AT, CA, CC, CG, CT, GA, GC, GG, GT, TA, TC, TG, TT] and k-mers of input genome are: AG, GT, TC, CT, TG, GC, CC, CT. Frequency decomposition of k-mers into the dictionary is [0, 0, 1, 0, 0, 1, 0, 2, 0, 0, 1, 0, 1, 0, 1, 1,0]. Re-scaled feature vectors is [0, 0, .125, 0,0, .125, 0, .25, 0, 0, .125, 0, .125, .125, 0] (Fig. 1).

## 2.3 Training

Finally proposed approach trained these feature vectors using multinomial Naïve-Bayes classifier [10]. For performance-valuation of the classifier, tenfold cross-validation is used. Thereafter, processing of these feature vectors is performed for efficient use by soft-computing approaches. Proposed approach de-scaled the calculated vectors to a variance that satisfy several statistical assumptions used by state-of-the-art classifiers algorithms. In addition, proposed approach also performed dimensional reduction using truncated singular value decomposition to minimize nonzero entries of the feature vectors [11, 12]. This greatly reduces running time for most of the classifiers while having a negligible effect on classification accuracy.

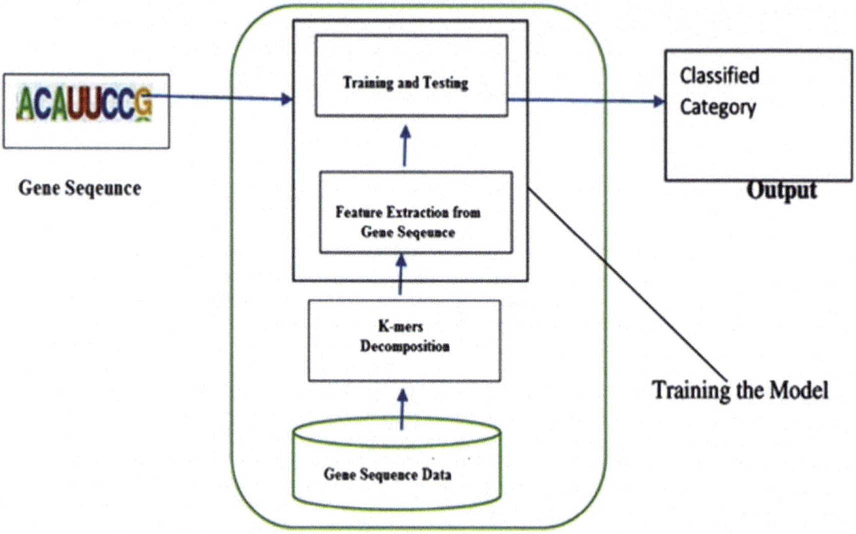


**Fig. 1** Architecture of gene sequence classification system

### 2.4 *Testing*

For testing proposed approach used the 80:20 ratios, in which 80% trained datasets and 20% are untrained datasets.

## 3 Result Discussions

Performance of proposed approach has been compared with state-of-the-art technique. To validate the performance of proposed method and state-of-the-art method tenfold cross-validation is used three times. From Table 1 and Fig. 2, it is confirmed that accuracy of proposed approach for various datasets is very high in comparison with state-of-the-art approach. Minimum improvement of accuracy achieved by proposed method over state-of-the-art method is 0.3% while improvement of maximum accuracy is 8% (Fig. 3).

## 4 Conclusion

In this paper, k-mer (substrings of length $k$) frequency decomposition and soft-computing-based approach is used to classify and identify the large set of unknown genes into some meaningful groups. It works by first taking a DNA sequence and

**Table 1** Performance comparison of proposed approach with state-of-the-art approach

| Datasets | Previous best accuracy (%) | Proposed approach accuracy (%) | | | Improvement (%) |
|---|---|---|---|---|---|
| | | Minimum | Maximum | Average | |
| H3 | 88.99 | 89.04 | 91.34 | 90.18 | 1.19 |
| H4 | 88.09 | 88.90 | 93.67 | 91.29 | 3.19 |
| H4ac | 78.84 | 81.64 | 88.37 | 85.01 | 6.16 |
| H3K4me1 | 74.42 | 78.85 | 84.56 | 81.71 | 7.29 |
| H3K4me2 | 74.20 | 79.32 | 85.34 | 82.33 | 8.13 |
| H3K4me3 | 71.50 | 75.62 | 82.31 | 78.97 | 7.47 |
| Splice | 96.18 | 96.25 | 98.10 | 97.18 | 0.99 |
| Promoter | 99.06 | 99.03 | 99.86 | 99.45 | 0.38 |

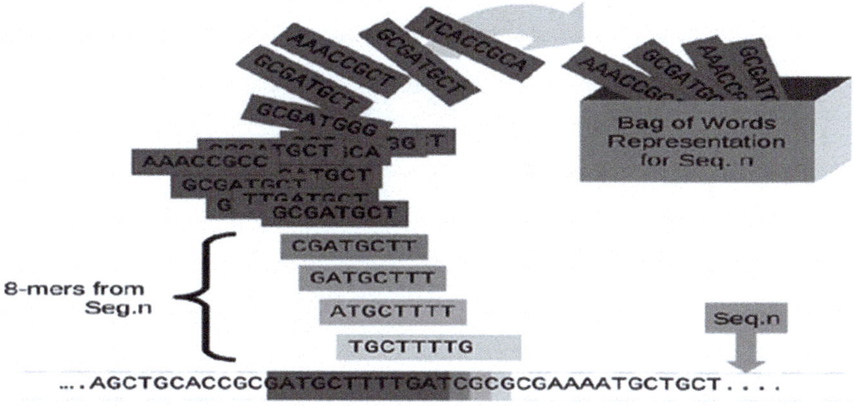


**Fig. 2** K-mer structure and bag of words representation of a DNA sequence

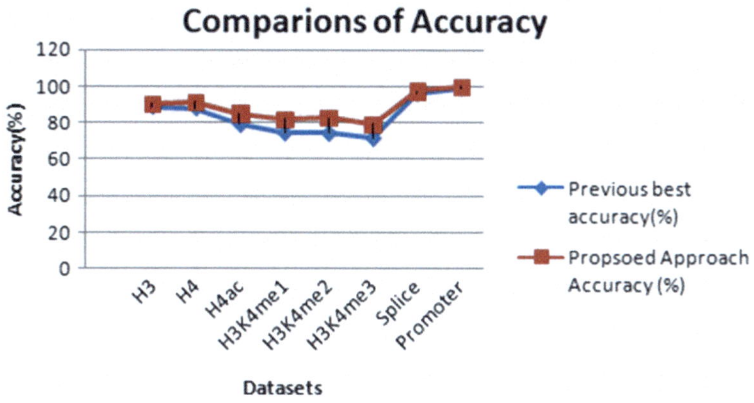


**Fig. 3** Comparison of accuracy of proposed approach with state-of-the-art approach

computing a vector of the proportions of every possible k-mer. These vectors are used as feature vectors, and a well-known supervised classification algorithm (multi-mode Naive Bayes classifier) is trained on the vectors. Experiments show that the proposed approach achieves the highest accuracy ($\geq$90%) along with the lowest running time in comparison with other state-of-the-art methods.

## References

1. Bokulich, N.A., et al.: Optimizing taxonomic classification of marker-gene amplicon sequences with QIIME 2's q2-feature-classifier plugin. Microbiome **6**(1), 90 (2018)
2. Nguyen, N.G., et al.: DNA sequence classification by convolutional neural network. J. Biomed. Sci. Eng. **9**(05), 280 (2016)
3. Eickholt, J., Cheng, J.: DNdisorder: predicting protein disorder using boosting and deep networks. BMC Bioinform. **14**(1), 88 (2013)
4. Leung, M.K.K., et al.: Deep learning of the tissue-regulated splicing code. Bioinformatics **30**(12), i121–i129 (2014)
5. Solis-Reyes, S., et al.: An open-source k-mer based machine learning tool for fast and accurate subtyping of HIV-1 genomes. PLoS One **13**(11), e0206409 (2018)
6. Ma, Jianmin, Nguyen, Minh N., Rajapakse, Jagath C.: Gene classification using codon usage and support vector machines. IEEE/ACM Trans. Comput. Biol. Bioinf. **6**(1), 134–143 (2009)
7. La Rosa, M., et al.: Genomic sequence classification using probabilistic topic modeling. In: International Meeting on Computational Intelligence Methods for Bioinformatics and Biostatistics. Springer, Cham (2013)
8. Mukhopadhyay, S., et al.: A comparative study of genetic sequence classification algorithms. In: Proceedings of the 12th IEEE Workshop on Neural Networks for Signal Processing. IEEE (2002)
9. Buldyrev, S.V., et al.: Long-range correlation properties of coding and noncoding DNA sequences: GenBank analysis. Phys. Rev. E **51**(5), 5084 (1995)
10. Sharma, T.K., Pant, M.: Opposition-based learning embedded shuffled frog-leaping algorithm. In: Soft Computing: Theories and Applications, pp. 853-861. Springer, Singapore (2018)
11. Mahajan, R.: Emotion recognition via EEG using neural network classifier. In: Soft Computing: Theories and Applications, pp. 429–438. Springer, Singapore (2018)
12. Shinde, S., Brijesh, I.: IoT-enabled early prediction system for epileptic seizure in human being. In: Soft Computing: Theories and Applications, pp. 37–46. Springer, Singapore (2020)
13. Kumar, S., Agarwal, S.: An efficient tool for searching maximal and super maximal repeats in large DNA/protein sequences via induced-enhanced suffix array. Recent Patents Comput. Sci. **12**(2), 128–134 (2019)

# Deep CNN Model for Identification of Fruit Disease

**Rishi Agrawal**

**Abstract** In today's scenario, a lot of problems are being solved by using different deep learning techniques. This paper used deep convolution neural networks to propose a model for orange fruit disease detection. Traditionally, fruit disease detection was a very time-consuming and manual process but with the advancement of computer and machine learning techniques, the process is becoming automated and faster. Many juice companies like Tropicana and Real continuously used different advanced techniques to distinguish between fresh and diseased fruits. The paper uses a visual geometry group named convolution neural network. Today machine learning techniques with the use of high-speed computing hardware devices have given us an opportunity to use different algorithms that automatically detect objects in an image. The paper used a dataset taken from Kaggle.com which contains three different images of fruits but the model used only oranges images for experimental purpose.

**Keywords** Deep convolution neural networks · Kaggle · Machine learning

## 1 Introduction

Traditionally, fruit disease detection was a very time-consuming and manual process but with the advancement of computer and machine learning techniques, the process is becoming automated and faster. Many juice companies like Tropicana and Real continuously used different advanced techniques to distinguish between fresh and diseased fruits. The paper uses a visual geometry group named convolution neural network. Today machine learning techniques with the use of high-speed computing hardware devices have given us an opportunity to use different algorithms that automatically detect objects in an image. Fruit disease detection is a machine learning model that will be used to label fruits as diseased or healthy. This model trains on the train dataset made for orange fruit. It consists of more than 3000 images out of which

R. Agrawal (✉)
GLA University, Mathura, India
e-mail: rishi.agrawal@gla.ac.in

 
T. K. Sharma et al. (eds.), *Soft Computing: Theories and Applications*,
Advances in Intelligent Systems and Computing 1381,
https://doi.org/10.1007/978-981-16-1696-9_18

1800 images are used for training, and the rest are used for testing and validation. This dataset will ensure that the model trains on an adequate number of images so that new images could be labeled with greater accuracy. The model will replicate the structure of the model used in ILSVRC 2014 competition. Machine learning applications have been developed by a few research centers around the world to detect fruit diseases. These techniques have been used for small- to medium-scale scientific purposes. The techniques required for the accurate and efficient detection of fruit diseases are yet in the developmental phase. These techniques could very well one day help in segregating fruits from one another on the basis of health. This research determines whether machine learning techniques can be used to improve pattern recognition and analysis of fruit image datasets gathered from the Internet.

## 2 Related Work

Deep convolution neural networks and object detection techniques are used to classify and identify tomato diseases. This approach is different than the approaches presented in [1–3] in the sense that it employs neural networks to extract features from the images rather than using image processing techniques. Two different object detection models are used in proposed system; the first one used is faster RCNN that is used to identify the species of the tomatoes. Then it locates the diseased part in those tomatoes. Second one uses the mask R-CNN to segment the shapes of the diseased part in those tomatoes. Region of interest extracted using these two object detection models are later fed into a deep convolutional neural network to extract features and map those features to the diseases. Four different deep convolutional networks were used in this study, namely ResNet-101, ResNet-50, VGG-16, MobileNet. The study finds that ResNet-101 performs better than all the other deep convolutional models but it also consumes the most amount of time to train on the dataset after object detection models are applied. ResNet-101 is found to have taken 23.82 h of time to complete the model training task. On the other hand, MobileNet is found to have taken 12.52 h and performed second to ResNet-50 which is itself second to ResNet-101. VGG-16 has the weakest performance and takes time comparable to ResNet-50 of about 20.38 h. Final mAP score for ResNet-101 stands highest at 99.64 with the best performance. The study [4] was done on the detection of tomato diseases. 11 different types of diseases were chosen to perform the classification. In [5], the study is done to evaluate the relative speeds of three different deep convolutional neural models GoogleNet, AlexNet, and CNN. All three models correctly do prediction on test data but task different duration of time to train due to the inherent complexity built into each of the models. In [6] very thorough study on detection of different types of banana diseases is done using deep convolutional neural networks. The dataset is gathered from Africa and Tamil Nadu with each image being labeled by experts and for particular disease that was present in that image. This dataset investigates five major banana diseases. The study underlines that for proper functioning of any machine learning model the dataset should be of good quality. To achieve this goal the study features, a dataset constructed with expert advice from two different continents. Each image is taken in different environmental conditions at different angles. It consists of images taken at different growing phases of the plant with dead/old leaves as well. This extensive effort to collect images and annotate each one led to a robust dataset for further investigation to take place. Three different machine learning architectures were used, namely ResNet50, InceptionV2, and MobileNetv1. Faster RCNN and single shot multibox (SSD) were used for object detection out of those available in TensorFlow. Then pre-trained models were trained with a COCO dataset. These architectures were then retrained using transfer learning. The metrics used for assessing the performance are loss function (includes weight of classification loss and localization loss), mAP score, and confusion matrix. Multi-SVM approach is having the highest efficiency among different methods [7, 8]. Linear and logistics regression methods can also be used for multiple analysis purpose [9]. The mean average precision (mAP) was used as the validation metric for banana disease and pest detection [10].

## 3 The Proposed Method

For the study in [1], total images used are 391 for apple disease detection. In [3], the dataset used is the same one used in [1]. Dataset used in [4] has 256 images with division in training, testing, and validation data ratio being 6:2:2. In [5], the dataset comprises 4900 images of 9 different types of fruits used to train each of the models. In [6], the dataset has 18,000 images of bananas with annotation to go with each type of architecture used in the investigation. This dataset is divided into following proportions: training (70%), validation (20%), testing (10%).

### *3.1 Convolutional Neural Network*

Convolutional neural network: A convolutional network is very similar to the artificial neural networks. And convolutional neural networks are widely used for tasks such as image recognition image classification and object detection tasks.

Benefits of using convolutional neural networks are.

1. ConvNets can do self-identification of regions of interest
2. This is the currently available most popular way of doing image-related work.

Limitations of using convolutional neural networks are.

1. Requires large dataset to work properly.
2. Overfitting or underfitting is common problems.
3. It is very specific to the task assigned for.

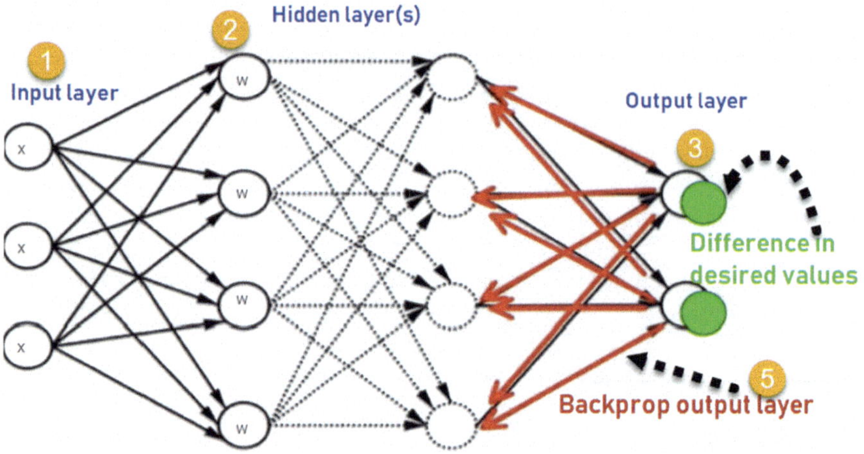


**Fig. 1** Simple perceptron model

Another issue that all the deep neural networks algorithms have is due to the backpropagation algorithm which is of vanishing and exploding gradients.

In [6], the study of plant/fruit disease detection of banana plants using convolution neural networks has been carried out. The following three different architectures were used to compare against each other: ResNet50, InceptionV2, MobileNetV1. In [6], dataset with 18,000 images taken from different regions of Africa and India(Tamil Nadu). To avoid the pitfalls of the poor performance produced by image datasets made in controlled environments, all these images used are taken in different environmental conditions. All the images used are labeled for feeding into CNN models Fig. 1.

## 4 Experimental Analysis

The data used for the detecting the fresh and rotten fruits was downloaded from Kaggle.com named "Fruits fresh and rotten for classification" created by Sriram Reddy Kalluri. This dataset is openly available on the online platform for data competitions. The dataset is of 3.6 GB which consists of the three folders namely—(i) dataset, (ii) train, and (iii) test. Each of the train and test folder contains six subfolder as follows: freshapples, rottenapples, freshbanana, rottenbanana, freshoranges, rotten oranges.

### *4.1 Changing of the Dataset*

For the purpose of use in this project, only the images of oranges have been used. The images have been separated in relabeled according to the following convention "dis. <image_number> .png" and "healthy. <image_number> .png" for diseased and healthy fruit images, respectively. These images are further divided into two folders, namely "train" and "test." Each of the train and test folders has two subfolders (a) healthy and (b) dis. Healthy folder contains all the healthy fruit images, and the other folder has all of the diseased fruit images for training and testing.

### *4.2 Analysis of the Three-Layer Implementations*

In the analysis of three-layer implementation, the result achieved with one layer of ConvNet was better than the ones achieved with three layers with around 92% accuracy.

- With one layer of ConvNet of size 32, activation function "ReLU" and padding to preserve the original dimension of the image after application of 3 × 3 filter. This model with a single layer gave an accuracy of around 80% (subject to change ±-2% on different runs of the same programs). This model overfits and between the 10th and 12th epoch, the same can be verified by looking at the graph where the two graph lines diverge (Fig. 2).
- With two layers of ConvNet, one of size of 32, and the other of size 64, and a max pooling layer. This graph looks a lot smoother and shows gradual progress as there are very few sharp edges (changes) in the graph. This model shows the highest accuracy of 92–93% on almost every run of the program (Fig. 3).

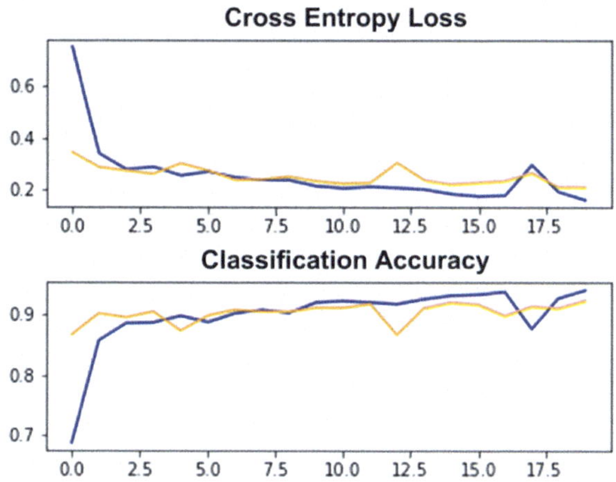


**Fig. 2** Result for single layer VGG

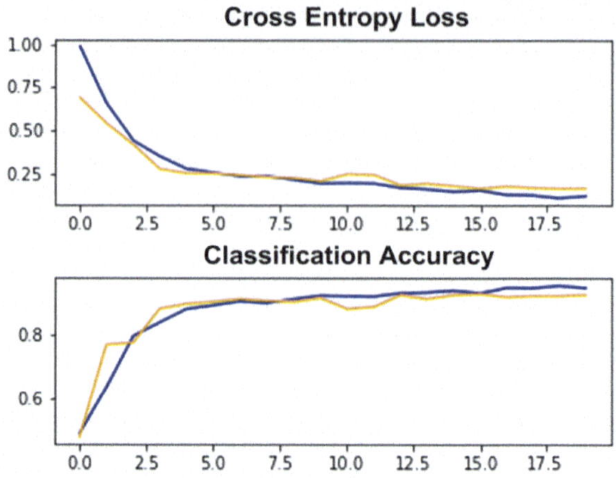


**Fig. 3** Result of using two layers of VGG

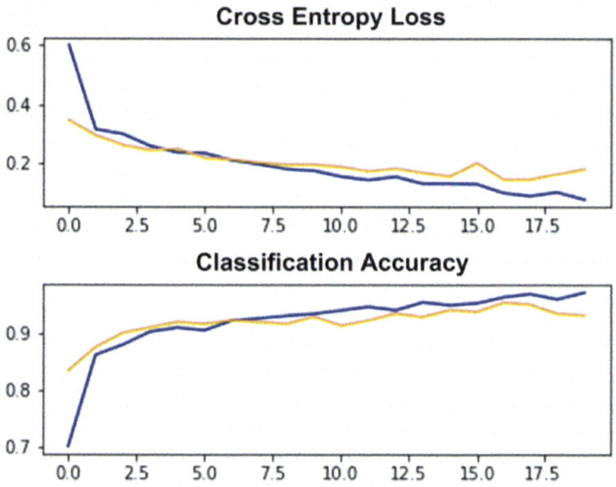


**Fig. 4** Result of using three layers of VGG

- With three layers of ConvNet: first one of size 32, second one of size 64, and third one of size 128, and all the other parameters being the same as earlier two models. This model shows 93% accuracy on the test data, but that is could be deceiving number because the model over fits really early, before the 7th epoch. This problem of overfitting is caused because the dataset only contains 3852 images. This shows that increasing the number of layers does not necessarily result in a better model (Fig. 4).

## 5 Conclusion

The result achieved with 1 layer of ConvNet was better than the ones achieved with three layers with around 92% accuracy. The models such as the one used in this study are trained on a dataset of images taken in a controlled environment. Such models lack generality and can easily perform poorly if images taken under different conditions are used. CNN is highly dependent on the quality of the dataset. Datasets must be designed very carefully so that no one feature or one specific case is more stressed than any other. With three layer of ConvNet, first one of size 32, second one of size 64, and third one of size 128, and all the other parameters being the same as earlier two models. This model show 93% accuracy on the test data, but that is could be deceiving number because the model overfits really early, before the 7th epoch. This problem of overfitting is caused because the dataset only contains 3852 images. This shows that increasing the number of layers does not necessarily result in a better model.

## References


1. Dubey, S.R., Jalal, A.S.: Fruit disease recognition using improved sum and difference histogram from images. Int. J. Appl. Pattern Recogn. **1**(2), 199–220 (2014)
2. Dubey, S.R.: Infected fruit part detection using K-means clustering segmentation technique. Int. J. Artif. Intel. Interact. Multimed. **2**(2), 65–72 (2013)
3. Dubey, S.R., Jalal, A.S.: Adapted approach for fruit disease identification using images. Int. J. Comput. Vis. Image Process. **2**(3), 51–65 (2012)
4. Wang, Q., Qi, F., Sun, M., Qu, J., Xue, J.: Identification of tomato disease types and detection of infected areas based on deep convolutional neural networks and object detection techniques. Comput. Intell. Neurosci **2**, 1–15 (2019)
5. Mohammed, N.A.: Evaluation of CNN, Alexnet and GoogleNet for fruit recognition Indonesian. J. Electr. Eng. Comput. Sci. **12**(2), 468–475 (2018)
6. Selvaraj, M.G., Vergara, A., Ruiz, H., Safari, N., Elayabalan, S., Ocimati, W., Blomme, G.: AI-powered banana diseases and pest detection. Plant Methods **3**, 15–20 (2019)
7. Suresh Raikwar, R.A.P.V.P.G.D.J.: Plant disease analysis and identification using multi SVM approach. Int. J. Adv. Sci. Technol. **29**(3), 4848–4856 (2020)
8. Agrawal, R.: Healthy leaves identification using multi SVM approach. Int. J. Adv. Sci. Technol **29**(3), 5992–5998 (2020)
9. Agrawal, R.: Identification of minimum support price using linear and logistic regression. Int. J. Adv. Sci. Technol. **29**(05), 2429–2435 (2020)
10. Geetharamani, G., Arun Pandian, J.: Identification of plant leaf diseases using a nine-layer deep convolutional neural network. Comput. Electr. Eng. **76**, 323–338 (2019)

# A Survey on Potential Techniques and Issues in 6G Communication

**Sweta**

**Abstract** While 5G is commercially introduced, worldwide researchers and academia have started looking beyond 5G and started working on 6G to support green communication and provide high quality of service and improved energy efficiency. In order to meet increasing demand of future application, there is need of improvement in existing architecture. 6G researchers are working on integrating traditional mobile network underwater, space, and aerial network to provide all time and everywhere network access. In this paper, potential techniques, terahertz, visible light communication, blockchain, molecular, quantum, and fundamental techniques are discussed for seamless and sustainable 6G network.

**Keywords** Terahertz · Visible light communication · Blockchain · Molecular communication · Quantum communication

## 1 Introduction

With the complete deployment of 5G in 2019, wireless communication has initiated beginning of true digital society with increased number of connected devices in comparison to 4G, high latency, data rate, and mobility. Looking back at the deployment of previous generation, it is found that it take about a decade for complete deployment of an existing generation and equal time is required for the development of succeeding generation. That is why when a generation enter into commercial utilization, research for next generation begin. Since 5G is in its initial stage of commercialization its high time for launch of successor of 5G, some countries already proposed their strategic plans for the development of 6G communication. Low carbon emission, energy efficiency, and increasing mobile traffic are the driving force behind evolution 6G wireless network. As 5G is expected to reach its limit in decade so in order to meet increasing demand latency and throughput. The main technical objectives for 6G communication are:

Sweta (✉)
Department of Electronics and Communication, GLA University, Mathura, India
e-mail: sweta.gla@gla.ac.in

 
T. K. Sharma et al. (eds.), *Soft Computing: Theories and Applications*,
Advances in Intelligent Systems and Computing 1381,
https://doi.org/10.1007/978-981-16-1696-9_19

- Improved global network coverage
- Secure and intelligent network connectivity
- Energy efficient network for advance devices
- Ultra-high data rate
- Ultra-low latency.

Many countries have already started to work on their strategic projects on 6G communication, Finland, in 2018, announced their 8 years 6G flagship program with an objective to design a 6G ecosystem [1]. USA is focusing on designing a terahertz-based 6G communication network, and UK and Germany are working on quantum technology-based 6G network [2]. In 2017, European Union launched three layer project to study basic of 6G technologies, in which main focus is on next generation forward error correction coding, channel modulation, and advance channel coding techniques for terabit wireless network.

## 2 Evolution of Mobile Communication Network

From the first analog communication in 1980s, communication industry has witnessed tremendous advancement in communication networks. It has been a long path and for researchers and academician to move from 1 to 5G and now beyond that. It took generations, standards, techniques that made it possible to attain from a simple voice transmission to full connected intelligent system. It take almost 10 year for a new generation to come into commercial utilization [3] (Fig. 1).

### *2.1 1G: First Generation*

First generation mobile network come in picture in 1980s, designed to provide voice service at a limited data rate of 2.4 kbps, based on analog signal transmission. 1G communication is the monopoly of state, and services are provided by state-owned operators only. Although voice quality in 1G is good, there are other drawback associated, no security, low transmission efficiency, and problem in hand off.

### *2.2 2G: Global System for Mobile Communication*

Along with better voice service 2G offer short message service (SMS) at a data rate of 64 kbps. Second generation of mobile communication was dominated by GSM, designed over digital modulation techniques as TDMA, CDMA.

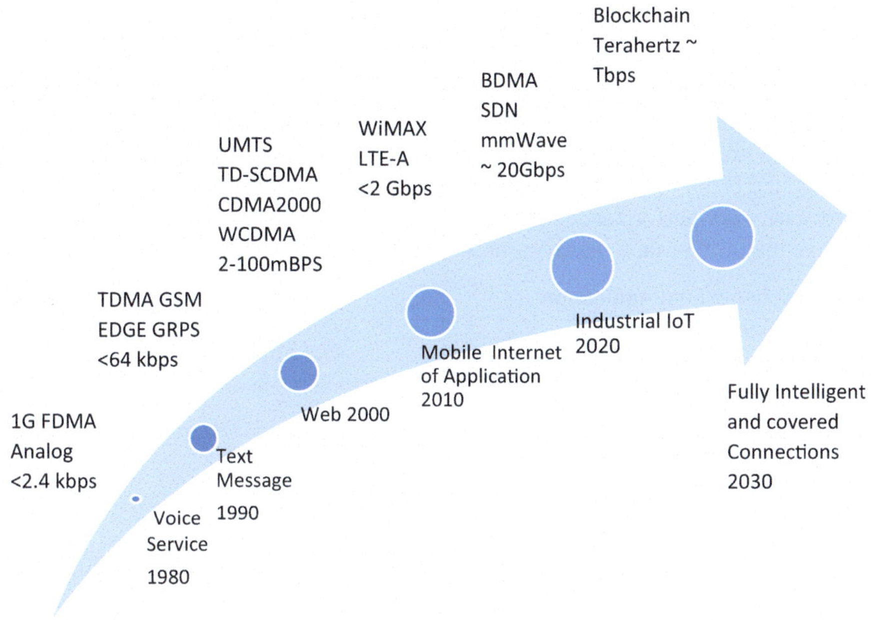


**Fig. 1** Evolution of mobile wireless communication

## 2.3 3G: Universal Mobile Telecommunication System

Third generation was proposed in 2000 with an objective of high-speed data transmission, and it provides data rate up to 2 Mbps. Additional service offered by 3G are high-speed Internet, web browsing, online TV streaming, navigational map as well as video service. To support global roaming, third generation partnership program (3GPP) was launched to decide technical standards, specification, and mobile system [4].

## 2.4 4G: Long-Term Evolution

Introduce in late 2000, it offer data rate up to 1 Gbps for downlink and 500 Mbps for uplink, improved spectral efficiency, reduced latency, and advanced application like video chat, high definition television, video broadcasting and provide terminal mobility to provide connectivity anywhere everywhere. LTE-A and WiMAX are the 4G standards, and it integrated advance technologies such as MIMO, OFDM, and CoMP.

### 2.5 5G: Green Communication

Main task of 5G communication is to provide massive connectivity, high data rate, latency, reliability, and energy-efficient network. 5G introduces use of millimeter wave which greatly increased data rate up to 10 Gbps. It integrated emerging technologies like D2D, millimeter wave, massive MIMO, spectrum sharing, and IOT to improve network performance.

Three major 5G usage scenarios proposed by IMT in 2020 are:

- Ultra-reliable communication
- Low latency
- Massive machine-type communication
- Enhanced mobile broad band.

## 3 6G Green Communication: Vision

6G is expected to advance the wireless technology by providing ultra-high speed, ultra-low latency, greater capacity, and nearly unlimited connectivity. 6G is expected to be available for commercial utilization by 2030 with high data rate at least 100–1000 times faster than 5G [5]. Moreover with rise in smart cities, society, buildings and house, there is increased demand for machine-to-machine and human-to-machine communication [2].

## 4 Promising Techniques for 6G Communication

See Fig. 2.

### 4.1 Spectrum Communication Technique

From the beginning of mobile communication in 1980s, spectrum usage played a vital role in fulfilling forever increasing demand for rise in data rate. With every next generation, we have seen tremendous expansion in spectrum resources. One of the main objective of 6G communication is to provide Tbps bit rate, and it is required to operate at higher frequencies at available spectrum and bandwidth. Terahertz and visible light are the most promising candidate for spectrum [6].

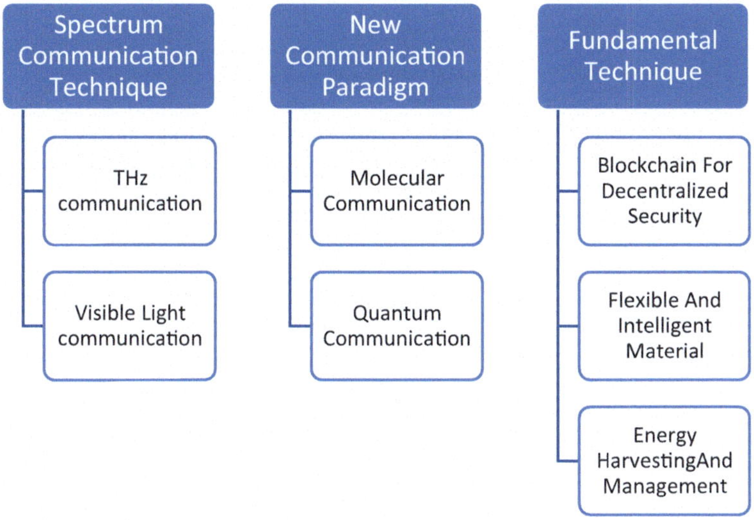


**Fig. 2** Promising techniques in 6G communication

### 4.1.1 Terahertz Communication

Terahertz operates for frequency range between 0.1 and 10 THz, and it is a spectral band between microwave and optical band. Three motivational characteristics of THz are:

- It supports data rate up to 100 Gbps with tens of GHz bands available in THz spectrum.
- Narrow beam and short pulse duration in THz wave offer more secure communication, limiting eavesdropping.
- Due to its ability to penetrate, some materials with small attenuation it find application in ultra-high speed and space communication.

There are tremendous challenges in mature application of THz wireless communication, and still there are aspect that need further research:

- Ultrahigh-frequency transistors and hardware material. With high thermal, electrical conductivity, and plasmonic effect, graphene emerged as most promising material to be considered for the same.
- Robust beamforming and scanning algorithm.
- Low complexity and power efficient hardware circuits.
- Designing of THz channel and noise modeling.
- Designing of low density channel code and energy-efficient modulation scheme.
- Role of nanoantennas in designing of ultra-massive MIMO system.
- Efficient synchronization scheme [7].

**Table 1** Data rate of VLC technology

| Lighting technology | Data rate |
|---|---|
| VLC based on phosphorus coated on blue LED | Up to 1 Gb/s |
| VLC based on RGB LED | Up to multi-Gb/s |
| VLC based on micro-LED | More than 10 Gb/s theoretically |
| Expectation from 6G | Hundreds of Gb/s or even Tb/s |

### 4.1.2 Visible Light Communication

Optical wireless communication is considered as one of the promising solution for RF-based communication and frequency range that includes infrared, visible light, and ultraviolet spectrum. Among all three visible light emerge as most promising spectrum with frequency range of 430–790 THz for optical wireless communication due to technological advantage and adoption of LED. High switching speed of LED at different light intensity makes it easy to encode data via emitted light in a variety of ways. For short range, VLC provides variety of advantages over conventional radio communication, Firstly, ultra-high bandwidth (THz) with free and unlicensed spectrum. Secondly, inability to penetrate opaque obstruction limit information transmission within an area that reduces inter-cell interference and enhances security. Thirdly, it does not generate harmful electromagnetic radiation thus made it suitable aircraft and hospitals where health is major concern [8] (Table 1).

## *4.2 New Communication Paradigm*

### 4.2.1 Molecular Communication

Molecular communication is a nature inspired new communication paradigm which uses biochemical signals for information transmission. Biochemical signals consist of lipid vesicles and particles of few nanometers to micrometers in size and are capable to propagate in aqueous or gaseous medium. Limitation of antenna size in radio communication at nanoscale dimension can also be overcome by molecular communication which is an ideal intrabody nanonetwork. Being biocompatible MC consumes a very small amount of energy for generation and propagation. Along with proposed advantages, MC has to overcome two major challenges, i.e., firstly, interface development between chemical and electrical domain. Secondly, security enhancement methods [9].

#### 4.2.2 Quantum Communication

In quantum communication, data is encoded using photon and quantum particles in quantum state which cannot be accessed without tampering it due to quantum principles like inalienable law and correlation of entangled particles. Superposition nature of qubits enhance data rate in QC. There are various branches of QC are:

- Quantum secure direct communication
- Quantum key distribution
- Quantum teleportation
- Quantum secret sharing.

Repeaters in QC divide long distance into segments and correct both photon loss and operational error, results in improved long distance communication [10].

### *4.3 Fundamental Techniques*

#### 4.3.1 Blockchain for Decentralized Security

Blockchain is considered as one of the most promising techniques for security, decentralized tamper resistance and anonymity made it suitable for various applications. Secure access to critical data and untamable distributed ledger containing all data provide stronger security in communication network. Blockchain-based decentralized control mechanism helps to establish direct communication link between network entities reducing administrative cost. It allows roaming across different networks and operators by integrating individual systems of different operators and proving unified authentication and authorization mechanism and billing system. Also, spectral efficiency of a communication system can be improved by integrating blockchain in spectrum sharing [11].

#### 4.3.2 Flexible and Intelligent Materials

For ultrahigh-speed communication, there is need of material that can support high frequency and temperature characteristics of emerging technologies. Novel materials such as indium phosphide, silicon germanium, gallium nitride, and graphene are already in que to be used in next generation communication. Beside these fluid materials are considered for designing of frequency reconfigurable antenna to improve flexibility and metasurface and metamaterial are an option to be considered in radio-controllable wireless communication.

### 4.3.3 Energy Harvesting and Management

With increasing demand for AI and IoT, there is very much need of energy-efficient communication techniques and equipment's. One of the emerging technique in energy harvesting is symbiotic radio which integrate passive backscatter devices with active transmission system [12]. Another promising technique is smart energy management that maintain a balance between energy requirement and supply [13].

# 5 Problems to Be Resolve in 6G Communication and Their Promising Solutions

As 6G communication is at its infant stage their will be multiple issues that it has to face and suppose to provide possible solutions for those problems. Among all persistent issue, three major issues that require prime focus of researchers in any network are power supply, hardware design, and network issues. All efforts are made to improve these three parameters in every next generation network (Fig. 3).

## *5.1 Power Supply Issue*

With increase in number of users, there is expectation from 6G to provide flexible and energy-efficient networks to support trillions of users. In present scenario, very few smartphones are capable of providing battery life of more then one day which is the bottleneck of existing communication network. For continuous growth of mobile devices in 6G communication, it is necessary to design an energy-efficient and simplified signal processing architecture. To deal with this new mobile architectures integrated with energy harvesting and wireless power transfer techniques can be developed. Moreover, integration of optimized power supply methods and energy-efficient wireless transmission technique is a promising approach for 6G network.

## *5.2 Network Security*

Network security is one of the critical issues for existing as well as 6G wireless network. Extended versions of physical layer security techniques employed in 5G such as LDPC-based secure massive MIMO and mmWave techniques can be employed in promising techniques of 6G like THz band application. Proper implementation and management of physical layer security along with network layer security are essential for next generation communication. For STIN distributed key

**Fig. 3** Problems and their possible solution in 6G communication

management mechanism jointly considering multicast and unicast communication, key management can be considered as promising solution [14].

### 5.3 *Hardware Design Issue*

For high-frequency communication network like terahertz band, there is need to attend miniaturization level upto micrometer or nanometer of antenna to be used. Integration of optical and electronics module, i.e., optoelectronics, can be considered as a promising solution for optimize utilization of on-chip integration, advance antenna techniques, and high-speed semiconductor for designing of high-frequency communication system.

## 6 Conclusion

Increase in number of users with passing time and need for high data rate with energy-efficient devices and wireless communication network, which leads to next wireless evolution toward 6G. High data rate, improved QoS, high network security, and energy-efficient device and network are the desired features of 6G. We started with evolution of wireless communication and discussed various features offered by individual generation and development with each successive generation. Then we focused on promising and emerging technologies that can be considered in 6G communication like THz and visible light communication, molecular and quantum communication, backscatter communication, blockchain, flexible and intelligent materials are explained that has the potential to fulfill all expectation from 6G communication. Challenges in the path of emerging communication network like hardware design, power supply, and network security are also discussed with their possible solutions. This survey is expected to enlighten path of future research work in 6G communication.

## References

1. Katz, M., et al.: 6 Genesis flag ship program: building the bridges towards 6G enabled wireless smart society and ecosystem. In: Proceedings of IEEE 10th Latin-American, conference communication (LATINCOM). pp. 1–9 (2018)
2. Yang, P., et al.: 6G wireless communication: vision and potential techniques. IEEE Netw. **33**(4), 70–75 (2019)
3. David, K., et al.: 6G vision and requirements: Is there is any need for beyond 5G. IEEE Veh. Technol. Mag. **13**(3), 72–80 (2018)
4. Gupta, A., et al.: A survey of 5G network: architecture and emerging technologies. IEEE Access **3**, 1206–1232 (2015)

5. Cacciapuoti, A.S., et al.: Beyond 5G: THz-based medium access protocol for heterogeneous network. IEEE Commun. Mag. **56**, 110–115 (2018)
6. Strinati, E.C., et al.: 6G: The next frontier: from holographic messaging to artificial intelligence using subterahertz and visible light communication. IEEE Veh. Technol. Mag. **14**(3), 42–50 (2019)
7. Tekbyk, K., et al.: Terahertz band communication systems: challenges, novelties and standardization efforts. Phys. Commun. **35**, Article no. 100700 (2019)
8. Cho, S., et al.: Enhancement of physical layer security with simultaneous beamforming and jamming for visible light communication systems. IEEE Trans. Inf. Forensic Secur. **14**(10), 2633–2648 (2019)
9. Nakano, T., et al.: Methods and applications of mobile molecular communication. Proc IEEE **107**(7), 1142–1456 (2019)
10. Gyongyosi, L., et al.: A survey on quantum channel capacities. IEEE Commun. Surv. Tuts. **20**(2), 1149–1205 (2018)
11. Ling, X., et al.: Blockchain radio access network (B-RAN): towards decentralized secure radio access paradigm. IEEE Access **7**, 9714–9723 (2019)
12. Long, R., et al.: Full-duplex back scatter communication in symbiotic radio systems. IEEE Access **7**, 21597–21608 (2019)
13. Zhou, Z., et al.: Software defined machine to machine communication for smart energy management. IEEE Commun. Mag. **55**(10), 52–60 (2017)
14. Yao, H., et al.: The space terrestrial integrated network: an overview. IEEE Commun. Mag. **56**(9), 178–185 (2018)

# Impact of COVID-19 in Architecture: A Survey-Based Method for Measuring and Understanding

**Bindu Agarwal, Sangita Rawat, Itishree Rana, Simran Agarwal, Anubhuti Aggarwal, and Promila Sharma**

**Abstract** As we crawl into the second decade of the twenty-first century, economy of several countries has hit rock bottom due to widespread of the pandemic "COVID-19". It is crucial to understand and comprehend that architecture is a powerful visual and objective tool that influences the economy of a region. Hence this research paper's objective is to unfold the impact of COVID-19 on architectural education and industry. To achieve this objective, a survey was conducted among more than 100 architecture students, academicians and professionals focusing on various factors such as finances, mental health, technology etc. In this paper, the relation between two variables is measured by Statistical Package for the Social Sciences (SPSS) software. The survey result depicts the major reasons recorded affecting the architectural/construction industry that were lack of labor, coordination issues, unfavorable government policies, and unavailability of material. Architectural students and academicians both strongly believed that education should not continue via online mode. It was noted that majority were affected by mental health issues/depression due to unemployment and financial losses.

**Keywords** COVID-19 · Mental health · Architecture · Construction · Unemployment · Financial loss

---

B. Agarwal (✉) · S. Rawat · I. Rana
School of Planning & Architecture, Graphic Era Hill University, Dehradun 248001, Uttarakhand, India
e-mail: bagarwal@gehu.ac.in

S. Agarwal
Shreeves Chaman Vihar, Phase-II, Dehradun 248001, Uttarakhand, India

A. Aggarwal
Graphic Era Hill University, Dehradun 248001, Uttarakhand, India

P. Sharma
Department of Life Sciences, Graphic Era University, Dehradun 248001, Uttarakhand, India

 
T. K. Sharma et al. (eds.), *Soft Computing: Theories and Applications*,
Advances in Intelligent Systems and Computing 1381,
https://doi.org/10.1007/978-981-16-1696-9_20

## 1 Introduction

World and its economy are always seen running at a faster pace with each day, but all stopped. It felt that everything had a pause, but not the virus. The novel coronavirus disease is believed to spread at the fastest pace with humans as its carrier. More than 22 million people across the world got infected and 788,000 people died (as on 20/08/2020). COVID-19 is an emerging, rapidly evolving situation. This pandemic has been reshaping the lifestyle of all around the globe. It has reshaped the ways we live, the ways we work. All the jolly smiles in this pandemic are now hidden behind masks and locked inside the rooms. In such scenario, we are forced to lock ourselves in the spaces which we were neglecting for so long. The only vaccine available to us is our closed spaces dwelled with such intricacies that we live, work and enjoy in those spaces only. Our houses are presumably the safest places now. Rethinking traditional building and structure designs has engaged architects and designers worldwide. The desire to prepare our homes for calamities or pandemics is no longer a surprise. Homes are now not just housing, but a multifunctioning structure, converted into offices, schools, parks, and exercise area. The form is not following function only, but also fear of infection. It is the scary contamination that is giving us the kind of spaces we want to live in. Going back to 1930s, we would find that rigid geometric buildings with long and light-colored walls, wide roof terraces, were then adopted for treating tuberculosis. Buildings were believed to function as a medical instrument. We were always focusing on architecture influencing environment or vice versa, but now the alarming time asks for architecture influencing health or vice versa. Architecture is beyond artistic prudence; it forms a bigger system by being a fundamental part of it.

The virus traveled from Wuhan, China to the entire world in no time and forced people to maintain social distance; “2 gaj ki duri”. Thus, every country decided for lockdown to minimize the spread of this disease. The lockdown affected all the sectors but education sector was impacted majorly. According to the UNESCO report, it had affected more than 90% of total world’s population during mid-April of the year 2020. More than 32 crores of students and youth in India and 120 crores worldwide have been affected by the nationwide lockdown. No toddlers were seen hopping in the playgrounds, no friend group was doing group assignments in college cafeterias. Silence prevailed, no husting and bustling was there. The pandemic held all classes, examinations, college semesters, internships at stake. Various initiatives were taken to regulate online classes for students and various e-learning software were launched. Zoom, Google Meet, Skype somehow replaced the blackboards and became a medium of learning which teachers across the nation adopted. Architecture education always had its roots to numerous workshops (Fig. 1), site-visits, design studios, discussions, (Fig. 2) lectures, and much more. While learning about architecture, nothing can replace practical experience, but the current pandemic situation took architectural education under attack. This is the creative trait of education which was emphasized on the studio format—student focused on independent projects, and

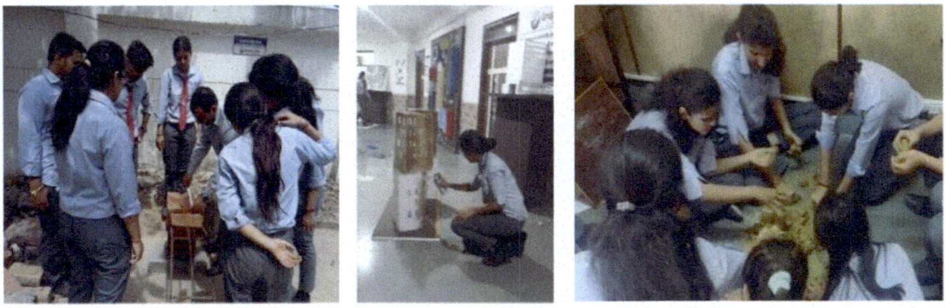

**Fig. 1** Architecture students' workshops during normal days

**Fig. 2** Jurys and discussions of Studio subjects during normal days of Bachelor of Architecture students

spent a far more time discussing it with their guides/professors, eventually producing a self-justified design.

Where architectural education inspired students to think out of the box, pandemic made them stick to their cubicles and their screens only. Tracing sheets, pencils, markers, colors were seen no longer lurking on the tables, but only on the screens.

## 2 Literature Review

The year 2020 came with the outbreak of COVID-19 and raised several questions before architects and urban planners regarding their role to make a more sustainable built environment in the post-pandemic world. This pandemic situation has not only impacted people's daily habits and behavior but also challenged our cities, buildings, and interiors to evolve and sustain in the "**New Normal World**".

The COVID-19 pandemic acted as a catalyst to boost people's awareness and concern about climate change and environment. The reduced air pollution level and change in environment during COVID-19 shutdown/lockdown in different parts of the world gathered the attention of all and gave us time to introspect.

The researcher also analyzes the impact of common pathogen and transactions of microbes and also provides details in their research paper so that the owners, administrators, and individual due to COVID-19 disease closes their schools, educational institutes, colleges, universities, corporate, organization government and non-government and do complete lockdown and social distancing is the only situation for the problem.

## 3 Materials and Methods

The scope of this study is limited to architectural/construction industry of India. This is an electronic survey-based study done via Google Forms and more than 100 responses have been recorded. The participants are majorly students, academicians/faculties, and professional from architectural field (49% are academicians and students while 51% are practicing architects). In this paper, the relation between two variables is measured by Statistical Package for the Social Sciences (SPSS) software.

## 4 Results

The results are divided into three sections: practicing architects, architectural students, and architectural academician/faculties, respectively. Since being tech-savvy is a challenge for most of the people in a developing country like India, more than 50% population comes from rural background. The result of respondents is to show the knowledge of software to work from home on the rating scale of 1–5 (5 for extremely good and 1 for extremely bad). The survey result reveals that maximum respondents have chosen 4 and minimum have chosen 1 as their response (Fig. 3),

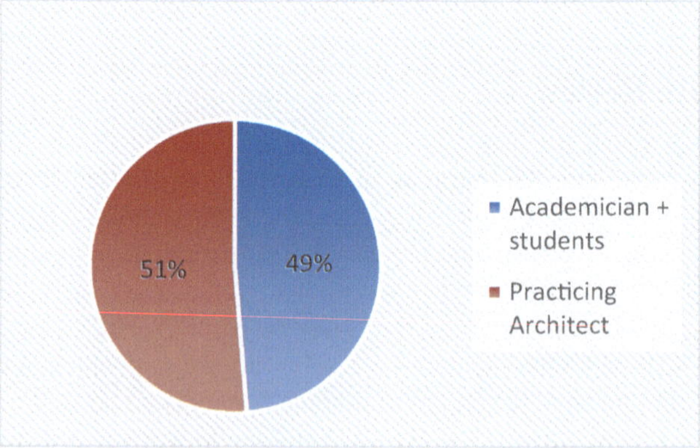


**Fig. 3** Jurys ratio of respondents

i.e., most of the people are tech-friendly and able to work from home. The study/work from home is not possible without the knowledge of software and technology and maximum people join hands with technology and learn software to perform as much as possible from there side in difficult situation, i.e., Corona.

## *4.1 Practicing Architects Opinion*

For practicing architects, the survey was done on a scale from 1 to 5, to check whether the pandemic became a boon or bane. The scale was ranged from strongly disagree to strongly agree. 45% practicing architects strongly believe that the business was impacted majorly, while 2% strongly disagree that it was not impacted, whereas 20% architects say that it maybe (Fig. 4).

We also calculated if the students felt depressed during this pandemic. It was done on a scale of 1–5, ranging from strong disagreement to strong agreement. 21.6% students strongly agree that they felt depressed, while 10.8% strongly disagree. However, 24.3% students took this with a balanced approach, with no agreement or disagreement (Fig. 5).

During the nationwide lockdown, the worst situations were seen. There was shortage of labor, materials, and supervision. The survey highlights all the factors which affected the practice during the pandemic. 11.9% people say that it was the shortage of construction materials, while 26.9% say that it was shortage of labor that impacted the practice. Certain guidelines were set by both the central and the state governments to enforce the lockdown and stop the spread of pandemic, 23.9% say that these policies affected the practice. Work from home was another trend that people all over the world adopted, so 22.4% believe that coordination with staff became a tedious task and then became a cause. All the factors were in some or other way responsible, 44.8% say that all the above-mentioned factors collectively affected the architectural practice (Fig. 6).

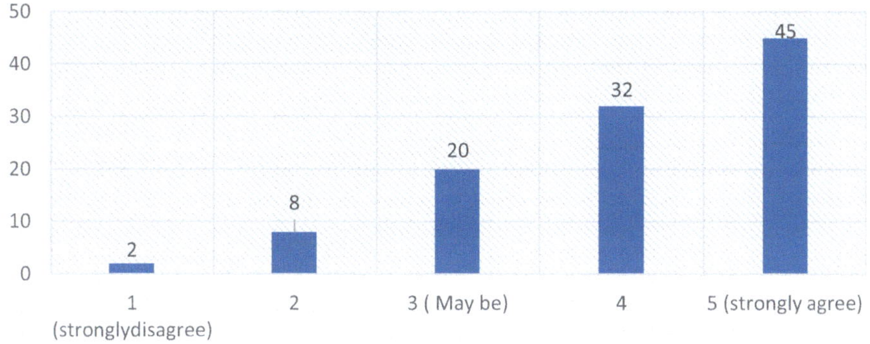


**Fig. 4** Impact on practicing architects' profession

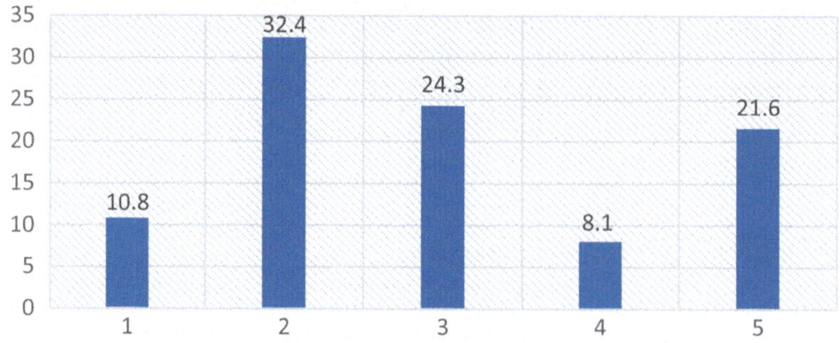


**Fig. 5** People feel depressed

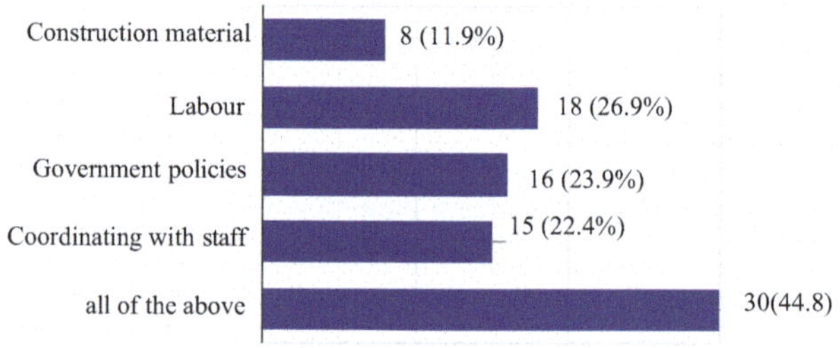


**Fig. 6** Issues affecting the practicing architects business

## 4.2 *Bachelor of Architecture Students' Feedback*

Students are so used to of exploring and visiting sites, attending studios and lectures. But pandemic stopped all and made students stay at home for a long. It majorly impacted the education system. Survey highlights a number of issues that students faced during this time. 39.5% students say that lack of discussions affected their studies, and 26.3% say that unavailability of printouts were the cause. More than 47% students say that network issues were the main cause of the discomfort. Also, 47.4% students believe that not going to sites and having on site experience was the main issue, whereas 55.3% students say that all the above factors were faced during this time (Fig. 7).

Architectural education is one trait that requires a lot of practical understanding. Survey examined if technology can replace the conventional educational system, 50% students strongly disagree and say that it cannot be continued over technology and requires studios and lectures, though 27.8% students say that yes it can be continued over technology (Fig. 8).

Architecture is one such profession, in which nothing can ever replace the practical experience and design studios. Endless discussions and sleepless nights pay a long way in architectural journey. The switch to technology was both friendly and tedious. But students were asked if pandemic elevated their architectural understanding since

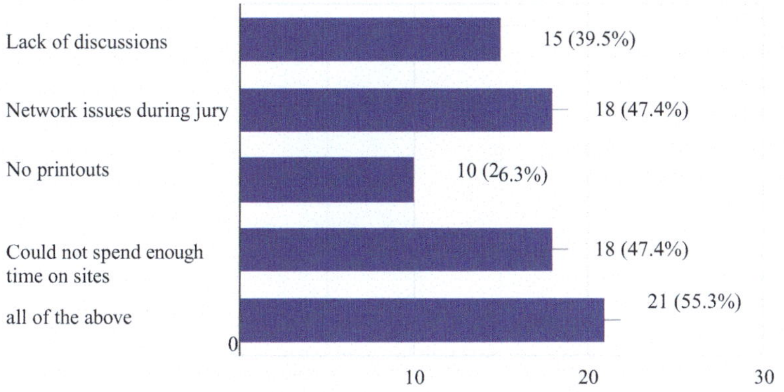


**Fig. 7** Major issues faced by Bachelor of Architecture students during the COVID Studio classes

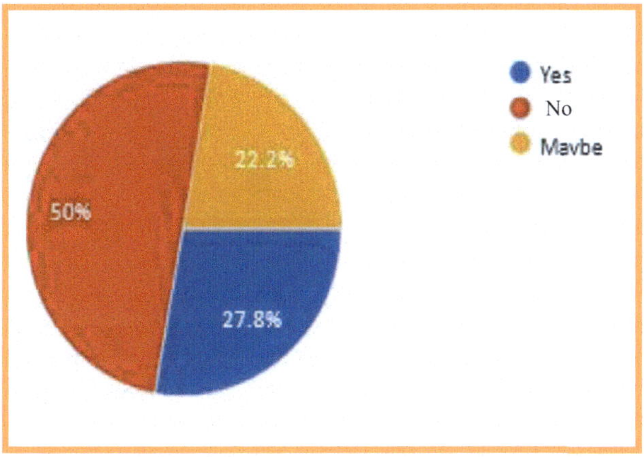


**Fig. 8** Students opinion for the architectural education can be continued over technology or it requires studio and lecture

they could explore a lot of literature and try their hands, or the understanding got decreased since no practical experience was there. 56% students firmly believe that their understanding got decreased while only 16% say that the understanding got increased. 28% students that it remained unaffected (Fig. 9).

## *4.3 Architectural Academicians View Point*

Faculties were asked about their views on class strength in online class. The survey was done on a scale of 1–5. 41.2% say that it was more during online classes and 11.8% say that it was very less (Fig. 10).

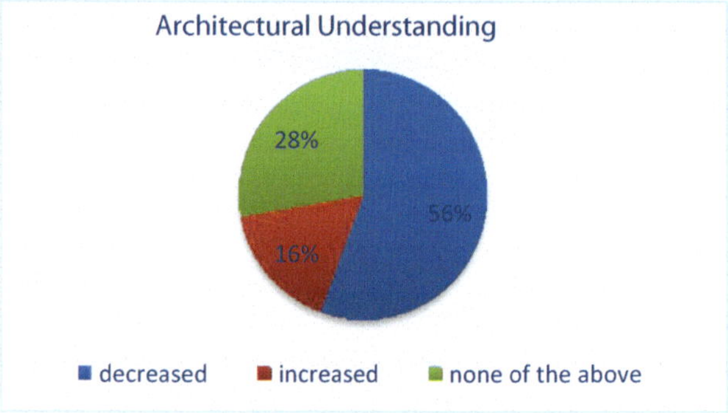


**Fig. 9** Students' opinion about architectural understanding

**Fig. 10** Architectural faculties review about student strength

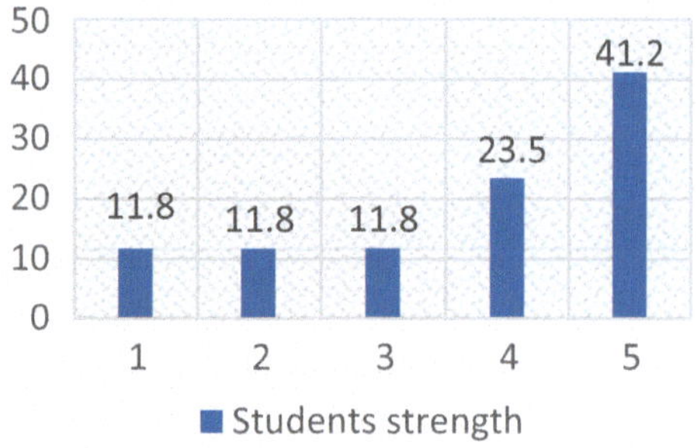


Students were given assignments and they were asked to submit on given dates just like in regular classes. Academicians were asked if the students did late submissions. The scale was set ranging from 1 to 5, from strongly disagree to strongly agree. 25% strongly believe that they did, whereas 18.8% strongly disagree for the same. Only 12.5% say that the submissions given by students were fairly on time (Fig. 11). Maximum is given to 4 rating, i.e., 37.5% and respondents agree with the submission and they believe students submitted on time.

**Fig. 11** Architectural faculties review student's submission

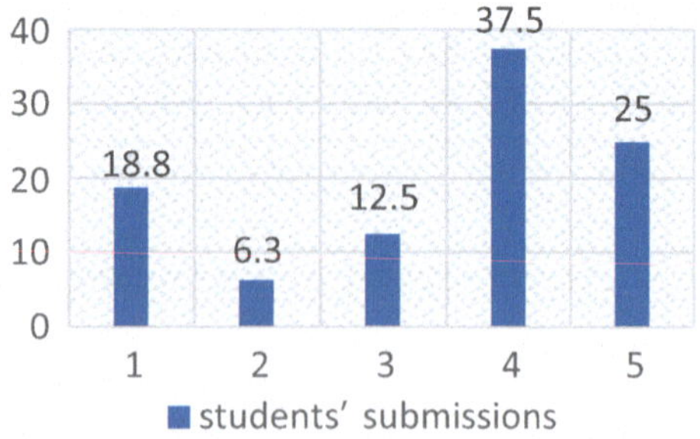

Academicians had to deliver classes and lectures over online platforms such as Zoom, Teams, and Google Meet. They must have faced certain difficulties, since they were all new to this, and every teacher has a different teaching methodology. The survey highlighted the scale ranging from 1 to 5, from bad to good. The analysis depicts that only 5.9% say that it was extremely good and they faced no difficulty where as 17.6% say that they faced a lot of difficulty. 29.4% people say that it was neither too difficult nor too easy.

Survey highlighted whether the academicians were happy or unhappy with increased work and decreased salaries, in other words can say deduction in salary but in work by the employers. On scale of 1–5, from unhappy to happy, the maximum 38% respondents choose 1 and strongly disagree with salary deduction because work load is same and work from home is also not easy. The work doing from home and technology load on the employer is also increased. Many of employers have to buy laptops and broadband connections (Fig. 12).

Technology was not always a friend to all, bit by bit all were adopting and learning it. Survey highlights how technology treated all ranging on a scale of 1–5 from bad to good. 35.1% say that the switch from conventional practices to technology was fairly good and only 2.7% say that it was not. 29.7% say that the switch was mid-way, neither too good nor too bad (Fig. 13).

Survey enquired the academicians about the problems they faced with students and their work outputs, three choices have been given bad, neither good nor bad. The results achieved were fairly balanced, as 11.1% believe that the work produce was

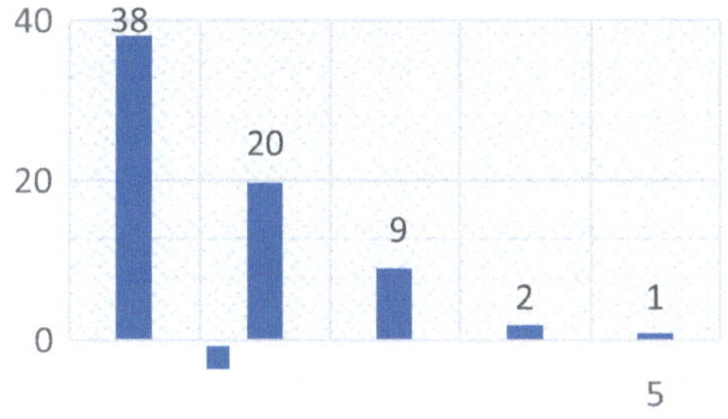


**Fig. 12** Faculties opinion for salary deduction

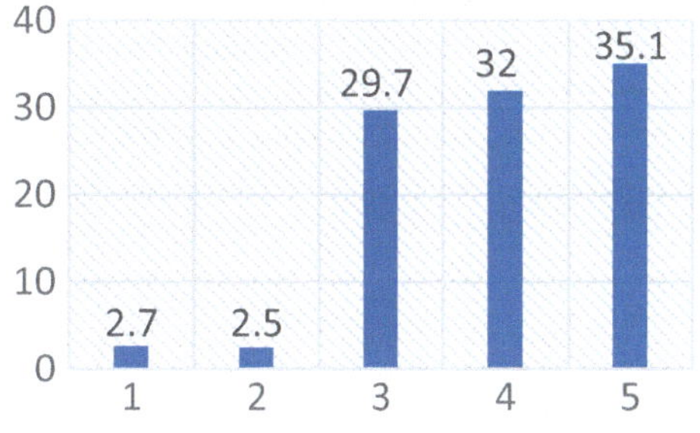


**Fig. 13** Architectural faculties opinion for online teaching

**Fig. 14** Students submission

good and 11.1% say that it was bad. 22.2% marked it on a mid-level and say that it was neither good nor bad, but was acceptable (Fig. 14).

## 5 Conclusion

Architecture is always believed to be an amalgamation of art and science, but now the pandemic also made technology integrate with it. However, 50% students think that architecture education is difficult to be continued over technology, it must follow the conventional "guru-shishya parampara", in which students come from far places to their study places and spend innumerable hours with their teachers and professors, imbibing the learnings and practices of the noble profession. Architecture is a collaborative process in which the fraternity cannot stay static.

Using SPSS–IBM software, the Pearson correlation between students view of architectural education affect during the lock down due to no workshops/seminars/visits/lectures/exhibitions and Corona has increased or decreased architectural understanding. The result show Perfect (Table 1) the value is near $\pm 1$, then it said to be a perfect correlation: as one variable increases, the other variable tends to also increase (if positive) or decrease (if negative). Results are high degree and if the coefficient value lies between $\pm 0.50$ and $\pm 1$, it is said to be a strong correlation.

**Table 1** Table captions should be placed above the tables

| Correlations | | | |
|---|---|---|---|
| | | Understanding | Result |
| Understanding | Pearson correlation | 1 | 0.449** |
| | Sig. (2-tailed) | | 0.000 |
| | N | 108 | 108 |
| Result | Pearson correlation | 0.449** | 1 |
| | Sig. (2-tailed) | 0.000 | |
| | N | 108 | 108 |

**Correlation is significant at the 0.01 level (2-tailed)

**Acknowledgements** Authors acknowledge the immense help received from the scholars whose articles are cited and included in references to this manuscript. The authors are also grateful to authors/editors/publishers of all those articles, journals, and books from where the literature for this article has been reviewed and discussed.

# References

1. Tate, J.E., Burton, A.H., Boschi-Pinto, C., Steele, A.D., Duque, J., Parashar, U.D., et al.: Global, regional, and national estimates of rotavirus mortality in children <5 years of age, 2000–2013. Clin. Infect. Dis. **62**(12), S96–S105 (2016)
2. El- Senousy, W.M.: Monoplex nested and semi-nested reverse transcription polymerase chain reaction for G and P genotyping of human rotavirus group A in clinical specimens and environmental samples. Int. J. Current Res. Rev. ISSN: 2231-2196 (Print), ISSN: 0975-5241(Online) Accepted: 07.08.2020, Published: 22.08.2020
3. Commercia Design: Impact of Covid-19 on Architecture Industry. https://www.commercialdesignindia.com/6020-impact-of-covid-19-on-architecture-industry
4. Munster, V.J., Koopmans, M., Van Doremalen, N., Van Riel, D., De Wit, E.: A novel coronavirus emerging in China-key questions for impact assessment. New England J. Med. (Posted: 2020)
5. Novel Coronavirus (2019-nCoV) Situation Reports: Organization WH (Posted: 2020)
6. Zhang, Y.: Hubei raises health emergency response to highest level. China Daily (Posted: 24 Jan 2020)
7. Rosenbaum, P.R., Rubin, D.B.: The central role of the propensity score in observational studies for causal effects. Biometrika **70**(1), 41–55 (Posted: 1983)
8. Internet: https://www.commercialdesignindia.com/6020-impact-of-covid-19-on-architecture-industry
9. Internet from https://www.iosrjournals.org
10. Student papers from 20-Aug-2011 Submitted to Assumption University on 20 Aug 2011. Internet. https://www.tandfonline.com/doi/pdf/10.1080/00220485.1991.10844710 Sources: Wikipedia and The New York Times
11. Dietz, L., Horve, P., Coil, D., Fretz, M., Wymelenberg, K. (2019). Novel Coronavirus (COVID-19) pandemic: built environment considerations to reduce transmission (Jan 2020) (2020).

Retrieved from American Society for Microbiology: https://msystems.asm.org/content/5/2/e00245-20
12. Harrouk, C.: Shift Architecture Urbanism Creates Hyperlocal Micro Markets that Operate During COVID-19 Shutdowns (2020). Retrieved from Archdaily: https://www.archdaily.com/936856/shift-architecture-urbanism-creates-hyperlocalmicro-markets-that-operate-during-covid-19-shutdowns
13. Honey-Roses, J.A.: The Impact of COVID-19 on Public Space: A Review of the Emerging Questions (2020)
14. The Covid-19 Crisis and Clean Energy Progress, Impact on Sectors and Technologies IEA (2020). https://www.iea.org/reports/the-covid-19-crisis-and-clean-energyprogress/buildings
15. Kronenburg, R.: Portable Architecture (1996)
16. Lubell, S.: Commentary: past pandemics changed the design of cities. Six ways COVID-19 could do the same. Los Angeles Times (2020). Retrieved from the planning report
17. Wu, X., Nethery, R.: A National Study on Long-Term Exposure to Air Pollution and COVID-19 Mortality in the United States (2020). Retrieved from Harvard. Retrieved from Harvard.edu: https://projects.iq.harvard.edsu/covid-pm

# Classification of Human Activities Using Support Vector Machine—A Review

**Sarvesh Kumar Swarnakar, Harshit Agrawal, and Ankita Goel**

**Abstract** This paper presents work on human activity detection dependent on smartphones with installed sensors and learning algorithms that was done not long ago. Perceiving human activity has earned a great amount of interest and has brought noteworthy research in recognizing logical data helpful to human activity recognition. Smartphones have installed inbuilt sensors that are fit to detect logical data of its clients, having a wide range capability of network connections. Installed sensors which include gyroscope sensor, tri-axial linear accelerometer, tri-axial accelerometer and direction sensors are utilized for movement information assortment. In this paper, we will talk about the classification algorithms used for classification of human activity and existing innovations relating to smartphone-based research in which human activity recognition will be examined.

**Keywords** Human activity recognition · Machine learning · Classification · Support vector machine

## 1 Introduction

Expansion of life expectancy, a causative factor for humanity's endurance to old age, is basically influenced by advancement in medical science and in treatment of ailments. It is assessed that by 2050 more than 2.01 billion individuals will be over 60 years [1]. With age, individuals become powerless against motor and physical inabilities, limiting their support in everyday activities. Seniors having social and physical inactivity are increasingly vulnerable to chronic diseases and disconnection with society, which cause negative results on mental, heart, lungs, therefore, mortality. Thus, on long-term care, large expenditure is spent; hence, social and healthcare administrations are searching for choices to adapt to the full-scale monetary difficulties related to an expanding ageing populace.

S. K. Swarnakar (✉) · H. Agrawal · A. Goel
Department of Information Technology, Galgotias College of Engineering and Technology, 1, Knowledge Park – III, Greater Noida, Uttar Pradesh, India
e-mail: sarvesh.swarnkar@galgotiacollege.edu



T. K. Sharma et al. (eds.), *Soft Computing: Theories and Applications*, Advances in Intelligent Systems and Computing 1381,
https://doi.org/10.1007/978-981-16-1696-9_21

A plenty of studies have proposed different strategies for activity monitoring problems using wearable sensors [2], remote sensor systems [3] and video recorders. Some used computer vision-based techniques for the classification of human activities but faced various restrictions related to environment like variable lighting. Some included recognition using body-fixed wearable sensors offering a more feasible and relatively cheap method to gauge human movement. In Ambient Assisted Living (AAL), activity detection and patient monitoring using smartphones has given more promising answers because of different features of cell phones like versatility, low cost, Bluetooth, Wi-Fi facility and 3G/4G communication feature.

Human activity recognition (HAR) has evoked significant enthusiasm for the different research zones and applications because of its potential use in proactive processing. Proactive processing is an innovation that expert—effectively foreseen individual's need in a set of circumstances, for example, medicinal services or life-care and takes suitable actions for their sake. Human activity recognition is also used to encourage physical workout and normal exercises for correction of different postures like children's sitting gesture and identification of different behaviours for early detection of any kind of illness. In [4], many more instances of applications are observed. We are in a need to have frameworks to perceive the human exercises so as to enhance the life's quality and screen the well-being of patients by recording the different exercises people do in day-to-day life in order to improve their well-being and give cautions in emergency. Distinguishing human activities from smartphones, data set has proved to be a mind-boggling task because of its enormous components.

Recognition can be cultivated by using the data recovered from different sources, for example, natural or body-worn sensors. A few methodologies have adapted devoted sensors for movement in various body parts, for example, waist, hands, chest, also thighs accomplishing great classification performance. These sensors are generally intolerable for the regular client and do not give a long-haul answer for activity monitoring (e.g. sensor repositioning posterior of dressing). Smartphones are raising new research open doors for applications that are dedicated to humans, where the client is the major source of collecting data and the cell phone is the first-hand detecting apparatus. Most recent gadgets accompany inserted sensors, for example, double cameras, microphones, accelerometers, gyroscope and many more. The utilization of mobile phones having inertial sensors is an elective answer for human activity recognition. These mass-advertised gadgets give an adaptable, reasonable and independent answer for consequently and subtly screen Activities of Daily Living (ADL) while additionally giving communication services.

## 2 Literature Review

All over the world, the human activity recognition, that uses sensors to perceive human activities, has been read for quite some while to deliver a much more basic framework with better accuracy. To create a preferred integrated framework having full-time and real-time supervision smartphone is an apt solution, but still there are finite number of tasks that explore HAR framework manufactured using the smartphone. The human activity detection system in advanced mobiles vows to begin another course in monitoring and medicinal services as well as in different fields. Consequently, medical establishments are effectively looking for cheaper arrangements. The need for the therapeutic partner is no longer required and patients can be remotely checked. So as to build the practicality of such applications, there are inquiries going on about human activity recognition incorporating walking, up, down, sitting, standing and lying. These days, smartphones are working with numerous sensors to enhance the communication ability for clients, incorporating accelerometer and gyroscope sensors. In this way, the possibility of use of inertial sensors to make a smartphone application is becoming increasingly reasonable for detection of human activities.

In the past few years, development related to the stream of deep learning has been witnessed, which also resulted in applications of HAR dealing with deep learning algorithms attaining terrific efficiencies. Main advantage of deep learning algorithms is its stacking structure which makes algorithms learn representative features automatically but it also has a drawback of demanding a large collection of labelled data for training of the model. Not only this, these algorithms have high computational load making it difficult to be used in real-time action recognition systems and applications. The enormous gathered information gives numerous significant information to perceive the human physical activities. There are other methods too where the information gathered is utilized in different supervised learning algorithms for the classification process. In [3, 5–7] algorithms like $K$-nearest neighbour (KNN), random forest, Naive Bayes algorithm and support vector machine that are applied for classification. Where dynamic activities can be effectively characterized, non-dynamic activities still present misclassification covers and low precision. Accuracy varies from 56 to 90% for all the activities for different algorithms. A basic design of human activity recognition system in shown in (Fig. 1).

## 3 Methods

### *3.1 Smartphone Sensors*

A wide range of sensory gadgets are used to decide the position and direction of an object. Gyroscope and the accelerometer are the most well-known of these sensors. Despite the fact that they are alike in purpose, they measure various things which are not the same. Thus, when consolidated into a sole gadget, they can make a ground-breaking cluster of data.

**Accelerometer**—An accelerometer is one of the fundamental devices among the latest smartphone technologies. The main objective is to recognize the modifications in the direction of smartphones regarding data and alter the introduction to suit the examination edge of the customer, for example, while inspecting for a site page

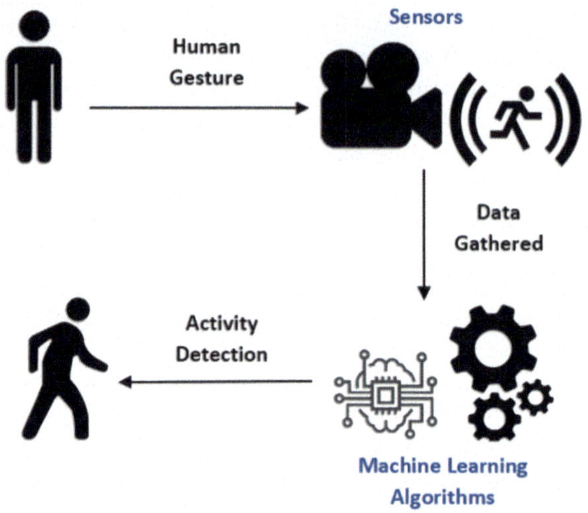


**Fig. 1** Design of human activity recognition system

with extended width, can get this scene to see from changing the inclination of smartphone to level. Exceptional way camera mode furthermore changes picture to scene when there is a modification in the direction of gadget/camera. This sensor also distinguishes the alteration by 3D (X, Y and Z turn) estimation of the accelerating speed.

**Gyroscope**—To determine direction, earth's gravity is used by a device known as gyroscope. Its ability is to keep up and control the position, level or direction depending on the principle of angular momentum, exactly when 'Gyros' used with an accelerometer recognizes movement from six axes, for instance right, left, up, down, forward and in backward. It furthermore identifies the spin, pitch and yaw movements. The three angular moments that can be seen from three axes X, Y and Z are yaw, spin and pitch. Using microelectrical and mechanical system (MEMS) advancement, gyroscopic sensors help in course reason and recognizing the gesture recognition systems utilized in smartphones and tablets.

### *3.2 Data Set*

Openly accessible data sets give an unreserved accessible source of information over various conducts to different scientists in the field. In the HAR analysis structure, some data sets have been discharged to the open space, few of them are mentioned below:

1. A Public Domain Data set for Human Activity Recognition Using Smartphones—This data set [8] focuses on the recognition of six diverse human

activities and this data set is having inertial information from smartphone gyroscope and accelerometers collected from around 30 individuals acting as subject performing daily life activities having the mobile phone attached to their waist. Activities performed in this data set are: Walking, walking up the stairs, walking down the stairs, laying, standing and sitting.

2. Data set—RealWorld (HAR)—This data set [9] is built to identify the different positions of the body where the devices can be attached to attain better results in the field of activity detection. Smartphones were attached to body parts like arm, waist, shin, chest, head and thigh and acceleration was recorded. Activities performed by 15 individuals in this data set are walking up the stairs, walking down the stairs, running, standing, jumping, sitting, walking in forward direction and laying. Sztyler and Stuckenschmidt [9] contain the information which depicts that this data set is transparent, true and has all the necessary data required for activity recognition.

## 3.3 Feature Extraction

Crude information collected from gadgets and sensors contain many concealed data and disturbance called noise. Helpful concealed data from the raw data can be discovered from feature extraction. In addition, it can take out the noise in the crude information from information assortment procedures or detectors. Choosing an appropriate feature will diminish the measure of memory and time required by the process of classification. Therefore, feature extraction is one of the most important steps for improving the classification algorithm's performance for implementation of human activity recognition. Various papers include feature extraction in their research for the same, to limit the noise in their data sets and to diminish the measure of memory and time for classification. Research papers using the phenomenon of feature extraction promise better results than others.

## 3.4 Classification Algorithms

Opting the right classification algorithm is a significant part in activity recognition which is used to classify various actions and exercises depending on the client's inputs. Most analysts utilize supervised classification algorithms. To create a classification model, the algorithms should be prepared with labelled samples. Commonly known supervised classification algorithms are Naïve Bayes, k-nearest neighbour, decision trees and support vector machine (SVM).

**Support Vector Machine**—To accomplish classification or regression tasks, support vector machine algorithm, a supervised classifier is commonly used. In

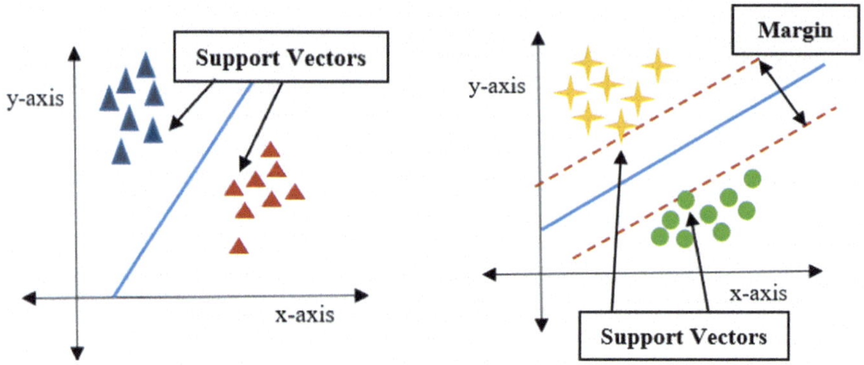


**Fig. 2** Support vector machine

this algorithm, the information is handled by mapping the element point in a n-dimensional space, where n is the features in the data set. When the plotting procedure is done, a classification procedure occurs to recognize the various categories of data and their labels. To classify the data into different classes, there is a need to determine and find the hyperplane that separates the various classes. SVM targets to discover a hyperplane separating two classes which augment the separation to the nearest focuses from each class. These nearest points are known as the support vectors as appeared in Fig. 2. To avoid mistakes and to improve accuracy, large margin separation is very important for better classification as much as possible. Therefore, SVMs search for the hyperplane which has the biggest edge or huge margin.

## *3.5 Applications of Human Activity Recognition*

Behind many fascinating applications, activity detection is a central component. The applications based on smartphone activity recognition could be classified as:

(1) Systems for the ultimate consumers such as collapse detection, fitness tracking, health monitoring, behaviour-based context-awareness applications, in automation of work and home.
(2) Systems for the masses and groups for example, activity-based crowdsourcing and social networking applications.
(3) Systems for the third person for instance, information collection research platforms, targeted advertising and accounting.

Nowadays, to enhance individuals' quality of life, a significant amount of study is taking place in human-focused zones that is human activity recognition becoming a noticeable research field. Applications in daily life monitoring are being developed to get a healthy lifestyle. There are sensor and system characteristics for improved user acceptance and are deployed in a large-scale care plan. Technologies like ambient intelligence, assistive automation, artificial intelligence and pervasive computing utilize HAR systems as a tool that gives data about individuals' conduct or activities which is ordinarily done by collecting and gathering signals from surrounding and sensors that can be worn externally on different body parts like waist, chest, knees and many more or using smartphones kept in pockets, then this data is used for classification of human actions using different machine learning algorithms. Numerous applications are present where human activity systems are utilized, for example, for crime prevention used in automatic observation of public places, for elderly and the youth care, applications consisting sleeping status monitoring of infants and determinations and prediction of demands for meal and other different stuff, intelligent interfaces for games and localization.

## 4 Discussion

When we talk about classification of data, we come across several classification algorithms in supervised learning. While survey the algorithms used for recognition of human physical actions include Naïve Bayes, decision tree algorithm, k-nearest neighbour, random forest and support vector machines (SVMs). In spite of the fact that there is not any study that can discover the best technique for human actions classifications, SVMs have been giving better results than other algorithms in many researches related to handwriting, speech and activity recognition.

First discussed the research on human activity recognition using different wearable sensors. In [2], classification of 10 human actions—types of 3D walking, 2D walking, static activities and plane motion—is done using a hierarchical detection method, data collected using tri-axial acceleration. The algorithms used are—decision tree (DT), artificial neural network (ANN), least square support vector machine (LS-SVM) and KNN. To eliminate the computational complexity and unclassifiable space, a small number of classifiers are used through multiple binary classification which was generated by decomposing the LS-SVM algorithm. Numerous experiments demonstrate that SVM is a higher-level classifier when accuracy is considered to be the main feature as SVM classifications have more ideal prediction performances compared to ANN classification. The average accuracy obtained using ANN is 92% where jumping has only 77% accuracy and LS-SVM obtains 94.5% accuracy. As KNN is more acceptable for small data sets, performance of KNN algorithm decreases as size of data set increases. KNN has 91% accuracy where activity walking downstairs has least accuracy of 86% whereas average accuracy of DT is 93% which is better than ANN and KNN but less than LS-SVM. Table 1 shows the summarized result of [2]. The recognition performance of the five walking actions is lower than others because the activities of 3D and 2D walking are almost the same. In this paper, the position to place the accelerometer sensor on the body can be changed to improve the accuracy.

Researchers also proposed methods to detect abnormal gestures by recognizing the daily life normal actions. Elimination method is adopted by many researches

**Table 1** Summary of algorithms in [2]

| Algorithms used for classification | Best algorithm | Accuracy % |
|---|---|---|
| DT | LS-SVM | 94.5% |
| ANN | | |
| LS-SVM | | |
| KNN | | |

**Table 2** Table of comparison for performance

| | Accuracy % |
|---|---|
| SVM conventional method | 95.7 |
| SVM proposed method | 94.3 |

which eliminate possible basic activities to detect abnormal activities like falling, instant change in basic activities, etc. Research paper [10] also used the techniques to detect abnormal as well as normal activities of daily life. Data set used wearable sensors—gyroscope and accelerometer with three-axis. A number of features extracted from the sensor's data were seven that were then used to classify nine activities—standing, walking, sitting, running, laying, falling, walking up, walking down and jumping. For classification of these nine discrete activities, multiclass SVM is used where kernel function is utilized for data set training. For shortening of computational power support vector machine algorithm, make use of a transition table that benefits in avoiding the unreachable actions. Accuracy for correct classification of activities using conventional approach is 95.7% where accuracy attained using the proposed method is 94.3%. The results achieved in [10] shown in (Table 2) clearly have lower computational time than the previous approaches but compromising the performance because real-time applications require instant results with reasonable accuracy. Not only this, but this approach also achieved accuracy near to the conventional algorithms and classifiers.

There are papers who have proposed methods which include data collected from wearable sensors and they try to aim for better efficiencies and accuracies for the human activity recognition field by utilizing the ensemble algorithm which is made by combining better classifiers for classification. Use of this method for classification can be seen in [11] where classification of ten activities of daily lives is done. By using this approach, novel detection model is applied to merge the performances of experiments conducted using diverse machine learning algorithms and classification models such as multilayer perceptron model, logistic regression techniques, KNN, random forest and SVM by making use of voting algorithms for HAR. All these machine learning algorithms mentioned in [11] have one or the other demerits of their own. Therefore, this study applied a voting rule for assembling different base models to attain one ideal predictive model. This machine learning technique is basic but a very strong technique used behind the idea to cluster various base learners for enhancing the performance giving 94% accuracy for MHEALTH data set used, and

**Table 3** Training set instances

| Type of activities | Number of Instances |
|---|---|
| Running | 968 |
| Walking | 487 |
| Sitting | 463 |
| Total | 1918 |

accuracy of 86% when USCHAD data set is used for training the model. The results depicted that the performance can be enhanced by merging finer machine learning algorithms to create ensemble learning models.

Now discussed researches on human activity recognition are those based on smartphones having embedded sensors. In [6], classification of three human actions—walking, running and sitting—has been finished making use of various algorithms having training set as shown in (Table 3), a multilayer perceptron classifier is used to sets aside the effort to create a model. Here, three classification algorithms are used for classification—SVM, J48 and Naive Bayes. When SVM classifiers are used, it is crucial to pick the proper features. In the event that the radial basis function is utilized as the kernel type giving 73.94% accuracy. J48 and Naive Bayes are straightforward and proficient algorithms for classification however can be a disadvantage when there are exceptions in the data set. On the other hand, if all the six activities are considered including walking, running, sitting, walking upstairs, walking downstairs, laying are classified through J48 and SVM, the correctness for J48 to recognize climbing stairs and climbing downstairs are just 81.2 and 74.2%, while support vector machine arrives at 96.3 and 85%. Thus, in [6] it is expressed that the performance of support vector machines could be improved if parameters in the data set are tuned. Regardless of whether J48 is a well-known and effective algorithm, it experiences issues in displaying countless complex activities like walking upstairs and walking downstairs.

In [5], the raw information was gathered from 10 to 11 volunteers with normal medical history of age between 11 and 27 with mobile phones in their pockets. The information is gathered from 10,933 examples as shown in (Table 4). Classification of 6 activities—sitting, standing, laying, walking, climbing stairs and walking downstairs. Human activity recognition is used as the principal form of the investigation for processed data sets collected using smartphones, and Anguita's exploration is

**Table 4** Samples collected from volunteers

| Type of activities | Number of samples |
|---|---|
| Standing | 4251 |
| Sitting | 3540 |
| Lying | 890 |
| Walking upstairs | 359 |
| Walking downstairs | 259 |
| Walking | 1634 |

utilized for testing information. According to this paper, the accuracy achieved for recognition of 2752/2957 examples was 89.1%, which was then improved to 93.39%, which implies higher than the past research result using support vector machine. On the other hand, when the data set of 561 features was reduced to the subsequent version with 248 features, the detection rate was higher than 89% as stated above but the drawback was that the laying movement had a low recognition rate 57% and the act was observed to be sitting. There is a scope of improvement in this examination by improving the component extraction for support vector machine and can be done using more balanced data having almost the same samples for better prediction.

There are various applications of human activity recognition like in medical research and human survey systems. In [7], classification of six basic human activities: walking, stumbling, running, walking upstairs and walking downstairs is done using a mobile phone-based recognition system. Time arrangement signals were gathered by the system utilizing an inherent accelerometer, created 30 features in both recurrence area and time, and afterwards the feature dimensionality was diminished to enhance the performance of the system. The action information was trained and tested utilizing three learning techniques: k-nearest neighbour algorithm, quadratic classifier and support vector machine. In the experiment in [7], the most suitable classification support vector machine.

The outcomes exhibit that SVM is a superior algorithm to work with for the classification of human activities. To diminish the measure of memory and time needed by the classification process, choosing reasonable features is important. Consequently, the performance of the classification algorithm will be improved.

## 5 Conclusion

It is vital to obtain a clearer image of the flow patterns of research in the zone of human activity recognition. This is achieved in this survey paper in regard to the classification algorithms utilized in human activity recognition utilizing smartphones. Also, it assisted with knowing different works being done in the region. The outcomes additionally show that SVM is a superior algorithm to work.

However, systems despite everything have some specific limitations. Level of recognition is low in some activities. Later on, there ought to be more research work to enhance the performance and increment the recognition capacities of the system. Rooms for upgrades additionally exist: the credit for the effectively classified dynamic actions goes to the features presented in the released data set [8], despite everything there is misclassification overlaps present in non-dynamic activities. Therefore, it requires further investigation of accessible sources of information and amendment of the HAR procedure pipeline stages. Future work will assess whether extra features are important to improve classifier performance, without adding computational complexity to the algorithm.

## References

1. WHO: World population ageing. Technical Report, UN World Health Organization, vol. 374, pp. 1–95 (2013)
2. Zheng, Y.: Human activity recognition based on the hierarchical feature selection and classification framework. J. Electr. Comput. Eng. **2015**, 1–9 (2015).
3. Corchado, J.M., Bajo, J., Tapia, D., Abraham, A., et al.: Using heterogeneous wireless sensor networks in a telemonitoring system for healthcare. IEEE Trans. Inform. Technol. Biomed. **14**(2), 234–240 (2010)
4. Ann, O.C., Theng, L.B.: Human activity recognition: a review. In: Control System, Computing and Engineering (ICCSCE), IEEE International Conference (2014)
5. Tran, D.N., Phan, D.D.: Human Activities Recognition in Android Smartphone Using Support Vector Machine. University of Information Technology Ho Chi Minh, Vietnam (2016)
6. Yin, X., Shen, W., Samarabandu, J., Wang, X.: Human activity detection based on multiple smart phone sensors and machine learning algorithms. In: 2015 IEEE 19th International Conference on Computer Supported Cooperative Work in Design (CSCWD). University of West em Ontario London, Ontario, Canada
7. Rasekh, A., Chen, C.A., Lu, Y.: Human Activity Recognition Using Smartphone. Texas A&M University, Fall CSCE666 Project Report (2011)
8. Anguita, D., Ghio, A., Oneto, L., Parra, X., Reyes-Ortiz, J.L.: A public domain dataset for human activity recognition using smartphones. In: 21st European Symposium on Artificial Neural Networks, Computational Intelligence and Machine Learning, ESANN 2013. Bruges, Belgium, 24–26 April 2013
9. Sztyler, T., Stuckenschmidt, H.: On-body localization of wearable devices: an investigation of position-aware activity recognition. In: IEEE International Conference on Pervasive Computing and Communications (PerCom), Sydney, NSW, pp. 1–9 (2016)
10. Palaniappan, A., Bhargavi, R., Vaidehi, V.: Abnormal human activity recognition using SVM based approach. In: ICRTIT-2012 IEEE International Conference (2012)
11. Nguyen, H.-D., Tran, K.P., Zeng, X., Koehl, L.: Wearable Sensor Data Based Human Activity Recognition Using Machine Learning: A New Approach (2019)

# Smartphone Inertial Sensors-Based Human Activity Detection Using Support Vector Machine

**Sarvesh Kumar Swarnakar, Harshit Agrawal, and Ankita Goel**

**Abstract** Perceiving human activity has earned a great amount of interest and has brought noteworthy research in recognizing logical data helpful to human activity recognition. It is the most essential technology behind different applications including health care, survey systems, surveillance, and many more. This paper presents work on human activity detection based on data collected from smartphones inertial sensors. Smartphones have installed inbuilt sensors that are fit to detect logical data of its clients, having a wide range capability of network connections. Installed sensors which include gyroscope sensor, tri-axial linear accelerometer, tri-axial accelerometer, and direction sensors are utilized for movement information assortment. Dataset used includes six activities: sitting, standing, laying, walking, walking upstairs and walking downstairs. In this paper, support vector machine learning algorithm with Gaussian radial basis kernel to classify all the six activities uniquely is used which resulted in better performance rate and less error rate as compared to the previous works already done in the area of HAR. Overlapping of dynamic and non-dynamic activity is improved with the overall accuracy of 98% which is more than the previous work results.

**Keywords** Human activity recognition · Smartphone · Inertial sensors · Machine learning · Classification · Support vector machine

## 1 Introduction

Expansion of life expectancy, a causative factor for humanity's endurance to old age, is basically influenced by advancement in medical science and in treatment of ailments. It is assessed that by 2050 more than 2.01 billion individuals will be over 60 years [1]. With age, individuals become powerless against motor and physical inabilities, limiting their support in everyday activities. Seniors having social and

S. K. Swarnakar (✉) · H. Agrawal · A. Goel
Department of Information Technology, Galgotias College of Engineering and Technology, 1, Knowledge Park – III, Greater Noida, Uttar Pradesh, India
e-mail: sarvesh.swarnkar@galgotiacollege.edu



T. K. Sharma et al. (eds.), *Soft Computing: Theories and Applications*,
Advances in Intelligent Systems and Computing 1381,
https://doi.org/10.1007/978-981-16-1696-9_22

physical inactivity are increasingly vulnerable to chronic diseases. Thus, on long-term care, large expenditure is spent; hence social and healthcare administrations are searching for choices to adapt to the full-scale monetary difficulties related to an expanding aging populace.

A plenty of studies have proposed different strategies for activity monitoring problems using wearable sensors [2–4], remote sensor systems [5], and video recorders. Some used computer vision-based techniques for the classification of human activities but faced various restrictions related to environment like variable lighting. Some included recognition using body-fixed wearable sensors offering a more feasible and relatively cheap method to gauge human movement. In Ambient Assisted Living (AAL), activity detection and patient monitoring using smartphones has given more promising answers because of different features of cell phones like versatility, low cost, Bluetooth, WIFI facility and 3G/4G communication feature.

One of the most helpful tools of our day-to-day life are cell phones. With the propelling innovations they are eventually fulfilling client issues and desires. Smartphones accompany distinctive inserted sensors which makes it conceivable to gather immense measures of data about the client's everyday life and exercises. Sensors have a major job in making smartphones increasingly utilitarian and mindful of nature. To make these devices progressively practical and incredible, developers add new modules and gadgets to the equipment. Gyroscope and accelerometer sensors are among these gadgets. Researches related to HAR based on accelerometer sensor are [6].

For most of the smartphone producers, an accelerometer has been a standard equipment. As its name proposes accelerometer gauges the variation in speed; not just the pace. Gyroscope is another sensor that has been a standard equipment for cell phones, which estimates direction by utilizing gravity. Smartphones are raising new research open doors for applications that are dedicated to humans, where the client is the major source of collecting data and the cell phone is the first-hand detecting apparatus. The utilization of mobile phones having inertial sensors is an elective answer for human activity recognition. These mass-advertised gadgets give an adaptable, reasonable and independent answer for consequently and subtly screen Activities of Daily Living (ADL) while additionally giving communication services.

## 2 Related Work

Human activity recognition (HAR) has evoked significant enthusiasm for the different research zones and applications because of its potential use in proactive processing. Proactive processing is an innovation that expert—effectively foreseen individual's need in a set of circumstances, for example, medicinal services or life-care and takes suitable actions for their sake. Human activity recognition is used to encourage physical workout and normal exercises for correction of different postures like children's sitting gesture and identification of different behaviors for early detection of any kind of illness. As in [7, 8] it is also used to detect any abnormal activity like falling in elderly people. In [9], many more instances of applications are observed. Bülbül et al. [10] present the detailed survey on HAR. Different machine learning procedures have been already utilized to recognize human activities.

The enormous gathered information gives numerous significant information to perceive the human physical activities. There are other methods too where the information gathered is utilized in different supervised learning algorithms for the classification process. In [5, 11–13], algorithms like ($K$-nearest neighbor) KNN, random forest, Naive Bayes algorithm, and SVM are applied for classification, where dynamic activities can be effectively characterized, non-dynamic activities still present misclassification covers and low precision. Accuracy varies from 56 to 90% for all the activities for different algorithms. In spite of the fact that there is not any study that can discover the best technique for human actions classifications, SVMs have been giving better results than other algorithms in many researches related to handwriting, speech and activity recognition.

Few researches on human activity recognition those based on smartphones having embedded sensors are discussed here. In [12], classification of three human actions—walking, running, and sitting—has been finished. Here, three classification algorithms are used for classification—SVM, J48, and Naive Bayes. When SVM classifiers are used, system gives 73.94% accuracy. On the other hand, if all the six activities are considered including walking, running, sitting, walking upstairs, walking downstairs, laying are classified through J48 and SVM, the correctness for J48 to recognize climbing stairs and climbing downstairs is just 81.2 and 74.2%, while support vector machine arrives at 96.3 and 85%. Thus, in [12], it is expressed that the performance of support vector machines could be improved if parameters in the dataset are tuned.

In [11], the raw information was gathered from 10,933 examples. Classification of six activities are done. According to this paper, the accuracy achieved for recognition of 2752/2957 examples was 89.1%, which was then improved to 93.39%, which implies higher than the past research result using support vector machine but the drawback was that the laying movement had a low recognition rate 57% and the act was observed to be sitting.

The outcomes of all the above-discussed researches exhibit that SVM is a superior algorithm to work with for the classification of human activities. To diminish the measure of memory and time needed by the classification process, choosing reasonable features is important. However, systems despite everything has some specific limitations, level of recognition is low in some activities. Later on, there ought to be more research work to enhance the performance and increment the recognition capacities of the system.

## 3 Proposed Approach

This paper proposed a methodology to build up a application program interface (API) using the best machine learning techniques; focused on the classification of six distinct activities, incorporating walking, upstairs, downstairs, laying, standing, and sitting; gathered a huge informational index for model training and testing. While performing each one of these activities, smartphones were kept in various directions and readings from mobile phone inertial sensors were gathered. The dataset therefore gathered consisted of features from the data that have not been investigated in past examinations, using support vector machine classifier for the detection of human activities achieving better results with less computational time.

This system consists of two parts—training and classification. Training is a part where the model is prepared using training dataset which is processed and scaled before model formation, and SVM supervised learning algorithm with RBF kernel is used for modeling and then the model prepared is evaluated by giving different sets of test data. When the final model is formed then the classification part starts where new set of data taken from user is given as input to the classification model to generate the result. The general architecture is depicted in Fig. 1.

Working of API starts where user inputs the test data and then chooses an activity to be classified, then the API gathers the data related to that activity and saves it in json format, then when user click on Predict button API will load both the model and test data json file and generate the result which is shown to the user with recall percentage and then result is saved in json format so that user can use it further. User has the option to choose from six activities: sitting, standing, laying, walking, walking upstairs, and walking downstairs, and this makes the API flexible and more efficient and it can also be integrated with different applications or systems working on human activity recognition.

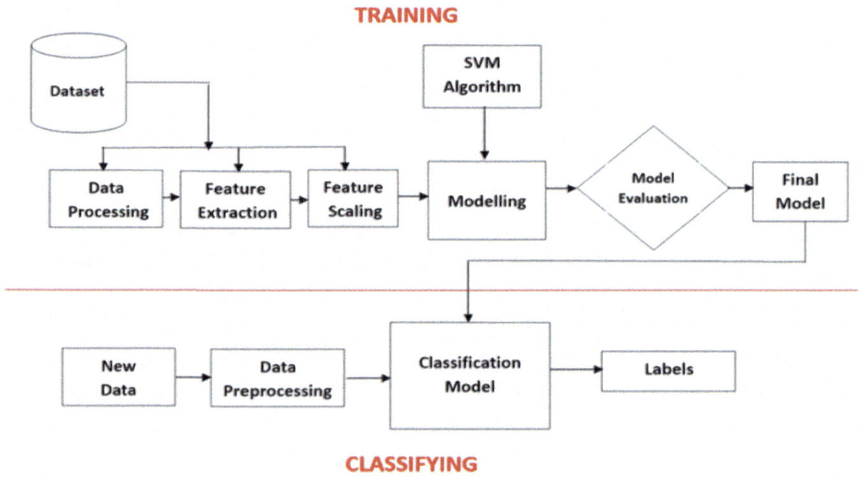


**Fig. 1** Architecture of the system

## 3.1 Smartphone Sensors

There are many approaches through which input can be taken for HAR systems. Earlier in many researches mode of input was through wearable sensors but due to the its disadvantages, this paper used smartphones having motion sensors for collecting the input. Gyroscope and the accelerometer are the most well-known of these sensors.

**Accelerometer**—An accelerometer is one of the fundamental devices among the latest smartphone technologies. The main objective is to recognize the modifications in the direction of smartphones regarding data and alter the introduction to suit the examination edge of the customer, for example, while inspecting for a site page with extended width, can get this scene to see from changing the inclination of smartphone to level. Exceptional way camera mode furthermore changes picture to scene when there is a modification in the direction of gadget/camera.

**Gyroscope**—To determine direction, earth's gravity is used by a device known as gyroscope. Exactly when "Gyros" used with an accelerometer recognizes movement from six axes, for instance, right, left, up, down, forward, and in backward. It furthermore identifies the spin, pitch, and yaw movements. The three angular moments that can be seen from three axes X, Y, and Z are yaw, spin, and pitch. Using microelectrical and mechanical system (MEMS) advancement, gyroscopic sensors help in course reason and recognizing the gesture recognition systems utilized in smartphones and tablets.

## 3.2 Dataset

In the HAR analysis structure, some datasets have been discharged to the open space, for example [14], focusing on the recognition of six diverse human activities as shown in Fig. 2, this dataset is having inertial information from smartphone gyroscope and accelerometers. Activities performed are walking activity, walking upstairs, walking downstairs, laying, standing, and sitting activity.

To acquire the dataset, a set of analysis is executed by a group of 29–30 selected subjects having age between 20 and 47 years where every subject followed a specific set of protocol of activities having waist-mounted phone. The whole process was carried out in the laboratory conditions but for gathering the more realistic data each participant was instructed to perform the chain of activities freely.

The dataset formed consists of 10,299 rows and 561 features which is partitioned into training and testing where randomly selected 70% of the dataset is taken as training set and 30% is taken as testing set. The data distribution is almost equal for all activities in both training dataset and testing dataset, they are closely balanced as depicted in Table 1.

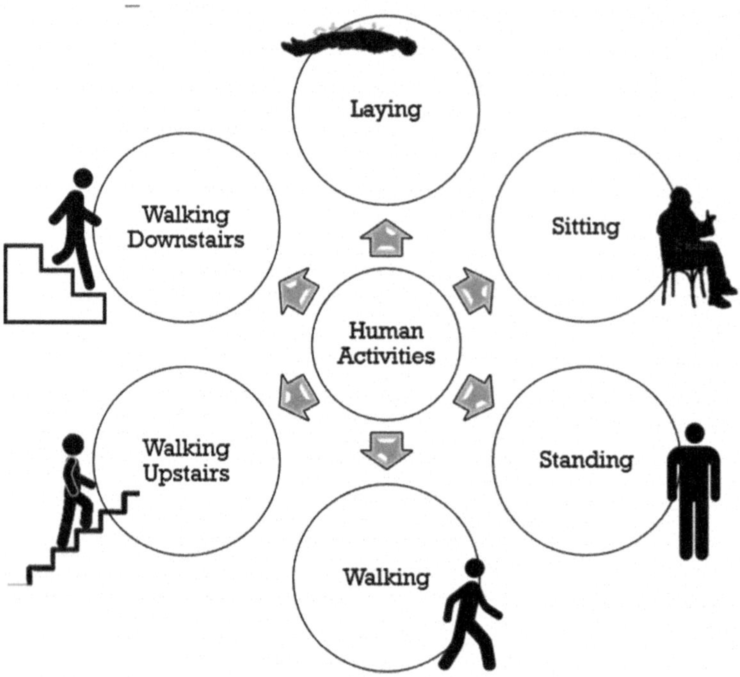


**Fig. 2** Different human activities

**Table 1** Distribution of training data

| Training data | |
|---|---|
| Activity | Count |
| Laying | 1407 |
| Sitting | 1286 |
| Standing | 1374 |
| Walking | 1226 |
| Walking_Downstairs | 986 |
| Walking_Upstairs | 1073 |

## *3.3 Preprocessing of Dataset*

Filtering of the data, replacing the undermined qualities and removal or selection of highlighted features is the main target of preprocessing. Raw or crude data is consistently fragmented which can create errors and so cannot be used for training of the model. That is the reason this paper also preprocessed our dataset before sending through a model. Here, three preprocessing techniques are used in this paper that is checking missing data, encoding categorical data, and feature scaling.

**Table 2** Samples collected from volunteers

| Activity | Encoded value |
|---|---|
| Laying | 0 |
| Sitting | 1 |
| Standing | 2 |
| Walking | 3 |
| Walking_Downstairs | 4 |
| Walking_Upstairs | 5 |

1. When data is not recorded properly, it may contain some values which are missing because of various reasons like data corruption. As algorithms of machine learning does not allow data having missing values that is why handling and checking missing data is essential. So applied methods check and replace or remove the missing values from datasets before actual training of data so that there is no missing data from training or testing dataset which needs to be corrected.
2. A large portion of the algorithms is unable to deal with categorical variables except if we transform those variables into numerical values. Dataset used for training and testing has "Activity" label which contains non-numeric data that is six activities which are classified in this paper which needs to be transformed from non-numerical labels into numerical labels, also did transformation of non-numerical labels into numerical labels using encoding categorical data method. Table 2 depicts training dataset labels which are transformed from non-numerical to numerical labels using encoding of categorical data method.
3. Prior to building any model, one of the major steps of data preprocessing is feature scaling. Some features might tend to dominate other features in the dataset because of the variance of magnitude more than the variance of other features, which is not something wanted in model that is why also done feature scaling of training and testing datasets before training of the model.

### *3.4 Classification Algorithm—Support Vector Machine*

Opting the right classification algorithm is a significant part in activity recognition which is used to classify various actions and exercises depending on the client's inputs; used SVM classifier for the detection of human activities. To accomplish classification tasks, SVM algorithm, a supervised classifier is commonly used. In this algorithm, the information is handled by mapping the element point in a n-dimensional space, where n tells the absolute number of elements or features in the dataset. When the plotting procedure is done, a classification procedure occurs to recognize the various categories of data and their labels. To classify the data into different classes, there is a need to determine and find the hyperplane that separates the various classes. SVM targets to discover a hyperplane separating two classes

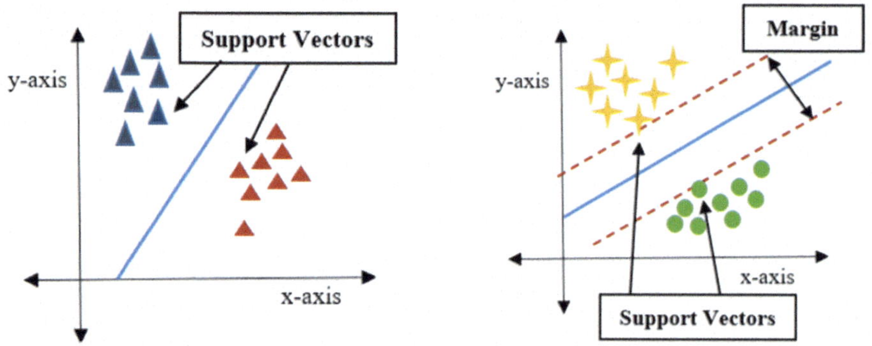


**Fig. 3** Support vector machine

which augment the separation to the nearest focuses from each class. These nearest points are known as the support vectors as appeared in Fig. 3. The hyperplane making the largest margin between the support vectors of different classes is chosen to be the best. To avoid mistakes and to improve accuracy, large margin separation is very important for better classification as much as possible.

Support vector machine is classified into different kinds based on kernel functions used. There are various types of kernel function, for example, linear kernel, polynomial kernel, radial basis function (RBF) kernel, and many more. This paper used SVM classifier with RBF kernel for human activity detection. It is not only a good kernel for nonlinear data points but have also proved to be a good kernel function for multiclass classification. One of the well-liked kernel function is radial basis function kernel or RBF kernel used in many machine learning algorithm mainly in SVM classifications. Example of RBF kernel is Gaussian kernel.

$$k(x, y) = \exp(-(||x-y||^2)/2s^2) \tag{1}$$

Equation (1) represents the RBF kernel with two samples $x$ and $y$, where $||x-y||^2$ is the square of the Euclidean distance between two vectors and the free parameter is s. It can also be written as Eq. (2) where $\gamma = (1/2\sigma^2)$.

$$k(x, y) = \exp(-g||x-y||^2) \tag{2}$$

RBF kernel possesses various advantages like it has low numerical problems and so used this kernel in our algorithm because of the nonlinear data and to get the best classification result as using other kernels will not result in the most effective hyperplane for our dataset. After the training of the model using above classifier achieved the overall accuracy of 98%.

**Table 3** Confusion matrix of the model

| | Laying | Sitting | Standing | Walking | WD | WU |
|---|---|---|---|---|---|---|
| Laying | 538 | 0 | 0 | 0 | 0 | 0 |
| Sitting | 2 | 439 | 16 | 0 | 0 | 1 |
| Standing | 0 | 11 | 533 | 0 | 0 | 0 |
| Walking | 0 | 0 | 0 | 489 | 4 | 6 |
| Walking downstairs | 0 | 0 | 0 | 6 | 406 | 9 |
| Walking upstairs | 0 | 0 | 0 | 15 | 2 | 470 |

## 4 Results

An API for human activity detection using support vector machine classifier was built where the dataset was collected using inertial smartphone sensors like accelerometer and gyroscope. The training dataset and testing dataset have a total of 7352 and 2947 observations, respectively, and have a total of 561 time and frequency features. Classification of these activities is done using SVM classifier with RBF kernel function.

To report the performance of model used for classification, a table or matrix is used known as confusion matrix based on a set of test data where the correct values are already familiar. Four distinct union of actual and predicted values form a confusion matrix where the count of correct and incorrect predictions is represented for each and every label or class. Confusion matrix of the model is represented in Table 3. We can conclude four quantities from confusion matrix which are true positive (TP), false negative (FN), true negative (TN), and false positive (FP). Using all these four quantities, precision, recall, and F1-score are calculated. To represent precision and recall, both simultaneously used F-measure. Equation (3) is the formula for F-measure.

$$\text{F - measure} = (2 * \text{Recall} * \text{Precision})/(\text{Recall} + \text{Precision}) \quad (3)$$

Using above formulas, performance matric is created as shown in Table 4. Overall accuracy of the model results in 98% which is much higher than the previous researches in human activity recognition field.

## 5 Conclusion

In this research, the general architecture utilized is presented to build HAR systems and emphasize the design issues such as selection of sensors, obtrusiveness, etc., which are independently evaluated based on the kind of system which is being developed. All the results in literature review also demonstrate that SVM is a best algorithm

**Table 4** Performance matric

| | Precision | Recall | F1-score | Support |
|---|---|---|---|---|
| Laying | 1.00 | 1.00 | 1.00 | 538 |
| Sitting | 0.98 | 0.96 | 0.97 | 458 |
| Standing | 0.97 | 0.98 | 0.98 | 544 |
| Walking | 0.96 | 0.98 | 0.97 | 499 |
| Walking downstairs | 0.99 | 0.96 | 0.97 | 421 |
| Walking upstairs | 0.97 | 0.97 | 0.97 | 487 |
| Avg/total | **0.98** | **0.98** | **0.98** | 2947 |

to work with in the field of HAR. Therefore, SVM is implemented to classify and recognize human activities.

This research overcomes certain restrictions which other system possessed. In many systems, percentage of recognition is low in some action with less computation time and in others if better accuracy is achieved they faced the problem of higher computation time and achieved both better accuracies and less computation time. The features chosen are by far the best set of features using limited sensors which can classify all dynamic and non-dynamic activities efficiently. API implemented perfectly with the overall accuracy of 98% and achieved 100% accuracy in classifying the laying activity and the least accuracy is 97% which is of classifying Sitting and all dynamic activities. This API can be integrated to other systems and applications and can be used as a base in different applications of human activity recognition. Future work will investigate whether feature addition is needed to enhance the classification model performance, without appending computational complexity to the algorithm. Human activity detection systems which exist are classifying activities by an individual person only which can be expanded in the direction of detecting activity trends having a group of individuals using social networks.

## References


1. WHO: World Population Ageing. Technical Report, UN World Health Organization, vol. 374, pp. 1–95 (2013)
2. Zheng, Y.: HAR Based on the Hierarchical Feature Selection and Classification Framework. J. Electr. Comput. Eng. (2015)
3. Nguyen, H.-D., Tran, K.P., Zeng, X., Koehl, L.: Wearable Sensor Data Based Human Activity Recognition Using ML: A New Approach (2019)
4. Wang, L., Gu, T., Chen, H., Tao, X., Lu, J.: Real-time activity recognition in wireless body sensor networks: from simple gestures to complex activities. In: 6th IEEE International Conference on Real-Time Computing Systems and Applications (2010)
5. Corchado, J.M., Bajo, J., Tapia, D., Abraham, A., et al.: Using heterogeneous wireless sensor networks in a telemonitoring system for healthcare. IEEE Trans. Inform. Technol. Biomed. **14**(2), 234–240 (2010)

6. Curone, D., Bertolotti, G.M., Cristiani, A., Secco, E.L., Magenes, G.: A real-time and self-calibrating algorithm based on triaxial accelerometer signals for the detection of human posture and activity. IEEE Trans. Inf Technol. Biomed. **14**(4), 1098–1105 (2010)
7. Palaniappan, A., Bhargavi, R., Vaidehi, V.: Abnormal human activity recognition using SVM based approach. In: ICRTIT-2012 IEEE International Conference (2012)
8. Shin, J.H., Lee, B., Park, K.S.: Detection of abnormal living patterns for elderly living alone using support vector data description. IEEE Trans. Inform. Technol. Biomed. **15**(3), 438–448 (2011)
9. Ann, O.C., Theng, L.B.: Human activity recognition: a review. In: Control System, Computing and Engineering (ICCSCE), 2014 IEEE International Conference (2014)
10. Bülbül, E., Çetin, A., Doğru, İ.A.: HAR Using Smartphones. IEEE (2018)
11. Tran, D.N., Phan, D.D.: Human Activities Recognition in Android Smartphone Using Support Vector Machine. University of IT Ho Chi Minh, Vietnam (2016)
12. Yin, X., Shen, W., Samarabandu, J., Wang, X.: Human Activity Detection Based on Multiple Smart Phone Sensors and Machine Learning Algorithms. University of West em Ontario London, Canada. 2015 IEEE 19th International Conference on (CSCWD)
13. Rasekh, A., Chen, C.A., Lu, Y.: Human Activity Recognition Using Smartphone. Texas A&M University, Fall CSCE666 Project Report (2011)
14. Anguita, D., Ghio, A., Oneto, L., Parra, X., Reyes-Ortiz, J.L.: A Public Domain Dataset for Human Activity Recognition Using Smartphones. 21st ESANN 2013. Bruges, Belgium 24–26 April 2013

# 475K Dataset

**Mangesh Wadwekar**

**Abstract** A lot of progress has been made in recent times on optical character recognition (*OCR*) like OpenCV, Tesseract OCR, Amazon Web Services Textract, Google Cloud Platform OCR, Microsoft Azure OCR to name a few, which are some platforms/libraries to convert image to text and image datasets like National Institute of Standards and Technology (NIST) and Google Street View House Number are available for printed text, handwritten text, and text from natural scenes. However, there is limited data available for printed text with *lower case*, *upper case*, *digits*, and *special characters* with varied *formatting* (bold, italic, underline, and strikethrough). In this paper, we are presenting dataset "475K Chars" that has 400+ synthesized *fonts* via computer with various combinations of text *formatting* like bold, italic, etc. Each image in the presented dataset is unique, and we have presented more than 475,000 images in our dataset. The dataset is open source and can be used to train deep learning neural networks, and images in this dataset have different text *formatting* with text format label for each character. Most of the characters have 16 different combinations of images of a font. To the best of our knowledge, this is the first attempt to present such a wide dataset publicly for scientific community. This dataset is publicly available and has been moderately tested for accuracy.

**Keywords** OCR · Latin characters · Digits · Special characters · Fonts · Formatting · Upper case · Lower case

## 1 Introduction

There are lot of datasets available for training and testing out models, but very few are publicly available datasets. We have open-source datasets like NIST [1], Chars74K [2], Google SVHN [3], ICDAR03-CH [4] etc., but these datasets lack in special character images and text formatting labels. In printed text, we can find the usage of special characters like asterisk (*), comma (,), period (.), exclamation (!), etc.,

M. Wadwekar (✉)
Independent Researcher, Boston, USA



T. K. Sharma et al. (eds.), *Soft Computing: Theories and Applications*,
Advances in Intelligent Systems and Computing 1381,
https://doi.org/10.1007/978-981-16-1696-9_23

quiet frequently. An annotation to the special character would improve the usability of the dataset. Further, on several occasions, in written text emphasis on keyword or specific word is put with the help of text formatting such as using bold, underline, or italic. Moreover, to highlight the changes in the text, people use the strikethrough that further help them to compare the old text with updated text (e.g., Apple Orange). Moreover, we are providing images for special characters as we see usage of comma (,), period (.), single and double quotes (',"), asterisk (*), percent (%), and many other special characters which come in day-to-day reading.

In the current work, we present a novel dataset "475K Chars" that covers all the above-mentioned text formatting with suitable labels and special characters. We have collected font images of 94 characters, and more than 400 fonts have been considered with different text formatting as it is a widely accepted norm to catch reader's attention, and all the images in our dataset are unique. Hence, inclusion of these special formatting in the dataset would ease the programmers or scientific community.

We do not have special characters in currently available datasets, so if a network is trained on these datasets, then it would fail to recognize special characters and punctuation marks. It might classify special characters as some other characters may hamper meaning of the whole sentence, e.g., if there is a Twitter keyword in the text, saying "Please support us on twitter with #example-keyword," then hashtag (#) can be classified as some other characters while converting image to text. Another example for punctuation could be "Let's eat Grandpa" vs "Let's eat, Grandpa". The whole meaning of the sentence is changed. Hence, training networks with punctuations and special characters could lead us in getting one step closer to automating comprehension analysis with image to text conversion, and there may be other uses as well. The goal of this dataset is to present a consolidated dataset of textual images. The 475K Chars dataset includes images of different fonts. Cartesian product of 94 characters with 407 fonts was considered with 16 different text formatting combinations, and some of the images were dropped as the usage of special characters is not seen with text formatting. Formatting options covered are bold (e.g., mango), italic (e.g., mango), underline (e.g., mango), and strikethrough (e.g., mango); with these four options, 16 combinations can be generated for each class. Typically, a single font contains 1168 images, which leads us to 475,376 images in the dataset, and these images have a size of $50 \times 50$ pixels.

Images are categorized by characters, i.e., all images of upper case "A" are in a single folder with all text formatting options. More has been described in upcoming sections on accessing dataset. This dataset can be used at places where we find printed text in images, e.g., image of a form, screenshot of a tweet or message, etc. Some of the work that already exists in this domain includes images of handwritten characters and digits in NIST [1] database; in Chars74K [2] and ICDAR03-CH [4], we can find images of printed text found in natural scene, i.e., picture taken of billboards or hoardings on roads or buildings. Google SVHN [3] has images of numbers in house addresses, so only numbers from natural scene can be trained with this dataset.

In rest of the paper, Sect. 2 presents summary of the field, Sect. 3 presents metadata of dataset, Sect. 4 presents dataset validation, Sect. 5 presents possible extension in future of the dataset, and finally, Sect. 6 concludes the work.

There could be numerous applications of this kind of OCR, like intelligent AGV [5] reading signs from environment and taking next steps accordingly. Another application is to read pages with watermark or reading checks used in banking; as they have different formatting, this dataset would provide help with that. There are new techniques for watermark such as robust digital image watermarking [6], but still there is a need for its OCR. In addition to that, watermarks are also used as source to verify authenticity, as seen in secure watermark scheme [7].

## 2 Related Work

There has been plethora of work done in OCR in the field of deep learning by introducing various algorithms and datasets for its use. Our work is to extend the data available publicly for people to use for any purpose. Already existing datasets for characters and numbers are NIST [1], MNIST [8], Chars74K [2], Google SVHN [3], ICDAR03-CH [4], etc. Out of these datasets, images are taken from various sources like handwritten text images, printed text images, and images from natural scene. Here we are extending existing datasets available by presenting synthesized image dataset which includes alphanumeric, case sensitive, with special characters and text formatting, with labels for all characters with their text format, e.g., bold and italic.

NIST [1] database has around 800K images of handwritten characters, and MNIST [10] is a subset (60K images) containing only digits from NIST [1]. NIST [1] has a wide variety of images for each character but does not have special characters. Since it is handwritten, so formatting is not applicable.

Chars74K [2] and ICDAR03-CH [4] datasets have images of characters and digits in printed format taken from natural scene, handwritten text, and synthesized computer fonts. These datasets also do not contain special characters, although they have text formatted images (bold, italic), but they lack formatting labels.

Google SVHN [3] has cropped images only of digits, taken from streets, natural scene; these images lack formatting option/labels around them.

## 3 Understanding Dataset

This dataset contains around 475K images divided by classes. Dataset contains 94 classes, out of which class 1–66 have 6512 images per class and classes 67–94 have 1628 images per class. Labels and classes are given same name and are classified as follows:

- Classes 1–10 are *digits* (1, 2, 3…0)
- Classes 11–62 are alphabets (*lower case* and *upper case*)
- Classes 63–66 are *special characters* having 6512 images per class (‘?’,’&’,’$’,’%’)
- Classes 67–94 are rest of the *special characters* having 1628 images per class, and detailed list is included in Table 1 in this section.

**Metadata**

- All images are 50 × 50 pixels in size.
- Folder containing images can act as label for that image. For example, all images for *lower case* ‘a’ are in folder named ‘11’.
- Image files follow naming format: <unique-number> - < formatting >
    - o For example, in folder ‘11’, first file name is ‘11-b.png’
    - o Here ‘11’ in ‘11-b’ is a unique sequence across dataset
    - o ‘b’ represents text *formatting* as bold
- Text format options are used with their naming symbols as follows:
    - o Bold as ‘b’, e.g., **Sample Text**
    - o Italic as ‘i’, e.g., *Sample Text*
    - o Underline as ‘u’, e.g., Sample Text
    - o Strikethrough as ‘s’, e.g., Sample Text
- There are 16 combinations made from these four *formatting* options as listed below (Fig. 1).

| Formatting | Symbol | Image |
|---|---|---|
| Bold | b |  |
| Bold, italic | bi |  |
| Bold, italic, strikethrough | bis |  |
| Bold, italic, strikethrough, underline | bisu |  |

(continued)

(continued)

| Formatting | Symbol | Image |
|---|---|---|
| Bold, italic, underline | biu |  |
| Bold, strikethrough | bs |  |
| Bold, strikethrough, underline | bsu |  |
| Bold, underline | bu |  |
| Italic | i |  |
| Italic, strikethrough | is |  |
| Italic, strikethrough, underline | isu |  |
| Italic, underline | iu |  |
| No formatting (normal font) | n |  |
| Strikethrough | s |  |

(continued)

(continued)

| Formatting | Symbol | Image |
|---|---|---|
| Strikethrough, underline | su |  |
| Underline | u |  |

### 3.1 *Dataset Details*

Dataset can be found over here:

https://github.com/mangeshw/validation_475k/tree/master/datasets/printed_alphabets

## 4 Validation

Dataset is validated using a subset of 76,000 images, 188 images from each class. Only unformatted or normal font images and bold font images were taken to evaluate the model, as including different formatting styles like underline and italic decreased the model accuracy. A better model with more layers for feature detection could be used for images with text formatting. Following is the code for model, and more details of code snippet can be found at following GitHub repository:

https://github.com/mangeshw/validation-475k

**Table 1** Class–label mapping

| Symbol | Label | Symbol | Label | Symbol | Label |
|---|---|---|---|---|---|
| 1 | 1 | w | 33 | $ | 65 |
| 2 | 2 | x | 34 | % | 66 |
| 3 | 3 | y | 35 | # | 67 |
| 4 | 4 | z | 36 | ~ | 68 |
| 5 | 5 | A | 37 | ‘ | 69 |
| 6 | 6 | B | 38 | ^ | 70 |
| 7 | 7 | C | 39 | ! | 71 |
| 8 | 8 | D | 40 | * | 72 |
| 9 | 9 | E | 41 | ( | 73 |
| 0 | 10 | F | 42 | ) | 74 |
| a | 11 | G | 43 | - | 75 |
| b | 12 | H | 44 | _ | 76 |
| c | 13 | I | 45 | = | 77 |
| d | 14 | J | 46 | + | 78 |
| e | 15 | K | 47 | { | 79 |
| f | 16 | L | 48 | } | 80 |
| g | 17 | M | 49 | [ | 81 |
| h | 18 | N | 50 | ] | 82 |
| i | 19 | O | 51 | ; | 83 |
| j | 20 | P | 52 | : | 84 |
| k | 21 | Q | 53 | ‘ | 85 |
| l | 22 | R | 54 | “ | 86 |
| m | 23 | S | 55 | , | 87 |
| n | 24 | T | 56 | < | 88 |
| o | 25 | U | 57 | > | 89 |
| p | 26 | V | 58 |  | 90 |
| q | 27 | W | 59 | @ | 91 |
| r | 28 | X | 60 | / | 92 |
| s | 29 | Y | 61 | \ | 93 |
| t | 30 | Z | 62 | \| | 94 |
| u | 31 | ? | 63 |  |  |
| v | 32 | & | 64 |  |  |

```
class
MyModel(Model):
def __init__(self):
    super(MyModel, self).__init__()
    self.conv1 = Conv2D(50, 3, activation='relu')
self.flatten = Flatten()
    self.d1 = Dense(128,
activation='relu')      self.d2 =
Dense(95, activation='softmax')
   def call(self,
x):      x =
self.conv1(x)
x =
self.flatten(x)
x = self.d1(x)
return self.d2(x)

# Create an instance of the
model model = MyModel()
```

Listing 1—Snippet of the validation code

## 5 Future Scope

This dataset was synthesized with 407 fonts, including digits, lower case and upper case characters and special characters with different text formatting options. Accuracy of this dataset has not been tested on full dataset, only normal and bold font images were trained, so that it can be scoped further to have a model with better accuracy for all text formats.

There is scope of introducing special characters with formatting. Formatting options were left for some special characters as they were confusing, e.g., symbol "less than" (<) with underline becomes (<) "less than equal to" or "minus" sign (-) with underline (-) appear to be somewhat "equals" sign or "asterisk" symbol (*) with strikethrough (*) looks like an underline to asterisk. Similarly, there could be more scenarios, which are left in this dataset, but could be picked up later and trained.

*Limitation*: Only Latin characters are used, and some special characters with formatting have been left out. Characters like "Rupee" (₹) are not present as of now, and as mentioned above characters like "less than equal to" (<) and "minus" sign (-) do not have images with formatting like strikethrough and underline.

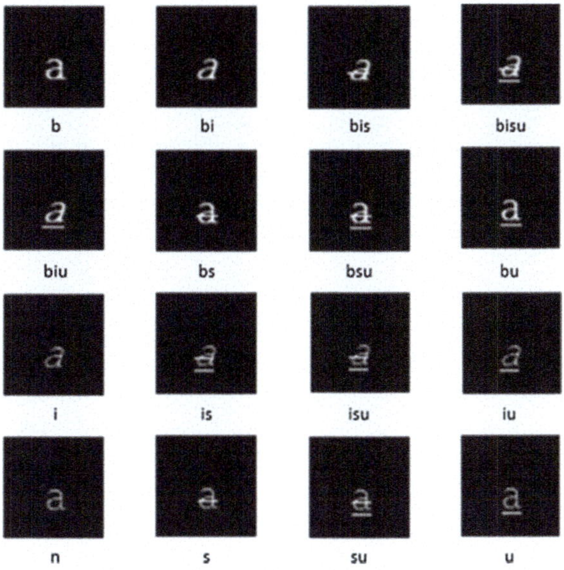


**Fig. 1** Display image of lower case alphabet "a" with different formatting options with labels

## 6 Conclusions

475K Chars are a synthesized computer font image dataset of lower case, upper case, digits, and special characters with different text formats with their labels. This dataset has been tested, and results look promising, further optimization may lead to better results. This dataset could act as a replacement/extension to the existing datasets for OCR. This would extend resources available in open-source community for training datasets.

**Acknowledgements** We would like to thank Dr. Rohit Verma (Manipal Institute of Technology, MAHE, Manipal, India) for providing us motivation, guidance on writing, and initial reviewing of this research paper.

## References

1. NIST: Available: https://www.nist.gov/
2. Chars74K: Available: https://www.ee.surrey.ac.uk/CVSSP/demos/chars74k/
3. Google SVHN: Available: https://ufldl.stanford.edu/housenumbers
4. Lucas, S.M., Young, R., Wong, S., Tang, A., Sosa, L., Panaretos, A.: ICDAR03-CH (2003)
5. Sheth, S., Ajmera, A., Sharma, A., Patel, S., Kathrecha, C.: Design and development of intelligent AGV using computer vision and artificial intelligence. In: Pant, M., Ray, K., Sharma, T., Rawat, S., Bandyopadhyay, A. (eds.) Soft Computing: Theories and Applications. Advances in Intelligent Systems and Computing, vol. 583. Springer, Singapore (2018). https://doi.org/10.1007/978-981-10-5687-1_31

6. Yadav, B., Kumar, A., Kumar, Y.: A robust digital image watermarking algorithm using DWT and SVD. In: Pant, M., Ray, K., Sharma, T., Rawat, S., Bandyopadhyay, A. (eds.) Soft Computing: Theories and Applications. Advances in Intelligent Systems and Computing, vol. 583. Springer, Singapore (2018). https://doi.org/10.1007/978-981-10-5687-1_3
7. Somwanshi, D., Asthana, M., Agarwal, A.: An OTSU-discrete wavelet transform based secure watermark scheme. In: Pant, M., Ray, K., Sharma, T., Rawat, S., Bandyopadhyay, A. (eds.) Soft Computing: Theories and Applications. Advances in Intelligent Systems and Computing, vol. 583. Springer, Singapore (2018). https://doi.org/10.1007/978-981-10-5687-1_25
8. LeCun, Y., Cortes, C., Burges, C. J.: MNIST (2012). Available: https://yann.lecun.com/exdb/mnist/

# Docker Security: Architecture, Threat Model, and Best Practices

**Sushant Chamoli and Sarishma**

**Abstract** With our ever-growing demands, virtualization is the technology to cater to our computing needs, thereby enabling users to leverage the full prowess of their computing resources. Virtualization, simply said, results in efficient usage of our resources, and containerization is the most prominent way to implement it. Docker is a container-based technology that helps to achieve virtualization. On the one hand, Docker helps us to provide a central point of control for various containers, however, on the other hand, it is a source of many security attacks too—if not configured properly. In this paper, we will be focusing on the Docker architecture, which is essential to understand how and from where attacks might originate. We will then dwell upon Docker threats along with various attack scenarios and steps we can take to eliminate such attacks. Further, we will shed light on some good practices to secure the Docker containers.

**Keywords** Docker · Security · Container · Architecture

## 1 Introduction

Virtualization has become the need of the hour in today's world. It is a cost-efficient approach to remove the under-utilization of resources. With cloud services becoming increasingly popular among the masses [1], virtualization has provided an approach to share resources among multiple users. Virtualization is achieved by two prominent technologies—virtual machines and containers. Virtual machines allow us to run different operating systems on a single host machine thus achieving hardware-level virtualization. However, since VMs have their own operating system, they are heavy on host machines and consequently takes time to load.

S. Chamoli (✉)
Graphic Era Hill University, Dehradun, India

Sarishma
Graphic Era University, Dehradun, India



T. K. Sharma et al. (eds.), *Soft Computing: Theories and Applications*,
Advances in Intelligent Systems and Computing 1381,
https://doi.org/10.1007/978-981-16-1696-9_24

Containers, on the other hand, are a lightweight alternative to VMs and have quicker startup times. Containers allow the sharing of a host kernel and thus achieves operating system-level virtualization. With the advent of Docker technology, the containerization of applications has been embraced by a whole lot of vendors [2]. However, since multiple containers share the same host kernel, security issues arise as a result. A malicious application can break out of the container and unethically access host kernel resources thereby compromising not only the host system but also all other containers running on it.

Bui [3] has given a detailed insight into Docker technology while focusing on the internal security aspect of Docker as well as external security modules that can be associated with Docker. Chelladhurai et al. [4] have discussed denial-of-service attacks on Docker architecture and proposed a memory-limit approach to prevent malicious containers from exhausting all the resources. Souppaya et al. [5] have given a detailed explanation of various possible attack surfaces in Docker technology and approaches to deal with them. Best practices have been defined for images, registries, orchestrators, containers, and host OS. Zerouali et al. [6] have evaluated the frequency of outdated packages used in Docker images by examining official Docker Hub images using the concept of technical lag. Amith Raj et al. [7] have discussed some basic Docker security features of Namespace and CGroup while also focusing on Linux Security Modules, SELinux, and AppArmor. Yasrab [8] has discussed security attacks like denial of service, ARP spoofing, and kernel exploits that can occur in a containerized environment while also discussing possible solutions to them.

Docker containers are most often deployed on cloud and [9] has described some of the most common attacks on cloud and cryptography as a means of security practice against data stealing. DDOS attacks are common on Docker architecture and [10] has discussed intrusion detection approaches by applying SVM, KNN, genetic Algorithm, and fuzzy neural network classifiers to segregate malicious from normal traffic.

Applications themselves have vulnerabilities which may in turn lead to infecting the whole system. Manhas and Taterh [11] have described popular Google Chrome browser which itself shipped with a lot of vulnerabilities in the past which lead to attacks ranging from DOS to remote code execution.

Liu et al. [12] have showcased how Docker Hub, the biggest repository of Docker images, can be exploited to carry out attacks on a machine. Using run commands, malicious images, and vulnerable packages existing in the image, a system can be attacked. Duarte and Antunes [13] have focused on vulnerabilities, mentioned in CVE databases, which affect Docker container security. The study deduces that main cause for these vulnerabilities is incorrect permission management, and static code analyzers are ineffective in detecting these vulnerabilities.

In this paper, we provide an insight into Docker architecture, discussing its vital components. We propose a threat model where we focus on the interactions a container has with the outside world. We focus on some security issues that Docker containers might face and the best security practices to mitigate them. Finally, we propose a case study to highlight how a docker misconfiguration might prove to be fatal.

## 2 Docker Architecture

Docker is a container-based virtualization tool [14]. It provides a centralized place for managing all container-related activities and operations. Docker architecture basically consists of Docker Daemon, Docker Client, and Docker Hub. Let us discuss all these components in a bit more detail:

### *2.1 Docker Daemon*

A Docker Daemon manages all the containers. Containers are runtime versions of images and contain the application and all its dependencies. With the ease with which containers can be created and destroyed, it becomes essential to provide a central controlling authority over these containers. A Docker Daemon receives a REST API request from Docker Client and operates accordingly on the containers.

### *2.2 Docker Client*

Docker Client allows users to fire Docker commands to interact with Docker containers. Since REST API is used by Docker Client to interact with Docker Daemon, Docker Client and Docker Daemon can exist on different machines. Docker Client allows users to input basic commands for pulling images, building containers out of them, starting and stopping containers apart from a plethora of other commands for managing containers.

### *2.3 Docker Registry*

Docker provides a registry feature where we can upload and download docker images. By default, Docker Hub is the registry with which all Docker Daemons communicate to pull and push images, however, we can configure other registries as well. Some popular Docker registries are Docker Hub, Quay, Amazon Elastic Container Registry, and Google Container Registry.

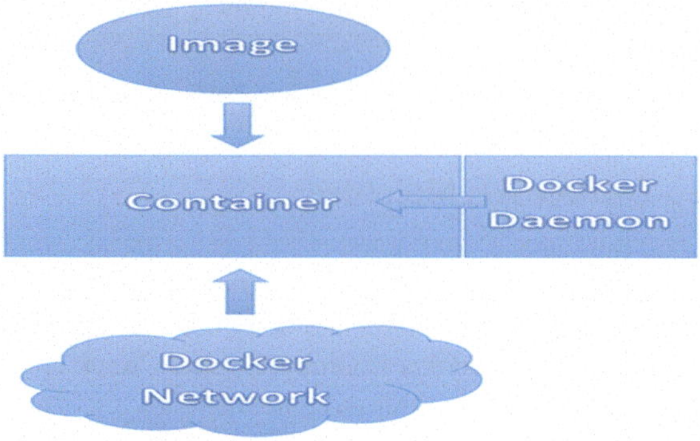


**Fig. 1** Docker architecture

## 3 Threat Model

The Docker system in itself is pretty secured if configured properly. If misconfigured, Docker containers can be subjected to a variety of attacks that might disrupt the entire application workflow. Hence, we classify a threat model based on types of attacks that can happen on a Docker system from the outside world (Fig. 1).

1. An attacker might inject malicious code in an image and upload it on Docker repository like Docker Hub, or an attacker can make use of an outdated library version(with a known vulnerability) to compose a docker image and then upload it on Docker repository. Since Docker containers are nothing but runtime versions of images, when a Docker image is downloaded and run, the container is vulnerable to attacks ranging from DDOS to whole host system meltdown.
2. An attacker might take advantage of a misconfigured Docker container and gain access to other containers and host system. For instance, a container created with root-level privilege might be compromised and consequently an attacker will get root privileges on the host system.
3. An attacker might infect a network to which Docker container is connected and then attack the containers. Unless configured, all containers are connected to a default network, therefore once the network is compromised, the containers are too.

## 4 Docker Security Issues and Solution

Docker Security is one of the foremost issues that come to mind while an organization uses Docker. While most of the security issues arise out of misconfiguration, others arise out of some limitations of the Docker framework. Let us discuss these issues in detail:

## 4.1 Issues Arising Out of Images

1. *Untrusted images* Docker containers are runtime versions of images. An image consists of an application and its dependencies. Docker has provided its official repository, Docker Hub—which contains thousands of trusted and verified images for users to use, as the default source for obtaining images, however, we can also pull images from some other source. Pulling images from untrusted sources and creating containers out of them can be dangerous to the host system [15]. Always images should be downloaded from a trusted source.
2. *Outdated images* Images in repositories may contain outdated packages [6], which might have some serious vulnerabilities. For enhancing features and fixing security bugs, new image versions are released by developers regularly in the Docker repository. However, if the same is not updated in our local machine, we might have an outdated version of an application running in our machine with probable loopholes known to the outside world. Images should be checked for their freshness consistently in our local machine by comparing it with the corresponding image tag in the Docker repository. With the ease with which containers can be built, using the new image version should not be an issue at all.
3. *Full-blown images* Docker images contain the application and its dependencies. These images are built on a base operating system. Having a complete Linux distro as the base for your application will prove to be a menace. Firstly the image size would drastically increase and secondly, a full-blown operating system would provide a larger attack surface to exploit. Various versions of Linux are available to use, like Alpine and Debian. These are very small in size and provide the bare essentials for application to run and thus cannot be exploited so easily.
4. *Confidentials in image* A Docker image is meant to be shared and distributed among various users. In certain circumstances, we need credentials to run a container from an image to use it. For instance, running a container from a MySQL image would result in an error until credentials are provided at runtime. Notice that the MySQL image does not contain any credentials and is thus safe to distribute. Including the secrets in the image would lead to secrets being embedded in the lower layers of image and which can be easily tampered with by a malicious user. Do not store credentials inside an image, instead, we can pass them at runtime using Orchestration tools like Kubernetes. Docker Secrets can be used to centrally store this data and securely transmit it to only those containers that need access to it [16].

## 4.2 Issues Arising Out of Docker Misconfiguration

1. *Root access to Containers* Docker Daemons need to run with root access to provide some features like port binding, mounting file systems, etc. By default, all containers created by Docker Daemons will have root access. However,

containers should never be run with root access as it provides an opportunity for an attacker to infect the container and consequently obtain root access on the host system. Always run the container in non-root mode [17]. Try running Docker containers with a non-root user:
*Docker run—user.*

2. *Allowing modifications to the Filesystems* Docker container often writes to the underlying file system, like most will be writing temp files. However, in some scenarios, they do not need any access. Creating containers in the default way allows them to modify the file system and thus could lead to a potential attack. Try running Docker container with read-only file systems:
*Docker run—read-only ubuntu.*
3. *No Memory constraint* Docker containers, if infected, can consume the whole memory of the host system and thus causing denial-of-service attacks onto other containers [4]. CGroup is a Linux kernel feature that Docker employs to restrict the amount of resources a container can use. The best practice is to allocate limited memory to Docker containers at startup:
*Docker run-it—m = 512 mb ubuntu.*

### *4.3 Issues Arising Out of Network*

1. *Same Network connectivity* By default, all Docker containers are connected to the same Docker network (Docker0). Containers might be working on sensitive information and thus such containers need to be isolated in a separate network. Use separate networks for containers according to the level of security needed. Try connecting sensitive containers to a user-defined separate network:
*Docker run—net = my_bridge alpine.*
2. *Unencrypted data in Overlay Network* By default, data transmitted between Docker hosts running on two different systems on overlay network is not encrypted. As such, a malicious container can oversee the communication between two containers which might jeopardize their security. We should try to encrypt network communication between containers, to be sure that even if data is sniffed upon, it is safe from prying eyes. While creating a user-defined network, encrypt all the communication using:
*Docker network create—opt encrypted—driver overlay.*

## 5 Case Study

Container Security is of paramount importance, especially in a cloud environment, where services are provided to users worldwide. A small misconfiguration could prove to be a target for attackers to exploit, which could lead to a whole system meltdown. Example of a misconfigured Docker attack is the popular Kinsing malware that targets open ports of Docker API and runs cryptominer in systems. These attacks have been going on for months, with thousands of attempts taking place nearly on a daily basis [18]. We will present a couple of scenarios to show how a misconfigured Docker container can be subjected to attacks and how to protect containers from such attacks.

## *5.1 System Configuration*

We have a Docker installed atop a virtual machine—allocated with 3 GB of RAM and 20 GB of disk space. We download Damn-Small-Vulnerable-Web (DSVW) image from Docker Hub, to perform a DOS attack on a Docker container.

## *5.2 Attack Scenario: 1*

We use Damn-Small-Vulnerable-Web (DSVW) Docker image, which is a pre-built image present in the Docker Hub and helps test a Web application against some popular vulnerabilities. We create a container out of this image and run it.

**Steps Involved**

1. We create and run the container from the DSVW image.
   *Docker run—p 1234:8000-it—name dsvw appsecco/dsvw.*
2. We open our browser and try to access localhost: 1234, so as to access the vulnerable Web application, DSVW.
3. We click *exploit* link at the bottom—under category—*denial of service (memory).*
4. We can see the entire system freeze within 10 s of our attack (Figs. 2 and 3).

**Protection from the Attack**

1. Try to limit the memory consumption of a container using memory and memory-swap parameters while creating the container.
   *Docker run—p 1234:8000—memory 100 m—memory-swap 100 m appsecco/dsvw.*

This way, we are trying to limit the memory used by the container to 100 MB. After this limit, the container stops running and the system is protected from memory exhaustion. We believe such memory restrictions can help prevent attackers from bringing down cloud containers and thus saving businesses money by preventing downtime (Figs. 4 and 5).

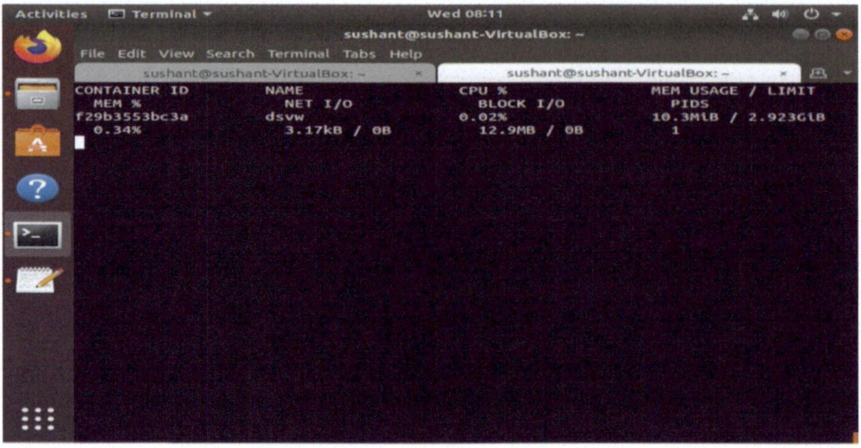


Fig. 2 Before DOS attack on container

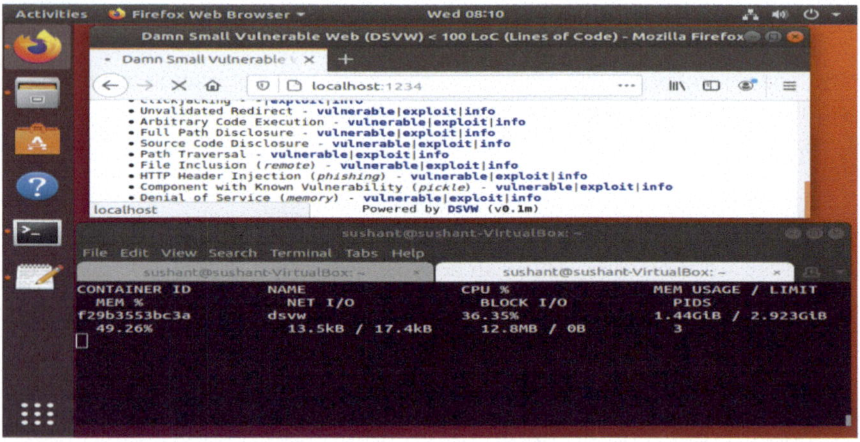


Fig. 3 After DOS attack on container

## 5.3 *Attack Scenario: 2*

### Steps Involved

1. We create two containers on the same network. We assume container1 to be running a normal application while container2 is running a data-sensitive application. Both are by default connected to the same Docker Bridge network.
   *Docker run—name container1 alpine/bin/bash*
   *Docker run—name container2 alpine/bin/bash*

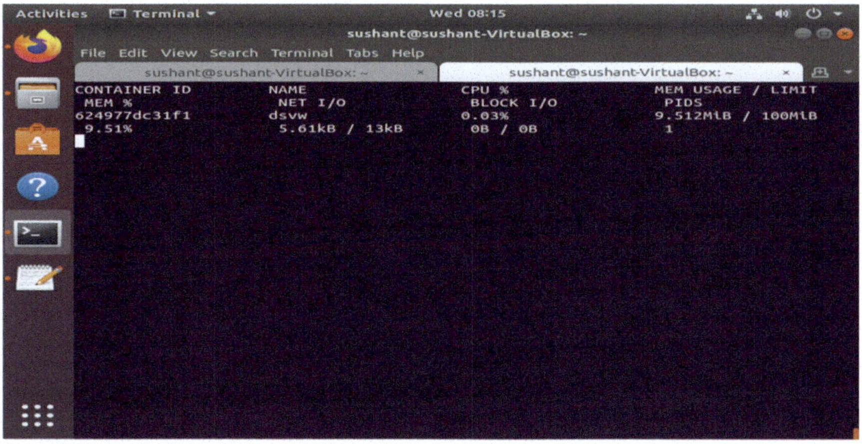


**Fig. 4** Before DOS attack on container (with memory limit applied)

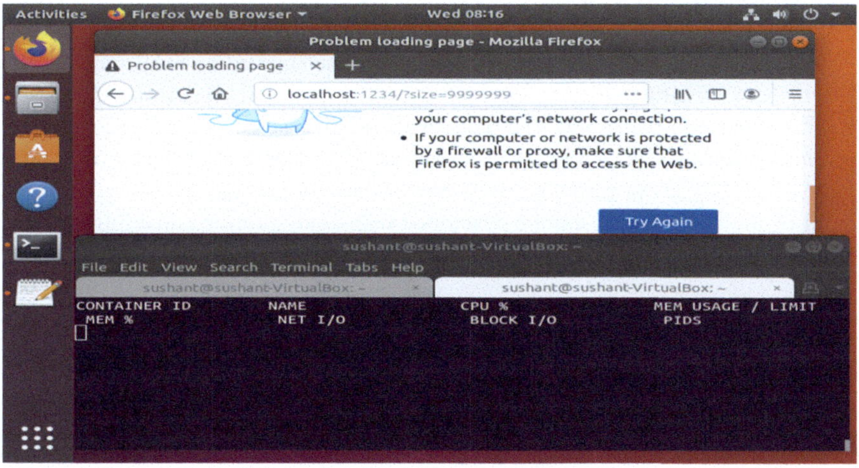


**Fig. 5** After DOS attack on container (with memory limit applied)

2. We try to ping container2 from container1, which is possible as both containers are on the same network. A malicious container can carry out an ICMP flood attack against container2 and bring it down.
   *ping—i 0.1—s 65,000 172.17.0.4*
   The above statement continuously pings a container, in a time interval of 0.1 s, with a large packet size of 65,000 bytes. Processing such huge packets at small intervals consumes system memory and ultimately brings the system down.

**Protection from the Attack**

1. Try to put sensitive containers in a separate network.

*Docker network create secureNetwork*
*Docker run—name container1 alpine /bin/bash*
*Docker run—name container2—net = secureNetwork alpine /bin/bash*
This way, we have container1 connected to default Docker Bridge network, while container2 is connected to secureNetwork. Now we cannot ping container2 from container1 as both are on different networks and it ensures customer privacy.

## 6 Conclusion

This research work has given us an insight into Docker technology, currently the most popular container-based virtualization tool in the market. We have discussed Docker components and proposed a threat model to identify the attack surfaces in the Docker architecture. We have segregated security loopholes into three main categories—issues arising out of images, docker misconfiguration, and network. We have dwelled into these security issues and suggested approaches on how to mitigate the attacks possible.

The issues described in this study needs to be refined according to business needs. Also, the use of Linux security modules like SELinux and AppArmor should be encouraged in the organization. The future work of this research would be developing an architecture that could enforce these best practices by default so that the user does not have to worry about missing out on any of these important security guidelines. An out-of-the-box solution can be developed with configuration capabilities to suit the organization's needs.

## References

1. Cloud Market Share: A look at the cloud ecosystem in 2020. In: Kinsta. https://kinsta.com/blog/cloud-market-share/. Accessed 29 Sep 2020
2. MarketsandMarkets: Application Container Market Size, Share and Global Market Forecast to 2023. In: Marketsandmarkets.com. https://www.marketsandmarkets.com/Market-Reports/application-container-market-182079587.html. Accessed 25 Sep 2020
3. Bui, T.: Analysis of Docker Security (2015). arXiv preprint arXiv:1501.02967
4. Chelladhurai, J., et al.: Securing docker containers from denial of service (dos) attacks. In: 2016 IEEE International Conference on Services Computing (SCC), IEEE (2016)
5. Souppaya, M., et al.: Application Container Security Guide (2nd Draft). National Institute of Standards and Technology (2017)
6. Zerouali, A., et al.: On the impact of outdated and vulnerable javascript packages in docker images. In: 2019 IEEE 26th International Conference on Software Analysis, Evolution and Reengineering (SANER), IEEE (2019)
7. Amith Raj, M.P., et al.: Enhancing security of docker using linux hardening techniques. In: 2nd International Conference on Applied and Theoretical Computing and Communication Technology (iCATccT), IEEE (2016)
8. Yasrab, R.: Mitigating Docker Security Issues (2018). arXiv preprint arXiv:1804.05039

9. Mishra, N., et al.: Secure Framework for Data Security in Cloud Computing. Soft Computing: Theories and Applications, pp. 61–71. Springer, Berlin (2018)
10. Aldubai, A.F., et al.: Analytical Study of Intruder Detection System in Big Data Environment. Soft Computing: Theories and Applications, pp. 405–416. Springer, Berlin (2018)
11. Manhas, S., Taterh, S.: A Comparative Analysis of Various Vulnerabilities Occur in Google Chrome. Soft Computing: Theories and Applications, pp. 51–59. Springer, Berlin (2018)
12. Liu, P., et al.: Understanding the security risks of docker hub. In: European Symposium on Research in Computer Security, Springer, Berlin (2020)
13. Duarte, A., Antunes, N.: An empirical study of docker vulnerabilities and of static code analysis applicability. In: 2018 Eighth Latin-American Symposium on Dependable Computing (LADC), IEEE (2018)
14. ZDNet: What is Docker and Why Is It So Darn Popular? In: ZDNet. https://www.zdnet.com/article/what-is-docker-and-why-is-it-so-darn-popular/. Accessed 22 Sep 2020
15. Sysdig: Docker Security Vulnerabilities and Threats. In: Sysdig. https://sysdig.com/blog/7-docker-security-vulnerabilities. Accessed 21 Sep 2020
16. Manage Sensitive Data with Docker Secrets. In: Docker Documentation. https://docs.docker.com/engine/swarm/secrets/. Accessed 19 Sep 2020
17. Docker Best Practices: Docker Security Best Practices. In: Snyk. https://snyk.io/blog/10-docker-image-security-best-practices/. Accessed 19 Sep 2020
18. Threat Alert: Kinsing Malware Attacks Targeting Container Environments. In: Blog.aquasec.com. https://blog.aquasec.com/threat-alert-kinsing-malware-container-vulnerability. Accessed 19 Sep 2020

# Breast DCE-MRI Segmentation for Lesion Detection Using Clustering with Multi-verse Optimization Algorithm

**Bikram Kar and Tapas Si**

**Abstract** The highest number of deaths among all types of cancers in women is caused by breast cancer. Therefore, early detection and diagnosis of breast cancer are very much needed for its treatment. Dynamic contrast-enhanced magnetic resonance imaging (DCE-MRI) is widely used for breast cancer diagnosis. In this paper, a segmentation method using modified hard-clustering technique with multi-verse optimizer (MVO) is proposed for the detection of breast lesion in DCE-MRI. The proposed method is termed as CMVO in this paper. First, MR images are denoised, and intensity inhomogeneities are corrected in the preprocessing steps. Then clustering technique is used in segmentation of the MR images. Finally, from the segmented images, lesions are extracted in the postprocessing step. The results of CMVO are compared with that of K-means algorithm and PSO-based hard clustering. The CMVO performs better than other methods in lesion detection in breast DCE-MRI.

**Keywords** Breast cancer · DCE-MRI · Segmentation · Clustering · Multi-Verse Optimizer

## 1 Introduction

### *1.1 Background*

According to World Health Organization (WHO)'s report [1], it is estimated that 627,000 women in the world have been died from breast cancer, and it is approximately 15% of all types of cancer deaths among women in 2018. Breast cancer is the most common cancer in women in India and accounts for 14% of all cancers

B. Kar
Department of Computer Science and Engineering, Sanaka Educational Trust's Group of Institutions, Malandighi, Durgapur, West Bengal 713212, India

T. Si (✉)
Department of Computer Science and Engineering, Bankura Unnayani Institute of Engineering, Bankura 722146, West Bengal, India



T. K. Sharma et al. (eds.), *Soft Computing: Theories and Applications*, Advances in Intelligent Systems and Computing 1381,
https://doi.org/10.1007/978-981-16-1696-9_25

in women [2, 3]. Organized and opportunistic screening programs in the developed countries result in a significant decrease in mortality caused due to breast cancer [4]. Recently, DCE-MRI is widely used for breast cancer detection, diagnosis and treatment planning or surgery.

## *1.2 Related Works*

This paper deals with the breast DCE-MRI segmentation for lesion detection. So many research contributions have been given in that area. Most of them are based on artificial inelegance and soft computing techniques. Here some related works on this fields are discussed. Wu et al. [5] proposed a method for segmentation of a lesion by breast DCE-MRI using Markov random field (MRF) model. In this method, breast image is taken first after that breast tissues are analyzed using contrast agent. This method develops an interactive segmentation process which identifies a region of interest from the whole breast MRI. ROI contains the suspected lesion. In this method, pre-contrast image is subtracted from the post-contrast image, and after that it will check any abnormality found or not in the subtracted image. Azmi and Norozi [6] proposed an advanced version of MRF method for the lesion segmentation in breast MR images. In this method to estimate the previous distribution, ratio of the two conditional probability distributions is applied. To measure the similarity between the pixels, texture features are used. Using this improved MRF model, we can solve computational complexity and sensitivity-related weakness. Jayender et al. [7] proposed a unique statistical learning algorithm to segment tumor. This algorithm used hidden Markov model (HMM), and it is very useful to identify and predict the particular pattern from noisy data. This HMM-based statistical learning method automatically segments the tumor cell from the healthy cell. Ribes et al. [8] devised an automatic breast MRI segmentation method using MRF. To implement this automatic segmentation method, two steps are used: adaptive denoising and statistical segmentation. For denoising purpose, a nonlinear filter anisotropic diffusion process is used, and a statistical model Markov random field was used for segmentation purpose. Shokouhi et al. [9] proposed an automatic computer-aided detection (CAD) system to detect benign and malignant breast lesion segmentation in DCE-MRI. Here, an advanced seeded region growing algorithm is used for segmentation purpose, and this region growing algorithm was developed using vesselness filter and fuzzy C-means (FCM) clustering method. In this system, at first, the motion artifacts are corrected, and then the breast segmentation is taken. Yao et al. [10] proposed a texture features and wavelet transform-based DCE-MRI technique to analyze breast tumor. In this system, co-occurrence matrix and run-length matrix are used to calculate the texture feature, and then wavelet transformation is used to extract the frequency feature. McClymont et al. [11] proposed a breast lesion segmentation procedure using mean-shift clustering algorithm and graph cut for automatic segment of doubtful breast lesions in multimodal breast MRI data. In this procedure, mean-shift clustering is used for feature extraction, and the graph cut is used to detect the clusters of similar suspicious cell. Wang et al. [12] proposed computerized 3D breast DCE-MRI system to segment and characterize the tumor using kinetic information and morphological features. In this proposed method, develop a computerized system to segment and characterized tumor with the help of kinetic information and morphological features of 3D breast DCE-MR images. In this method, an integrated color map is created by intersecting kinetic, and to detect suspected breast lesions, color map is used. An advanced FCM method was used to segment the tumor. Gubern-Mérida et al. [14] proposed a novel computer-aided system which automatically identifies lesion in breast DCE-MRI. In this method, at first correct the motion artifacts of the image and segment, and then blob and relative enhancement voxel features are used on the breast area to locate breast lesion candidate after that false positive reduction is used with morphological and kinetic features from the candidate region. Khalvati et al. [15] devised an atlas-based segmentation technique to segment the breast boundary cell in three-dimensional MRI. This proposed algorithm combines probablistic atlas and atlas selection approaches in one framework. Tan et al. [16] proposed an automatic segmentation method of breast MRI. First, the binary images are generated from the MR images to separate the background. Second, the undesired artifacts and noise are removed using multiple morphological operations. Finally, the chest contour masks are generated to separate the breast region from the chest. Al-Faris et al. [19]proposed an improved seeded region growing algorithm-based automatic tumor segmentation method in breast magnetic resonance imaging. This CAD system is able to generate the region of interest (ROI), initial seed and threshold values for seeded region growing without user interaction. This proposed methodology is divided into three stages. In the first stage, the axial images are split, and after that the noise is reduced. In the second stage, detection and deletion of breast skin are done with the help of two algorithms: level set active contour and morphological thinning. In the third stage ,tumor segmentation is done using improved seeded region growing method. Pandey et al. [18] proposed an automatic and quick segmentation method to detect ROI and density of breast. In this proposed method to segment the breast region of interest K-means clustering, adaptive filtering and for morphological operation local adaptive thresholding are used .For breast density segmentation, fuzzy C-means thresholding is used with mean value histogram. This method is very effective for breast MRI with high density. Recently, deep learning (DL) gains the popularity in segmentation of breast MRI. Xu et al. [17] devised a segmentation method for breast DCE-MRI using convolutional neural network (CNN), and a 2D U-Net was trained in supervised learning mode with training dataset and tested with test dataset. Benjelloun et al. [20] proposed deep learning-based automatic breast tumor segmentation method in DCE-MRI images. In this proposed method for image segmentation purpose, we use deep neural network which is based on U-Net architecture. This architecture generates good results on small dataset also. Arjmand et al. [21] proposed a method to segment the breast tumor using K-means clustering and cuckoo search optimization technique. Here K-means clustering algorithm is used to segment the input MRI grayscale images into three clusters: background, normal tissues and tumor cell, and this segmentation procedure is done on the basis of intensity level of the tissue. In K-means clustering algorithm, the primary centroid is selected randomly for this reason, the different final answer occurs for different choice, and another problem in K-means clustering is local minimum problem. So, cuckoo search optimization technique is used to find the optimal value of initial centroids. Ting et al. [22] proposed an CNN-based breast cancer detection algorithm. Previously used manual segmentation procedure to detect malignant cell was very time consuming, and it depends on machine and expertise of the operator. But this proposed CNNI-BCC algorithm used supervised deep learning techniques to classify breast cancer lesion. Zheng et al. [23] proposed a deep learning-based efficient AdaBoost algorithm to detect the malignant cell in breast. For early stage diagnosis and treatment, it is very important to classify the malignant cell from the benign cell. In this proposed method, deep learning-based CNN is used for the classification of tumor using breast masses, and in CNN, the segmentation is performed in each and every pixel. And in this method, efficeint AdaBoost algorithm is used to detect the breast cancer. Priya and Krishnaveni [24] proposed biomedical-based monitoring system 'LabView' using watershed algorithm to detect the cancer cell or tumor cell from the normal cell, and it is a very important part to detect cancer. Here watershed algorithm is used to detect the cancer cell from the medical images taken from MRI or CT scan. At first, the noise is reduced from the given images, then region of interest (ROI) is selected, and then watershed segmentation procedure is applied to separate the normal cell from the malignant cell.

### *1.3 Objectives*

The objective of this paper is the segmentation of the breast DCE-MRI for lesion detection. That is why modified hard-clustering techniques with MVO [29] algorithm are proposed to segment the lesion in breast DCE-MRI in this paper. This paper also demonstrates the novel application of MVO algorithm in medicine. The experimental results of the proposed method are compared with the K-means algorithm and PSO-based hard-clustering technique. The empirical results show that the proposed CMVO outperforms other methods used in the comparative study.

## 2 Materials and Methods

### *2.1 DCE-MRI Dataset*

Total 10 Sagittal T2-weighted DCE-MRI slices of two patients have been taken from Cancer Genome Atlas Breast Invasive Carcinoma (TCGA-BRCA) [25, 26]. All MRI slices having a size greater than $256 \times 256$ are resized to $256 \times 256$. All the breast DCE-MRI slices are given in Fig. 1.

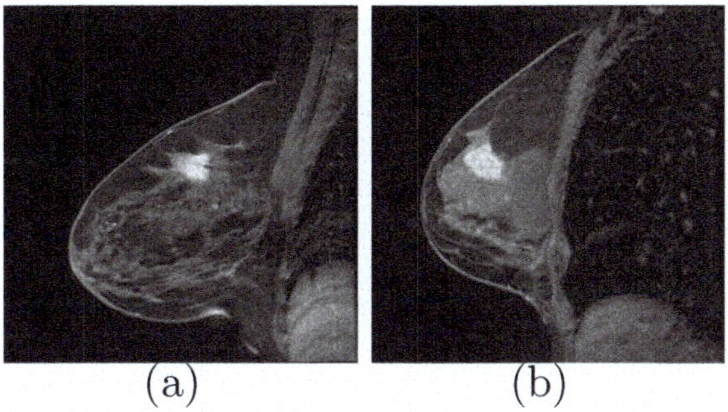


**Fig. 1** Breast DCE-MR images (representative). **a** MRI slices of patient-1, and **b** MRI slices of patient-2

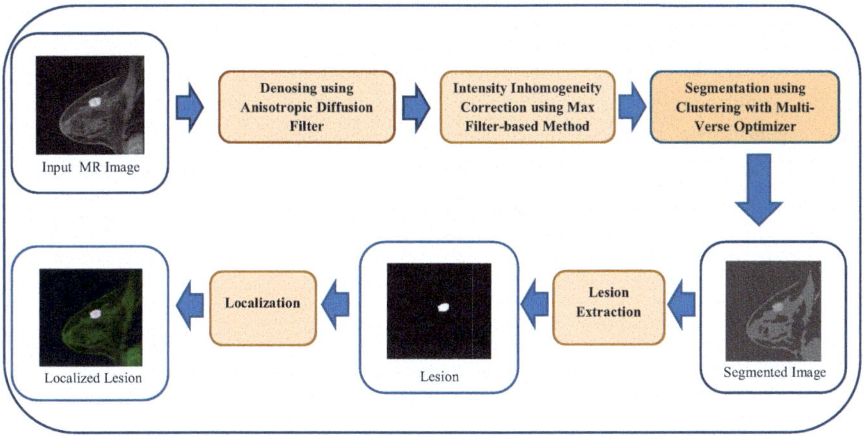


**Fig. 2** Outline of the proposed method

## 2.2 *Proposed Method*

The proposed methods have three steps as follows: (1) preprocessing, (2) segmentation and (3) postprocessing. The outline of the proposed FWA-based segmentation method is given in Fig. 2.

### 2.2.1 Preprocessing

The difficulties due to the presence of noise and intensity inhomogeneities (IIH) in the MR images are faced by segmentation process. Therefore, noise is removed using anisotropic diffusion filter [27]. Then IIH is corrected using max filter-based method [28].

### 2.2.2 Segmentation Using Clustering with MVO

In this paper, hard (i.e., partitional)-clustering technique with MVO is applied to cluster the breast MR images for segmentation. Clustering is an unsupervised learning technique widely applied to biomedical data such as MR images [32–34]. K-means algorithm [35] is a well-known clustering technique, but its main drawback is being suffering from premature convergence in the local optima. That is why metaheuristc-based clustering technique has drawn the attention because it performs better clustering than K-means algorithm. In metaheuristic-based clustering, metaheuristic such as particle swarm optimizer (PSO) and fireworks algorithm (FWA) are used to search the optimal cluster centers in the data. In this work, MVO-based clustering technique, each universe in MVO represents a clustering solutions, i.e., cluster centers, and a predefined objective function is evaluated as fitness function, i.e., inflation rate of the universe. Let now assume $C_i(x_{i1}, x_{i2}, \ldots, x_{id})$ is the cluster center where $d$ is the dimension of the data. The cluster centers are arranged in a linear array by placing one center after one another like $\{C_1, C_2, \ldots, C_i, \ldots, C_K\}$ where $K$ is cluster number in the data and $2 \leq K < N$. A pictorial representation of this is given in Fig. 3.

In [30], hard clustering with MVO was used for binarization of handwritten documents, and the intra-cluster spread (ICS) was used as an objective function defined as follows:

$$\text{ICS} = \sum_{k=1}^{\mathcal{K}} \sum_{\mathbf{x}_i \in C_k} \| \mathbf{x}_i - \mathbf{m}_k \|^2 \tag{1}$$

In a good clustering solution, a good *trade-off* between compactness (i.e., intra-cluster distance) and separation (i.e., inter-cluster distance) is needed. In ICS, only intra-cluster distance in all cluster is considered, and it is minimized using MVO algorithm in that study. Separation is also needed to be maximized for obtaining good clustering solutions. In this paper, MVO-based clustering technique is modified by incorporating *DB-index* [31] as objective function in it in place of ICS. *DB-index* is the ratio of intra-cluster compactness and inter-cluster separation. It is discussed in Sect. 2.2.

*MVO Algorithm* In this paper, for breast DCE-MRI segmentation, MVO algorithm [29] is used for hard clustering. This MVO algorithm is based on the big bang theory. Three most important ideas of the multi-verse concept are as follows:

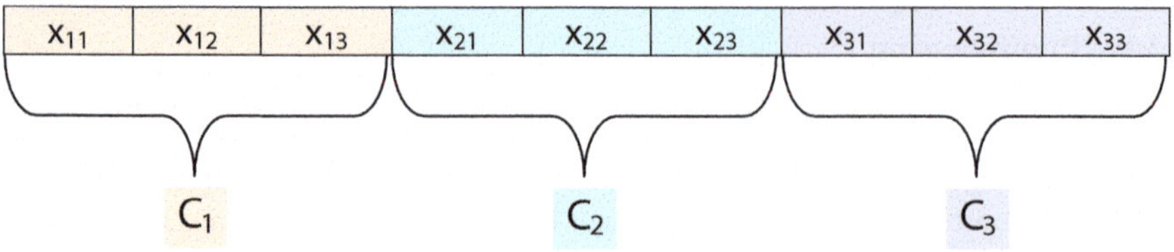


**Fig. 3** Universe representation in MVO

white holes, black holes and wormholes, and these are the base of MVO algorithm. MVO is a population-based optimization algorithm. Here the white hole and black holes help in exploration of the search spaces in MVO, and the wormholes assist in exploitation. At the time of MVO optimization process, the following rules are applied to the universe(solution): (1) If the inflation rate will be high, then the chances of white hole will be high; (2) If the inflation speed will be high, then the chances of black holes will be low; (3) Solution with high inflation speed tends to transfer the object through white holes; (4) Solution with low inflation rate receives more object through black holes, and (5) Object in all universe (solution) is moved across the best universe(solution) through wormholes.

Let assume $\mathcal{X}_i = (x_{i1}, x_{i2}, \ldots, x_{i\mathcal{D}})$ is the $i$th universe in the space, and the $j$th object of the $i$th universe is represented by each component $x_{ij}$ where $\mathcal{D}$ is the dimension of the universe. The inflation rate ($I$) of the $i$th universe is represented by the fitness, i.e., the objective function value $f(\mathcal{X}_i)$. For object movement, first the $k$th universe $\mathcal{X}_k$ has been selected using roulette wheel selection strategy, and then to move the $j$ objects of the $k$th universe to the $i$th universe, the following equation is used:

$$x_{ij} = \begin{cases} x_{kj} & \text{if } r_1 < \mathrm{NI}(\mathcal{X}_i) \\ x_{ij} & \text{if } r_1 \geq \mathrm{NI}(\mathcal{X}_i) \end{cases} \tag{2}$$

where $r_1$ is the uniformly distributed random number in (0,1) and $\mathrm{NI}(\mathcal{X}_i)$ is the normalize inflation rate of the $i$th universe. In order to make local changes for each universe through wormholes for exploitation or local search, the following equation is used:

$$x_{ij} = \begin{cases} \begin{cases} x_{ij} + \mathrm{TDR} \times (\mathcal{X}_{\min} + (\mathcal{X}_{\max} - \mathcal{X}_{\min}) \times r_4) & r_3 < 0.5 \\ x_{ij} - \mathrm{TDR} \times (\mathcal{X}_{\min} + (\mathcal{X}_{\max} - \mathcal{X}_{\min}) \times r_4) & r_3 \geq 0.5 \end{cases} & r_2 < \mathrm{WEP} \\ x_{ij} & r_2 \geq \mathrm{WEP} \end{cases} \tag{3}$$

where $r_2$, $r_3$ and $r_4$ are uniformly distributed random number in (0, 1). The traveling distance rate (TDR) and the wormhole existence probability (WEP) are two important parameters. TDR is defined as follows:

$$\mathrm{TDR} = 1 - \frac{t^{\frac{1}{p}}}{T^{\frac{1}{p}}} \tag{4}$$

where $T$ is the maximum number of iterations, $t$ is the current iteration, and $p$ is the exploitation accuracy over iterations. The higher $p$ makes the MVO to have faster and more accurate exploitation during the search. WEP is defined as follows:

$$\mathrm{WEP} = \mathrm{WEP}_{\min} + (\mathrm{WEP}_{\max} - \mathrm{WEP}_{rmmin}) \times \frac{t}{T} \tag{5}$$

$\mathrm{WEP}_{\min}$ and $\mathrm{WEP}_{\max}$ are minimum and maximum range of WEP, respectively.

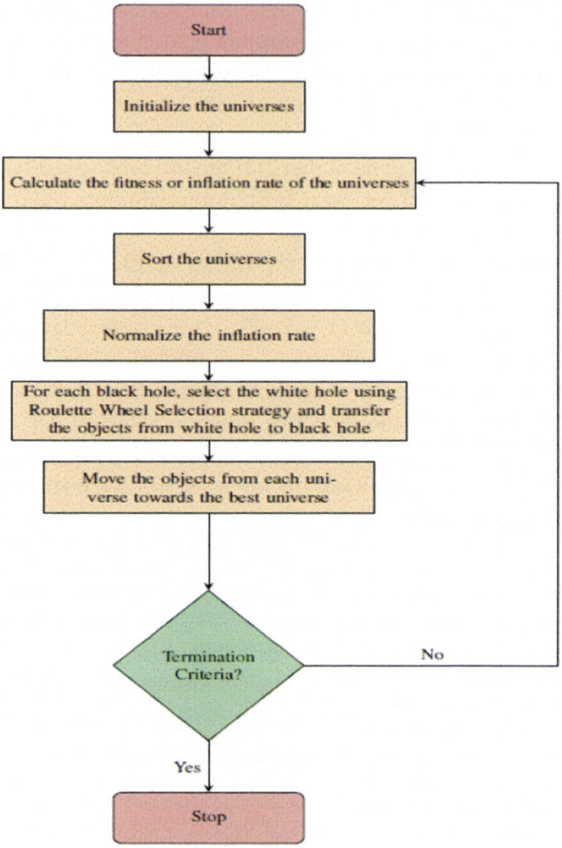


**Fig. 4** Flowchart of MVO algorithm

The flowchart of MVO is given in Fig. 4.

*Objective Function* In the current work, the objective function in the MVO-based clustering technique (CMVO) is *DB-index*. Previously, *DB-index* was used to validate the clustering performance in the segmentation of brain MRI [32–34]. *DB-index* is the ratio of intra-cluster compactness and inter-cluster separation. $i$th intra-cluster compactness is computed as follows:

$$S_{i,t} = \left[ \frac{1}{N_i} \sum_{\mathbf{X} \in C_i} \|\mathbf{X} - \mathbf{m_i}\|^t \right]^{\frac{1}{t}} \tag{6}$$

Inter-cluster distance, i.e., separation between $i$th and $j$th cluster is computed as follows:

$$d_{ij,p} = \left[ \sum_{i=1}^{d} |m_{i,k} - m_{j,k}|^p \right]^{\frac{1}{p}} = \|\mathbf{m_i} - \mathbf{m_j}\|^p \tag{7}$$

where $\mathbf{m_i}$ is the $i$th cluster centroid, $t, p \geq 1, t, p$ are integer, and $N_i$ is the number of elements in the $i$th cluster $C_i$. *DB–index* is finally measured by the following equation:

$$DB(K) = \frac{1}{K}\sum_{i=1}^{K} \max_{j \in K, j \neq i} \left\{ \frac{S_{i,t} + S_{j,t}}{d_{ij,p}} \right\} \quad (8)$$

A valid optimal clustering is indicated by the smallest *DB–index*. In the current work, *DB–index* is used as objective function, i.e., inflation rate for universe in MVO.

### 2.2.3 Postprocessing

After clustering the MR images using CMVO, the lesions are extracted from clustered images. The lesions in MR images have the hyperintensities, and pixels with hyper-intensities are selected from the clustered images. Finally, the lesions are localized in the original MR images.

## 3 Results and Discussion

At the outset, noises in the breast MR images are removed using anisotropic diffusion filter. Then intensity inhomogeneities are corrected. The proposed clustering method CMVO is applied to segment the preprocessed MR image with cluster number 3. The parameters of MVO are set as follows: number of universe ($N$) = 30, $\text{WEP}_{\text{min}}$ = 0.2, $\text{WEP}_{\text{max}}$ = 1, exploitation accuracy ($p$) = 6. The algorithm is terminated when it reaches maximum number of function evaluation = 3000, or there is no improvement in the best solution in successive 20 iterations. The performance of clustering is measured using *DB-index*. The same experiment is repeated for 51 runs independently. The results of CMVO is compared with CPSO and K-means algorithms. The mean and standard deviation (in parenthesis) of *DB-index* values are reported in Table 1. It is observed that the mean *DB-index* values of CMVO are lower than that of other methods which indicates the CMVO performs better clustering as well as segmentation than others. To test the statistical significance of the quantitative results, Wilcoxon signed rank test [36], the nonparametric test is conducted. The significance level ($\alpha$) of the test is 0.01, and the test results are given in Table 2. It is noticed that CMVO statistically outperforms both CPSO and K-means methods.

The clustered images obtained from the best run of each method are given in Fig. 5. The extracted lesions from the clustered images (Fig. 5) are given in Fig. 6. By observing the lesion images of CPSO and CMVO by open eyes, it cannot be differentiated, and it is difficult to say which one performs better segmentation based on the best results obtained by these two methods. But it can be easily observed that both methods detect a small number of healthy tissues as lesions in the MRI slice of

**Table 1** Mean and standard deviation (in parentheses) of DB-index values over 51 independent runs

| MRI# | CPSO | K-means | CMVO |
|---|---|---|---|
| 1 | 0.08696 (0.0011) | 0.14458 (9.42E-05) | **0.08524 (1.69E–05)** |
| 2 | 0.09125 (0.0012) | 0.14793 (0.0004) | **0.08940 (1.04E–05)** |
| 3 | 0.09272 (0.0008) | 0.14885 (0.0012) | **0.09113 (8.73E–06)** |
| 4 | 0.09132 (0.0015) | 0.15166 (1.41E-16) | **0.08939 (2.05E–05)** |
| 5 | 0.08833 (0.0008) | 0.14857 (1.06E-06) | **0.08728 (6.98E–05)** |
| 6 | 0.08821 (0.0006) | 0.14252 (2.05E-05) | **0.08480 (0.0002)** |
| 7 | 0.08842 (0.0009) | 0.14459 (0.0002) | **0.08284 (4.45E–05)** |
| 8 | 0.08823 (0.0011) | 0.14703 (9.29E-06) | **0.08157 (0.0003)** |
| 9 | 0.08802 (0.0005) | 0.14346 (0.0003) | **0.08237 (9.98E–05)** |
| 10 | 0.08832 (0.0013) | 0.13813 (2.25E-06) | **0.08330 (0.0002)** |

Results in bold face indicates better

**Table 2** Wilcoxon signed ranks test statistics on mean *DB-index* over 51 independent runs. $R^+$: sum of positive ranks, $R^-$: sum of negative ranks

| S. No. | Comparison | $R^+$ | $R^-$ | $Z$ | $p$(2-tailed) | Winner |
|---|---|---|---|---|---|---|
| 1 | CPSO versus CMVO | 55 | 0 | −2.803060 | 0.005062 < 0.01 | CMVO |
| 2 | K-means versus CMVO | 55 | 0 | −2.803060 | 0.005062 < 0.01 | CMVO |

the first patient (Fig. 6a, c). It is also observed that CMVO detects lesions better than K-means. In the lesion images (Fig. 6b, e), K-means methods detect a large number of healthy tissues as lesions. The localized lesions in the original MR images are given in Fig. 7.

Robustness is an important performance criterion of metaheuristic algorithms, and it is very much needed in segmentation of breast MR images. The lower standard deviation of the performance measures over multiple runs indicate a higher robustness. The standard deviation of CMVO is very much lower than that of CPSO for all the images. But K-means has the lower standard deviation than CMVO, whereas its segmentation performance is very poor.

The experiment is conducted on a Desktop PC having Intel(R) Core(TM) i7-4770 @3.40 GHz CPU, 8 GB RAM Windows 7 Professional 64-bit OS and MATLAB 2016b software. The average computational time costs for CMVO, CPSO and K-means are 81.70, 63.22 and 0.1135 s, respectively. The average computational time of K-means is very low compared to CMVO and CPSO, but its performance is not satisfactory. Though the computational cost of CMVO is higher than CPSO, its performance is more satisfactory than CPSO.

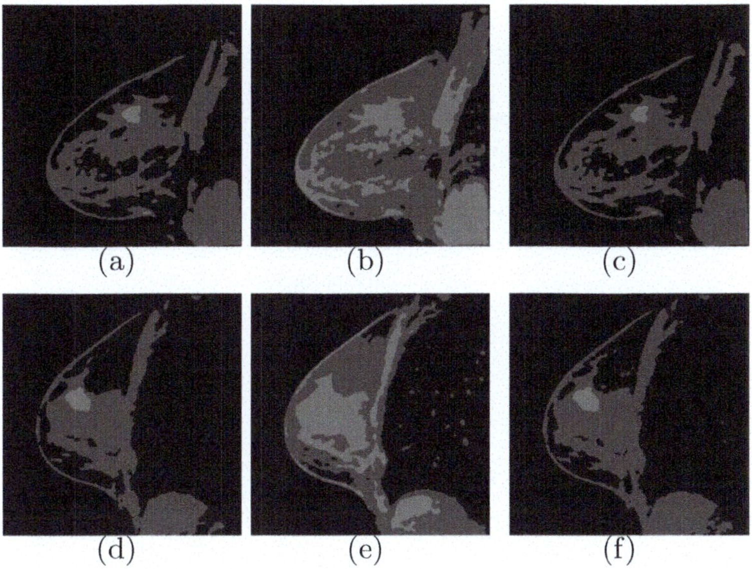


**Fig. 5** Clustered images generated using different methods for patient-1. First column: CPSO, second column: K-means, third column: CMVO

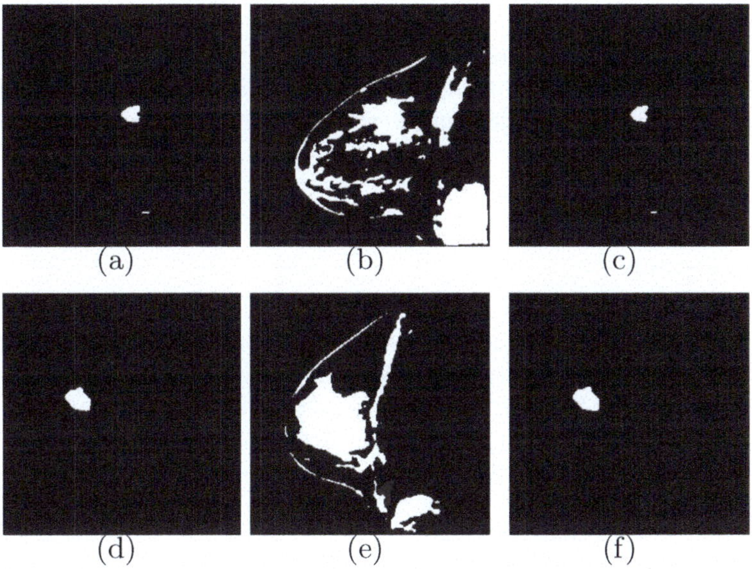


**Fig. 6** Extracted lesions from segmented images generated using different methods for patient-1. First column: CPSO, second column: K-means, third column: CMVO

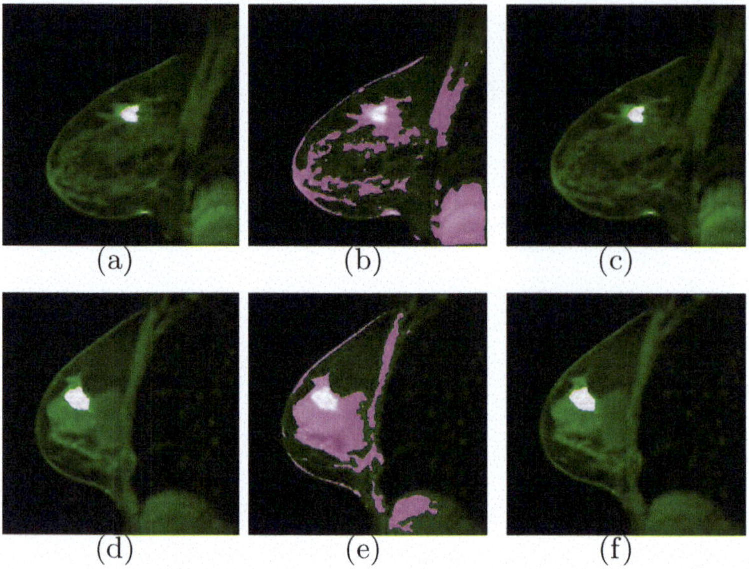


**Fig. 7** Localized lesions (bright-colored spot) in MR images for patient-1. First column: CPSO, second column: K-means, third column: CMVO

From the discussion of both quantitative (*DB-index*) and qualitative (visual) results, it may be concluded that the CMVO performs better than CPSO and K-means. It is efficient and effective in segmentation of breast MR images as well as lesion detection in breast.

## 4 Conclusion

A segmentation method is developed in this paper for breast DCE-MRI using MVO. In the proposed method, clustering with MVO is applied in the segmentation of the breast MR images for lesion detection. The segmentation results are compared with CPSO and K-means algorithm. The proposed methods statistically outperform other methods. The proposed method is efficient and effective in lesion detection in breast MR image. In this work, only gray levels are used as feature in clustering. In the future work, wavelet and texture features will be utilized in clustering leading to performance improvement.

## References


1. https://www.who.int/cancer/prevention/diagnosis-screening/breast-cancer/en/. Accessed 05 Dec 2019
2. Ferlay, J., Soerjomataram, I., Ervik, M., et al.: GLOBOCAN 2012 v1.0, Cancer Incidence and Mortality Worldwide: IARC CancerBase No. 11 [Internet]. F. Lyon, International Agency for Research on Cancer (2013)
3. Bray, F., Ren, J.S., Masuyer, E., et al.: Estimates of global cancer prevalence for 27 sites in the adult population in 2008. Int. J. Cancer **132**(5), 1133–45 (2013)
4. Shetty, M.K. (eds.) Breast Cancer Screening and Diagnosis—A Synopsis. Springer, New York (2015)
5. Wu, Q., Salganicoff, M., Krishnan, A., Fussell, D.S., Markey, M.K.: Interactive lesion segmentation on dynamic contrast enhanced breast MR using a Markov Model, Medical Imaging 2006: image processing. In: Reinhardt, J.M., Pluim,J.P.W. (eds.) Proceedings of SPIE, vol. 6144, 61444M (2006)
6. Azmi, R., Norozi, N.: A new Markov random field segmentation method for breast lesion segmentation in MR images. J. Med. Signals Sens. **1**(3) (2011)
7. Jayender, J., Gombos, E., Chikarmane, S., Dabydeen, D., Jolesz, F.A., Vosburgh, K.G.: Statistical learning algorithm for in situ and invasive breast carcinoma segmentation. Comput. Med. Imaging Graph. **37**, 281–292 (2013)
8. Ribes, S., Didierlaurent, D., Decoster, N., Gonneau, E., Risser, L., Feillel, V., Caselles, O.: Automatic segmentation of breast MR images through a Markov random field statistical model. IEEE Trans. Med. Imaging **33**(10), 1986–1996 (2014)
9. Shokouhi, S.B., Fooladivanda, A., Ahmadinejad, N.: Computer-aided detection of breast lesions in DCE-MRI using region growing based on fuzzy C-means clustering and vesselness filter. EURASIP J. Adv. Signal Process. **39** (2017)
10. Yao, J., Chen, J., Chow, C.: Breast tumor analysis in dynamic contrast enhanced MRI using texture features and wavelet transform. IEEE J. Sel. Top. Signal Process. **3**(1), 94–100 (2009)
11. McClymont, D., Mehnert, A., Trakic, A., Kennedy, D., Crozier, S.: Fully automatic lesion segmentation in breast MRI using mean-shift and graph-cuts on a region adjacency graph. J. Magn. Reson. Imaging **39**, 795–804 (2014)
12. Wang, T.C., Huang, Y.H., Huang, C.S., Chen, J.H., Huang, G.Y., Chang, Y.C., Chang, R.F.: Computer-aided diagnosis of breast DCE-MRI using pharmacokinetic model and 3-D morphology analysis. Magn. Reson. Imaging **32**, 197–205 (2014)
13. Sim, K.S., Chia, F.K., Nia, M.E., Tso, C.P., Chong, A.K., Abbas, S.F., Chong, S.S.: Breast cancer detection from MR images through an auto-probing discrete Fourier transform system. Comput. Biol. Med. **49**, 46–59 (2014)
14. Gubern-Mérida, A., Kallenberg, M., Mann, R.M., Martí, R., Karssemeijer, N.: Breast segmentation and density estimation in breast MRI: a fully automatic framework. IEEE J. Biomed. Health Inform. **19**(1), 349–357 (2015)
15. Khalvati, F., Ortiz, C.G., Balasingham, S., Martel, A.L.: Automated segmentation of breast in 3D MR images using a robust atlas. IEEE Trans. Med. Imaging **34**(1) (2015)
16. Tan, Y., Liu, L., Liu, Q., Wang, J., Ma, X., Ni, H.: Automatic breast DCE-MRI segmentation using compound morphological operations. In: Proceedings of 4th International Conference on Biomedical Engineering and Informatics (BMEI), pp. 147–150 (2011)
17. Xu, X., Fu, L., Chen, Y., Larsson, R., Zhang, D., Suo, S., Hua, J., Zhao, J.: Breast region segmentation using convolutional neural network in dynamic contrast enhanced MRI. In: Proceedings of of 40th Annual International Conference of the IEEE Engineering in Medicine and Biology Society (EMBC), Honolulu, HI, USA, pp. 750–753 (2018)
18. Pandey, D., Yin, X., Wang, H., Su, M.Y., Chen, J.H., Wu, J., Zhang, Y.: Automatic and fast segmentation of breast region-of-interest (ROI) and density in MRIs. Heliyon **4**, (2018). https://doi.org/10.1016/j.heliyon.2018.e01042

19. Al-Faris, A.Q., Ngah, U.K., Isa, N.A.M., Shuaib, I.L.: Computer-aided segmentation system for breast MRI tumour using modified automatic seeded region growing (BMRI-MASRG). J. Digit Imaging **27**(1), 133–144 (2014)
20. Benjelloun, M., Adoui, M.E., Larhmam, M.A., Mahmoudi, S.A.: Automated breast tumor segmentation in DCE-MRI using deep learning, In: Proceedings of 4th International Conference on Cloud Computing Technologies and Applications (Cloudtech) (2018). https://doi.org/10.1109/CloudTech.2018.8713352
21. Arjmand, A. , Meshgini, S., Afrouzian, R., Farzamnia, A.: Breast tumor segmentation using K-means clustering and cuckoo search optimization, In: Proceedings of 9th International Conference on Computer and Knowledge Engineering (ICCKE). https://doi.org/10.1109/ICCKE48569.2019.8964794
22. Ting, F.F., Tan, Y.J., Sim, K.S.: Convolutional neural network improvement for breast cancer classification. Expert Syst. Appl. **120**, 103–115 (2019)
23. Zheng, J., Lin, D., Gao, Z., Wang, S., He, M., Fan, J.: Deep learning assisted efficient AdaBoost algorithm for breast cancer detection and early diagnosis. IEEE Access **8**. https://doi.org/10.1109/ACCESS.2020.2993536
24. Priya, H., Krishnaveni, M.S.: LabVIEW implemented breast cancer detection using watershed algorithm. In: Proceedings of 5th International Conference on Communication and Electronics Systems (ICCES) (2020). https://doi.org/10.1109/ICCES48766.2020.9137945
25. Lingle, W., Erickson, B.J., Zuley, M.L., Jarosz, R., Bonaccio, E., Filippini, J., Gruszauskas, N.: Radiology data from the cancer genome atlas breast invasive carcinoma [TCGA-BRCA] collection. Cancer Imaging Arch. (2016) http://doi.org/10.7937/K9/TCIA.2016.AB2NAZRP
26. Clark, K., Vendt, B., Smith, K., Freymann, J., Kirby, J., Koppel, P., Moore, S., Phillips, S., Maffitt, D., Pringle, M., Tarbox, L., Prior, F.: The cancer imaging archive (TCIA): maintaining and operating a public information repository. J. Digit. Imaging **26**(6), 1045–1057 (2013)
27. Mohana, J., Krishnavenib, V., Guo, Y.: A survey on the magnetic resonance image denoising methods. Biomed. Signal Process. Control **9**, 56–69 (2014)
28. Balafar, M.A., Ramli, A.R., Mashohor, S.: A new method for MR grayscale inhomogeneity correction. Artif. Intell. Rev. **34**, 195–204 (2010)
29. Mirjalili, S., Mirjalili, S.M., Hatamlou, A.: Multi-verse optimizer: a nature-inspired algorithm for global optimization. Neural Comput. Appl. **27**, 495–513 (2016). https://doi.org/10.1007/s00521-015-1870-7
30. Elfattah, M.A., Hassanien, A.E., Abuelenin, S., Bhattacharyya, S.: Multi-verse optimization clustering algorithm for binarization of handwritten documents. In: Bhattacharyya, S., et al. (eds.) Recent Trends in Signal and Image Processing, Advances in Intelligent Systems and Computing, vol. 727. https://doi.org/10.1007/978-981-10-8863-6_17
31. Davies, D.L., Bouldin, D.W.: A cluster separation measure. IEEE Trans. Pattern Anal. Mach. Intell. **1**(2), 224–227 (1979)
32. Si, T., De, A., Bhattacharjee, A.K.: Brain MRI segmentation for tumor detection using Grammatical Swarm based clustering algorithm. In: Proceedings of IEEE International Conference on Circuits, Power and Computing Technologies (ICCPCT-2014), Nagercoil, India (2014)
33. Si, T., De, A., Bhattacharjee, A.K.: Grammatical swarm based segmentation methodology for lesion segmentation in brain MRI. Int. J. Comput. Appl. **121**(4), 1–8 (2015)
34. Si, T., De, A., Bhattacharjee, A.K. MRI brain lesion segmentation using generalized opposition-based glowworm swarm optimization. Int. J. Wavelets Multiresol. Inf. Process. **14**(5), 1650041 (2016) (29 p)
35. MacQueen, J.B.: Some methods for classification and analysis of multivariate observations. In: Proceedings of 5th Berkeley Symposium on Mathematical Statistics and Probability. 1, pp. 281–297. University of California Press (1967)
36. Derrac, J., Garcia, S., Molina, D., Herrera, F.: A practical tutorial on the use of nonparametric statistical tests as a methodology for comparing evolutionary and swarm intelligence algorithms. Swarm and Evolutionary Computation **1**, 3–18 (2011)

# Using Linguistically Non-local Punjabi Queries to Search the Global Web

**Harjit Singh**

**Abstract** Most part of the web is based on English language. The information, one requires may be available in English-based Web sites that are returned by the search engines based on some keywords extracted from the search query. Although, web searching is now allowing non-English queries also, but when non-English queries are used to search the web, the search engine returns Web sites related to the language that is used for search query. These local language Web sites may not provide relevant information to the user. We can call these queries as linguistically local queries. This paper presents a procedure to search the global English web using linguistically non-local Punjabi language queries. The input queries are processed through various natural language processing techniques such as tokenization, stemming, translation, transliteration, etc. The processing extracts and generates web search keywords, which are then used to reformulate the web search query to get the globally relevant information. It can further be enhanced to enable multi-language web search to get relevant information across the web.

**Keywords** Non-English web search · Natural language processing · Non-local web search in local language

## 1 Introduction

World Wide Web is a major source of information, and most of the knowledge resources are available in English language. Although with multi-lingual support, the local languages are now being used in Web sites, but as second-level languages [1]. It is preferred to develop bi-lingual Web sites in which English is considered as the first language [2]. So, in this information era, most of the information on the web is available in English language.

Search engines are used by the people to retrieve the information from the web. These search engines rank the Web sites using various parameters. If we search for

H. Singh (✉)
Punjabi University, Patiala, India



T. K. Sharma et al. (eds.), *Soft Computing: Theories and Applications*,
Advances in Intelligent Systems and Computing 1381,
https://doi.org/10.1007/978-981-16-1696-9_26

some information on the web, the results are displayed by the search engines as per the ranking of Web sites [3]. If we want to access global information about some topic, we still have to search the web in English language [4]. Although, the search engines are providing multi-lingual support, but searching the web in a local language provides results that are available in that local language [5]. In simple words, if we try to search for same information using an English query and a local language query, totally different results are returned by the search engine. The information returned by the search engine using local language query may not be the most relevant information [6]. The objective of an information retrieval system should always be to provide most relevant information to the user.

It means even by providing the multi-lingual support through search engines, the users are not able to get the relevant information using local language queries. Punjabi is a language of about 100 million people worldwide which belongs to Punjab state in India and migrated Punjabi people in other countries such as Canada, US, UK, UAE, etc. [7–9] Someone can search the web using Punjabi, but search results provided contain Web sites that provide information in Punjabi. Some popular Punjabi Web sites are https://www.punjabiuniversity.ac.in, https://epaper.punjabitribuneonline.com, https://www.rozanaspokesman.in, https://www.ajitjalandhar.com, https://jagbani.punjabkesari.in. But if somebody wants to search global information related to a topic using Punjabi query, that person will not get the proper information. Such queries may be called linguistically non-local Punjabi queries (LNL Punjabi queries).

This paper presents a way to search the web for linguistically non-local information using Punjabi queries. The used methodology is a combination of various natural language processing (NLP) techniques of such as tokenization, stemming, translation, transliteration, etc. Tokenization and stemming are used as primary processing in NLP research such as for preprocessing [10], text summarization [11, 12], sentiment analysis [13], user review analysis [14], etc.

## 2 Literature Review

Zhang and Lin investigated the support of multiple languages in search engines [15]. The authors conducted a survey of existing search engines providing support for multiple languages. They compared and characterized the search engines based on evaluation criteria and described the strengths along with the weaknesses with multiple language support. They found that the search engines Google, Onlinelink, and EZ2Find are providing the best support for multiple languages and revealed that this feature is limited at lexical level.

In 2007, a workshop was organized to promote the research on web searching for non-English queries [16]. The papers for South Asian languages were presented. The workshop proceedings concluded that the search engines can be made more effective if they consider more properties of non-English languages and stressed the need for more research on real search situations.

Lewandowski revealed the problems that are faced by the web searchers when finding information in foreign languages [5]. They examined the capability of search engines Google, MSN, Yahoo, and Ask against English and German language documents. They submitted 50 search queries of each language and investigated the first 20 results returned by each search engine. They revealed that there were no problems in getting results in the same language that was used in the interface; but when the results were confined to a foreign language, MSN and Google faced difficulties.

Gao et al. introduced a way to utilize English-based web page ranking to improve ranking of Chinese web pages [17]. Their relevance ranker first takes a Chinese query as input and translates it into equivalent English query. The researchers produced a cross-lingual graph as a relationship among English and Chinese pages and then used the ranking method developed by Qin et al. [18]. They showed improvements over the existing mono-lingual Chinese web page rankers.

Mequannint et al. [19] invented a bi-lingual search engine for Amharic-English. Their system used a number of components. They developed a translation system that used a bidirectional dictionary to translate Amharic to English and vice versa. The system is also supported by a transliteration module to translate proper nouns. The web search engine was able to search information in Amharic and English. They used the open-source search engine named Nutch experimentation which includes indexing and ranking.

Rózsa et al. [20] studied the behavior of a particular group of web search users. The non-English users still have to use queries in English to fulfill their information requirements. They revealed that when users search in a foreign language such English, they face many challenges which do not exist if they use their native language. They adopted qualitative research methods for their study and described the differences in the behavior of web search users when searching in English and in their native language.

Fu [21] analyzed the reformulation strategies adopted by web searchers for mixed language queries. Their study was focused on Chinese-English queries. They made use of query logs as well as semi-structured interviews to do the analysis. These results can help researchers to make improvements in existing search engines, so that the engines can provide more appropriate results for mixed language queries.

## 3 Methodology

The adopted methodology is a combination of various techniques of NLP such as tokenization, stemming, translation, transliteration, etc. The motive is to extract keywords from the Punjabi web query and use those keywords to reformulate the query. Figure 1 shows the model of the system.

The overall architecture is divided into two modules named keyword extraction module and query reformulation module. These modules are explained as follows.

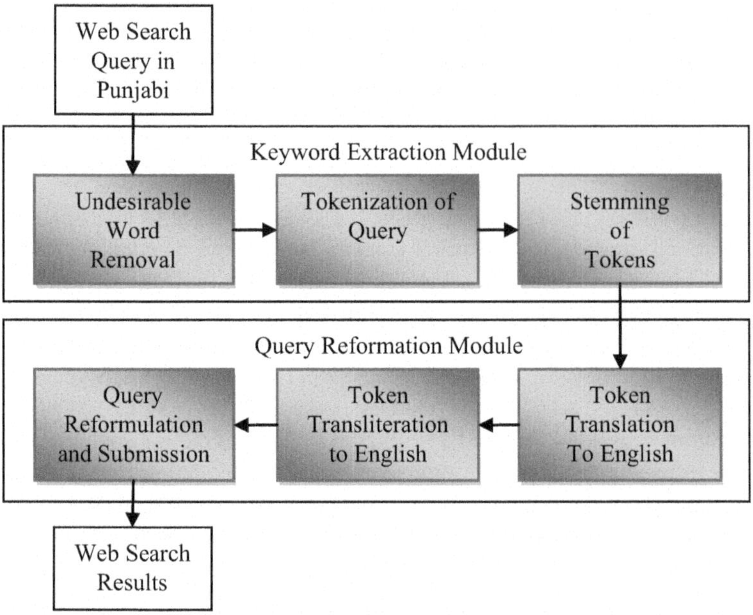


**Fig. 1** Model of the system

### *3.1 Keyword Extraction Module*

This module takes the web search query in Punjabi language as input and processes it to extract keywords from the query sentence. The Punjabi web query is processes to remove undesirable words. These are the common words that are used to make the sentence grammatically correct, but these are not desirable to search the web. For example, the helping verbs in a language are not required. It eliminates the unwanted words and shortens the web query. A list of undesirable words was generated by taking most of the data from the paper by Kaur et al. [22]. Any word in the web query that matches with a word in the list is removed from the query. The words that are not important to search engine are included in this list. The purpose is to keep only the important words to proceed further.

The shortened web query is split into individual words using white space as a delimiter. This process is called tokenization and individual words so obtained are called tokens [23]. These Punjabi tokens may be inflected words, i.e., the words may have suffixes attached to them [24]. For example, some words may have been made plurals by appending some suffix. These tokens are stemmed to remove the suffixes and keep the base words. A lightweight stemmer by Harjit et al. [24] is used to stem Punjabi tokens. The resultant set of tokens is the set of keywords extracted from the input Punjabi web query. These keywords are used by the search engine to produce its results.

## *3.2 Query Reformulation Module*

This module takes the list of Punjabi keywords as input. A translation engine is used to translate the individual keywords to English. The translation engine uses a bi-lingual dictionary prepared using the dataset from IndoWordNet [25]. The engine works by searching the Punjabi word in dictionary and if a match occurs, the equivalent English word is returned. The engine translates only dictionary words and common nouns. Since proper nouns are not available in dictionary, so those are left as it is. The proper nouns are transliterated to English equivalents by the transliteration module. It uses a rule-based algorithm by Singh and Oberoi [26]. The web query is reformulated by using the English tokens. The resultant tokens are concatenated by separating them with a white space to generate a new web query. The search query can be optimized for Google search engine by using search operators in between keywords. For example, AND (Google search operator) can be used in between the keywords to instruct Google search engine to search the information containing all these keywords [27]. This web query is then used to search the web. Algorithm 1 shows the working of the system.

**Algorithm 1: Working of the System**

```
Step 1: Let, Punjabi Web Search Query as Qp
       List of Undesirable Words as Lu
       Bi-Lingual Dictionary as Dbl
       List of Tokens as Lt
       White Space as Ws
       Token as T
       Punjabi Token as Tp
       English Token Te
        English Web Search Query as Qe
Step 2: Read Qp
Step 3: Match Words of Lu in Qp
Step 4: If Found then Replace with Ws
Step 5: Tokenize Qp and store tokens in Lt
Step 6: Open and Read Tp from Lt
Step 7: Do
           Stem Tp
           Update Tp in Lt
           Read Next Tp from Lt
       While Tp exists in Lt
Step 8: Open and Read Tp from Lt
Step 9: Do
           Search Tp in Dbl
           If Match Found Read Te Then
              Replace Tp with Te in Lt
           End If
           Read Next Tp from Lt
       While Tp exists in Lt
Step 10: Open and Read T from Lt
```

```
Step 11: Do
              If T is Punjabi Token Then
                 T_e=Transliterate_To_English (T)
                 Replace T with T_e in L_t
              End If
              Read Next T from L_t
          While T exists in L_t
Step 12: Concatenate Tokens of L_t to form Q_e
Step 13: Submit Q_e to Search Engine
Step 14: Display search results.
Step 15: End
```

## 4 Test and Results

The system was tested using a sample of 365 Punjabi web search queries. The queries were typed manually, and the results returned by the search engine were analyzed for relevance and as per expectations. Table 1 shows some example Punjabi queries with their equivalent reformulated web queries.

It was found that 323 search queries were processed by the system as expected, and the search engine returns the globally relevant results. So,

$$\begin{aligned}
&\text{Total number of test search queries} = 365\\
&\text{No. of queries processed by the system as expected} = 323\\
&\text{Accuracy} = (\text{No. of Correct Results/Total Number Inputs}) \times 100\\
&\quad = 323/365 = 88.49\%
\end{aligned}$$

**Table 1** Punjabi queries with equivalent reformulated web queries

| Punjabi query | Reformulated web query |
|---|---|
| ਬਿਲ ਗੇਟਸ ਬਾਰੇ ਜਾਣਕਾਰੀ ਦੇਵੋ | Bill gates |
| ਕੋਵਿਡ-19 ਦੇ ਸੰਸਾਰ ਭਰ ਵਿੱਚ ਕਿੰਨੇ ਕੇਸ ਹਨ | Covid-19 worldwide case |
| ਸਭ ਤੋਂ ਵੱਧ ਕੋਵਿਡ-19 ਕੇਸਾਂ ਵਾਲਾ ਦੇਸ਼ ਕਿਹੜਾ ਹੈ<br>ਨਾਸਾ ਕੀ ਹੈ | Highest covid-19 case country<br>NASA |
| ਦੁਨੀਆਂ ਦੇ ਸਭ ਤੋਂ ਅਮੀਰ ਵਿਅਕਤੀ ਕਿਹੜੇ ਹਨ | World richest |
| ਰੂਸ ਵਿੱਚ ਖੇਤੀ ਯੋਗ ਖੇਤਰਫਲ ਕਿੰਨਾ ਹੈ | Russia agriculture area |
| ਦੁਨੀਆਂ ਦੀ ਸਭ ਤੋਂ ਉੱਚੀ ਇਮਾਰਤ ਕਿਹੜੀ ਹੈ | World highest building |

**Fig. 2** Accuracy percentage of the system

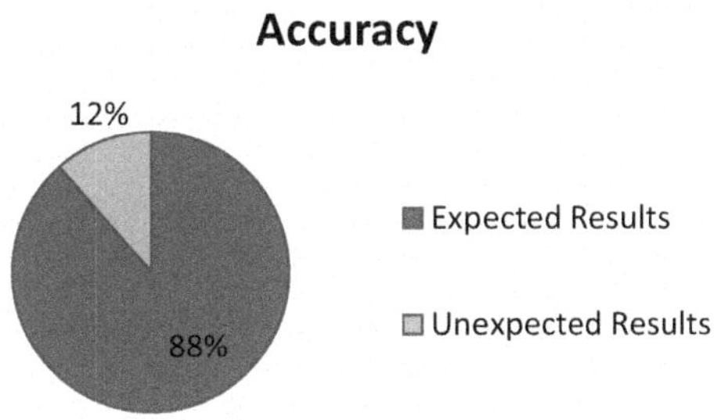


The 42 queries were processed by the system; but due to some limitations of the used processing techniques, the reformulated web search query was found incomplete. Figure 2 shows the accuracy percentage of the system.

## 5 Future Scope

The system was unable to produce expected results for 12% test queries. It was due to the lack of proper translation and transliteration of keywords. Generally, the people mix some English words in Punjabi language script to enter their query. In such cases, the system failed to reformulate the web query properly. In future scope, the system could be improved to handle these issues. Also with little more efforts, the system could be extended to handle other languages that are similar to Punjabi such Hindi. In this way, the system could be made multi-lingual.

## 6 Conclusion

When local language queries are used to search the web, the search engine returns results based on that language. The Punjabi language queries are processed by the search engine to display information available in Punjabi Web sites. If a user wants to get linguistically non-local information using a Punjabi web search query, he/she does not get the relevant information. This paper presented a system model that can be used to get global and relevant information using Punjabi language web search queries. The input queries are processed through various natural language processing techniques such as tokenization, stemming, translation, transliteration, etc. The processing extracts and generates web search keywords in English to get the globally relevant information. The system was tested using 365 Punjabi web search queries, and for 88.49% input queries, it produced the results as expected. Due to the lack of proper translation and transliteration of keywords, it was unable to produce expected results for 12% test queries. Generally, the people mix some English words in Punjabi language script to enter their query. In such cases, the system failed to reformulate the web query properly. In future, the system could be improved to handle these issues and can be extended to make it multi-lingual.

## References


1. Hillier, M.: The role of cultural context in multilingual website usability. Electron. Commer. Res. Appl. **2**(1), 2–14 (2003). https://doi.org/10.1016/S1567-4223(03)00005-X
2. Miraz, Dr., Ali, M., Excell, P.: Multilingual website usability analysis based on an international user survey. In: Proceedings of the Fifth International Conference on Internet Technologies and Applications (ITA 13), pp. 236–244. Creative and Applied Research for the Digital Society (CARDS), Glyndŵr University in Wrexham, North East Wales, UK (2013)
3. Honarvar, A.: Web page rank estimation in search engine based on SEO parameters using machine learning techniques. Int. J. Comput. Sci. Netw. Secur. **17**, 95–100 (2017)
4. Esmaeili Aliabadi, D., Avsar, B., Esmaeili, E., Yousefnezhad, R.: Investigating global language networks using Google search queries. Exp. Syst. Appl. **121**, 66–77 (2019). https://doi.org/10.1016/j.eswa.2018.12.016
5. Lewandowski, D.: Problems with the use of web search engines to find results in foreign languages. Online Inf. Rev. **32**(5), 668–672 (2008). https://doi.org/10.1108/14684520810914034
6. Mohamed, E., Elmougy, S., Aref, M.: Toward multi-lingual information retrieval system based on internet linguistic diversity measurement. Ain Shams Eng. J. **10**(3), 489–497 (2019) ISSN: 2090-4479. https://doi.org/https://doi.org/10.1016/j.asej.2018.11.009.
7. Tej, K.: Bhatia: major regional languages. In: Kachru, B.B., Kachru, Y., Sridhar, S.N. (eds.) Languages in South Asia, pp. 121–131. Cambridge University Press, Cambridge (2008)
8. Abbas Malik, M.G.: Punjabi machine transliteration. In: Proceedings of the 21st International Conference on Computational Linguistics and 44th Annual Meeting of the ACL, pp. 1137–1144. Association for Computational Linguistics, Sydney, Australia (2006)
9. Rahman, T: Soft power of Punjabi: language in the domain of pleasure. J. Punjab Sikh Stud. **24**(1&2: Spring-Fall 2017), 73–94 (2017)
10. Sunita, R.V.: An optimizing preprocessing algorithm for enhanced web content. In: Pant, M., Sharma, T., Verma, O., Singla, R., Sikander, A. (eds.) Soft Computing: Theories and Applications, Advances in Intelligent Systems and Computing, vol. 1053, pp. 63–71. Springer, Singapore (2020). https://doi.org/10.1007/978-981-15-0751-9_6
11. Soni, V., Kumar, L., Singh A.K., Kumar M.: Text summarization: an extractive approach. In: Pant, M., Kumar Sharma, T., Arya, R., Sahana, B., Zolfagharinia, H. (eds.) Soft Computing: Theories and Applications, Advances in Intelligent Systems and Computing, vol. 1154, pp. 629–637. Springer, Singapore (2020). https://doi.org/10.1007/978-981-15-4032-5_57
12. Kirmani, M., Manzoor Hakak, N., Mohd, M., Mohd M.: Hybrid text summarization: a survey. In: Ray, K., Sharma, T., Rawat, S., Saini, R., Bandyopadhyay, A. (eds.) Soft Computing: Theories and Applications, Advances in Intelligent Systems and Computing, vol. 742, pp. 63–73. Springer, Singapore (2019). https://doi.org/10.1007/978-981-13-0589-4_7
13. Sharma, P., Gupta, K., Bareja, M., Jain, V.K.: Effective approach for sentiment analysis on movie reviews. In: Pant, M., Sharma, T., Verma, O., Singla, R., Sikander, A. (eds.) Soft Computing: Theories and Applications, Advances in Intelligent Systems and Computing, vol. 1053, pp. 491–501. Springer, Singapore (2020). https://doi.org/10.1007/978-981-15-0751-9_46
14. Kumar, K., Bamrara, R., Gupta, P., Singh, N.: M2P2: movie's trailer reviews based movie popularity prediction system. In: Pant, M., Sharma, T., Verma, O., Singla, R., Sikander, A. (eds.) Soft Computing: Theories and Applications, Advances in Intelligent Systems and Computing, vol. 1053, pp. 671–681. Springer, Singapore (2020). https://doi.org/10.1007/978-981-15-0751-9_62

15. Zhang, J., Lin, S.: Multiple language supports in search engines. Online Inform. Rev. **31**(4), 516–532 (2007). ISSN: 1468-4527. https://doi.org/10.1108/14684520710780458
16. Lazarinis, F., Ferro, J.V., Tait, J.: Improving non-English web searching (iNEWS07). ACM SIGIR Forum **41**(2), 72–76 (2007). https://doi.org/10.1145/1328964.1328977
17. Gao, W., Blitzer, J., Zhou, M.: Using English information in non-English web search. In: Proceedings of the 2nd ACM Workshop on Improving Non English Web Searching (iNEWS '08), pp. 17–24. Association for Computing Machinery, New York, NY, USA (2008)
18. Qin, T., Liu, T.-Y., Zhang, X.-D., Wang, D.-S., Xiong, W.-Y., Li, H.: Learning to rank relational objects and its application to web search. In: Proceedings of the 17th international conference on World Wide Web (WWW '08), pp. 407–416. Association for Computing Machinery, New York, NY, USA (2008)
19. Munye, M., Atnafu, S.: Amharic-English bilingual web search engine. In: Proceedings of the International Conference on Management of Emergent Digital EcoSystems (MEDES '12), pp. 32–39. Association for Computing Machinery, New York, NY, USA (2012)
20. Rózsa, G., Komlodi, A., Chu, P.: Online searching in English as a foreign language. In: Proceedings of the 24th International Conference on World Wide Web (WWW '15 Companion), Article 86, pp. 875–880. Association for Computing Machinery, New York, NY, USA (2015)
21. Fu, H.: Analysis of Chinese-English mixed language query reformulation strategies and patterns during web searching. In: Proceedings of the Association for Information Science and Technology, pp. 1–5. American Society for Information Science, USA (2016)
22. Kaur, J., Saini, J.R.: Punjabi stop words: a Gurmukhi, Shahmukhi and Roman Scripted Chronicle. In: Proceedings of the ACM Symposium on Women in Research 2016 (WIR '16), pp. 32–37. Association for Computing Machinery, New York, NY, USA (2016)
23. Rai, A., Borah, S.: Study of various methods for Tokenization. In: Mandal, J., Muk-hopadhyay, S., Roy, A. (eds.) Applications of Internet of Things, Lecture Notes in Networks and Systems, vol. 137, pp. 193–200. Springer, Singapore (2021). https://doi.org/10.1007/978-981-15-6198-6_18
24. Singh, H., Oberoi, A.: Lightweight stemming approach for Punjabi language text: an NLIDB subsystem alternative. J. Res. Electron. Comput. Eng. (A Unit of I2OR) **7**(1), 2334–2337 (2019)
25. Bhattacharyya, P.: IndoWordNet. In: Dash, N., Bhattacharyya, P., Pawar, J. (eds.) The WordNet in Indian Languages, pp. 1–18. Springer, Singapore (2017). https://doi.org/10.1007/978-981-10-1909-8_1
26. Singh, H., Oberoi, A.: An efficient Romanization of Gurmukhi Punjabi proper nouns for pattern matching. Int. J. Recent Technol. Eng. **8**(3), 634–640 (2019). https://doi.org/10.35940/ijrte.B2467.098319
27. Google Issue Tracker (Search Query Language). https://developers.google.com/issue-tracker/concepts/search-query-language. Last accessed 12 Sept 2020

# A Supervised Learning Approach by Machine Learning Algorithms to Predict Diabetes Mellitus (DM) Risk Score

**Tarin Sultana Sharika, Abdullah Al Farabe, Ghalib Ashraf, Nahian Raonak, and Amitabha Chakrabarty**

**Abstract** The utilization of artificial intelligence (AI) has become a valuable part of medical research. Diabetes is one of the top maladies on the planet, and it is a long-term disease that happens if the body cannot utilize glucose appropriately. This paper indicates the demand for the right procedures, advancement, and usage of successful well-being observing strategies ought to be selected to battle against diabetes. So early discovery and treatment with the utilization of different procedures must be preferred. Our model consists of four machine learning algorithms which are K-Nearest Neighbor, Random Forest, Decision Tree, and Logistic Regression. These algorithms are used on a dataset of 15,000 diabetes mellitus patients along with nine features, where the results are compared to show which algorithm gives the best accuracy, including statistical analysis to get a more perfect result of our model. Among all four algorithms, the random forest gives the best accuracy of around 92%, where the rest of the algorithms give between 78 and 90%. Hopefully, this study could be very helpful in medical science to predict the risk score of diabetes mellitus (DM) early and to classify them.

**Keywords** Diabetes Mellitus (DM) · Linear Discriminant Analysis (LDA) · Logistic Regression · Random Forest · Decision Tree · KNN · Data analysis · Prediction · ROC-AUC curve · Statistical analysis

## 1 Introduction

The American Heart Association (AHA) has found that almost 68% of patients die due to diabetes [1]. Diabetes type-I occurs due to a shortage of insulin. The only procedure is to inject insulin on a daily basis to control glucose levels in the blood. Diabetes type-II mostly occurs in adults which is counted around 90% of diabetes

---

T. S. Sharika · A. Al Farabe (✉) · G. Ashraf · N. Raonak · A. Chakrabarty
Department of Computer Science and Engineering, BRAC University, 66 Mohakhali, Dhaka 1212, Bangladesh
e-mail: amitabha@bracu.ac.bd

 
T. K. Sharma et al. (eds.), *Soft Computing: Theories and Applications*, Advances in Intelligent Systems and Computing 1381,
https://doi.org/10.1007/978-981-16-1696-9_27

cases. Moreover, the body cannot make good use of the insulin that it produces. Inactivity of health or overweight can mainly be the causes of type-II diabetes. Physical work, maintaining food habits, oral drugs, and insulin are the treatment of diabetes type-II patients [2]. Diabetes is linked with other diseases too such as sight loss, kidney disease, heart disease, and brain stroke amputations. Nowadays, people get affected by diabetes at an early age, and worldwide it is estimated that around 79,000 children develop type-I diabetes [3]. Moreover, a diabetes patient has a high risk of having an operation, and it is said that the people with surgical cases having DM are assessed using bed days around 45% contrasted to patients without DM who are admitted to clinical wards [4]. In this research, the target is only on diabetes mellitus (DM) risk score. It does not matter it is type-I or type-II.

Nowadays, machine learning has become extremely popular for analysts to diagnose a disease in the beginning. With the breadth of information in the galenical field, it has become very easy to examine with more precision rates and better discoveries. Machine learning is a drifting methodology for the early expectations of various illnesses precisely dependent on the clinical information of the patients.

Our motivation is utilizing statistics and probability from medical information could be more supportive to develop classification systems. Expanding on the medical science field, it has become easy to search for prevention in the early stage of any illness. So, our principle aim of building a decision support framework is to determine and diagnose diabetes with higher prediction accuracy. Figure 1 shows the process of this research.

In this research, a dataset has been collected from kaggle.com containing fifteen thousand of information, consisting of consisting of nine inputs and one outcome. The inputs are PatientID, DiastolicBloodPressure, BMI, Pregnancies, Age, TricepsThickness, DiabetesPedigree, SerumInsulin, PlasmaGlucose, and Diabetic have a binary value, where 1 indicates the presence of diabetes and 0 otherwise [5]. We have checked co-relationship among the attributes. Heatmap is to show relationships among attributes. Moreover, it shows which attributes affect the output most. The strength of correlation among attributes determines the quality of the dataset. The first motive of this study is to reduce unnecessary data from the dataset to get

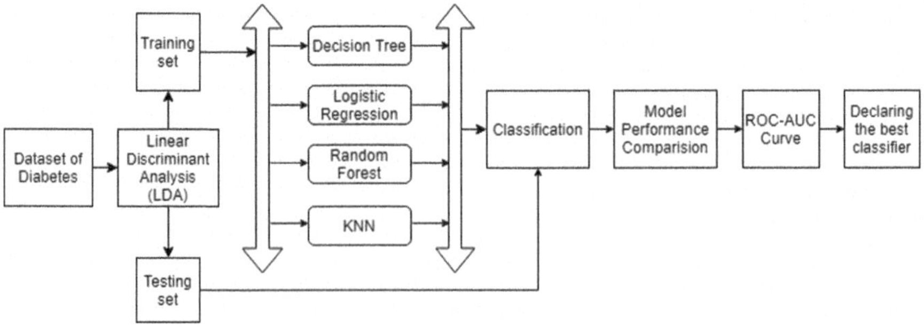


**Fig. 1** Overall architecture

more accurate results for which linear discriminant analysis (LDA) is used for data preprocessing [6]. For our research purpose, the following four machine learning algorithms—decision tree, random forest, KNN, and logistic regression—are used to get the best prediction classifier. After using these algorithms, the dataset is analyzed from the confusion matrix that has been found from the algorithms. Next, AUC-ROC curves show how much the model is able to characterize among classes. Moreover, the discussion about sensitivity (TPR), specificity, precision, $F1$ score, and Matthews correlation coefficient is to analyze the performance of our model. The research is also showing the beneficial positive effect of using how open-source prediction is getting stronger [7].

## 2 Literature Review

Contemporary researchers have been working with diabetes datasets to predict outcomes using various algorithms. Warke et al. [8] have used machine learning to develop a diabetes diagnosis system to predict whether the patient is suffering with DM or not. Moreover, the study has developed a classification model made by using various classifiers. This research has worked on a 15000 cases of DM dataset founded from different National Institute of Diabetes. It worked in four phases. In the first phase, scatter plot is used to visualize the dataset to see the distribution of data and finding correlation if any by using the seaborn library. Then four algorithms—SVM, decision tree, Naïve Bayes, and KNN—have been classified where Naive Bayes gives the best prediction of around 72% where the rest algorithms give 62–68% accuracy. So, in this research, it has declared that compared to other classifiers, Naive Bayes could be the best classifier to forecast.

Pande et al. [9] have worked on PIMA Indians Diabetes dataset in kaggle.com and have uploaded the whole procedure. For the data visualization, first Dist Plot is used to determine the probabilities of specific data values within the distribution. The next pair plot is used for explaining a relationship between two variables to understand the best set of features. Moreover, at different values, violin plots have shown the probability density of the data (in the simplest case this could be a histogram). When data distribution has more than one peak, violin plots are more effective than a boxplot (as it indicates only summary statistics). In addition, in this research, heatmap shows the correlation between features indicates a clear understanding of the relationships between variables. It could be useful in data analysis and modeling. At last, this research has predicted accuracy with three classifiers—logistic regression, decision tree, and gradient boosting. Among them, logistic regression gives the best accuracy of 80.7%, where decision tree and gradient boosting give 75.0 and 80.2%. This research aims only to show the distribution of their dataset by various plots and showing accuracy by using different classifiers [9].

Li et al. [10] have made a management system for diabetes research purposes. The paper has found those methods among the AI papers which are used to find out which steps have taken most common from all of them in diabetes education. SVM is the most common algorithm that has been used for prediction. ANN is used to guide diet exercises. Case-based reasoning is used to calculate insulin individually. Moreover, GA is recommendable to monitor blood sugar. In addition, ES is a system to data process by fuzzy logic, and DT is another learning algorithm which helps to combine patients to get better outcomes to diagnose diabetes.

From studying the above research in our paper, LDA has used to identify NaN values. Linear discriminant analysis (LDA) has been used for data preprocessing systems so that before the process of classification, it is possible to reduce the number of unnecessary data from the dataset. Next to look for correlation among the attributes, the Heatmap is used by using the seaborn function if there is any feature that could be removed from the dataset. Moreover, four ML algorithms are used such as random forest (RF), logistic regression (LR), KNN, and decision tree from where accuracy has been predicted for each and compared among them to determine the best classifier. Only based on the accuracy, it is not enough to say which algorithm is the best classifier so our statistical analysis has been carried out using by using ROC-AUC curve, confusion matrix, sensitivity (TPR), specificity, precision, $F1$-score, and Matthews correlation coefficient to show the performance of our model. So, by analyzing features, data preprocessing, and finding the accuracy and performance of the model, we have come to a result that which algorithm could be said to be the best classifier in our model.

# 3 Research Methodology

## *3.1 Data Description*

Our research is based on diabetes mellitus, and a supervised diabetes dataset from kaggle.com [5] has taken where 15000 data is included with ten different attributes. It is examining the supervised dataset containing both inputs and outputs features concluding with a result of the best classifier to upgrade the decision support framework. There were a total of ten attributes where "Patient ID" has been deleted from the dataset(unnecessary). So, now the dataset consists of eight attributes as inputs and one output which consists of "yes = 1" and "no = 0". "Yes" indicates the presence of diabetes, and "no" indicates the absence of diabetes. The dataset does not consist of any string values. The dataset contains ten attributes named PatientID (ID number of patients), DiastolicBloodPressure (pressure in the arteries when the heart rests), Pregnancies (number of pregnancies), Age (age of patient), TricepsThickness (skinfold thickness), DiabetesPedigree (function score of likeliness of DM in relation to relatives of the patient), SerumInsulin (2-hour serum insulin (mu U/ml)), PlasmaGlucose (concentrated 2 hours in an oral glucose resistance test), and Diabetic (output of diabetes (1) or not (0)).

## 3.2 *Feature Score*

**Heatmap**: A heatmap is based on data analysis which uses colors by showing a bar graph which uses width and height for better data visualization. The validity of the process is 0 up to 1, and if it crosses more than 1 or less than 0, then it should recheck the whole data if there is any wrong or not. Otherwise, no relation is depending on them [11]. We have used a heatmap on a 15,000 diabetes patient dataset to show the co-relationship among the inputs.

In Fig. 2, dark blue represents a high correlation, while light white/blue is a low correlation. A high absolute correlation between two features indicates that they are collinear with each other. If the correlation is greater than the threshold, it is considered a problem, and then it argues to remove one of the features between them

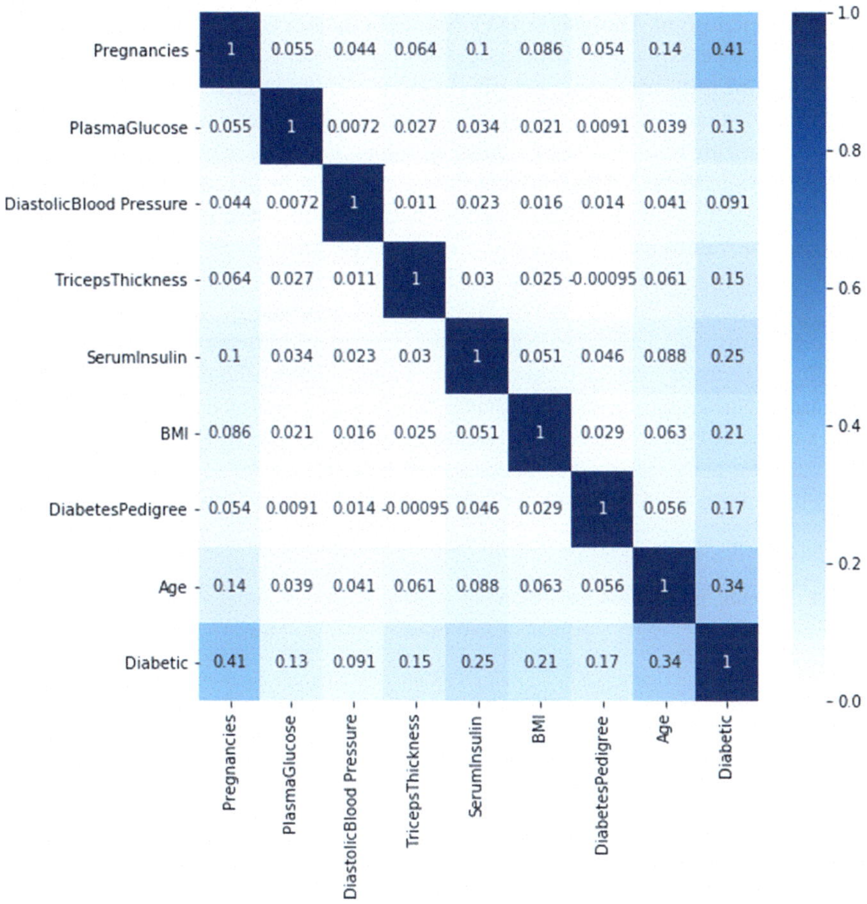


**Fig. 2** Showing heatmap by analyzing from the dataset

[12]. Seaborn's heatmap function is used to avoid one of the correlated features. If there is any relation, then one can be removed, and it will not affect so much in the result.

Figure 2 shows that among the inputs, the highest correlation result is around 0.14 (<0.5) only between Pregnancy and Age among the features. As the scale is less than 0.5, so all the features have taken to testify as the outcomes are depending on every input.

# 4 System Implementation

## *4.1 Missing Values imputation*

**Linear Discriminant Analysis (LDA)**: Since our dataset is large, there could be lots of NaN or unnecessary values which could be a big problem to get a perfect prediction. So that the linear discriminant analysis (LDA) algorithm is used to remove all those unnecessary values. From the dataset, the missing values are marked, the rows with missing values are removed, and the missing values are replaced with sensible values.

Missing values in a dataset can cause mistakes with some AI calculations, so LDA is helping to get rid of them. The purpose was to fill up these missing values with mean column values using the imputer function [13, 14].

Table 1 shows the counts of NaN and means values of every feature after using LDA on the dataset. Table 1 shows Pregnancy attribute has 4377 NaN values. The rest of the attributes have zero NaN values. Our main purpose is to use LDA for ignoring unnecessary values from our dataset but excluding the Pregnancy attribute, there are no NaN values in other attributes. So, mean imputation has been used for the "Pregnancies" to ignore those NaN values. The mean value of Pregnancies is used to replace the NaN values. The accuracy of the LDA model is 76.8%. After that, the features were used as inputs for our model and applied different classifications

**Table 1** Imputing missing values by LDA

| Features (Input) | Count of zero values | Mean |
|---|---|---|
| Pregnancies | 4377 | 3 |
| Glucose | 0 | 107 |
| BP | 0 | 71 |
| Skin | 0 | 28 |
| Insulin | 0 | 137 |
| BMI | 0 | 31.5 |
| Pedigree | 0 | 0.39 |
| Age | 0 | 30 |

such as KNN, LR, random forest, and decision tree. By applying these algorithms, accuracy has been measured for the performance evaluation and picked the best algorithm by comparing them with each other. In addition, we have used sensitivity, specificity, precision, $F1$-score, and Matthews correlation coefficient to get better performance through our models. Moreover, we have used the ROC-AUC curve for showing the performances of our processes. In our process, we have used Spyder (py35) and Jupyter. During the implementation of the prediction engine scikit-learn, NumPy and Matplotlib libraries have been used.

# 5 Algorithm Implementation

## 5.1 *Train/Test Split*

In machine learning models, a typical issue confronted is with overfitting and underfitting. Overfitting happens when the model gains from the information so well that it prompts no preparation mistake. It learns the commotions and deviations of the information as ideas and fits the model excessively well to the information. In any case, when the model is given new information, it neglects to make precise forecasts since it was overfitted with the clamors and deviations. Underfitting is the inverse, and it can neither fit the information accessible nor to new information [15].

The train/test split has been implemented by using packages like scikit-learn's model. The data was split into an 80:20 ratio, where 80% was used as a train and the rest was used 20% for testing. In most cases, this is the ideal ratio of splitting data. A 50:50 ratio was also considered but since the dataset did not contain a large number of instances, taking only 50% of the original data might have led to the underfitting of the model and then using our splitting data into the algorithms—KNN [16], random forest, logistic regression, and decision tree [17] one by one.

# 6 Result Analysis

In the KNN algorithm, different values of n have been used in Table 2.

**Table 2** Accuracy based on neighbor values

| K-value | Accuracy (%) |
|---|---|
| 1 | 84.6 |
| 2 | 83.4 |
| 3 | 86.8 |
| 4 | 86.2 |

From Table 2, it seems that for K=3 neighbor, KNN gives the best accuracy of 86.8%. So, in this case, neighbor 3 is considered to take for KNN. Accuracy is a performance measure of different machine learning algorithms. As the working mechanism is different, so the algorithms have given different accuracy.

## *6.1 Model Performance*

In this section, the confusion matrix has been measured from the four classifiers that have been used. From the confusion matrix TP, TN, FP, and FN have taken to calculate several equations such as sensitivity, specificity, precision, accuracy, $F1$-score, Matthews correlation coefficient, and AUC score for each of the algorithms of our model to predict DM risk score.

Table 3 shows that random forest gives 92.60%, decision tree gives 88.6%, KNN gives 86.80%, and logistic regression gives 78.6% accuracy. Now we can differentiate easily that random forest which gives 92.60% accuracy is the best machine algorithm for our prediction model.

**ROC and AUC**: AUC means the measurement of separability, and ROC means a probability curve. AUC-ROC helps the measurement of the quality of a model. Execution estimation is a basic task in machine learning. It shows the capability of a model to distinguish between classes. If AUC is near 1, it has a good prediction of separability. If the model is poor, then the AUC value will be near 0. Moreover, if the AUC is 0.5, that means that the model has no capability of separating classes [18, 19]. The curve plots have two parameters such as

- True Positive Rate (TPR)—indicates y-axis

$$\text{Recall/TPR/Sensitivity} = \frac{\text{TP}}{\text{TP} + \text{FN}} \tag{1}$$

- False Positive Rate (FPR)—indicates x-axis

$$\text{FPR} = \frac{\text{FP}}{\text{FP} + \text{TN}} \tag{2}$$

**Table 3** Accuracy of algorithms

| Algorithm | Accuracy (%) |
|---|---|
| Random forest | 92.6 |
| Logistic regression | 78.6 |
| Decision tree | 88.6 |
| KNN | 86.80 (for neighbor=3) |

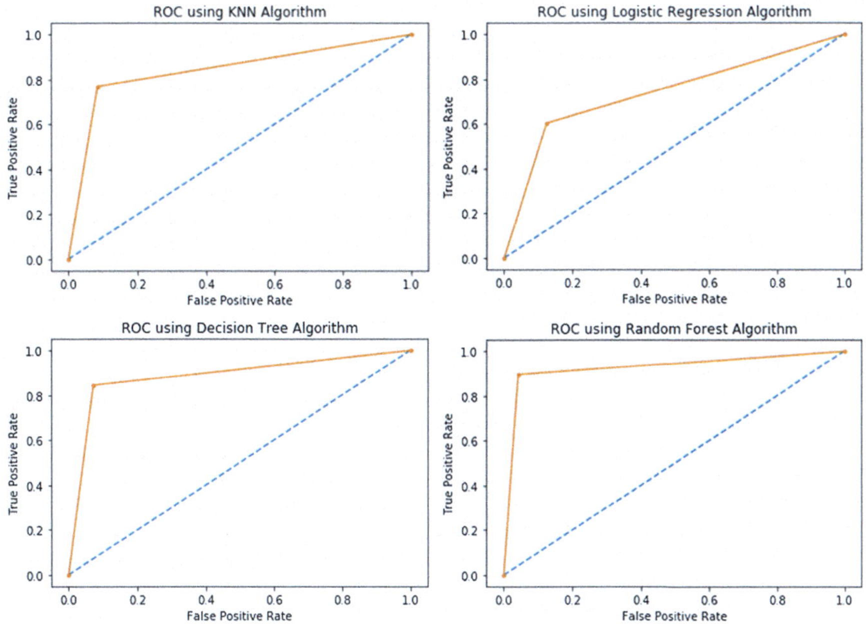


**Fig. 3** ROC of each algorithm

Figure 3 of ROC curves shows that the random forest curve is almost touching near to 1 which indicates a good classifier. Logistic regression is giving the worst prediction.

**Performance Model**: The performance model includes sensitivity(TPR), specificity, precision, $F1$-score, and Matthews correlation coefficient which are the measurements of performance for each algorithm by also using equations (1) and (2). With accuracy and ROC-AUC procedures, the performance model is also relevant to the prediction model to get more accurate results.

In Table 4, each of the algorithms gives more than 0.5 of AUC and near to 1 which indicates that this model has a good measure of separability. Random forest gives 0.906 which is the highest AUC of all the algorithms used. So, in this model, random forest has given the best job in classifying the DM. Confusion matrix of each algorithm has also calculated which consists of TP, FP, FN, and TN [20].

**Table 4** AUC score of machine learning algorithms

| Algorithm | Accuracy (%) |
|---|---|
| Random forest | 0.906 |
| Decision tree | 0.886 |
| KNN ($n = 3$) | 0.843 |
| Logistic regression | 0.739 |

**Precision**:

$$\text{PREC} = \frac{\text{TP}}{\text{TP} + \text{FP}} \quad (3)$$

**F1-score**:

$$F1 = \frac{2\text{TP}}{2\text{TP} + \text{FP} + \text{FN}} \quad (4)$$

**Matthews correlation coefficient (MCC):**

$$\text{MCC} = \frac{\text{TP} \times \text{TN} - \text{FP} \times \text{FN}}{\sqrt{(\text{TN} + \text{FP})(\text{TN} + \text{N})(\text{TP} + \text{FN})(\text{TP} + \text{FP})}} \quad (5)$$

**Sensitivity(SN/TPR):**

$$\text{SN} = \frac{\text{TP}}{\text{TP} + \text{FN}} \quad (6)$$

**Specificity:**

$$\text{SP} = \frac{\text{TN}}{\text{TN} + \text{FP}} \quad (7)$$

Table 5 has been calculated by using Eqs. 3–7 [21]. In Table 5, we have analyzed the performance of our model using some performance metrics. In this model, random forest gives a specificity of 0.9304 which means 93.4% of people without diabetes were correctly identified. Also, sensitivity is 0.924 which means 92.4% of people with diabetes were correctly identified by our model. Moreover, precision, $F1$-score, and Matthews correlation coefficient of random forest are 0.9678, 0.9457, and 0.8336 which is also higher than the rest of the algorithms used in this model. Table 5 is showing that for sensitivity, specificity, precision, $F1$-score, and Matthews correlation coefficient, random forest is giving the best and highest accuracy com-

**Table 5** Comparison of tabulated data in random forest, KNN, decision tree, and logistic regression

| Algorithm | Sensitivity (TPR) | Specificity | Precision | $F1$ Score | Matthews correlation coefficient |
|---|---|---|---|---|---|
| Random forest | 0.9245 | 0.9304 | 0.9678 | 0.9457 | 0.8336 |
| KNN | 0.8912 | 0.8169 | 0.9163 | 0.9036 | 0.6968 |
| Decision tree | 0.9239 | 0.8525 | 0.9279 | 0.9259 | 0.7746 |
| Logistic regression | 0.8625 | 0.5780 | 0.8451 | 0.8537 | 0.4486 |

pared to other algorithms used. In the end, based on accuracy, ROC-AUC curve, sensitivity, specificity, precision, $F1$-score, and Matthews correlation coefficient, the whole research considering that the random forest is the best classifier among all of the algorithms.

## 7 Conclusion and Future Works

In this model, four algorithms—KNN, random forest tree, logistic regression, and decision tree—have been used. LDA is valuable to see if there are any NaN values or not. Then the heatmap is used to see the correlation among inputs so that one of the correlated features could be removed from the dataset. Therefore, confusion metrics are also used to measure model performance in Table 5 such as sensitivity, specificity, precision, $F1$-score, and Matthews correlation coefficient. After analyzing all these results, the research has to come to the conclusion that among all of the four algorithms the random forest provides the best accuracy in the predicted model. Besides, random forest is the best algorithm for regression or classification-based problems, and it handles binary or categorical features easily. In a random forest, prediction speed is also faster than training speed, and so it will require less time as well. However, finding statistical analysis and accuracy is not enough to diagnose a diabetes patient, because it also depends on the depth of the data information. The data we have used from Kaggle might not be enough to get proper accuracy. It is just the beginning of our research, and our main goal is to develop a decision support framework model that predicts the risk score of DM. In addition, we are planning to work on the complexities that a diabetes patient can have side effects like cardiovascular diseases, kidney diseases, and lung disease. Image processing could be used by detecting spam information from a dataset [22]. We have collected a dataset from a hospital, and soon we will investigate to improve our results where deep learning is another choice. Hopefully, this prediction model could take a step further to predict the early stages of other diseases that are related to diabetes.

## References

1. Falvo, D., Holland, B.: Medical and Psychosocial Aspects of Chronic Illness and Disability (2017)
2. International Diabetes Federation—What is Diabetes. https://bit.ly/3dciMhX. Retrieved 30 Nov 2020 (2020)
3. Patterson, C., Guariguata, L., Dahlquist, G., Soltész, G., Ogle, G., Silink, M., Patterson, C., Guariguata, L., Dahlquist, G., Soltész, G., Ogle, G., Silink, M. (2014). Diabetes in the young—A global view and worldwide estimates of numbers of children with type 1 diabetes. Diabetes Res. Clin. Pract. **103**(2), 161–175. https://doi.org/10.1016/j.diabres.2013.11.005
4. Wukich, D.K.: Diabetes and its negative impact on outcomes in orthopaedic surgery. World J. Orthoped. **6**(3), 331–339 (2015). https://doi.org/10.5312/wjo.v6.i3.331

5. Diabetes from DAT263x Lab01. https://www.kaggle.com/fmendes/diabetes-from-dat263x-lab01. Retrieved 30 Nov 2020 (2020)
6. Sharland, D.E.: Davidson's principles and practice of medicine. Postgr. Med. J. **58**(677), 195 (1982)
7. Kakkar, M., Jain, S., Bansal, A., Grover, P.S.: Is open-source software valuable for software defect prediction of proprietary software and vice versa?. In: Pant, M., Ray, K., Sharma, T., Rawat, S., Bandyopadhyay, A. (eds.) Soft Computing: Theories and Applications. Advances in Intelligent Systems and Computing, vol. 583. Springer, Singapore (2018). https://doi.org/10.1007/978-981-10-5687-1_21
8. Warke, M., Kumar, V., Tarale, S., Galgat, P., Chaudhari, D.J.: Diabetes diagnosis using machine learning algorithms. Diabetes **6**(03) (2019)
9. PIMA Indians Diabetes (Beginner). https://www.kaggle.com/rishpande/pima-indians-diabetes-beginner/notebook. Retrieved 30 Nov 2020 (2020)
10. Li, J., Huang, J., Zheng, L., Li, X.: Application of artificial intelligence in diabetes education and management: present status and promising prospect. Front. Public Health **8**, 173 (2020). https://doi.org/10.3389/fpubh.2020.00173
11. Feature Selection Techniques in Machine Learning with Python. https://towardsdatascience.com/feature-selection-techniques-in-machine-learning-with-python-f24e7da3f36e. Retrieved 30 November 2020 (2020)
12. Linear Regression and its Assumptions. https://towardsdatascience.com/linear-regression-and-its-assumptions-ef6e8db4904d. Retrieved 30 November 2020 (2020)
13. Brownlee, J.: How to Handle Missing Data with Python. https://machinelearningmastery.com/handle-missing-data-python/. Retrieved 30 Nov 2020 (2020)
14. Senapti, R., Shaw, K., Mishra, S., Mishra, D.: A novel approach for missing value imputation and classification of microarray dataset. Procedia Eng. **38** (2012). https://doi.org/10.1016/j.proeng.2012.06.134
15. Brownlee, J.: Overfitting and underfitting with machine learning Aalgorithms. https://machinelearningmastery.com/overfitting-and-underfitting-with-machine-learning-algorithms/. Retrieved 30 Nov 2020 (2020)
16. Srivastava, T.: K Nearest Neighbor—KNN Algorithm—KNN in Python & R. https://www.analyticsvidhya.com/blog/2018/03/introduction-k-neighbours-algorithm-clustering/. Retrieved 30 Nov 2020 (2020)
17. Sabharwal, M.: The use of soft computing technique of decision tree in selection of appropriate statistical test for hypothesis testing. In: Soft Computing: Theories and Applications, pp. 161–169. Springer, Singapore (2018)
18. Brownlee, J.: How to use ROC curves and precision-recall curves for classification in python. https://machinelearningmastery.com/roc-curves-and-precision-recall-curves-for-classification-in-python/. Retrieved 30 Nov 2020 (2020)
19. Understanding AUC—ROC Curve. https://towardsdatascience.com/understanding-auc-roc-curve-68b2303cc9c5. Retrieved 30 November 2020 (2020)
20. Medium. https://towardsdatascience.com/understanding-confusion-matrix-a9ad42dcfd62. Retrieved 30 Nov 2020 (2020)
21. Medium. https://towardsdatascience.com/accuracy-recall-precision-f-score-specificity-which-to-optimize-on-867d3f11124. Retrieved 30 Nov 2020 (2020)
22. Kumar, J., Taterh, S., Kamnthania, D.: Study and comparative analysis of various image spamming techniques. In: Soft Computing: Theories and Applications, pp. 351–365. Springer, Singapore (2018)

# Predicting Road Accident Severity Due to Weather Conditions Using Classification Algorithms

**R. Harikrishnan and Benafsha Cyrus Postwala**

**Abstract** This paper explores various approaches taken to predict the severity of a road accident based on the prevalent weather conditions during the accident using accident data in the USA from February 2016 to June 2020. Accuracies of pre-existing algorithms such as decision tree classifiers and Gaussian Naïve Bayes along with the accuracy of an artificial neural network (ANN) developed from scratch in conjunction with voting ensemble techniques are compared. These models attempt to determine the extent of severity of an accident in the presence of independent variables such as weather conditions, visibility, wind speed and precipitation. Severity of an accident refers to the impact of the accident on traffic delay. Applications for this predictor lie in timely re-routing of traffic in the event of an accident. By successfully predicting the severity, traffic can be re-directed to alternate routes if required, thus facilitating the smooth flow of traffic without any massive traffic jams.

**Keywords** Prediction algorithm · Multi-layer perceptron · Accident severity · Classification · Machine learning

## 1 Introduction

The future of decision making relies heavily on prediction algorithms that are driven by extensive amounts of data. With the growing applications of data science and new machine learning algorithms being invented, most industries rely on predictors and classifiers to make decisions and project data. New strides in essential services such as health care, government, law enforcement, etc., are due to research and development in artificial intelligence for these specific application areas. Using deep learning (CNNs) [1] models to determine whether there are tumours in images of MRI

R. Harikrishnan (✉)
Symbiosis Institute of Technology (SIT), Symbiosis International Deemed University (SIDU), Pune, India

B. C. Postwala
University of East Anglia, Norwich, UK



T. K. Sharma et al. (eds.), *Soft Computing: Theories and Applications*, Advances in Intelligent Systems and Computing 1381,
https://doi.org/10.1007/978-981-16-1696-9_28

scans could help doctors identify abnormalities and facilitate timely treatment. It is usually seen that specific algorithms work best for certain specific applications. These include Naïve Bayes for language classification [2], CNNs for image classification [3], logistic regression for linearly separable data and many more. Developing a classifier for a new application involves extensive trials and experimentation on various existing algorithms to find the most suitable one. Such an application area is formulating a prediction algorithm that can predict the effect of a road accident on traffic congestion. While it may be extremely difficult to predict when a road accident will occur, steps can be taken to determine how the accident can affect delays in traffic such as road blockage. Every year, around 6 million accidents occur in the USA. This means, through 2016 and 2020, there have been approximately 30 million accidents in the country. By considering factors such as the weather conditions at the time of the accident, early prediction of severity is possible so that necessary steps can be taken to mitigate massive traffic congestions. Usually, minor accidents do not affect the traffic. However, major accidents with higher severity could adversely affect traffic and induce extended time delays. Time-critical scenarios involving ambulances and police vehicles could greatly benefit from a pre-emptive warning of a road blockage. Other similar works such as "Accident risk of road and weather conditions on different road types" [4] relate the effect of weather conditions on the road with the risk factor of accidents that take place in those conditions. This paper focuses on weather conditions, visibility, wind speed and precipitation to determine the severity of an accident which could be "more severe" or "less severe". The prediction is made with the help of an artificial neural network (multi-layer perceptron) [5] and other pre-existing algorithms such as decision trees and Gaussian Naïve Bayes specifically designed for our purpose. The accuracy is compared with individual pre-existing classification algorithm accuracies. The motivation of this paper is to focus on the application of a predictor after an accident takes place to assess the severity and determine if traffic blockage may occur so that timely re-routing can take place for essential vehicles, thus preventing further delays. This paper first describes the proposed work in Sect. 2 and then goes on to describe the various classification algorithms used in Sect. 3. The results are discussed in Sect. 4, and finally, the paper is concluded along with its future scope in Sect. 5.

## 2 Proposed Work

Development of this model involved researching existing work and machine learning algorithms. The data was then acquired, and extraction of independent and dependent variables was carried out. After this, various data pre-processing techniques were applied, and issues such as data imbalance and null values were rectified. The dataset was then divided into training and test data where the former constituted 80% of the total dataset and the latter, 20%. The training dataset was trained on the Gaussian Naïve Bayes and decision tree classifier algorithms. After this, an artificial neural network was built using a combination of rectified linear unit and sigmoid layers.

To further improve accuracy, a technique called voting ensemble was applied where two classifiers are combined to predict the data. The accuracies of each model were then compared to find which was most favourable.

**Part 1**: Initial tasks involved were research on existing technology, understanding and study of various classification models and techniques, data preparation methods and identification of suitable packages and software to accomplish the tasks.

**Part 2**: After this, procurement of data was done from Kaggle. In order to make the data more suitable, various pre-processing tasks such as handling missing values, elimination of outliers, extraction of required columns and rows of data, assignment of integers to denote names or "strings" as well as merging of similar classes were carried out.

**Part 3**: Following this, the cleaned data was then split into training and test sets. The training set was then trained on individual algorithms, namely, Gaussian Naïve Bayes, decision trees and a multi-layer perceptron, and the testing set was validated. Each model was evaluated based on its confusion matrix, accuracy score, logarithmic loss and F1 score. The performance of each model was then reported.

Figure 1 shows a more detailed workflow of the project, highlighting the intricate processes that were carried out along with the major ones.

From Fig. 1, we can see that each major process such as data pre-processing and training models involves several tasks.

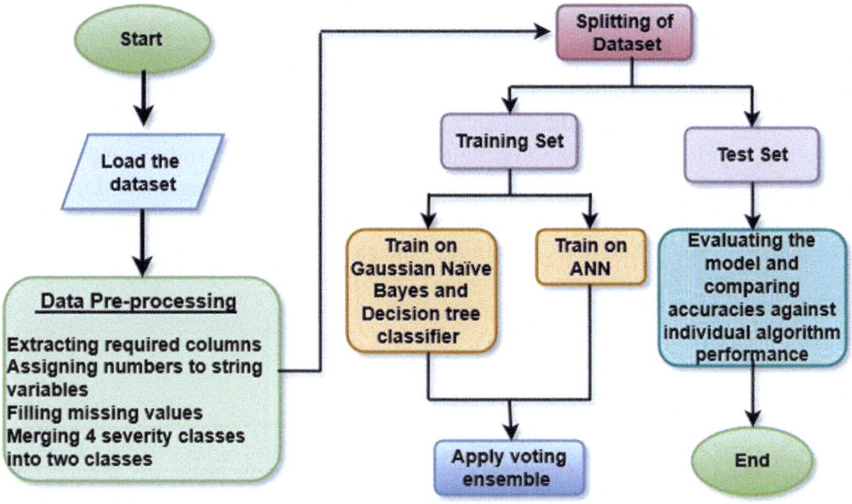


**Fig. 1** Workflow diagram of detailed process

**Table 1** Variable names, classes and their description

| Variable name | Classes/values | Description |
|---|---|---|
| Severity | Classes: 1, 2, 3, 4 | 1 = least severe, 4 = most severe |
| Visibility | Value from 0 to 140 miles | The values signify the distance till which a driver can see ahead of him/her |
| Wind speed | Value from 0 to 984 miles per hour | The values signify the speed of wind |
| Precipitation | Value from 0 to 25 inches | Deposition of rain, snow, sleet, etc. |
| Weather condition | Over 120 different types | Various weather conditions such as rain, snow and thunderstorm. |

### 2.1 Data Source

The data was acquired from a free dataset on the Kaggle Website comprising over 3.5 million records of accidents recorded from February 2016 till June 2020 in the USA. It consisted of 49 different columns from 49 American states [6]. This model only utilised the relevant data, namely severity, visibility, wind speed, precipitation and weather condition.

### 2.2 Data Description

Table 1 shows each variable name, its classes, types of values and a short description of each variable. Here, the dependent variable is severity of the accident, and independent variables are visibility, wind speed, precipitation and weather condition. Table 1 gives a clear idea of each variable that has been used. The dependent variable is "Severity", and the independent variables are "Visibility", "Wind Speed", "Precipitation" and "Weather Condition". The range of each variable is also given along with a short description.

### 2.3 Data Pre-processing

To make data usable for the model, various pre-processing tasks are carried out. These tasks vary with the nature of the variable. The type of pre-processing for each variable is given as follows.

Severity: This variable originally consisted of four distinct classes that denoted severity from least to most, i.e. 1–4, respectively. For ease of classification, the four classes were merged into two distinct classes, where severity levels 1 and 2 were denoted as 0 (less severe) and levels 3 and 4 were denoted as 1 (more severe).

Weather condition: The weather conditions were in the form of words or "string" form and required to be converted to numbers. There were more than 120 different types of weather conditions. As most of the weather conditions were similar in nature, they were grouped into eight levels, ranging from 0 to 7 in increasing order of adversity. They are shown in Table 2.

By classifying each type of weather condition at a certain level, similar conditions are grouped together for ease.

After assigning suitable values and eliminating null values, a class imbalance was discovered. The number of cases where severity was 0 was found to be more

**Table 2** Weather condition levels grouped with similar weather types

| Weather condition level | Condition names |
|---|---|
| 0 | Clear, fair, fair/windy, mostly cloudy/windy, funnel cloud, N/A precipitation |
| 1 | Light rain, overcast, mostly cloudy, light snow, scattered clouds, partly cloudy, cloudy, partly cloudy/windy, drizzle/wind |
| 2 | Rain, haze, light drizzle, fog, shallow fog, light freezing rain, drizzle, light rain showers, mist, patches of fog, light haze, light rain/windy, showers in the vicinity, light rain shower, cloudy/windy, light fog, fog/windy, partial fog, partial fog/windy, rain showers, rain shower |
| 3 | Snow, light freezing drizzle, heavy rain, smoke, light freezing fog, volcanic ash, blowing sand, blowing dust/windy, widespread dust, light rain with thunder, thunder in the vicinity, sand, snow grains, light snow/windy, wintry mix, haze/windy, light rain shower/windy, light drizzle/windy, light snow shower, freezing drizzle, light sleet, light snow grains, light snow showers, light hail, light ice pellets |
| 4 | Light thunderstorms and rain, hail, light thunderstorms and snow, snow showers, rain/windy, patches of fog/windy, drizzle and fog, wintry mix/windy, light freezing rain/windy, light snow and sleet, light snow and sleet/windy, freezing rain, sleet, light snow with thunder, widespread dust/windy, light blowing snow, heavy rain shower, snow/windy |
| 5 | Thunderstorms and rain, thunderstorm, T-storm, thunder, low drifting snow, small hail, thunder/windy, sand/dust whirlwinds, blowing dust, smoke/windy, snow and sleet, blowing snow/windy, rain and sleet, heavy sleet, snow and sleet/windy, heavy snow with thunder, light thunderstorm, heavy freezing drizzle, freezing rain/windy, thunder and hail, drifting snow, thunder and hail/windy, drifting snow, heavy rain showers, T-storm/windy, heavy snow/windy, heavy smoke |
| 6 | Heavy thunderstorms and rain, heavy T-storm, heavy T-storm/windy, blowing snow, heavy snow, ice pellets, squalls, heavy thunderstorms and snow, heavy ice pellets, heavy freezing rain, squalls/windy, sand/dust whirlwinds/windy, snow and thunder, thunder/wintry mix/windy, heavy blowing snow, heavy thunderstorms with small hail, thunderstorms and snow, dust whirls, sand/dust whirls nearby |
| 7 | Tornado |

than double the cases where severity was 1. In order to rectify this, a simple up-sampling technique was used which added more cases with severity equal to one, thus equalising the number of values in the two classes. All these steps were performed using the pandas, numpy and sci-kit learn libraries in Python.

### 2.4 Splitting of Dataset

After pre-processing the data, the dataset was split into two parts:

- Training set (80% of dataset).
- Test set (20% of dataset).

The model learnt the features corresponding to each class label class (severity) from the training set. After learning the features, the model is tested by predicting the severity of unseen data, i.e. test set data. This gives us an idea of how well the model can perform on unseen data.

## 3 Classification Algorithms

In order to determine the severity of an accident, various algorithms were considered. Following are the various models.

### 3.1 Gaussian Naïve Bayes

The Naïve Bayes algorithm is based on the Bayes theorem and assumes that the variables are not dependent on each other. This is useful while dealing with nonlinearly separable data that cannot be classified by simple methods such as logistic regression. The Gaussian Naïve Bayes [7] is simply the Naïve Bayes algorithm with a Gaussian distribution. Here, in addition to the probabilities of each term, we also require the mean and standard deviation of the terms.

Every attribute of the class can be defined by a Gaussian probability density function given by Eq. (1)

$$Xi \sim N(\mu, \sigma 2) \tag{1}$$

This type of probability function is in the form of a bell curve and is given by Eq. (2)

$$N(\mu, \sigma 2)(x) = \left(\frac{1}{\sqrt{2\pi\sigma 2}}\right) * \mathrm{e}^{-(x-\mu)^2/2\sigma 2} \tag{2}$$

Here, $\mu$ is the mean, and $\sigma^2$ is the variance.

For the accident severity prediction model, experimentation with the Gaussian Naïve Bayes [8] algorithm was carried out. Implementation of the Gaussian Naïve Bayes model was carried out on Python with the help of the sklearn library.

## *3.2 Decision Tree Classifier*

The decision tree classifier is another popular classifier that was experimented with, for this particular problem. Classification in a decision tree classifier is done by traversing from the root and consequent nodes, to the leaf of the tree, where the nodes define the instances and leaves signify the class. Again, like the Gaussian Naïve Bayes model, decision tree classifiers are particularly useful for classifying nonlinear data. The decision tree was implemented using the sklearn library in Python.

## *3.3 Voting Ensemble Method Using Multi-layer Perceptron and Decision Trees Classifier*

While existing algorithms are easy to experiment on, another approach to predict accident severity is the creation of an also known as a multi-layer perceptron. This gives more parameters to control and an in-depth understanding of how the model can carry out the classification. A multi-layer perceptron consists of an input layer, hidden layers and an output layer. A perceptron that does not contain any hidden layers can classify linear data, while hidden layers are introduced when the data is nonlinear.

The ANN [8] model consisted of three rectified linear unit (activation functions) layers and one final sigmoid layer. The first ReLu [9] layer consisted of eight neurons, the second had six neurons, and the 3rd had three neurons. The sigmoid layer had one neuron. These layers are cascaded, and each layer serves as an input to the next.

This neural network follows the backpropagation algorithm [10]. The backpropagation algorithm behaves similar to a feedback model that tries to minimise the cost function. Each neuron has variable weights and biases, which are adjusted to minimise the cost function.

Ensemble methods are often used to increase the accuracy of a machine learning model by combining two or more models. The two simple ensemble techniques are voting and averaging ensemble, where the former is used for classification tasks and the latter for regression. This model utilises a hard voting ensemble method that combined the predictions of decision tree classifier and artificial neural networks. A hard voting ensemble is essentially a method where the exact labels of a class are added, and the class with the most votes is selected. A voting ensemble is created using a multi-layer perceptron and a decision tree classifier.

# 4 Results and Discussion

## *4.1 Performance Metrics Used to Assess the Model*

While the accuracy score of a model is the most widely used parameter to judge a model's performance, other performance metrics [11] are also useful to evaluate a model on a case-to-case basis. Considering this problem statement, both false positives and false negatives pose issues in a real-life scenario. For example, consider a situation where a severe accident is reported and a new route is to be determined. If the model falsely classifies the accident as "less severe", a traffic jam could occur, and important vehicles like police jeeps and ambulances could face major problems. On the other hand, if a less severe accident has occurred and the model classifies it as "more severe", the re-routing of traffic could prove to be useless.

Confusion matrix: A confusion matrix is a matrix that denotes the actual and predicted values of the model. It is a convenient way of reporting the correct and incorrect predictions of the model. If the true positives and true negatives exceed the false positives and false negatives, respectively, the model is said to be more accurate. If the inverse is true, then the model is not as accurate. A binary classification model would have a 2 × 2 confusion matrix, while a model that classifies three classes would have a 3 × 3 confusion matrix.

Accuracy score: The accuracy score denotes the ratio of correctly made predictions to the total number of predictions. This can be derived from the confusion matrix as given in Eq. (3)

$$\frac{\text{True Positives} + \text{True Negatives}}{\text{True Positives} + \text{False Positives} + \text{False Negatives} + \text{True Negatives}} \tag{3}$$

Logarithmic loss: By calculating the logarithmic loss, we can clearly determine how many incorrect classifications our model makes. The value of log loss can range from 0 to infinity, where a value closer to 0 indicates a more accurate model. It is given by Eq. (4)

$$\text{LogarithmicLoss} = \frac{-1}{N}\sum_{i=1}^{N}\sum_{j=1}^{M} y_{ij} * \log(pij) \tag{4}$$

where there are "$N$" number of samples that belong to "$M$" number of classes.

"$y_{ij}$" stands for whether sample $i$ is a part of class $j$ or not.

"$pij$" shows the probability of sample $i$ being a part of class $j$.

F1 score: F1 [12] score is used to determine the correctly classified instances. It is another method of determining the accuracy of a model. It is the harmonic mean of precision and recall. Where precision is the ratio between the correctly classified positive values (true positives) and the total number of predicted positive values. Recall is the ratio of the correctly classified positive values (true positives) and the actual number of positive values. Thus, F1 is given by Eq. (5)

$$F1 = 2 * \frac{1}{\frac{1}{\text{precision}} + \frac{1}{\text{recall}}} \tag{5}$$

## 4.2 Results

Following are the results obtained for each model. Figures 2, 3 and 4 show the confusion matrices for Gaussian Naive Bayes, decision tree classifier [13] and ANN algorithm, respectively.

The following Table 3 shows each model [14] and its corresponding accuracy score, logarithmic loss and F1 score.

From Table 3, it is found that the decision tree classifier provides the best accuracy. When paired with the multi-layer perceptron in an ensemble model, the accuracy slightly improves.

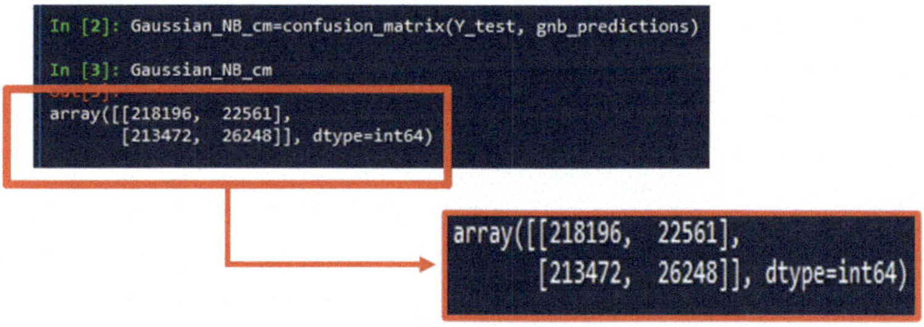


**Fig. 2** Gaussian Naive Bayes confusion matrix

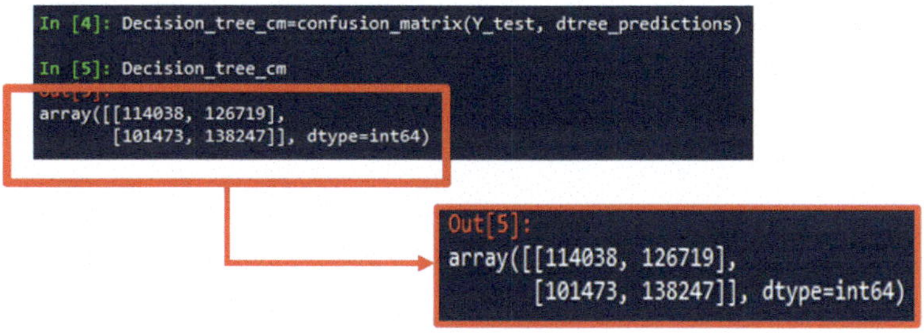


**Fig. 3** Decision tree classifier confusion matrix

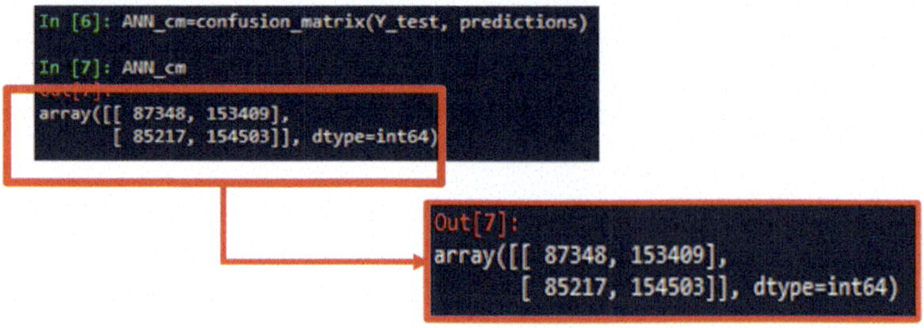


**Fig. 4** ANN confusion matrix

**Table 3** Performance metrics for each model

| Model name | Accuracy score | Logarithmic loss | F1 score |
|---|---|---|---|
| Gaussian Naïve Bayes | 0.5080 | 16.991 | 0.1797 |
| Decision tree classifier | 0.5269 | 16.3373 | 0.5495 |
| Artificial neural network | 0.5139 | 16.7893 | 0.4626 |
| Voting ensemble | 0.5278 | 16.31 | 0.5499 |

## 5 Conclusion and Future Scope

As weather conditions have little effect on the severity of an accident, predicting the dependent variable was difficult. Thus, the limitation of this model is the accuracy. The maximum accuracy received was 52.78% by using a voting ensemble method, which included a combination of a multi-layer perceptron and decision tree classifier as they perform well when there exists an imbalanced dataset. The logarithmic loss was also the least at 16.31 using this method. The next steps towards a successful "accident severity predictor" would be improving the model accuracy by implementing other algorithms or combinations of more algorithms. Additionally, unsupervised learning methods or clustering techniques could also be used to determine the variables that influence the severity considerably. By improving the model further, a robust system can be developed that could revolutionise the dynamicity of navigation systems.

## References

1. Siar, M., Teshnehlab, M.: Brain Tumor detection using deep neural network and machine learning algorithm. In: 9th International Conference on Computer and Knowledge Engineering, ICCKE, pp. 363–364 (2019)

2. Duan, L., Di, P., Li, A.: A new Naive Bayes text classification algorithm. TELKOMNIKA Indonesian J. Electr. Eng. Inst. Adv. Eng. Sci. **12**, 948–949 (2014)
3. Manjunath Jogin, M., Madhulika, M.S., Divya, G.D., Meghana, R.K., Apoorva, S.: Feature extraction using convolution neural networks (CNN) and deep learning. In: International Conference on Recent Trends in Electronics, Information & Communication Technology, IEEE, pp. 2320–2321 (2018)
4. Malin, F., Norros, I., Innamaa, S.: Accident risk of road and weather conditions on different road types. Accid. Anal. Prevent. **122**, 182 (2019)
5. Rana, A., Singh Rawat, A., Bijalwan, A., Bahuguna, H.: Application of multi layer (perceptron) artificial neural network in the diagnosis system: a systematic review. In: International Conference on Research in Intelligent and Computing in Engineering, IEEE, pp. 2–3 (2018)
6. Moosavi, S., Samavatian, M.H., Parthasarathy, S., Teodorescu, R., Ramnath, R.: Accident risk prediction based on heterogeneous sparse data: new dataset and insights. In: Proceedings of the 27th ACM SIGSPATIAL International Conference on Advances in Geographic Information Systems, ACM (2019)
7. Gayathri, B.M., Sumathi, C.P.: An automated technique using Gaussian Naïve Bayes classifier to classify breast cancer. Int. J. Comput. Appl. **148**, 16–21 (2016)
8. Gupta, K.K., Vijay, R., Pahadiya, P.: A review paper on feature selection techniques and artificial neural networks architectures used in thermography for early stage detection of breast cancer. In: Soft Computing: Theories and Applications, Advances in Intelligent Systems and Computing, Springer, Berlin, pp. 457–458 (2020)
9. Dahiya, S., Tyagi, R., Gaba, N.: Streamlining choice of CNNs and structure framing of convolution layer. In: Soft Computing: Theories and Applications, Advances in Intelligent Systems and Computing, Springer, pp. 707–709 (2020)
10. Zhixin, S., Bingqing, L.: Research of improved back-propagation neural network algorithm. In: 12th International Conference on Communication Technology Electronics, IEEE, pp. 763–765 (2010)
11. Malsa, N., Singh, P., Gautam, J., Srivastava, A., Singh, S.P.: Source of treatment selection for different states of India and performance analysis using machine learning algorithm for classification. In: Soft Computing: Theories and Applications, "Advances in Intelligent Systems and Computing, Springer, p. 240 (2020)
12. Ghori, K.M.U., Imran, M., Nawaz, A., Abbasi, R.A., Ullah, A., Szathmary, L.: Performance analysis of machine learning classifiers for non-technical loss detection. J. Ambient Intell. Hum. Comput. 10 (2020)
13. Ankitdeshpandey, Karthi, R.: Development of intrusion detection system using deep learning for classifying attacks in power systems. In: Soft Computing: Theories and Applications, Advances in Intelligent Systems and Computing. Springer, pp. 755–756 (2020)
14. Ochieng, N., Mwangi, W., Ateya, I.: Generalization performance comparison of machine learners for the detection of computer worms using behavioral features. In: Soft Computing: Theories and Applications, Advances in Intelligent Systems and Computing, Springer, pp. 677–678 (2020)

# Heuristics for Generalized Minimum Dominating Set Problem

**Mallikarjun Rao Nakkala and Alok Singh**

**Abstract** For any edge-weighted undirected graph $G = (V, E)$ with a fixed threshold value, a generalized dominating set $G_D$ is a subset of $V$ such that every node of $V$ either belong to $G_D$ or have at least one adjacent node in $G_D$ with an additional condition that each node not in $G_D$ must satisfy the threshold constraint, i.e., the sum of edge weights of a non-member node to the members of $G_D$ must be at least equal to the threshold value. The generalized minimum dominating set problem (GMDS) seeks a generalized dominating set with minimum cardinality. If every edge weight is larger than or equal to the threshold value, then the problem becomes the conventional minimum dominating set (MDS) problem. Being a generalization of MDS, GMDS is $\mathcal{NP}$-hard. GMDS finds application in text summarization. In this paper, we present six greedy heuristics to solve the GMDS problem. To our knowledge, these are the first heuristic approaches for GMDS. We have tested these approaches on a wide range of instances. Computational results show the effectiveness of our heuristics.

**Keywords** Combinatorial optimization · Dominating set · Generalized minimum dominating set · Heuristic

## 1 Introduction

Given an undirected graph $G = (V, E)$, a dominating set $D$ is a subset of $V$ such that each node $v \in V$ is either present in $D$ or adjacent to at least one node present in $D$. Nodes of $D$ are termed as dominating nodes or dominators, and all the other nodes are termed as non-dominating or dominatee or dominated nodes. The minimum dominating set (MDS) problem seeks a dominating set of minimum cardinality. The generalized minimum dominating set problem (GMDS) is an extension of MDS for

M. R. Nakkala · A. Singh (✉)
School of Computer and Information Sciences, University of Hyderabad, Hyderabad 500046, India
e-mail: alokcs@uohyd.ernet.in



T. K. Sharma et al. (eds.), *Soft Computing: Theories and Applications*, Advances in Intelligent Systems and Computing 1381,
https://doi.org/10.1007/978-981-16-1696-9_29

(edge) weighted graphs. Given a weight function $w : E \rightarrow \Re^+$ associated with edges of $G$ and a threshold value $T$, the GMDS seeks a dominating set with minimum cardinality such that the sum of edge weights from each non-dominating node to dominating nodes is at least $T$. This additional constraint on the dominating set is referred to as threshold constraint and a dominating set satisfying this constraint is referred to as a generalized dominating set (GDS). If $w(e) \geq T \; \forall e \in E$, then GMDS is equivalent to MDS. Hence, GMDS is $\mathcal{NP}$-hard as it can be considered as a generalization of MDS, which is known to be $\mathcal{NP}$-hard [4].

Figure 1 illustrates GMDS with the help of an example. Figure 1a is the input graph. Figure 1b, c shows the optimal solutions of GMDS with $T = 10$ and $T = 20$, respectively, on the input graph of Fig. 1a. In these two figures, dominating nodes are shown in dark gray color and dominated nodes are shown in light gray color. So optimal solution of GMDS with $T = 10$ is $\{a, b, d\}$ and with $T = 20$ is $\{a, b, c, d\}$. Please note that these optimal solutions are not unique as {e, f, g} is another optimal solution for $T = 10$, and {c, e, f, g} is another optimal solution for $T = 20$. Please note that {a, d} is an optimal solution for MDS on this graph. If $T \leq 6$, then GMDS on this graph is equivalent to MDS. As can be seen in this example, cardinality of GMDS on any given graph is a monotonically non-decreasing function of $T$.

GMDS was introduced by Xu and Zhou [13] in 2016 who cited the application of this problem in the context of text summarization, i.e., extracting a set of sentences from one or more documents which can best summarize the content of these documents. Earlier, Shen and Li [11] used MDS for text summarization, and GMDS was introduced as an extension of MDS for this task. From the sentences in the documents, a graph is constructed where a node is created for each sentence and an edge is created between two nodes if the similarity score between the two corresponding sentences is not less than a certain minimum value. Each sentence is pre-processed before computing the similarity score to remove the stop words and to convert each remaining word to its standard prototype. GMDS allows use of lower minimum similarity score while constructing the graph which yields a denser graph. Size of dominating set is inversely proportional to density of the graph. A higher threshold value while finding GMDS compensates the use of lower minimum similarity score while constructing the graph. Hence, use of GMDS is expected to obtain smaller set of sentences than MDS without affecting the quality of summarization.

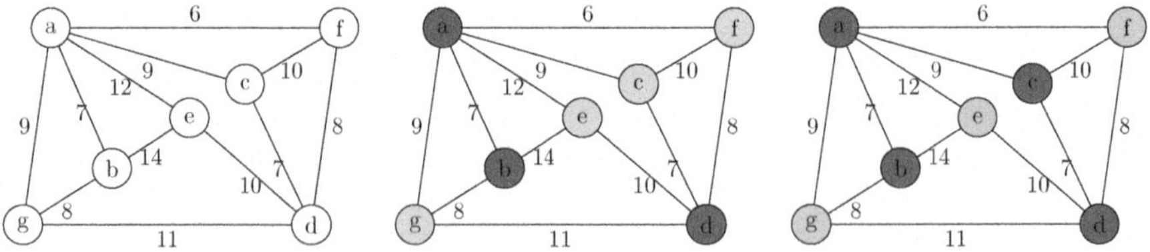

(a) Input graph (b) Optimal solution for $T$=10 (c) Optimal solution for $T$=20

**Fig. 1** Illustrating GMDS with two different values of $T$

Many heuristic approaches exist in the literature for the MDS and its extension on node-weighted graphs, viz. minimum weight dominating set problem (MWDS) [1–3, 5–10, 12]. However, GMDS, which is an extension of MDS on edge-weighted graphs, has not been studied so far from the heuristic perspective. While introducing the GMDS, Xu et al. [13] utilizing the concepts from statistical physics have proposed a replica symmetric (RS) spin glass theory model for this problem and a message-passing belief propagation-guided decimation (BPD) algorithm to solve this model. These are highly complex techniques and require knowledge of statistical physics to understand them. However, there exists no simple heuristic for this problem in the literature. This has motivated us to develop the heuristic approaches for GMDS.

In this paper, we have presented six greedy heuristics to solve the GMDS problem. These heuristics make use of the basic concepts like the number of adjacent nodes and the sum of edge weights of nodes to construct a GDS. These heuristics are augmented with a redundant node removal procedure to further improve the quality of solutions obtained. The performance of these six heuristics has been evaluated on the benchmark instances derived from standard benchmark instances available in the literature for MWDS. Computational results show the effectiveness of our heuristics.

The remainder of this paper is organized as follows: Sect. 2 presents our various heuristics for GMDS. Computational results and their analysis have been presented in Sect. 3. Finally, we conclude the paper in Sect. 4 by listing our contributions and some possible directions for future research.

## 2 Heuristics for GMDS

We have developed six greedy heuristics for GMDS. These heuristics will be named H1, H2, H3, H4, H5 and H6. These heuristics first construct a GDS, and then, redundant nodes are removed to improve the solution quality. Algorithm 1 presents the common framework for these six heuristics. To construct a GDS, the candidate solution ($G_D$) is initialized to empty set $\emptyset$, and then, all mandatory nodes are inserted into $G_D$. After that, an iterative process ensues where during each iteration a node is chosen as per a greedy policy from $V \setminus G_D$ and inserted into $G_D$. This iterative process continues till $G_D$ becomes a GDS. After that, redundant nodes are removed to further improve the quality of $G_D$. Mandatory nodes are those nodes that must be part of any feasible solution, and the function *mandatory_nodes()* identify such nodes. Clearly, any node whose sum of weights of incident edges is less than the given threshold value is a mandatory node. By adding these nodes to the candidate solution at the beginning can aid in obtaining better quality solutions. The function *greedy_policy()* chooses a dominating node for adding into the candidate solution as per a greedy policy, and its actual implementation varies from one heuristic to another. A dominating node $v$ belonging to a GDS $G_D$ is called redundant if $G_D \setminus \{v\}$ is also a GDS. The quality of a GDS can be improved by removing the redundant nodes. The redundant node removal process works iteratively. In every iteration, it first identifies the set of redundant nodes in $G_D$, and then, a node with a minimum

**Input**: An edge-weighted undirected graph $G := (V, E)$, $w(e) \in \Re^{+} \forall e \in E$.
**Output**: Best solution found.

$G_D \leftarrow \emptyset$;
$G_D \leftarrow G_D \cup$ mandatory_nodes($V$);
**while** $G_D$ *is not a GDS* **do**
  $v \leftarrow$ greedy_policy($V \setminus G_D$);
  $G_D \leftarrow G_D \cup \{v\}$;
**end**
$G_D \leftarrow$ redundant_removal($G_D$);
**return** $G_D$;

**Algorithm 1:** General structure of our heuristics

value of sum of its edge weights (value of $SW_{we}$ at the beginning of the heuristic) is selected from the set of redundant nodes (ties are broken arbitrarily) and removed from $G_D$. The process repeats till no redundant node remains in $G_D$. The method *redundant_removal()* implements the redundant node removal process.

Like the usual practice in describing a dominating set construction algorithm, we have followed a coloring scheme. It is assumed that initially, all nodes $v \in V$ are colored white. Whenever a node is added to partially constructed GDS, its color is changed to black, and colors of its those white adjacent nodes which satisfy the threshold constraint are changed to gray. Termination condition for our heuristics can be restated in terms of this coloring scheme, i.e., our heuristics terminate when no white node is left in the graph. So when the construction of GDS completes, all nodes can have either black or gray color. Further, dominating nodes have black color, and dominated nodes have gray color.

Four quantities, viz. CLR($u$), $D_w$, $SW_{we}$, $WT_{set}$, in addition to threshold $T$ are used by the presented heuristic approaches. These four quantities are defined in Table 1. All these four quantities are updated whenever a new node is selected as a dominating node.

As mentioned earlier, the greedy policy is different for different heuristics. H1 primarily depends on edge weights to white adjacent nodes ($SW_{we}$). The greedy policy of H2, H3 and H5 involves two steps, and first step is common to these three.

**Table 1** Notational convention

| Quantity | Definition | Mathematical expression |
|---|---|---|
| CLR($u$) | 1 in case vertex $u$ is white, otherwise 0 | – |
| $D_w(u)$ | Number of white nodes adjacent to $u$ | $\sum_{(u,v)\in E} CLR(v)$ |
| $SW_{we}(u)$ | Sum of edge weights of $u$ to the white nodes | $\sum_{(u,v)\in E} w(u, v) \times \mathrm{CLR}(v)$ |
| $WT_{set}(u)$ | Sum of edge weights of $u$ to the nodes in $G_D$ | $\sum_{((u,v)\in E)\wedge(v\in G_D))} w(u, v)$ |

In the first step, they check for a white node $u$ such that $D_w(u)$= 1 and $u$ satisfies the threshold constraint when its sole white adjacent node $v_i$ is selected as a dominating node. If such a node $u$ is found, then its sole white adjacent node $v_i$ is selected as a dominating node. If no such node is found in the first step, then only it enters into the second step. The reason behind first step is the empirical observation that for nodes like $u$, selecting $v_i$ immediately yields better solution in most cases. In the second step, H2 is exactly the same as H1. On the other hand, H3 in the second step primarily depends on the number of white adjacent nodes ($D_w$). The greedy policy of H4 primarily depends on the product of $SW_{we}$ and $D_w$. As mentioned already, greedy policy of H5 involves two steps, and second step is exactly same as the greedy policy of H4. H6 works differently from rest, and it uses a ranking mechanism to implement its greedy policy. The greedy policy used by these six heuristics is described in detail in the subsequent subsections. We have already described the common features of these six heuristics, and these will not be repeated while describing each of these heuristics.

## 2.1 *First Heuristic (H1)*

The greedy policy of our first heuristic referred to as H1 depends on $SW_{we}$ to select a dominating node, and it uses $D_w$ only as a tiebreaker whenever a tie occurs in $SW_{we}$ values. This greedy policy makes use of Eq. 1 to select a non-black node, i.e., it selects a non-black node having the highest $SW_{we}$ value. If two or more nodes have the same highest $SW_{we}$ value, then Eq. 2 is used to break the tie. If again a tie occurs, then one node among those involved in this second tie is selected uniformly at random. The basic idea behind this greedy policy is that by making a node with a highest value of $SW_{we}$, dominating can aid in making its adjacent white nodes dominated with lesser number of dominating nodes. This may result in GDS of smaller size.

$$v \leftarrow \arg\max_{u \in V \setminus G_D} (SW_{we}(u)) \tag{1}$$

$$v \leftarrow \arg\max_{u \in V \setminus G_D} (D_w(u)) \tag{2}$$

## 2.2 *Second Heuristic (H2)*

Our second heuristic referred to as H2 has a greedy policy that involves two steps. These two steps are already explained. The second step is exactly same as the greedy policy of H1.

## 2.3 Third Heuristic (H3)

The greedy policy of third heuristic named H3 works in the same way as H2 with one difference in the second step. To select a dominating node in second step, it uses Eq. 2 instead of Eq. 1. Here, Eq. 1 has been used as the tie-breaking rule. The basic idea behind the second step of this greedy policy is that by making a node with a highest value of $D_w$, dominating can aid in making more of its adjacent nodes dominated, thereby leading to a GDS of smaller size. The second step of $H3$ can be used as a standalone greedy policy like the greedy policy of $H1$. We have tried this approach, but it produces the worst results overall than the six heuristics described here, and hence, it is not included in this paper.

Figure 2 illustrates H2 and H3 with the help of an example. Figure 2a is the input graph and $T = 10$. The GDS computed by H2 and H3 is shown in Fig. 2b, c, respectively. We will trace the working of H2. Initially, all nodes are white, and $G_D$ is set to $\emptyset$. Node $a$ has the sum of edge weights as 8, which is less the value of $T$ (10). Hence, it is a mandatory node. So node $a$ is added to $G_D$ and colored black. All its adjacent nodes are violating the threshold constraint, so none of its neighbors are colored gray. Now, there are no mandatory nodes. Then, H2 begins the iterative process. In the first iteration, no white node exists with white degree one. So second step of greedy policy is used. Using Eq. 1, node $c$ with highest white edge-weighted sum (17) is selected as a dominating node and is added to $G_D$ by changing its color to black. As its adjacent node $b$ satisfies the threshold constraint, $b$ is made a dominated node, and its color is changed to gray. In the second iteration, again, no white node exists with white degree one. Using Eq. 1, node $f$ with the highest white edge-weighted sum is added to $G_D$ and is colored black. Its adjacent white node $h$ is made dominating and colored gray as $h$ satisfies the threshold constraint. In the third iteration, the white degree of two nodes $e$&$g$ is one, and if their only white adjacent node $d$ is made dominated, then these two nodes satisfy the threshold constraint. So the node $d$ is added to $G_D$ and colored black. Now, the white adjacent nodes $e$ & $g$ satisfy the threshold constraint and are made dominating and colored gray. Now,

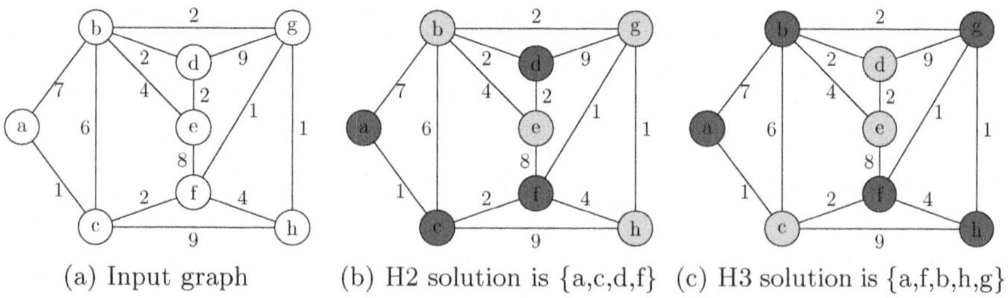


**Fig. 2** Illustrating H2 and H3 on a graph with 8 nodes, 14 edges and $T = 10$

there is no white node in the graph and H2 ends by returning the dominating set $G_D = \{a, c, f, d\}$ shown in Fig. 2b. Analogously, working of H3 can be traced to verify H3 returns the dominating set $G_D = \{a, f, b, h, g\}$ shown in Fig. 2c.

## *2.4 Fourth Heuristic (H4)*

This heuristic named as H4 works exactly same as H1 with only one difference. It uses Eq. 3 instead of Eq. 1 and 1 instead of Eq. 2 to break the tie. The greedy policy of this heuristic combines the ideas used in greedy policies of H1 and H3 by using the product of $D_{\mathrm{w}}$ and $\mathrm{SW}_{\mathrm{we}}$. The basic idea behind this greedy policy is that by making a node with a highest value of $D_{\mathrm{w}} \times \mathrm{SW}_{\mathrm{we}}$ dominating can aid not only in making its adjacent white nodes dominated with lesser number of dominating nodes but also in making more of its adjacent nodes dominated, thereby leading to a better GDS.

$$v \leftarrow \arg\max_{u \in V \setminus G_D} (D_{\mathrm{w}}(u) \times \mathrm{SW}_{\mathrm{we}}(u)) \tag{3}$$

## *2.5 Fifth Heuristic (H5)*

Our fifth heuristic referred to as H5 also uses a greedy policy involving two steps as mentioned already. The second step is exactly same as the greedy policy of H4.

## *2.6 Sixth Heuristic (H6)*

The sixth heuristic is named as H6. The greedy policy of H6 is based on ranks which are assigned to all non-black nodes using the Eq. 4 at the beginning of every iteration. Node with the highest rank is selected as a dominating node. If two or more nodes have the same highest rank, then a node with a minimum value of $\mathrm{WT}_{\mathrm{set}}$ is selected. If still a tie exists, then it is broken arbitrarily. With this ranking procedure, a dominating node is selected in such a way that the selected node will cover maximum number of nodes when they become dominated. Rank of a node $u$ depends on the threshold value $T$ and two quantities of its all white adjacent nodes $v_i$ $\forall (u, v_i) \in E$, viz. $w(u, v_i)$, $\mathrm{WT}_{\mathrm{set}}(v_i)$.

$$\mathrm{Rank}(u) \leftarrow x + y \tag{4}$$

where $x$ and $y$ are computed as follows:

$$x \leftarrow \sum_{((u,v_i)\in E)\wedge(\mathrm{CLR}(v_i)=1)\wedge(\mathrm{WT}_{\mathrm{set}}(v_i)+w(u,v_i)\geq T))} 1.0 \tag{5}$$

$$y \leftarrow \sum_{((u,v_i)\in E)\wedge(\mathrm{CLR}(v_i)=1)\wedge(\mathrm{WT}_{\mathrm{set}}(v_i)+w(u,v_i)<T))} \frac{(\mathrm{WT}_{\mathrm{set}}(v_i) + w(u, v_i))}{T} \tag{6}$$

## 3 Computational Results

We have evaluated the performance of our heuristics on three types of datasets derived from the benchmark datasets for the minimum weight dominating set (MWDS) available in the literature. MWDS datasets are three types, viz. unit disk graph (UDG), Type-I and Type-II graphs. All these graphs are node weighted. UDG graphs are generated by Potluri and Singh [10]. Dataset has six different node sizes, viz. 50, 100, 250, 500, 800 and 1000. These nodes are randomly distributed in an area of 1000 × 1000 units, and there is an edge between the two nodes if the distance between them does not exceed $R$. Two different values of $R$ (150 and 200) are considered. Type-I and Type-II datasets are generated by Jovanovic et al. [5]. The number of nodes in these two datasets varies from 50 to 1000, and the number of edges varies from 50 to 20,000. There exist 10 different graph instances for each combination of the number of nodes and the number of edges/range in these three datasets. There are 53 instance groups of such 10 instances in each of Type-I and Type-II datasets. For UDG dataset, number of such groups is 12. So a total of 1180 graph instances divided into 118 groups are there in these three datasets.

These MWDS datasets are converted to GMDS datasets by discarding the node weights and assigning the weights on every edge. For UDG and Type-I graph instances, the threshold value $T$ is fixed to 10, and the weight of each undirected edge is assigned an integer value randomly in the closed interval of [1, 9]. For Type-II graph instances, threshold value $T$ is fixed to 20, and the weight of each undirected edge is assigned randomly an integer value in the closed interval of [6, 15]. Actually, Type-I and Type-II datasets for MWDS are identical upon discarding the node weights. Hence, the same graph is tested with two different threshold values and two different intervals of edge weights. We also refer to these three datasets as UDG, Type-I and Type-II datasets.

Our six heuristics are implemented in C, and a Linux-based 2.0 GHz Core-i3-5005U system with 4 GB of memory has been used for their executions. We have reported the results before and after redundant node removal procedure. There is a possibility that the relative performance of heuristics may change before and after the redundant node removal procedure. We have reported the results for each of the 118 instance groups. For each group, reported results are the average of results on 10 instances belonging to that group. As already mentioned, these 10 instances have

**Table 2** Comparison of H1, H2, H3, H4, H5 and H6 heuristics' results for UDG graph instances

| $\|V\|$ | $R$ | Before redundant node removal | | | | | | After redundant node removal | | | | | |
|---|---|---|---|---|---|---|---|---|---|---|---|---|---|
| | | H1 | H2 | H3 | H4 | H5 | H6 | H1 | H2 | H3 | H4 | H5 | H6 |
| 50 | 150 | 31.1 | 29.9 | **29.7** | 30.9 | **29.7** | 30.5 | 29.7 | 29.5 | **29.4** | 29.7 | **29.4** | 29.6 |
| 50 | 200 | 25.5 | 24.4 | 24.6 | 25.7 | 24.5 | **24.1** | 23.9 | 23.2 | 23.3 | 24.0 | **23.0** | **23.0** |
| 100 | 150 | 48.9 | **45.0** | 45.7 | 49.2 | 45.6 | 48.7 | 44.6 | **43.1** | 43.2 | 44.9 | 43.3 | 44.4 |
| 100 | 200 | 30.6 | **29.3** | 31.1 | 31.6 | 30.5 | 31.0 | 28.5 | 28.6 | 29.2 | 29.1 | **28.4** | 29.2 |
| 250 | 150 | 57.3 | 56.5 | 58.0 | 58.4 | **56.1** | 58.0 | 53.7 | 53.8 | 53.7 | 53.9 | 53.5 | **53.4** |
| 250 | 200 | **36.3** | 36.6 | 37.9 | 36.5 | 36.5 | 36.4 | **34.5** | 35.0 | 36.4 | 34.9 | 34.6 | 34.7 |
| 500 | 150 | 64.4 | 64.8 | 65.6 | 64.8 | 65.7 | **62.8** | 61.2 | 62.1 | 61.6 | 61.1 | 63.1 | **61.0** |
| 500 | 200 | 41.0 | 41.4 | 42.5 | **40.7** | 41.6 | 41.6 | 38.7 | 39.5 | 40.7 | **38.5** | 39.9 | 40.4 |
| 800 | 150 | **66.8** | 67.8 | 70.3 | 67.9 | 68.8 | 68.9 | **64.1** | 65.6 | 67.2 | 65.0 | 65.5 | 66.1 |
| 800 | 200 | **42.8** | 43.9 | 44.4 | 43.4 | 44.3 | **42.8** | **41.3** | 42.1 | 42.7 | 42.0 | 42.4 | 41.8 |
| 1000 | 150 | **68.5** | 70.2 | 73.6 | 70.2 | 71.5 | 70.2 | **66.6** | 67.4 | 70.6 | 67.5 | 68.3 | 68.7 |
| 1000 | 200 | **44.0** | 45.1 | 46.7 | 44.5 | 45.3 | 44.3 | **42.6** | 43.3 | 44.0 | 42.8 | 43.8 | 43.5 |

the same number of nodes and edges/ranges. Execution time in almost all cases is less than 1 second, so we have not reported the execution times.

Table 2 shows the results obtained by six heuristics on UDG instances. In the same way, Tables 3 and 4 present results on Type-I and Type-II instances. In these tables, the number of nodes is represented as $|V|$, and the number of edges as $|E|$, and $|R|$ denotes the transmission range in case of UDG dataset. Summary of results before and after redundant node removal has been shown in Tables 5 and 6, respectively. In these tables, the performance of every heuristic on UDG, Type-I and Type-II instances, has been compared with other heuristics in terms of number of instance groups where heuristic on the left is better (B), equal (E) or worse (W) compared with the heuristic at the top.

Performance of the six heuristics varies depending on the type of instance group. By observing the results presented in various tables before and after redundant node removal, most of the heuristics have same relative performance. In all cases, the quality of the solutions has been improved after the redundant node removal procedure.

On UDG instances (Table 2), out of 12 instance groups H1, H2, H3, H4, H5 and H6 after redundant node removal have obtained best results on 5, 1, 1, 1, 3, 3 instance groups, respectively. If the graph has more nodes, then H1 has performed better than other heuristics on UDG instances. If a graph has less number of nodes, then H5 performed better than other heuristics. H6 also performs better on graphs with lesser number of nodes. H2, H3 and H4 have poor performance on UDG graphs.

On Type-I instances (Table 3), out of 53 instance groups H1, H2, H3, H4, H5 and H6 after redundant node removal have obtained better results on 6, 17, 16, 12, 19, 7 instance groups, respectively. The performance of H2 and H3 is better than other heuristics on sparse graphs (a graph with less number of edges). H4 and H5 have better performance on dense graphs (a graph with more number of edges). Occasionally,

**Table 3** Comparison of H1, H2, H3, H4, H5 and H6 heuristics' results for Type-I graph instances

| $\|V\|$ | $\|E\|$ | Before redundant node removal | | | | | | After redundant node removal | | | | | |
|---|---|---|---|---|---|---|---|---|---|---|---|---|---|
| | | H1 | H2 | H3 | H4 | H5 | H6 | H1 | H2 | H3 | H4 | H5 | H6 |
| 50 | 50 | 34.1 | **33.6** | **33.6** | 34.1 | **33.6** | **33.6** | 34.1 | **33.6** | **33.6** | 34.1 | **33.6** | **33.6** |
| 50 | 100 | 29.5 | 27.2 | **27.0** | 29.3 | 27.1 | 28.7 | 26.8 | **26.3** | 26.6 | 27.5 | 26.6 | 27.2 |
| 50 | 250 | 15.9 | 14.9 | 14.8 | 15.0 | **14.5** | 15.8 | 14.9 | 14.6 | **14.3** | 14.7 | **14.3** | 15.1 |
| 50 | 500 | 8.3 | 8.7 | 8.4 | **8.2** | 8.6 | 8.3 | **8.0** | 8.3 | 8.2 | **8.0** | 8.4 | 8.2 |
| 50 | 750 | 6.1 | 6.2 | 5.8 | 5.9 | 6.0 | **5.7** | 5.9 | 5.9 | 5.8 | 5.8 | 5.9 | **5.7** |
| 50 | 1000 | 4.6 | 4.6 | 4.9 | 4.5 | 4.5 | **4.4** | 4.4 | 4.4 | 4.9 | 4.3 | 4.3 | **4.1** |
| 100 | 100 | 69.5 | **68.6** | **68.6** | 69.5 | **68.6** | 68.8 | 69.3 | **68.6** | **68.6** | 69.3 | **68.6** | 68.8 |
| 100 | 250 | 52.6 | 48.5 | 48.4 | 52.6 | **47.8** | 51.7 | 48.4 | 46.9 | **46.6** | 48.6 | 46.8 | 48.0 |
| 100 | 500 | 30.7 | **30.4** | **30.4** | 31.6 | 30.6 | 31.2 | 29.8 | 29.3 | 29.4 | 29.5 | 29.4 | **29.0** |
| 100 | 750 | 22.4 | 22.4 | **21.5** | 22.2 | **21.5** | 22.8 | 21.9 | 21.6 | **21.1** | 21.8 | **21.1** | 22.0 |
| 100 | 1000 | 17.7 | 18.0 | 17.7 | **17.1** | 17.6 | 18.3 | 17.5 | 17.7 | 17.4 | **16.6** | 16.8 | 17.1 |
| 100 | 2000 | 9.7 | 9.8 | 10.0 | **9.2** | 9.6 | 9.7 | 9.5 | 9.7 | 9.9 | **9.2** | 9.6 | 9.6 |
| 150 | 150 | 105.4 | **104.4** | **104.4** | 105.4 | **104.4** | **104.4** | 105.0 | **104.4** | **104.4** | 105.0 | **104.4** | **104.4** |
| 150 | 250 | 92.3 | **85.4** | 86.7 | 93.0 | 86.0 | 89.2 | 87.4 | **84.6** | 85.2 | 87.6 | 84.7 | 86.3 |
| 150 | 500 | 63.7 | 58.9 | 59.1 | 63.9 | **58.7** | 62.1 | 59.5 | 56.6 | 57.0 | 59.5 | **56.4** | 57.8 |
| 150 | 750 | 49.0 | 46.0 | 46.3 | 48.0 | **45.2** | 49.1 | 46.0 | **44.6** | 45.0 | 45.8 | **44.6** | 46.6 |
| 150 | 1000 | 37.4 | 37.4 | 37.7 | 37.7 | **36.5** | 38.4 | 36.2 | 36.1 | 36.6 | 36.2 | **35.8** | 37.4 |
| 150 | 2000 | **20.6** | 20.8 | 21.4 | **20.6** | 20.8 | **20.6** | 20.4 | 20.6 | 21.0 | 20.5 | 20.7 | **20.1** |
| 150 | 3000 | 14.9 | 15.1 | 15.6 | **14.5** | 15.1 | 15.1 | 14.8 | 15.0 | 15.5 | **14.4** | 15.0 | 15.1 |
| 200 | 250 | 132.9 | **127.3** | **127.3** | 132.6 | **127.3** | 129.5 | 130.4 | **127.2** | **127.2** | 130.4 | **127.2** | 128.7 |
| 200 | 500 | 104.8 | 97.1 | 96.8 | 105.3 | **96.2** | 101.1 | 95.3 | 94.1 | **93.1** | 96.8 | 93.8 | 94.5 |
| 200 | 750 | 80.8 | 75.4 | **75.1** | 81.1 | 75.4 | 80.7 | 74.6 | **72.1** | 72.4 | 74.6 | 73.0 | 74.4 |
| 200 | 1000 | 64.3 | 60.3 | **59.5** | 63.3 | 60.4 | 63.7 | 60.8 | 59.1 | **57.9** | 60.0 | 58.4 | 60.7 |
| 200 | 2000 | 35.4 | 36.2 | 36.5 | **35.1** | 36.4 | 35.5 | 35.0 | 35.3 | 35.9 | 34.7 | 35.3 | **34.5** |
| 200 | 3000 | 25.9 | 25.7 | 26.2 | **25.1** | 25.8 | 26.0 | 25.4 | 25.3 | 25.8 | **25.0** | 25.7 | 25.8 |
| 250 | 250 | 173.6 | **172.1** | **172.1** | 173.4 | **172.1** | 172.3 | 173.3 | **172.1** | **172.1** | 173.1 | **172.1** | 172.3 |
| 250 | 500 | 145.5 | **132.0** | 132.4 | 144.8 | 133.3 | 139.9 | 134.0 | 129.0 | **128.6** | 133.4 | 129.5 | 133.6 |
| 250 | 750 | 118.6 | 108.8 | 110.3 | 116.4 | **108.4** | 117.5 | 107.7 | **104.2** | 105.4 | 108.9 | 104.5 | 108.2 |
| 250 | 1000 | 96.4 | 90.0 | 89.5 | 94.9 | **88.9** | 94.6 | 90.3 | **85.6** | 86.7 | 89.0 | 85.9 | 89.7 |
| 250 | 2000 | **52.7** | 52.8 | 53.8 | **52.7** | 53.0 | 53.2 | **51.1** | 51.6 | 52.4 | 51.7 | 52.2 | 52.0 |
| 250 | 3000 | **37.4** | 38.4 | 39.9 | 37.8 | 39.1 | 39.6 | **37.1** | 38.2 | 38.7 | 37.4 | 38.2 | 39.0 |
| 250 | 5000 | 25.1 | 26.0 | 26.3 | **24.9** | 25.5 | 25.6 | **24.9** | 25.6 | 25.8 | **24.9** | 25.3 | 25.5 |
| 300 | 300 | 204.9 | **201.8** | **201.8** | 204.9 | **201.8** | 202.3 | 204.5 | **201.8** | **201.8** | 204.5 | **201.8** | 202.3 |
| 300 | 500 | 187.0 | 171.2 | 169.9 | 185.8 | **169.6** | 178.6 | 172.4 | 167.7 | 167.3 | 173.6 | **167.0** | 171.3 |
| 300 | 750 | 159.6 | 144.0 | 143.4 | 159.4 | **143.1** | 154.1 | 144.4 | **138.1** | 138.7 | 143.9 | 138.6 | 143.3 |
| 300 | 1000 | 132.2 | **119.9** | 121.9 | 130.3 | 120.4 | 129.7 | 121.2 | **115.7** | 117.0 | 121.6 | 116.2 | 121.6 |
| 300 | 2000 | 74.8 | 73.7 | **72.9** | 73.5 | 74.0 | 75.0 | 72.3 | 72.0 | **70.5** | 71.0 | 71.6 | 71.6 |
| 300 | 3000 | 52.7 | 54.1 | 54.2 | **52.1** | 53.3 | 53.5 | 52.5 | 53.3 | 53.2 | **51.9** | 52.7 | 52.4 |
| 300 | 5000 | 34.7 | 36.5 | 36.2 | **34.5** | 36.3 | 35.3 | **34.5** | 35.8 | 35.8 | **34.5** | 35.8 | 35.1 |

(continued)

**Table 3** (continued)

| $\|V\|$ | $\|E\|$ | Before redundant node removal | | | | | | After redundant node removal | | | | | |
|---|---|---|---|---|---|---|---|---|---|---|---|---|---|
| | | H1 | H2 | H3 | H4 | H5 | H6 | H1 | H2 | H3 | H4 | H5 | H6 |
| 500 | 500 | 340.6 | **336.3** | **336.3** | 340.6 | **336.3** | 336.6 | 339.9 | **336.3** | **336.3** | 339.9 | **336.3** | 336.6 |
| 500 | 1000 | 291.6 | 263.8 | 264.8 | 289.3 | **263.2** | 277.6 | 267.3 | 258.0 | 258.5 | 266.8 | **257.2** | 262.6 |
| 500 | 2000 | 191.3 | 178.5 | **176.6** | 189.8 | 179.8 | 187.5 | 177.1 | 171.9 | **170.6** | 176.7 | 174.1 | 175.9 |
| 500 | 5000 | 89.2 | 90.2 | 90.5 | **87.9** | 89.1 | 90.6 | 88.1 | 88.6 | 88.7 | 87.0 | **86.9** | 88.4 |
| 500 | 10,000 | 50.2 | 52.2 | 51.9 | **50.0** | 52.6 | 51.2 | **49.7** | 52.0 | 51.4 | 49.9 | 51.8 | 50.8 |
| 800 | 1000 | 525.1 | **498.7** | 499.2 | 525.2 | 499.2 | 510.3 | 511.5 | **497.9** | 498.5 | 512.2 | 498.6 | 505.1 |
| 800 | 2000 | 417.7 | **381.9** | 382.7 | 416.7 | 383.0 | 404.8 | 384.2 | 371.1 | 370.0 | 384.7 | **369.7** | 377.4 |
| 800 | 5000 | 212.7 | 205.8 | 206.3 | 210.2 | **205.2** | 214.6 | 204.4 | 201.0 | 201.7 | 203.0 | **200.3** | 205.9 |
| 800 | 10,000 | 118.0 | 121.6 | 122.5 | **117.8** | 121.2 | 120.9 | 117.4 | 120.2 | 120.4 | **117.0** | 119.2 | 118.9 |
| 1000 | 1000 | 683.2 | **674.3** | **674.3** | 683.4 | **674.3** | 674.8 | 680.8 | **674.3** | **674.3** | 680.9 | **674.3** | 674.8 |
| 1000 | 5000 | 324.5 | 308.5 | 307.1 | 320.9 | **305.7** | 320.4 | 306.7 | 299.2 | 297.9 | 306.4 | **295.9** | 305.0 |
| 1000 | 10,000 | 179.3 | 180.2 | 183.7 | **176.9** | 181.4 | 181.2 | 175.4 | 178.1 | 179.3 | **174.0** | 178.1 | 177.2 |
| 1000 | 15,000 | 128.4 | 132.5 | 132.8 | **126.8** | 131.6 | 129.6 | 127.4 | 130.8 | 130.7 | **125.8** | 130.3 | 127.7 |
| 1000 | 20,000 | 100.9 | 105.4 | 105.6 | **99.9** | 104.5 | 102.9 | 100.5 | 104.5 | 104.5 | **99.2** | 103.6 | 102.2 |

H1 and H6 also performed better. Overall, H2 and H5 have better performance on Type-I instances. However, H4 consistently performed better on three largest group of instances.

On Type-II instances (Table 4), out of 53 instance groups H1, H2, H3, H4, H5 and H6 after redundant node removal have obtained better results on 11, 14, 21, 13, 21, 3, respectively. Before redundant node removal, H5 has the highest performance. After the redundant node removal procedure, H3 and H5 performed better than other heuristics. H2 performs better on a graph with the number of edges is nearly the same as the number of nodes. As the number of edges increases relative to the number of nodes, then H3 and H5 performed better. If the number of edges is very high compared with the number of nodes, then H1 and H4 performed better. H6 has poor performance on these graphs. The performance of heuristics varies even on graphs with the same number of nodes and the different number of edges. If the degree of connectivity is low, then H2 performed better. As the degree of connectivity increases, H3 and H5 performed better. If the degree of connectivity is very high, then H1 and H4 performed better. In very few cases, H6 performed better.

## 4 Conclusion

To solve the generalized minimum dominating set (GMDS) problem, we have presented six greedy heuristics, viz. H1, H2, H3, H4, H5 and H6. For constructing a GDS, all these heuristics add the mandatory nodes at the beginning and then work iteratively utilizing a greedy policy. Greedy policy used in these heuristics varies from

**Table 4** Comparison of H1, H2, H3, H4, H5 and H6 heuristics' results for Type-II graph instances

| $\|V\|$ | $\|E\|$ | Before redundant node removal | | | | | | After redundant node removal | | | | | |
|---|---|---|---|---|---|---|---|---|---|---|---|---|---|
| | | H1 | H2 | H3 | H4 | H5 | H6 | H1 | H2 | H3 | H4 | H5 | H6 |
| 50 | 50 | 32.4 | **31.3** | **31.3** | 32.3 | **31.3** | **31.3** | 32.2 | **31.3** | **31.3** | 32.1 | **31.3** | **31.3** |
| 50 | 100 | 27.5 | 25.3 | **24.8** | 27.4 | **24.8** | 26.6 | 25.5 | 24.9 | **24.3** | 25.7 | **24.3** | 25.2 |
| 50 | 250 | 14.2 | 13.6 | 14.0 | 14.3 | **13.5** | 14.4 | 13.5 | **13.2** | 13.5 | 13.6 | **13.2** | 13.7 |
| 50 | 500 | **7.5** | 7.9 | 8.0 | 7.6 | 8.0 | 7.7 | **7.3** | 7.7 | 7.8 | **7.3** | 7.8 | 7.5 |
| 50 | 750 | **5.5** | 5.6 | **5.5** | 5.7 | 5.7 | 5.6 | **5.4** | 5.6 | **5.4** | 5.5 | 5.5 | 5.6 |
| 50 | 1000 | 4.2 | 4.3 | 4.1 | **4.0** | **4.0** | 4.2 | **4.0** | **4.0** | 4.1 | **4.0** | **4.0** | **4.0** |
| 100 | 100 | 64.8 | **63.2** | **63.2** | 64.8 | **63.2** | 63.3 | 64.3 | **63.2** | **63.2** | 64.3 | **63.2** | 63.3 |
| 100 | 250 | 48.5 | **44.2** | 44.6 | 48.6 | 44.6 | 47.1 | 45.4 | **42.4** | 42.7 | 45.8 | 42.7 | 45.1 |
| 100 | 500 | 29.2 | 28.8 | **27.9** | 28.8 | 28.5 | 28.7 | 27.8 | 27.9 | **26.8** | 27.9 | 27.7 | 27.5 |
| 100 | 750 | 20.1 | 20.0 | 20.4 | **19.9** | 20.5 | 21.0 | 19.6 | **19.5** | 19.9 | 19.7 | 20.1 | 20.8 |
| 100 | 1000 | **15.7** | 16.4 | 16.1 | **15.7** | 16.5 | 16.4 | **15.6** | 16.2 | 15.9 | **15.6** | 16.3 | 16.1 |
| 100 | 2000 | **8.8** | 9.5 | 9.4 | 9.3 | 9.4 | 9.2 | **8.8** | 9.3 | 9.3 | 9.1 | 9.3 | 9.1 |
| 150 | 150 | 96.4 | **93.9** | **93.9** | 96.4 | **93.9** | 94.0 | 95.8 | **93.9** | **93.9** | 95.8 | **93.9** | 94.0 |
| 150 | 250 | 89.0 | 81.2 | **80.5** | 89.3 | **80.5** | 86.0 | 82.7 | 79.2 | **78.7** | 82.8 | **78.7** | 82.0 |
| 150 | 500 | 60.2 | **54.8** | 55.9 | 59.3 | 56.0 | 58.9 | 56.2 | **53.4** | 54.1 | 56.0 | 54.2 | 55.5 |
| 150 | 750 | 44.0 | 42.9 | 42.9 | 43.2 | **42.3** | 43.6 | 42.2 | 41.6 | 41.6 | 42.3 | **41.4** | 42.3 |
| 150 | 1000 | 34.0 | 34.4 | 34.4 | **33.8** | 33.9 | 34.9 | 33.6 | 33.5 | 33.6 | 33.3 | **32.9** | 33.6 |
| 150 | 2000 | **19.3** | 19.6 | 20.2 | 19.5 | 20.2 | 19.9 | 19.3 | 19.6 | 20.0 | **19.2** | 19.9 | 19.6 |
| 150 | 3000 | 13.7 | 14.8 | 14.7 | **13.5** | 14.6 | 14.2 | 13.6 | 14.3 | 14.6 | **13.5** | 14.4 | 14.0 |
| 200 | 250 | 124.7 | 115.4 | **115.1** | 124.4 | **115.1** | 120.4 | 120.5 | 115.1 | **114.8** | 120.1 | **114.8** | 118.7 |
| 200 | 500 | 97.7 | 89.7 | 88.5 | 98.8 | **88.2** | 96.2 | 89.9 | 86.8 | 86.2 | 89.9 | **85.7** | 89.1 |
| 200 | 750 | 72.9 | 70.0 | 68.9 | 72.6 | **68.6** | 73.3 | 69.4 | 67.4 | **66.5** | 69.5 | 66.6 | 70.2 |
| 200 | 1000 | 58.9 | 56.3 | 55.7 | 58.0 | **55.6** | 59.6 | 57.1 | 54.7 | **53.9** | 56.3 | 54.1 | 56.5 |
| 200 | 2000 | **32.4** | 33.7 | 33.5 | 32.9 | 33.6 | 33.5 | **32.3** | 33.1 | 33.0 | 32.5 | 33.3 | 32.9 |
| 200 | 3000 | 23.3 | 24.1 | 23.7 | **22.8** | 24.6 | 24.3 | 23.2 | 24.1 | 23.7 | **22.6** | 24.2 | 24.2 |
| 250 | 250 | 160.0 | **155.7** | **155.7** | 160.0 | **155.7** | **155.7** | 158.7 | **155.7** | **155.7** | 158.7 | **155.7** | **155.7** |
| 250 | 500 | 138.1 | 124.5 | 123.8 | 137.3 | **123.6** | 131.5 | 126.4 | 121.1 | 120.8 | 125.6 | **120.6** | 124.9 |
| 250 | 750 | 111.3 | 102.9 | **101.7** | 112.1 | 102.0 | 110.5 | 102.5 | 98.2 | **97.6** | 102.5 | 98.0 | 101.7 |
| 250 | 1000 | 88.5 | **83.7** | 85.0 | 89.5 | 84.5 | 88.9 | 83.6 | **81.3** | 82.3 | 84.1 | 81.8 | 84.2 |
| 250 | 2000 | 49.0 | 50.1 | 50.0 | **48.5** | 49.5 | 50.6 | 48.1 | 49.6 | 49.2 | **47.7** | 48.5 | 49.5 |
| 250 | 3000 | 35.0 | 36.2 | 36.0 | **34.5** | 36.1 | 35.5 | 34.9 | 35.9 | 35.6 | **34.4** | 35.6 | 35.0 |
| 250 | 5000 | 23.3 | 25.3 | 24.0 | **22.6** | 24.3 | 24.2 | 23.3 | 24.8 | 23.6 | **22.6** | 23.9 | 24.1 |
| 300 | 300 | 192.6 | **187.6** | **187.6** | 192.6 | **187.6** | 187.8 | 192.1 | **187.6** | **187.6** | 192.1 | **187.6** | 187.7 |
| 300 | 500 | 174.9 | 158.7 | **158.3** | 176.1 | **158.3** | 168.1 | 162.8 | 156.0 | **155.7** | 163.5 | 155.8 | 161.3 |
| 300 | 750 | 149.3 | 134.6 | 134.8 | 149.6 | **133.8** | 146.3 | 137.2 | 130.4 | 131.1 | 137.5 | **130.1** | 135.0 |
| 300 | 1000 | 121.8 | 112.5 | 112.9 | 122.6 | **112.4** | 120.9 | 114.4 | 109.9 | 109.7 | 115.4 | **109.4** | 112.8 |
| 300 | 2000 | **67.2** | 68.0 | 68.2 | 68.3 | 68.5 | 70.0 | **65.7** | 66.6 | 67.2 | 66.9 | 67.1 | 66.7 |
| 300 | 3000 | **47.9** | 50.1 | 50.9 | 48.6 | 50.3 | 49.1 | **47.2** | 49.3 | 49.6 | 47.9 | 49.3 | 48.5 |
| 300 | 5000 | 31.9 | 33.4 | 33.8 | **31.6** | 33.1 | 32.6 | 31.6 | 33.1 | 33.5 | **31.5** | 32.9 | 32.4 |

(continued)

**Table 4** (continued)

| $\|V\|$ | $\|E\|$ | Before redundant node removal | | | | | | After redundant node removal | | | | | |
|---|---|---|---|---|---|---|---|---|---|---|---|---|---|
| | | H1 | H2 | H3 | H4 | H5 | H6 | H1 | H2 | H3 | H4 | H5 | H6 |
| 500 | 500 | 322.5 | **315.7** | **315.7** | 322.5 | **315.7** | 315.8 | 321.0 | **315.7** | **315.7** | 320.9 | **315.7** | 315.8 |
| 500 | 1000 | 281.6 | 249.7 | 249.5 | 280.7 | **249.4** | 267.7 | 257.4 | 243.9 | **243.5** | 257.7 | 244.0 | 253.0 |
| 500 | 2000 | 176.2 | 166.8 | 166.2 | 175.4 | **166.0** | 176.5 | 167.3 | 163.0 | **161.3** | 168.4 | **161.3** | 167.4 |
| 500 | 5000 | **80.4** | 84.0 | 83.4 | 81.4 | 82.9 | 83.8 | **79.9** | 82.7 | 82.3 | 80.4 | 81.9 | 82.8 |
| 500 | 10,000 | 46.2 | 49.2 | 48.1 | **45.7** | 48.0 | 47.8 | 46.1 | 48.6 | 47.8 | **45.6** | 47.7 | 47.8 |
| 800 | 1000 | 504.4 | 468.6 | **468.4** | 503.9 | **468.4** | 483.7 | 486.3 | 467.5 | **467.2** | 485.8 | **467.2** | 477.8 |
| 800 | 2000 | 397.3 | 359.9 | **357.9** | 397.9 | 359.2 | 383.7 | 365.5 | 348.8 | **348.5** | 365.6 | 349.9 | 361.7 |
| 800 | 5000 | 193.0 | 192.1 | **191.0** | 191.9 | 191.7 | 196.5 | 189.2 | **187.7** | **187.7** | 188.5 | 187.8 | 190.4 |
| 800 | 10,000 | **108.5** | 114.1 | 114.1 | 109.3 | 113.6 | 113.4 | **107.8** | 112.7 | 112.7 | 108.6 | 112.3 | 111.1 |
| 1000 | 1000 | 651.7 | **637.5** | **637.5** | 651.7 | **637.5** | 637.8 | 648.6 | **637.5** | **637.5** | 648.6 | **637.5** | 637.8 |
| 1000 | 5000 | 293.6 | 285.0 | 284.5 | 293.1 | **280.4** | 296.9 | 281.9 | 277.4 | 277.8 | 281.8 | **275.4** | 281.5 |
| 1000 | 10,000 | 164.9 | 168.2 | 168.0 | **163.6** | 169.2 | 168.1 | **162.4** | 165.6 | 166.3 | 162.5 | 167.5 | 165.5 |
| 1000 | 15,000 | 116.6 | 122.9 | 123.7 | **116.0** | 122.9 | 120.7 | 115.9 | 122.0 | 122.4 | **115.6** | 121.7 | 120.0 |
| 1000 | 20,000 | 93.1 | 99.5 | 98.4 | **91.7** | 97.7 | 95.5 | 92.7 | 98.4 | 97.2 | **91.6** | 96.7 | 95.3 |

**Table 5** Summary table (before redundant node removal)

| | | H1 | | | H2 | | | H3 | | | H4 | | | H5 | | | H6 | | |
|---|---|---|---|---|---|---|---|---|---|---|---|---|---|---|---|---|---|---|---|
| | | B | E | W | B | E | W | B | E | W | B | E | W | B | E | W | B | E | W |
| Type-I | H1 | – | – | – | 18 | 3 | 32 | 19 | 1 | 33 | 9 | 8 | 36 | 14 | 0 | 39 | 21 | 3 | 29 |
| | H2 | 32 | 3 | 18 | – | – | – | 26 | 9 | 18 | 31 | 0 | 22 | 15 | 11 | 27 | 37 | 3 | 13 |
| | H3 | 33 | 1 | 19 | 18 | 9 | 26 | – | – | – | 33 | 1 | 19 | 12 | 10 | 31 | 33 | 2 | 18 |
| | H4 | 36 | 8 | 9 | 22 | 0 | 31 | 19 | 1 | 33 | – | – | – | 20 | 1 | 32 | 25 | 1 | 27 |
| | H5 | 39 | 0 | 14 | 27 | 11 | 15 | 31 | 10 | 12 | 32 | 1 | 20 | – | – | – | 39 | 3 | 11 |
| | H6 | 29 | 3 | 21 | 13 | 3 | 37 | 18 | 2 | 33 | 27 | 1 | 25 | 11 | 3 | 39 | – | – | – |
| Type-II | H1 | – | – | – | 22 | 0 | 31 | 21 | 1 | 31 | 19 | 7 | 27 | 21 | 0 | 32 | 29 | 1 | 23 |
| | H2 | 31 | 0 | 22 | – | – | – | 13 | 10 | 30 | 30 | 1 | 22 | 12 | 8 | 33 | 33 | 4 | 16 |
| | H3 | 31 | 1 | 21 | 30 | 10 | 13 | – | – | – | 32 | 0 | 21 | 14 | 16 | 23 | 39 | 3 | 11 |
| | H4 | 27 | 7 | 19 | 22 | 1 | 30 | 21 | 0 | 32 | – | – | – | 21 | 2 | 30 | 28 | 0 | 25 |
| | H5 | 32 | 0 | 21 | 33 | 8 | 12 | 23 | 16 | 14 | 30 | 2 | 21 | – | – | – | 34 | 2 | 17 |
| | H6 | 23 | 1 | 29 | 16 | 4 | 33 | 11 | 3 | 39 | 25 | 0 | 28 | 17 | 2 | 34 | – | – | – |
| UDG | H1 | – | – | – | 7 | 0 | 5 | 9 | 0 | 3 | 10 | 0 | 2 | 7 | 0 | 5 | 7 | 1 | 4 |
| | H2 | 5 | 0 | 7 | – | – | – | 11 | 0 | 1 | 6 | 2 | 4 | 9 | 0 | 3 | 6 | 1 | 5 |
| | H3 | 3 | 0 | 9 | 1 | 0 | 11 | – | – | – | 5 | 0 | 7 | 1 | 1 | 10 | 2 | 1 | 9 |
| | H4 | 2 | 0 | 10 | 4 | 2 | 6 | 7 | 0 | 5 | – | – | – | 6 | 1 | 5 | 2 | 1 | 9 |
| | H5 | 5 | 0 | 7 | 3 | 0 | 9 | 10 | 1 | 1 | 5 | 1 | 6 | – | – | – | 5 | 1 | 6 |
| | H6 | 4 | 1 | 7 | 5 | 1 | 6 | 9 | 1 | 2 | 9 | 1 | 2 | 6 | 1 | 5 | – | – | – |

**Table 6** Summary table (after redundant node removal)

| | | H1 | | | H2 | | | H3 | | | H4 | | | H5 | | | H6 | | |
|---|---|---|---|---|---|---|---|---|---|---|---|---|---|---|---|---|---|---|---|
| | | B | E | W | B | E | W | B | E | W | B | E | W | B | E | W | B | E | W |
| Type-I | H1 | – | – | – | 17 | 2 | 34 | 19 | 0 | 34 | 14 | 12 | 27 | 16 | 1 | 36 | 22 | 0 | 31 |
| | H2 | 34 | 2 | 17 | – | – | – | 26 | 10 | 17 | 33 | 0 | 20 | 15 | 15 | 23 | 33 | 2 | 18 |
| | H3 | 34 | 0 | 19 | 17 | 10 | 26 | – | – | – | 32 | 1 | 20 | 11 | 13 | 29 | 31 | 4 | 18 |
| | H4 | 27 | 12 | 14 | 20 | 0 | 33 | 20 | 1 | 32 | – | – | – | 19 | 1 | 33 | 25 | 1 | 27 |
| | H5 | 36 | 1 | 16 | 23 | 15 | 15 | 29 | 13 | 11 | 33 | 1 | 19 | – | – | – | 35 | 4 | 14 |
| | H6 | 31 | 0 | 22 | 18 | 2 | 33 | 18 | 4 | 31 | 27 | 1 | 25 | 14 | 4 | 35 | – | – | – |
| Type-II | H1 | – | – | – | 21 | 1 | 31 | 21 | 3 | 29 | 23 | 10 | 20 | 21 | 1 | 31 | 27 | 2 | 24 |
| | H2 | 31 | 1 | 21 | – | – | – | 17 | 11 | 25 | 31 | 2 | 20 | 15 | 11 | 27 | 32 | 5 | 16 |
| | H3 | 29 | 3 | 21 | 25 | 11 | 17 | – | – | – | 31 | 0 | 22 | 16 | 16 | 21 | 35 | 4 | 14 |
| | H4 | 20 | 10 | 23 | 20 | 2 | 31 | 22 | 0 | 31 | – | – | – | 20 | 2 | 31 | 25 | 3 | 25 |
| | H5 | 31 | 1 | 21 | 27 | 11 | 15 | 21 | 16 | 16 | 31 | 2 | 20 | – | – | – | 34 | 4 | 15 |
| | H6 | 24 | 2 | 27 | 16 | 5 | 32 | 14 | 4 | 35 | 25 | 3 | 25 | 15 | 4 | 34 | – | – | – |
| UDG | H1 | – | – | – | 9 | 0 | 3 | 8 | 1 | 3 | 9 | 1 | 2 | 7 | 0 | 5 | 7 | 0 | 5 |
| | H2 | 3 | 0 | 9 | – | – | – | 9 | 0 | 3 | 6 | 0 | 6 | 6 | 0 | 6 | 7 | 0 | 5 |
| | H3 | 3 | 1 | 8 | 3 | 0 | 9 | – | – | – | 4 | 0 | 8 | 2 | 1 | 9 | 2 | 1 | 9 |
| | H4 | 2 | 1 | 9 | 6 | 0 | 6 | 8 | 0 | 4 | – | – | – | 6 | 0 | 6 | 5 | 0 | 7 |
| | H5 | 5 | 0 | 7 | 6 | 0 | 6 | 9 | 1 | 2 | 6 | 0 | 6 | – | – | – | 7 | 1 | 4 |
| | H6 | 5 | 0 | 7 | 5 | 0 | 7 | 9 | 1 | 2 | 7 | 0 | 5 | 4 | 1 | 7 | – | – | – |

one heuristic to another. Solutions obtained by these heuristics are further improved by deleting the redundant nodes. To evaluate the performance of these heuristics, we have used three benchmark datasets, viz. UDG, Type-I and Type-II. These datasets are derived from the standard benchmark datasets of minimum weight dominating set problem (MWDS). Performance of these heuristics varies depending on type of dataset, number of nodes and density of graphs. On UDG instances, H1 performed better, whereas on Type-I and Type-II instances H5 performed better. For largest group of instances, H1 and H4 performed better than others. To our knowledge, these are first heuristic approaches for GMDS. Hence, these heuristics will serve as baseline approaches for any future attempt to solve this problem.

Performance of our heuristics can be improved further at the expense of increased computational time by incorporating local search strategies such as repeatedly exchanging two dominating nodes with a dominated node whenever such an exchange is possible. As a future work, we plan to develop metaheuristic approaches for GMDS utilizing the ideas presented in this paper.

**Acknowledgements** The first author acknowledges the financial assistance received in the form of a Senior Research Fellowship from the Council of Scientific and Industrial Research (CSIR), Government of India.

## References


1. Albuquerque, M., Vidal, T.: An efficient matheuristic for the minimum-weight dominating set problem. Appl. Soft Comput. **72**, 527–538 (2018)
2. Chalupa, D.: An order-based algorithm for minimum dominating set with application in graph mining. Inf. Sci. **426**, 101–116 (2018)
3. Chaurasia, S.N., Singh, A.: A hybrid evolutionary algorithm with guided mutation for minimum weight dominating set. App. Intell. **43**(3), 512–529 (2015)
4. Garey, M.R., Johnson, D.S.: Computers and Intractability: a guide to the theory of NP-completeness. W. H. Freeman, San Francisco (1979)
5. Jovanovic, R., Tuba, M., Simian, D.: Ant colony optimization applied to minimum weight dominating set problem. In: Proceedings of the 12th WSEAS International Conference on Automatic Control, Modelling and Simulation (ACMOS'10), pp. 322–326. World Scientific and Engineering Academy and Society (WSEAS), Stevens Point, Wisconsin, USA (2010)
6. Lin, G., Zhu, W., Ali, M.M.: An effective hybrid memetic algorithm for the minimum weight dominating set problem. IEEE Trans. Evol. Comput. **20**(6), 892–907 (2016)
7. Lin, G., Guan, J.: A binary particle swarm optimization for the minimum weight dominating set problem. J. Comput. Sci. Technol. **33**(2), 305–322 (2018)
8. Nitash, C., Singh, A.: An artificial bee colony algorithm for minimum weight dominating set. In: 2014 IEEE Symposium on Swarm Intelligence, pp. 1–7. IEEE (2014)
9. Potluri, A., Singh, A.: Two hybrid meta-heuristic approaches for minimum dominating set problem. In: Proceedings of the Second International Conference on Swarm, Evolutionary, and Memetic Computing (SEMCCO 2011), Part II. Lecture Notes in Computer Science, vol. 7077, pp. 97–104. Springer (2011)
10. Potluri, A., Singh, A.: Hybrid metaheuristic algorithms for minimum weight dominating set. Appl. Soft Comput. **13**(1), 76–88 (2013)
11. Shen, C., Li, T.: Multi-document summarization via the minimum dominating set. In: Huang, C., Jurafsky, D. (eds.) Proceedings of the 23rd International Conference on Computational Linguistics COLING 2010. pp. 984–992. Tsinghua University Press (2010)
12. Shetgaonkar, S., Singh, A.: Hybridization of artificial bee colony algorithm with estimation of distribution algorithm for minimum weight dominating set problem. In: Proceeding of the Sixth International conference on ICT for Sustainable Development (ICT4SD 2020). Advances in Intelligent Systems and Computing, vol. 1270, pp. 1–16. Springer (2020)
13. Xu, Y.Z., Zhou, H.J.: Generalized minimum dominating set and application in automatic text summarization. J. Phys.: Con. Ser. **699**, 012014 (2016). https://doi.org/10.1088/1742-6596/699/1/012014

# Assessing the Supervised Machine Learning Algorithms for Human Activity Recognition Using Smartphone

**Peyush Kumar, Drishya Lal Chuke, Parul Kalra Bhatia, and Deepti Mehrotra**

**Abstract** Smart phones have come all the way up to become an essential part of our lives compared to any other device. People use it during their free time, some people work on their smart phones, for social media, photos, retrieving information, etc. As it can be utilized in many different fields, human body medical and fitness are also a very important field that smart phones can be used to simplify the process of medication and human body surveillance. Imagine a mobile application constantly analyzing the movement of the body through the help of sensors present in our smart phone in real time, and looking through these values, it can detect the current state of a human body like walking, sitting, laying, standing, etc., altogether human activities. In this paper, we aim to test and analyze various machine learning algorithms such a way that it recognizes the current state of the body and predicts the activity being performed with maximum accuracy. The trained models can further be deployed for real-time use in either mobile devices or on computer using a website as interface. Most of the smart phone contains accelerometer and gyroscope which are the sensors built-in with the smart phone hardware. We have taken the dataset from the UCI machine learning repository.

**Keywords** Machine learning · Accelerometer · Gyroscope · Mobile device

## 1 Introduction

Human activity recognition is a field of study that deals with detecting the precise action or movement of a person which is done with the help of data collected from sensors. Movement involves activities performed indoor such as sitting, walking, running, laying, walking upstairs or downstairs, etc. The reading is recorded in many ways like within video radar, sensors, or other wireless method. Human activity recognition has been studied since the 1980s [1], and many extensive research and studies have been carried out till the present year. It was a very important and useful

P. Kumar (✉) · D. L. Chuke · P. K. Bhatia · D. Mehrotra
Amity School of Engineering and Technology, Amity University Uttar Pradesh, Noida, India



T. K. Sharma et al. (eds.), *Soft Computing: Theories and Applications*,
Advances in Intelligent Systems and Computing 1381,
https://doi.org/10.1007/978-981-16-1696-9_30

field to research, as it is connected to many different fields of study such as medicine, human–computer interaction, and sociology. After the invention of smart phone, human activity recognition in a smart phone was a great topic of research, as it had great advantage over other computer vision-based devices. As smart phones are portable, more compatible, it did not require infrastructure support like computer vision-based device, e.g., requirement of video camera installations in the nursing homes to have a real-time and full-time supervision. This paper attempts to analyze various supervised machine learning algorithms. Most of the smart phone contains accelerometer and gyroscope which are the sensors built-in with the smart phone hardware that can be used to detect human activity [2]. Human activity recognition is one of the main components of software being developed for assisted living. Assisted living is useful for people who suffer from medical conditions like Alzheimer's disease or patients at risk of stoke as they need help in monitoring their day to day activities. Moreover, it can be used for monitoring elderly people in the nursing homes [3]. In this paper, we attempt to analyze machine learning algorithms on the basis of various parameters like precision, training time, testing time, recall, $f1$-score, etc. For our study, first we performed data pre-processing techniques on our dataset. After that exploratory data analysis was done to visualize the dataset for analysis. This was proceeded by training and then testing the models to obtain the results for comparison. Various parameters are used to compare the trained models.

## 2 Literature Review

Attempt to recognize human activity has been going around since the 1980s with the help of machine learning; various attempts have been made to develop a model to accurately detect human activity for applications in the field of health care, old age supervision, and active and assisted living (AAL) [4]. One of the challenging tasks has been to collect precise and resourceful data for human activity recognition for the supervised machine learning models to be trained with. An alternate approach was provided to use unsupervised learning models to overcome the limitation of generating training data and the number of activities that could be detected [5]. Another unsupervised approach was taken using the classification method called MCODE (finds clusters in a network) for activity recognition [6]. However, with the availability of accurate and meaningful training data, supervised learning algorithms generally attain greater accuracies as compared to the unsupervised learning. One such public domain dataset was used in this paper which took into consideration six activities and carried out a set of experiments on thirty volunteers to obtain precise results [7].

# 3 Methodology

## *3.1 Model Design*

Following steps were followed to implement the machine learning algorithms and obtain the results:

- *Collecting Data*: It is the first step to compile the data on which the algorithm would be trained. Collected data was obtained from the dataset using the read_csv() method of the python pandas library. Except the rows containing activity and subject, all the remaining features, i.e., 561 were used to test and train the models.
- *Cleaning the Data*: In this stage, you make sure that your data has all useful fields, so there are no meaningless fields that you do not require. The dataset did not require cleaning.
- *Analyze Data:* After cleaning the data, you analyze your data, i.e., analyze everything and find patterns and similarities in the dataset. Dimensionality reduction was done, and the data was visualized using the t-SNE algorithm. Figure 1 was used to make the observation that the siting and standing activities overlapped, while the other four activities can be easily separated.

From Fig. 2, we can observe that the stationary and moving activities are linearly separable by motion. From Fig. 3, we can observe that if angle($X$, gravityMean) > 0, the activity is laying.

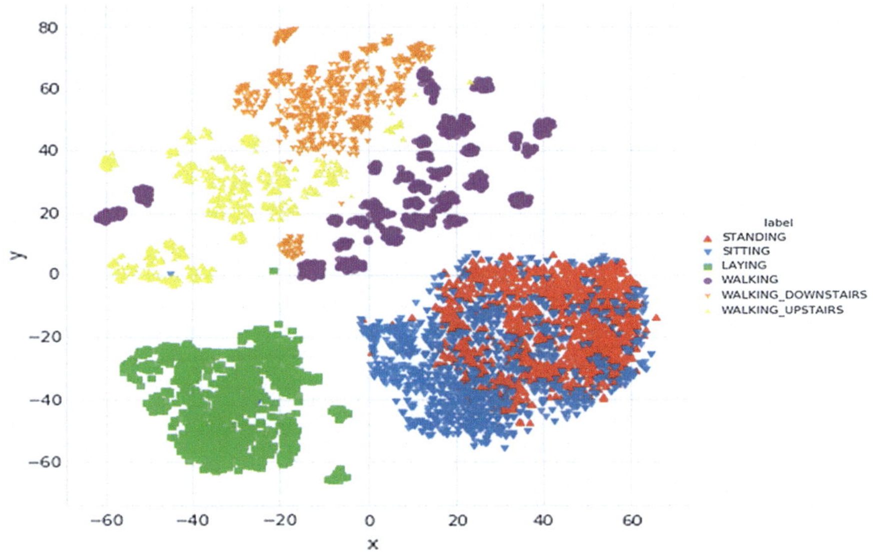


**Fig. 1** Data visualization after applying t-SNE

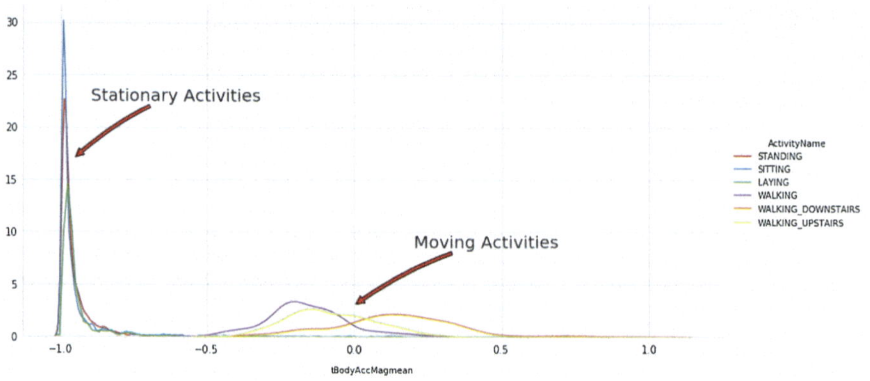


**Fig. 2** Time space plot of mean values of magnitude of acceleration

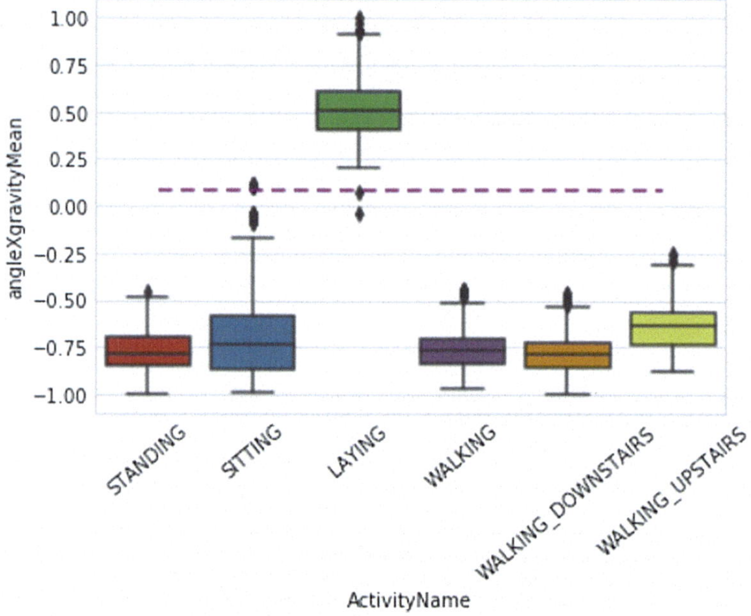


**Fig. 3** Box plot for angle between $x$ axis and gravity mean

- *Train the Algorithm*: When the analyzing of data is completed, it will train the algorithm based on that analysis.
- *Test the Algorithm*: After training the algorithms, it will test the algorithm how it is going to perform by providing certain set of data. Actual output is already known and is compared with the results to check the precision.

**Table 1** Dataset description

| | |
|---|---|
| 1 | Features are normalized and bounded within [−1, 1] |
| 2 | From each window, a feature vector was obtained by calculating variables from the time and frequency domain |
| 3 | Each feature vector is a row on the text file |
| 4 | The signals that we obtained are tBodyGyroMag, tBodyGyroJerkMag, fBodyAcc-XYZ, fBodyAccJerk-XYZ, fBodyGyro-XYZ, fBodyAccMag, fBodyAccJerkMag, fBodyGyroMag, fBodyGyroJerkMag, tBodyAcc-XYZ, tGravityAcc-XYZ, tBodyAccJerk-XYZ, tBodyGyro-XYZ, tBodyGyroJerk-XYZ, tBodyAccMag, tGravityAccMag, and tBodyAccJerkMag |
| 5 | These signals are then used to estimate the variables: mean(), std(), skewness(), curtosis(), maxInds(), meanFreq(), skewness(), iqr(), correlation(), entropy(), energy(), std(), etc. |
| 6 | We can obtain some other vectors by taking the average of signals: gravityMean, tBodyAccMean, tBodyAccJerkMean, tBodyGyroMean, and tBodyGyroJerkMean |

## 3.2 *Dataset Description*

The dataset was taken form UCI machine learning repository. The analyses have been completed with a gathering of 30 volunteers inside an age section of 19–48 years. Every individual performed six exercises (WALKING, WALKING_UPSTAIRS, WALKING_DOWNSTAIRS, SITTING, STANDING, LAYING) wearing a cell phone (Samsung Galaxy S II) on the midriff. The dataset description is given in Table 1 and the supervised algorithms applied are described in Table 2.

## 3.3 *Machine Learning Algorithms Applied*

The program is based on supervised machine learning, where we have used seven machine learning algorithms. Scikit-learn library is used for all the seven algorithms.

## 3.4 *Analysis*

Table 3 shows that logistics regression is the best algorithms as it is giving the lowest error among all the others models. It boasts precision, recall, and $f$-score of 0.94. Also, it takes significantly less training time and testing time as compared to other algorithms. It is the suitable model to be deployed for further use in predicting the activities on the basis of real-time data collection. The next best algorithms on the basis of accuracy score are gradient boosting and random forest with accuracy of 91.53% and 90.95%, respectively. However, training time of gradient boosting is significantly more than that taken by random forest classifier. Support vector classifier too gives desirable results with 89.43% accuracy score with training and testing time

**Table 2** Supervised algorithms applied

| | |
|---|---|
| Support vector machine (SVC) | It is mainly used for classification purpose where in you want to classify a data into two different segments depending upon the features of the data |
| Logistic regression | Logistic regression is a prescient demonstrating procedure, and it generally includes expectation. It gauges the connection between a needy (target) and an autonomous variable (indicator) |
| K-nearest neighbors | K-nearest neighbors, it stores all the available cases and classifies new cases based on this similarity measure. The "K" in KNN algorithm is the nearest neighbors we wish to take vote from |
| Random forest | Classification is based on dividing the dataset into different categories or groups by adding label or a technique of categorizing the observation into different category. We classify it to perform predictive analysis on it. Most of the time, random forest is trained with bagging method |
| Decision tree | Decision tree is based on classification. Classification is based on dividing the dataset into different categories or groups by adding label or a technique of categorizing the observation into different category |
| Gradient boosting | In gradient boosting, base learners are generated sequentially in such a way that the present base learner is always more effective than the previous one. Basically, the overall model improves sequential with each iteration |
| Gaussian Naïve Bayes | It is a classification technique which is based on Bayes' theorem. It assumes that any particular feature in a class is completely unrelated to the presence of any other feature |

**Table 3** Algorithms testing results

| Algorithm used | Training time (mm:ss:ms) | Testing time (mm:ss:ms) | Accuracy | Precision | Recall | $F1$-score |
|---|---|---|---|---|---|---|
| SVC | 00:01:96 | 00:09:08 | 89.43% | 0.90 | 0.90 | 0.89 |
| Logistic Regression | 00:02:17 | 00:00:01 | 94.17% | 0.94 | 0.94 | 0.94 |
| KNN | 00:00:11 | 00:16.79 | 87.09% | 0.87 | 0.87 | 0.087 |
| Random Forest | 00:07:00 | 00:00:19 | 90.95% | 0.91 | 0.91 | 0.91 |
| Decision Tree | 00:01:13 | 00:00:01 | 83.97% | 0.84 | 0.84 | 0.83 |
| Gradient Boosting | 01:29:98 | 00:00:31 | 91.53% | 0.91 | 0.92 | 0.91 |
| Gaussian NB | 00:00:06 | 00:00:46 | 71.46% | 0.77 | 0.73 | 0.69 |

of 1.96 s and 9.08 s, respectively. Out of all the algorithms, Gaussian Naïve Bayes gives the worst accuracy score of 71.46% compared to other algorithms. To further analyze performance of logistic regression, a classification report was obtained.

From Table 4, it is observed that logistic regression comparatively struggles to accurately identify activities like sitting and walking upstairs. Siting activity comes out to be classified with lowest precision 0.88 followed by walking upstairs with a precision of 0.90. Standing constitutes for the lowest recall of 0.88 followed by 0.93 of walking and sitting. However, it boasts accurate detection of laying activity. Classification report for SVC and random forest was also obtained.

From the classification report of SVC and random forest in Tables 5 and 6, respectively, we can again observe that they comparatively struggle to identify activities like sitting, standing, and walking upstairs. However, they depict nearly accurate

**Table 4** Classification report for logistic regression

| Activity | Precision | Recall | $F1$-score | Support |
|---|---|---|---|---|
| Laying | 1.00 | 1.00 | 1.00 | 1407 |
| Sitting | 0.88 | 0.93 | 0.90 | 1286 |
| Standing | 0.94 | 0.88 | 0.91 | 1374 |
| Walking | 0.99 | 0.93 | 0.86 | 1226 |
| Walking downstairs | 0.95 | 0.97 | 0.96 | 986 |
| Walking upstairs | 0.90 | 0.95 | 0.93 | 1073 |

**Table 5** Classification report for SVC

| Activity | Precision | Recall | $F1$-score | Support |
|---|---|---|---|---|
| Laying | 0.99 | 1.00 | 0.99 | 1407 |
| Sitting | 0.80 | 0.84 | 0.82 | 1286 |
| Standing | 0.85 | 0.80 | 0.83 | 1374 |
| Walking | 0.98 | 0.88 | 0.93 | 1226 |
| Walking downstairs | 0.92 | 0.92 | 0.92 | 986 |
| Walking upstairs | 0.84 | 0.94 | 0.88 | 1073 |

**Table 6** Classification report of random forest

| Activity | Precision | Recall | $F1$-score | Support |
|---|---|---|---|---|
| Laying | 0.99 | 1.00 | 0.99 | 1407 |
| Sitting | 0.82 | 0.92 | 0.87 | 1286 |
| Standing | 0.91 | 0.81 | 0.86 | 1374 |
| Walking | 1.00 | 0.88 | 0.93 | 1226 |
| Walking downstairs | 0.87 | 0.95 | 0.91 | 986 |
| Walking upstairs | 0.87 | 0.951 | 0.89 | 1073 |

**Table 7** Classification report of gradient boosting

| Activity | Precision | Recall | $F1$-score | Support |
|---|---|---|---|---|
| Laying | 0.99 | 1.00 | 0.99 | 1407 |
| Sitting | 0.86 | 0.92 | 0.89 | 1286 |
| Standing | 0.93 | 0.86 | 0.89 | 1374 |
| Walking | 0.98 | 0.86 | 0.92 | 1226 |
| Walking Downstairs | 0.88 | 0.95 | 0.88 | 986 |
| Walking Upstairs | 0.86 | 0.91 | 0.92 | 1073 |

identification for laying. Identifying sitting and walking upstairs have low precision for gradient boosting as well as observed from Table 7.

## 4 Results and Discussion

From the supervised machine learning algorithms, logistic regression came forward with the best possible accuracy as shown in Fig. 4.

The results were obtained without the use of in-depth domain knowledge and any data mining techniques. It is possible to enhance the ability to recognize the activities with much higher precision by combing the machine learning algorithms with other models. For example, SVM and hidden Markov model can be combined to overcome the drawbacks when used separately [8]. However, without any enhancements, logistic regression proves to best suited for further use for human activity recognition. One challenge that may arise in activity recognition is that every person has diverse movement actions which may affect the results as different people are used for collecting the data and testing it later [9]. Real-time data can be accessed using

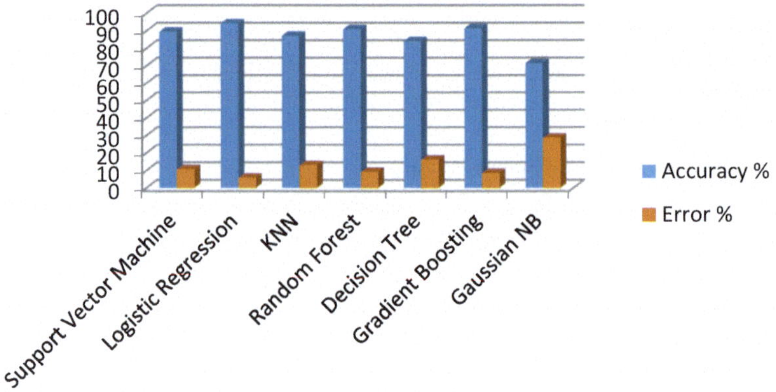


**Fig. 4** Accuracy of algorithms used

the mobile sensors such as gyroscope and accelerometer and convolutional neural networks for local feature selection [10]. In the latest development, bidirectional LSTM networks are being used for HAR with inertial sensors in smartphones [11]. Hidden Markov models and neural networks have also been studied to recognize human activities [12].

## 5 Conclusion

We can conclude that among supervised machine learning algorithms, regression provides the highest precision for human activity recognition. For future scope, the trained models that provide desirable results can be either be freezed or used to make an API to be deployed for real-time use in either as a mobile application or a website. Our study has only taken into account the basic machine learning algorithms and obtaining higher accuracy may be possible using ensemble models, combining multiple machine learning models. Further, deep leaning algorithms may also prove to be more accurate.

## References

1. Johansson, G.: Visual perception of biological motion and a model for its analysis. Percept. Psychophys. **14**, 201–211 (1973). https://doi.org/10.3758/BF03212378
2. Kaur, S., Dinesh Kumar, J., Chaudhary, G.: Human Activity Recognition Using Tri-Axial Angular Velocity. Springer, Berlin (2020). https://doi.org/10.1007/978-981-15-4992-2_47
3. Sebestyen, G., Stoica, I., Hangan, A.: Human activity recognition and monitoring for elderly people. In: IEEE 12th International Conference on Intelligent Computer Communication and Processing (ICCP) (2016)
4. Ranasinghe, S., Al Machot, F., Mayr, H.C.: A review on applications of activity recognition systems with regard to performance and evaluation. Int. J. Distrib. Sensor Netw. (2016)
5. Kwon, Y., Kang, K., Bae, C.: Unsupervised learning for human activity recognition using smartphone sensors. Exp. Syst. Appl. (2014). https://doi.org/10.1016/j.eswa.2014.04.037
6. Yonggang, L., Ye, W., Li, L., Jun, Z., Letian, S., Ye, L.: Towards unsupervised physical activity recognition using smartphone accelerometers. Multim. Tools Appl. (2017). https://doi.org/10.1007/s11042-015-3188-y
7. Anguita, D., Ghio, A., Oneto, L., Parra, X., Reyes-Ortiz, J.L.: A public domain dataset for human activity recognition using smartphones. In: European Symposium on Artificial Neural Networks, Computational Intelligence and Machine Learning. Bruges, Belgium (2013)
8. Wu, H., Pan, W., Xiong, X., Xu, S.: Human activity recognition based on the combined SVM and HMM. In: IEEE International Conference on Information and Automation (ICIA) (2014)
9. Ismael, A.A., Jayabalan, M., Al-Jumeily, D.: A study on human activity recognition using smartphone. Int. J. Soft Comput. Eng. (2020)
10. Ignatov, A.: Real-time human activity recognition from accelerometer data using convolutional neural networks. Appl. Soft Comput. (2017). https://doi.org/10.1016/j.asoc.2017.09.027
11. Yu, S., Qin, L.: Human activity recognition with smartphone inertial sensors using Bidir-LSTM networks. In: 2018 3rd International Conference on Mechanical, Control and Computer Engineering (ICMCCE), Huhhot (2018)

12. Bhardwaj, R., Dang, K., Gupta, S.C., Kumar, S.: Review on human activity recognition using soft computing. In: Pant, M., Ray, K., Sharma, T., Rawat, S., Bandyopadhyay, A. (eds,) Soft Computing: Theories and Applications. Advances in Intelligent Systems and Computing, vol 583. Springer, Singapore (2018). https://doi.org/10.1007/978-981-10-5687-1_70

# Support Vector Machine Technique to Prognosis Breast Cancer

**Virendra Singh Kushwah, Aradhana Saxena, Janki Sharan Pahariya, and Sandip Kumar Goyal**

**Abstract** Breast cancer is one of the most inescapable infections among ladies in the India and around the world. Early determination is a critical advance in recovery and treatment. Nonetheless, it is anything but a simple one because of a few vulnerabilities in discovery utilizing mammograms. AI (ML) strategies can be utilized to plan and create instruments for doctors that can be utilized as a powerful system for early discovery and analysis of bosom malignancy which will significantly improve the endurance pace of patients. This paper utilizes support vector machine calculation to anticipation the bosom malignancy. The Wisconsin Diagnosis Breast Cancer informational index was utilized as to gauge exactness. The outcomes acquired are serious and can be utilized for identification and treatment.

**Keywords** Breast cancer · Support vector machine · Prognosis

## 1 Introduction

Breast cancer is the most well-known malignant growth in ladies in India. It starts when cells in the chest begin to grow wild. These cells regularly structure tumors that can be seen through X-bar or felt as knocks in the chest area. Early end basically grows the chances of perseverance. The key troubles against its area is the best approach to mastermind tumors into hazardous (unsafe) or benign (non-dangerous). A tumor is seen as undermining if the cells can form into including tissues or spread to far away locales of the body. A positive tumor does not assault near to tissue nor spread

V. S. Kushwah (✉)
VIT Bhopal University, Sehore, India

A. Saxena · J. S. Pahariya
Rustamji Institute of Technology, Gwalior, India

S. K. Goyal
Maharishi Markandeshwar (Deemed to be University), Mullana, Ambala, India



T. K. Sharma et al. (eds.), *Soft Computing: Theories and Applications*,
Advances in Intelligent Systems and Computing 1381,
https://doi.org/10.1007/978-981-16-1696-9_31

to various bits of the body way where threatening tumors can. Yet, generous tumors can be not kidding in the event that they push on crucial structures, for example, veins or nerves [1].

AI procedure can significantly improve the degree of analysis in bosom malignant growth. Breast cancer is the most well-known malignancy in many urban areas and the second generally normal in rustic India. Breast malignant growth represents 25–32% of female diseases in all urban areas across India. In India, one lady is determined to have bosom malignant growth at regular intervals. One lady kicks the bucket of bosom malignant growth, at regular intervals in India. An expected 70,218 ladies kicked the bucket of bosom malignant growth in India, in the year 2012, the most noteworthy on the planet for that year. Just 60% of ladies who are treated for bosom malignant growth get by for in any event five years post-treatment in India when contrasted with 89% in the USA. The high death rate is because of the absence of mindfulness and postponement in screening and finding. Over 50% of bosom malignant growth patients in India are experiencing stages 3 and 4, where the odds of endurance are incredibly low. India sees a spray in the instances of breast cancer in the age gathering of 30 to 50, and the equivalent is probably going to increment. Breastfeeding frequently interferes with periods, which prompts less feminine cycles and less estrogen introduction, prompting about a 4.3% decrease in the general danger of bosom malignant growth for at regular intervals of breastfeeding. As indicated by the National Institute of Health, breast cancer survivors are at an expanded danger of osteoporosis. Examination shows that accomplished doctors can distinguish malignant growth by 79% precision, while a 91% (now and again up to 97%) exactness can be accomplished utilizing machine learning strategies [2].

## 2 Support Vector Machine (SVM

### *2.1 Background*

SVM is one of the directed ML grouping methods that is generally applied in the field of malignancy finding and anticipation [3]. SVM works by choosing basic examples from all classes known as help vectors and isolating the classes by producing a direct capacity that partitions them as extensively as conceivable utilizing these help vectors [4]. Subsequently, it tends to be said that a planning between an information vector to a high dimensionality space is made utilizing SVM that intends to locate the most appropriate hyperplane that partitions the informational index into classes [5]. This straight classifier expects to boost the separation between choice hyperplane and the closest information point, which is known as the minimal separation, by finding the most appropriate hyperplane [6]. Support vector machine is overseen learning model. It is supported as a result of its portrayal execution. In SVM count, every data thing is plotted as an encourage in the $n$-dimensional space, here $n$ is supreme number of features which are being used for request and the assessment of every part is addressed by the bearings of the data point [7]. SVM contains a decision hyperplane that is used to parts the data reasons for different classes using most extraordinary edge. Datacenters which misrepresentation close the hyperplane are called as support vectors. This request cycle makes non-direct decision cutoff points and gatherings datacenters which do not have vector space depiction [8].

SVM is one of the most standard AI request techniques that is used for the representation and finish of ailment. As shown by SVM, the classes are disconnected with hyperplane that is involved assistance vectors that are essential models from all classes [9]. The hyperplane is a separator that is recognized as decision limit among the two model packs. SVM can be used for gathering tumors as agreeable or undermining subject to patient's age and tumors size [10]

## 2.2 *Support Vector Machine*

For a choice hyperplane $\mathbf{x}^T\mathbf{w} + b = 0$ to isolate the two classes $P = \{(\mathbf{x}_i, 1)\}$ and $N = \{(\mathbf{x}_i, -1)\}$, it needs to fulfill

$$y_i(\mathbf{x}_i^T\mathbf{w} + b) \geq 0 \tag{1}$$

for both $\mathbf{x}_i \in P$ and $\mathbf{x}_i \in N$. Among all such planes satisfying this condition, we have to find the ideal one $H_0$ that disconnects the two classes with the maximal edge (the partition between the decision plane and the closest test centers).

The ideal plane should be in the two classes, so the great ways from the plane to the closest point on either side is the same. We describe two additional planes $H_+$ and $H_-$ that are comparing to $H_0$ and experience the direct closest toward the plane on either side:

$$\mathbf{x}^T\mathbf{w} + b = 1, \quad \text{and} \quad \mathbf{x}^T\mathbf{w} + b = -1 \tag{2}$$

All focuses $\mathbf{x}_i \in P$ on the positive side ought to fulfill

$$\mathbf{x}_i^T\mathbf{w} + b \geq 1, \quad y_i = 1 \tag{3}$$

and all focuses $\mathbf{x}_i \in N$ on the negative side ought to fulfill

$$\mathbf{x}_i^T\mathbf{w} + b \leq -1, \quad y_i = -1 \tag{4}$$

These can be consolidated into one disparity:

$$y_i(\mathbf{x}_i^T\mathbf{w} + b) \geq 1, \quad (i = 1, \ldots, m) \tag{5}$$

The fairness holds for those focuses on the planes $H_+$ or $H_-$. Such focuses are called *uphold vectors*, for which

$$\mathbf{x}_i^T \mathbf{w} + b = y_i \tag{6}$$

i.e., the accompanying holds for all help vectors:

$$b = y_i - \mathbf{x}_i^T \mathbf{w} = y_i - \sum_{j=1}^{m} \alpha_j y_j (\mathbf{x}_i^T \mathbf{x}_j) \tag{7}$$

What is more, the acceptable ways from the origin to the three equivalent planes $H_-$, $H_0$ and $H_+$ are, independently, $|b - 1|/||\mathbf{w}||$, $|b|/||\mathbf{w}||$, and $|b + 1|/||\mathbf{w}||$, and the partition between planes $H_-$ and $H_+$ is $2/||\mathbf{w}||$.

We will probably increase this detachment, or, equivalently, to restrict the standard $||\mathbf{w}||$. Directly the issue of finding the ideal decision plane with respect to $\mathbf{w}$ and $b$ can be figured as:

$$\begin{aligned}
&\text{minimize :}\\
&\frac{1}{2}\mathbf{w}^T\mathbf{w} = \frac{1}{2}||\mathbf{w}||^2 \text{ (objective function) subject to :}\\
&y_i(\mathbf{x}_i^T \mathbf{w} + b) \geq 1, \quad \text{or} \quad 1 - y_i(\mathbf{x}_i^T \mathbf{w} + b) \leq 0, \qquad (i = 1, \ldots, m)
\end{aligned}$$

Since the objective work is quadratic, this obliged progression issue is known as a quadratic program (QP) issue. (If the objective work is straight rather, the issue is an immediate program (LP) issue.) This QP issue can be enlightened by Lagrange multipliers system to restrict the going with

$$L_p(\mathbf{w}, b, \alpha) = \frac{1}{2}||\mathbf{w}||^2 + \sum_{i=1}^{m} \alpha_i (1 - y_i(\mathbf{x}_i^T \mathbf{w} + b)) \tag{8}$$

regarding $\mathbf{w}$, $b$ and the Lagrange coefficients $\alpha_i \geq 0$ $(i = 1, \cdots, \alpha_m)$. We let

$$\frac{\partial}{\partial W} L_p(\mathbf{w}, b) = 0, \quad \frac{\partial}{\partial b} L_p(\mathbf{w}, b) = 0 \tag{9}$$

These lead, individually, to

$$\mathbf{w} = \sum_{j=1}^{m} \alpha_j y_j \mathbf{x}_j, \quad \text{and} \quad \sum_{i=1}^{m} \alpha_i y_i = 0 \tag{10}$$

Subbing these two conditions once more into the outflow of $L(\mathbf{w}, b)$, we get the *double problem* (regarding $\alpha_i$) of the above-mentioned *base problem*:

$$\text{and maximize and } L_d(\alpha) = \sum_{i=1}^{m} \alpha_i - \frac{1}{2}\sum_{i=1}^{m}\sum_{j=1}^{m} \alpha_i \alpha_j y_i y_j \mathbf{x}_i^T, \mathbf{x}_j$$

$$\text{Subject to and } \alpha_i \geq 0, \quad \sum_{i=1}^{m} \alpha_i y_i = 0$$

The double issue is identified with the base issue by:

$$L_d(\alpha) = inf_{(\mathbf{w},b)} L_p(\mathbf{w}, b, \alpha) \tag{11}$$

i.e., $L_d$ is the best lower bound (infimum) of $L_p$ for all $\mathbf{w}$ and $b$.

Taking care of this double issue (a simpler issue than the basic one), we get $\alpha_i$, from which $\mathbf{w}$ of the ideal plane can be found.

Those focuses $\mathbf{x}_i$ on both of the two planes $H_+$ and $H_-$ (for which the balance $y_i(\mathbf{w}^T\mathbf{x}_i + b) = 1$ holds) are called *uphold vectors*, and they relate to positive Lagrange multipliers $\alpha_i > 0$. The preparation relies just upon the help vectors, while all different examples from the planes $H_+$ and $H_-$ are not significant.

For a help vector $\mathbf{x}_i$ (on the $H_-$ or $H_+$ plane), the obliging condition is

$$y_i\left(\mathbf{x}_i^T\mathbf{w} + b\right) = 1 \quad (i \in sv) \tag{12}$$

here $sv$ is a set of all indices of support vectors $\mathbf{x}_i$ (corresponding to $\alpha_i > 0$). Substituting

$$\mathbf{w} = \sum_{j=1}^{m} \alpha_j y_j \mathbf{x}_j = \sum_{j \in sv} \alpha_j y_j \mathbf{x}_j \tag{13}$$

we get

$$y_i(\sum_{j \in sv} \alpha_j y_j \mathbf{x}_i^T \mathbf{x}_j + b) = 1 \tag{14}$$

Note that the summation only contains terms corresponding to those support vectors $\mathbf{x}_j$ with $\alpha_j > 0$, i.e.,

$$y_i \sum_{j \in sv} \alpha_j y_j \mathbf{x}_i^T \mathbf{x}_j = 1 - y_i b \tag{15}$$

For the optimal weight vector $\mathbf{w}$ and optimal $b$, we have:

$$\begin{aligned}||\mathbf{w}||^2 &= \mathbf{w}^T\mathbf{w} \\ &= \sum_{i\in sv} \alpha_i y_i \mathbf{x}_i^T \sum_{j\in sv} \alpha_j y_j \mathbf{x}_j \\ &= \sum_{i\in sv} \alpha_i y_i \sum_{j\in sv} \alpha_j y_j \mathbf{x}_i^T \mathbf{x}_j \\ &= \sum_{i\in sv} \alpha_i (1 - y_i b) \\ &= \sum_{i\in sv} \alpha_i - b \sum_{i\in sv} \alpha_i y_i \\ &= \sum_{i\in sv} \alpha_i\end{aligned}$$

The last equality is due to $\sum_{i=1}^{m} \alpha_i y_i = 0$ shown above. Recall that the distance between the two margin planes $H_+$ and $H_-$ is $2/||\mathbf{w}||$, and the margin, the distance between $H_+$ (or $H_-$) and the optimal decision plane $H_0$, is

$$\frac{1}{||\mathbf{w}||} = \left(\sum_{i\in sv} \alpha_i\right)^{-1/2} \tag{16}$$

# 3 Material and Methods

## *3.1 Data Source*

This paper utilizes the Breast Cancer Wisconsin (Diagnostic) database, we can make a classifier that can help analyze patients and anticipate the probability of a bosom malignancy. We are utilizing support vector machine (SVM) strategy for forecast of the order of bosom malignant growth to discover execution. Backing vector machine is managed learning model. It is favored [11] because of its arrangement execution. In SVM count, every data thing is plotted as an encourage in the n-dimensional space, here n is finished number of features which are being used for portrayal and the assessment of every segment is addressed by the bearings of the data point. SVM contains a decision hyperplane that is used to parts the data motivations behind different classes using most noteworthy edge. Datacenters which misrepresentation close the hyperplane are called as support vectors. This portrayal cycle produces non-direct decision cutoff points and gatherings datacenters which do not have vector space depiction.

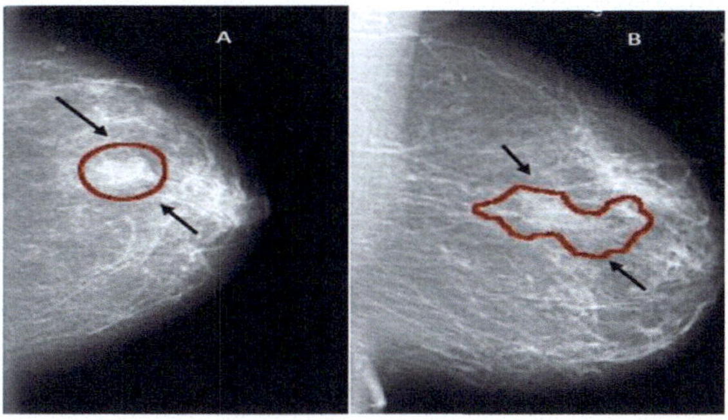


**Fig. 1** Examples of mammograms: **a** Mass; **b** MCs

Mammography is presently one of the significant strategies to identify breast cancer early. The attractive reverberation imaging (MRI) is the most appealing option in contrast to mammogram. In any case, the MRI test is done when the radiologists need to affirm about the presence of the tumor. The downside of the MRI is that the patient could build up an unfavorably susceptible response to the differentiating specialist, or that a skin contamination could create at the spot of infusion. It might cause claustrophobia. Masses and micro-calcifications (MCs) are two significant early indications of the sickness as shown in Fig. 1.

A mass can be either kindhearted or dangerous. The distinction among generous and threatening tumors is that the kindhearted tumors have round or oval shapes, while dangerous tumors have an incompletely adjusted shape with an unpredictable diagram. Furthermore, the dangerous mass will seem more white than any tissue encompassing it [12].

## 3.2 *Related Work*

We should utilize our prepared model to make a forecast utilizing our testing information [13]. It is a table with two lines and two sections that report the quantity of bogus positives FP, bogus negatives FN, genuine positives TP, and genuine negatives TN with:

1. True Positive (TP): Sick patients effectively recognized as debilitated
2. False Positive (FP): Healthy patients inaccurately recognized as debilitated
3. True Negative (TN): Healthy patients accurately recognized as sound
4. True Negative (FN): Sick patients mistakenly recognized as sound.

Table 1 gives a relationship between positive and negative with true and false.

**Accuracy** The precision speaks to the extent of genuine outcomes, both genuine positives and genuine negatives, among the complete number of cases analyzed [13]. The classifier precision is a proportion of how well the classifier can effectively foresee cases into their right classification. It is the quantity of right forecasts partitioned

**Table 1** Relationship between positive and negative with true and false

| Predicted output/labels (Ground truth) | Algorithm outputs/labels | |
|---|---|---|
| | Patient without disease (Negative/0) | Patient with disease (Positive/1) |
| Patient without disease (False/0) | TN | FP |
| Patient with disease (True/1) | FN | TP |

by the absolute number of cases in the informational collection. It is important that the exactness is exceptionally reliant on the limit picked by the classifier and can in this way change for various testing sets. Along these lines, it isn't the ideal technique to look at changed classifiers yet may give a review of the class [14].

$$\text{Accuracy} = \frac{\text{TP} + \text{FN}}{\text{TP} + \text{FN} + \text{FP} + \text{TN}} \tag{17}$$

**Recall** Recall known as sensitivity in general terms might be characterized as the proportion of legitimately decided positive examples to the all perceptions. Review might be viewed as a measure for the adequacy of the framework in foreseeing positives and deciding expenses.

$$\text{Recall} = \frac{\text{TP}}{\text{TP} + \text{FN}} \tag{18}$$

**Precision** The degree of correctness in deciding the positive results might be characterized as accuracy. It is fundamentally the proportion between evident positives and the general arrangement of positives. This portrays the taking care of limit of the framework for positive qualities however does not give understanding into the negative qualities.

$$\text{Precision} = \frac{\text{TP}}{\text{TP} + \text{FP}} \tag{19}$$

**F1-Score** It is the weighted average of precision and recall. This measure thus thinks about both kind of bogus qualities. $F1$-score is viewed as ideal when at 1 and is a complete disappointment when at 0.

$$F1\ \text{Score} = \frac{2 * (\text{Precision} * \text{Recall})}{(\text{Precision} + \text{Recall})} \tag{20}$$

# 4 Result and Discussion

## 4.1 Dataset Columns

There are 31 columns into the dataset which is being used to evaluate the result.

In the AI field and all the more explicitly the factual arrangement issue, a disarray grid is a specific table design that awards representation of the execution of a calculation, regularly an administered learning one. Every segment speaks to the occasions in a real class while each line of the network speaks to the examples in a anticipated class (or the other way around) [15].

## 4.2 Performance Evaluation

Table 2 shows the relationship between our features. It has been also observed that there are 357 Benign and 212 Malignant, we have in our dataset as you can see Fig. 3. Number 1 represents the Benign and 0 represents the Malignant, respectively. Note that 1.0 (Orange) = Benign (No Cancer) 0.0 (Blue) = Malignant (Cancer) (Fig. 2)

Now let us check the correlation between our features. So, it has been noted down that there is a strong correlation between the mean radius and mean perimeter, mean area and mean perimeter. Now it is time to train the model and to make a prediction using our testing data and get the classification report to know that how many healthy patients have cancer?

The above is confusion matrix and denoting that 40 patients are predicted healthy while 66 are suffered by cancer as shown in Fig. 4.

As we can see in Table 3, our model did not do a very good job in its predictions. It predicted that 48 healthy patients have cancer.

# 5 Conclusion and Perspectives

In this paper, we zeroed in on a hazardous illness that causes demise for some ladies over the world which is the breast cancer. Bosom malignant growth expectation is critical in the region of Medicare and Biomedical. In this paper, we zeroed in on building a classifier which targets anticipating the most extreme malignant growth known as breast cancer. In this, we proposed a contributed strategy to determine this ailment and give data about the patient status. This paper portrays the bosom malignant growth model as a characterization task and portrays the usage of the support vector machine (SVM) strategy to characterize bosom malignant growth as kindhearted or threatening. The consequences of SVM comprise of exactness and accuracy. To sum up the created technique, the underlying step, in view of information social event of patients as text/csv document. Presently remove the non-

**Table 2** Relationship between features

| Mean radius | Mean texture | Mean perimeter | Mean area | Mean smoothness | Mean compactness | Mean concavity | Mean concave points | Mean symmetry | Mean fractal dimension |
|---|---|---|---|---|---|---|---|---|---|
| 0 17.99 | 10.38 | 122.80 | 1001.0 | 0.11840 | 0.27760 | 0.3001 | 0.14710 | 0.2419 | 0.07871 |
| 1 20.57 | 17.77 | 132.90 | 1326.0 | 0.08474 | 0.07864 | 0.0869 | 0.07017 | 0.1812 | 0.05667 |
| 2 19.69 | 21.25 | 130.00 | 1203.0 | 0.10960 | 0.15990 | 0.1974 | 0.12790 | 0.2069 | 0.05999 |
| 3 11.42 | 20.38 | 77.58 | 386.1 | 0.14250 | 0.28390 | 0.2414 | 0.10520 | 0.2597 | 0.09744 |
| 4 20.29 | 14.34 | 135.10 | 1297.0 | 0.10030 | 0.13280 | 0.1980 | 0.10430 | 0.1809 | 0.05883 |

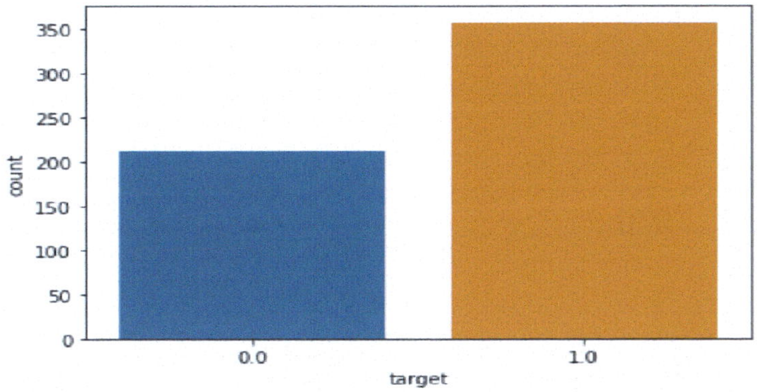


**Fig. 2** Represent benign by 0.0 and malignant by 1.0

**Fig. 3** Confusion matrix-1

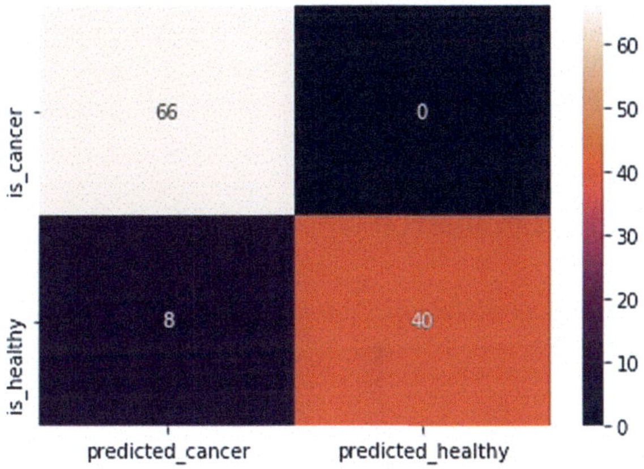


**Fig. 4** Confusion matrix-2

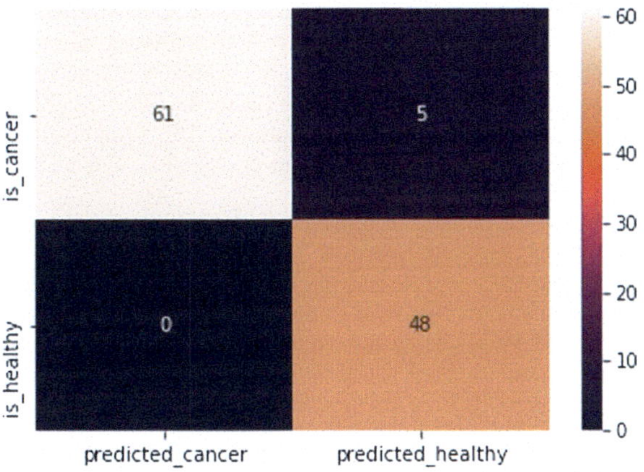

**Table 3** Classification report of results

| | Precision | Recall | $f1$-score | Support |
|---|---|---|---|---|
| 0.0 | 1.00 | 0.83 | 0.91 | 48 |
| 1.0 | 0.89 | 1.00 | 0.94 | 66 |
| Accuracy | | | 0.93 | 114 |
| Micro avg. | 0.95 | 0.92 | 0.93 | 114 |
| Weighted avg. | 0.94 | 0.93 | 0.93 | 114 |

significant component like id and other. At long last, the SVM classifier is utilized for order, which train models to sort disease patients as indicated by their analysis. Exploratory outcomes show that the viability of model. SVM accomplishes 96.09% grouping precision on test subsets.

**Acknowledgements** Authors are thankful to School of Computing Science and Engineering (SCSE) at Vellore Institute of Technology (VIT) Bhopal University for giving highly motivational supports.

## Conflict of interest

The authors declare that there is no conflict of interest in this paper.

## References

1. Chakraborty, S.: Bayesian kernel probit model for micro array based cancer classification. Comput. Stat. Data Anal. **12**, 4198–4209 (2009)
2. Guyon, I., Weston, J., Barnhill, S., Vapnik, V.: Gene selection for cancer classification using support vector machines. Mach. Learn. **46**, 389–422 (2002)
3. McCarthy, J.F., Marx, K., Hoffman, P.E.: Applications of machine learning and high-dimensional visualization in cancer detection, diagnosis, and management. Ann. N.Y. Acad. Sci. **62**, 10201259 (2004)
4. Tan, A.C., Gilbert, D.: Ensemble machine learning on gene expression data for cancer classification. Appl. Bioinform. **2**, 75–83 (2003)
5. Abreu, P.H., et al. Predicting breast cancer recurrence using machine learning techniques: a systematic review. ACM Comput. Surv. (CSUR) **49**(3), 52 (2016)
6. Hall, P., Park, B.U., Samworth, R.J.: Choice of neighbor order in nearest-neighbor classification. Ann. Stat. 2135–2152 (2008)
7. Van't Veer, L.J., et al.: Gene expression profiling predicts clinical outcome of breast cancer. Nature **415**(6871), 530–536 (2002)
8. Sarkar, S.K., Nag, A.: Identifying patients at risk of breast cancer through decision trees. Int. J. Adv. Res. Comput. Sci. **08**, 88–96 (2017)
9. UCI Breast Cancer Wisconsin (Original) Dataset. https://archive.ics.uci.edu/ml/datasets/Breast+Cancer+Wisconsin

10. Huang, M.W., Chen, C.W., Lin, W.C., Ke, S.W., Tsai, C.F.: SVM and SVM ensembles in breast cancer prediction. PloS one **12**(1) (2017)
11. Abreu, P.H., et al.: Predicting breast cancer recurrence using machine learning techniques: a systematic review. ACM Comput. Surv. (CSUR) **49**(3), 52 (2016)
12. Tang, J., Rangayyan, R.M., Xu, J., El Naqa, I., Yang, Y.: Computer-aided detection and diagnosis of breast cancer with mammography: recent advances. IEEE Trans. Inf. Technol. Biomed. **13**(2), 236–251 (2009)
13. Powers, D., Ailab : Evaluation: from precision, recall and F-measure to ROC, informedness, markedness & correlation. J. Mach. Learn. Technol. **2**, 2229–3981 (2011). https://doi.org/10.9735/2229-3981
14. Prasad, D., Goyal, S.K., Sharma, A., Bindal, A., Kushwah, V.S.: System model for prediction analytics using k-nearest neighbors algorithm. J. Comput. Theor. Nanosci. **16**(10), 4425–4430 (2019)
15. Ting, K.M.: Confusion matrix. In: Sammut, C., Webb, G.I. (eds.) Encyclopedia of Machine Learning and Data Mining. Springer, Boston, MA (2017)

# Time Dilation in Lorentzian and Modified Euclidean Metrics of Space–time

**Talal Al-Ameen and Imad Muhi**

**Abstract** In the relativistic Lorentzian metric transformation, Minkowski (The Principle of Relativity, 1908 [1]) realised the possibility for a two-dimensional representation of the interrelationships between bodies having large relative velocities. This realisation paved the way for the concept of spacetime in Einstein's Special Theory of Relativity. Hartle and Hawking's proposal (Phys. Rev. D 28:2960, 1983 [2]) employed Wick's rotation transformation to transform the Lorentzian metric to a Euclidean metric in an attempt to resolve the issue of singularity in the light cone. This work is an attempt to show that the concept of spacetime, involving light cone in two dimensions, based on the Lorentzian and Euclidean metrics is rather a mathematical artifact than a physical reality and may lead to conflicting results as far as time dilation is concerned.

**Keywords** Light cone · Lorentz metric · Euclidean metric · Time dilation

## 1 Introduction

A light cone (or null cone), in Special Relativity, is the surface describing the temporal evolution of a flash of light in Minkowiski space–time [1]: This, in 2-D, can be depicted as a Pythagoras triangle (Fig. 1).

The time dilation for the above light cone can be expressed as:

T. Al-Ameen (✉) · I. Muhi
Department of Mathematics and Natural Sciences, Prince Mohammad Bin Fahd University (PMU), Al Khobar, Kingdom of Saudi Arabia
e-mail: tameen@pmu.edu.sa

I. Muhi
e-mail: imuhi@pmu.edu.sa

 
T. K. Sharma et al. (eds.), *Soft Computing: Theories and Applications*,
Advances in Intelligent Systems and Computing 1381,
https://doi.org/10.1007/978-981-16-1696-9_32

**Fig. 1** Showing the light cone in 2-D, where $i = \sqrt{-1}$, $t$ (the coordinate time) is the time interval spanned by the space–time interval ($t_r$), and $x$ ($x = ct$, where $c$ is the speed of light) is the distance in space along the corresponding axes ($x$-axis)

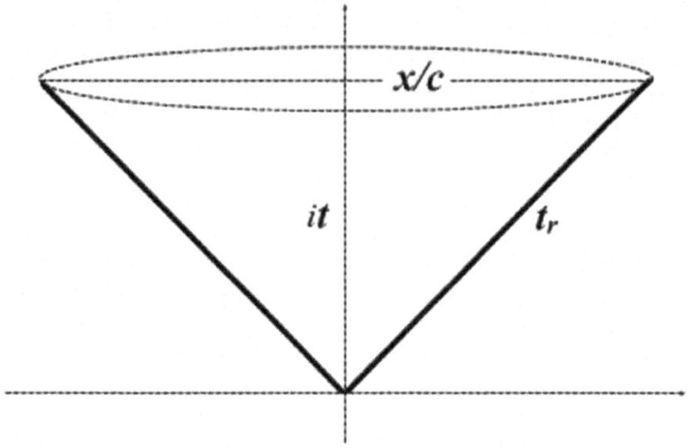


$$t_r^2 = (x^2/c^2) - t^2$$
$$t_r^2 = -(t^2 - (x^2/c^2))$$
$$-t_r^2 = t^2 - (x^2/c^2)$$
$$-t_r^2/t^2 = 1 - (v^2/c^2) \quad \text{where} \quad v = x/t.$$

$|-t_r{}^2/t^2| = |1 - (v^2/c^2)|$ or $t_r{}^2/t^2 = 1 - (v^2/c^2)$ on the assumption that $v^2/c^2 \leq 1$, where | | represents an absolute value.

But this result implicitly implies $(-t_r{}^2/t^2) = t_r{}^2/t^2$ which is false! Nevertheless, this implication has been assumed in finding $t_r = t\,(1 - (v^2/c^2))^{½}$ or $t_r = t/\gamma$ where $\gamma = 1/(1 - (v^2/c^2))^{1/2}$ is the Lorentz factor [3–5].

Alternatively, we may arbitrary assume $t_r{}^2 = |(x^2/c^2) - t^2|$ or $t_r{}^2/t^2 = |(v^2/c^2) - 1|$. If so, then we will have either $t_r{}^2/t^2 = -((v^2/c^2) - 1)$ or $t_r{}^2/t^2 = ((v^2/c^2) - 1)$.

Rejecting $t_r{}^2/t^2 = (v^2/c^2) - 1$, because it demands tachyons (objects outside the light cone moving with $v > c$), $t_r{}^2/t^2 = -((v^2/c^2) - 1)$, or $t_r{}^2/t^2 = (1 - (v^2/c^2))$, will result in $t_r = t/\gamma$ where $\gamma = 1/(1 - (v^2/c^2))^{1/2}$. However, this arbitrary assumption remains mathematically ad hoc even if it leads to a valid physical result from the relativity point of view.

## 2 The Light Cone as Per Lorentzian Metric of Space–time

The configuration of the light cone in the above (Fig. 1) is mathematically acceptable but only if time is considered to be imaginary ($it = \sqrt{-1}\ t$). This approach to the light cone may need to be reassessed for a number of reasons. First, while it may be mathematically justified to represent time as imaginary, this should also be supported by physical evidence. Simply stating that time can be imaginary is not enough to justify conceiving of physical time as imaginary. Time is real; as such, the assumption that time is imaginary for the purpose of depicting the light cone seems rather ad hoc.

Secondly, the model of the light cone that has been established in Standard Special Relativity, which is a model of space–time, presupposes that the speed of light determines what can be observed (or not) in space–time. The presumption in the model is that the speed of light ($c$) cannot be exceeded. Surely, a model of space–time should be represented by a space–time cone where $c$ is not the determinant factor. If $c$ should be exceeded, then a model determined by c would become redundant. For instance, to avoid abandoning our present model of the light cone it is generally denied that quasar jets are actually exceeding c as observations suggest. Most astrophysicists prefer to suppose that what is being witnessed are optical illusions—less a phenomenon is admitted that is incompatible with the theory of special relativity. If the light cone model of Standard Special Relativity is correct then these jets exist outside the light cone and should not be detectable. The presumption in the model that c cannot be exceeded means that superluminal observations by the Hubble Telescope have to remain inexplicable. In order to protect the integrity of the model of the light cone it would surely be better to set the speed of the moving object ($v$), rather than $c$, as the determinant factor.

A third reason to reassess this model of the light cone is that it violates the basis of the Pythagorean theorem, i.e. that the hypotenuse ($t_r$) should be greater than either the adjacent ($x/c$) or the opposite ($t$). Assigning either $x/c$ or $t$ as imaginary, as is the case in Standard Special Relativity, would make the hypotenuse smaller. By synthesising the imaginary to the real, Standard Special Relativity presents us with a rather unusual model of the light cone that can neither be considered real nor imaginary—but what might be referred to as a "complex" entity. The *absolute value* (or modulus or magnitude) of a complex entity $s = x + iy$ is $|s|^2 = x^2 + y^2$. Choosing $y = ict$, which ironically causes the imaginary to become real, will lead to $|s|^2 = x^2 - c^2t^2$, not $|s|^2 = c^2t^2 - x^2$ as shown above in ${t_r}^2/t^2 = 1 - v^2/c^2$ (or $c^2{t_r}^2 = c^2t^2 - x^2$). To achieve $|s|^2 = c^2t^2 - x^2$, as Standard Special Relativity imposes upon physics, is mathematically ad hoc. There is no mathematical justification to assume time-like dimension ($x$) necessarily to be smaller than space-like dimension ($ct$): $c^2t^2 > x^2$; quite the opposite. It seems that $c^2t^2 > x^2$ is an imposed dogma rather than a mathematical fact. This assumption may well be the reason why Einstein, and his peers rejected the possibility of superluminal speeds; as it would require exiting the boundaries and constraints of the light cone (and/or space–time) resulting in cause and effect conundrums e.g. time travel.

The imposition of $c^2t^2 > x^2$ in Standard Special Relativity assumes that objects possessing superluminal speeds must necessarily exit outside space–time (existing in the imaginary). However, mathematic demands that $c^2t^2 < x^2$, which means that superluminal objects remain within the confines of space–time.

Rather than being prepared to interpret the possibility of superluminal speeds as spelling the end of Standard Special Relativity mainstream physicists have found explanations to delay its replacement by a better model to explain the kinematics of the universe. For example, the extraordinary speeds of quasars that are celestial bodies postulated as travelling at superluminal velocities at the edge of the universe are rationalised by mainstream physicists to be an effect of distorted regions of space–time; yet, Einstein's Relativity demands a space–time that is at large homogeneous and isotropic. Similarly, experiments in 2011 by CERN researchers that seemed to show neutrinos travelling faster than the speed of light were later dismissed as unreliable due to technical errors. Even those who admitted the possibility of superluminal neutrinos claimed that this did not necessarily constitute a violation of Einstein's Relativity but rather the first evidence of the existence of extra dimensions (New Scientist, 28 September, 2011). As this argument, which seems to be a desperate attempt to defend Standard Special Relativity, also constitutes a violation of the Standard approach—as Einstein never postulated extra dimensions; it would perhaps be more appropriate to simply accept that Standard Special Relativity lacks the ability to explain the possibility of either superluminal neutrinos or anything else, other than photons, travelling at the speed of light or at even greater speeds.

Explaining away experimental evidence demonstrating neutrinos travelling at the speed of light by appealing to the existence of extra dimensions does little to save Einstein's Special Relativity but rather, ends up revealing another one of its weaknesses—the solution to the problem will either be to explain superluminality or to explain extra dimensions.

The three above queries that highlight ad hoc mathematical procedures in the light cone model presented in Fig. 1 are serious enough to cause doubt in the general approach taken by Standard Special Relativity, but there are further reasons why this model of the light cone needs to be questioned.

A fourth concern is that a careful examination of the relation, $-t_r^2 = t^2 - x^2/c^2$, would inform us that after taking the square root of both sides, $(-t_r^2)^{½} = (t^2 - x^2/c^2)^{½}$, we will be left with the surprising result of an imaginary value on the left side of the equation and a real value on the right: $it_r = (t^2 - x^2/c^2)^{½}$. Now the hypotenuse becomes imaginary.

A further problem, a fifth concern, is that the relation: $-t_r^2/t^2 = 1 - v^2/c^2$, where $v = x/t$, which is transformed into the absolute value $|-t_r^2/t^2| = |1 - v^2/c^2|$ thereby resulting in $t_r^2/t^2 = 1 - v^2/c^2$, has similarly been calculated in an ad hoc manner. The purpose of this particular mathematical step ensures that $v^2/c^2$ remains smaller than 1, i.e. that $c > v$. If the absolute value had not been introduced in such a manner, then $v^2/c^2$ would have been greater than 1, i.e. $v > c$. Employing ad hoc mathematical procedures to ensure $c > v$ can perhaps be understood within the context of Maxwell's of light was maintained as the maximum speed in nature. Such pervading principles in physics formed the consensus view when Lorentz and Einstein were formulating their approach to Standard Special Relativity. To remain faithful to the prevailing view that the speed of light was a censor in nature, Minkowiski, Lorentz and Einstein formulated a model of the light cone that ensured $c > v$.

Another area of concern, sixth now, is that the model only holds when $x/c$ is smaller than $t$, however, this assumption creates two different times: $x/c$, which can be called $t'$ (on the horizontal), and coordinate time $t$ (on the vertical). If the light cone in two dimensions initiated from the premise where the proper relativistic distance s can be expressed as $s = ict + x$ then one should ask what $ct$ actually is? The space-like dimension is represented by $ct$ and the time-like dimension is represented by $x$. Now if $ct = x$ then the time-like dimension and the space-like dimension become equivalent. This condition $x = ct$ (or $ct = x$) presents a grave problem in Standard

Special Relativity as it would nullify the model of the light cone completely: $s^2 = -x^2 + x^2 = 0$. By assuming that $x/c$ is smaller than $t$ (or vice versa) Standard Special Relativity has once again brought about another ad hoc mathematical justification for their model of the light cone. A model that must assume that the Pythagorean triangle is derived from a rectangular space–time rather than a space–time that is square. This is because time dilation only occurs within the light cone when depicted in rectangular space–time, i.e. where the adjacent and opposite sides are not equal, thus, allowing values for $x/c$ to be different from $t$ or ($it$). To represent the light cone with the adjacent and opposite sides being of equal length (and/or time) would have $x/c = t$ and would nullify the light cone, as already stated.

A model of the light cone established upon rectangular space–time causes another concern, the seventh, in respect to the Big Bang. Rectangular space–time suggests that the first flash of light from the Big Bang covered in the same time different distances in two-perpendicular directions. This would violate Standard Special Relativity's own claim that the speed of light is constant in any direction.

An eighth concern is that while Minkowiski, Lorentz and Einstein established a hyperbolic (non-Euclidean) model of the light cone that rejected the idea that rigidity is a principle in nature they, nevertheless, founded a Special Relativity with a starting point that begins with a rigid body in motion; a model that applies to rigid bodies—not applicable to non-rigid bodies. This contradiction in Special Relativity seems to have gone unnoticed: having rigid bodies after embracing non-Euclidean mathematics seems rather ad hoc.

It is interesting that the heirs of Standard Special Relativity, a century later, while remaining loyal to the fundamental premises established by the model of the light cone attempted to reinterpret it in parabolic (Euclidean) terms. One such attempt is known as *The No-Boundary Proposal.* The appeal to Euclidean mathematics was in order to overcome the issue of what existed before the Big Bang, i.e. the nature of space–time during and before the singularity assumed before the Big Bang; however, we shall see that this reinterpretation in Euclidean terms results in problems to the concept of time dilation.

## 3 The Light Cone According to the No-Boundary Proposal

In an attempt to resolve the issue of singularity in the light cone, James Hartle and Stephen Hawking proposed a theory based on the idea that the universe, like the Earth, did not have a boundary. In order to propose this, they transformed the Lorentzian metric into a Euclidean one [2].

The difference between the Lorentzian and the Euclidean metric is that the "time" element becomes negative in the Lorentzian metric:

$$\text{Lorentzian metric: } s^2 = -c^2t^2 + x^2$$
$$\text{Euclidean metric: } s^2 = c^2t^2 + x^2$$

Hartle and Hawking's proposal employed Wick's rotation transformation to modify the time axis. In the Wick rotation, the time axis is multiplied by the imaginary unit $i$ (or replacing $t$ in the Lorentzian metric with $it_{\text{img}}$). Thus, conversion of the Lorentzian metric in Euclidean terms is:

$$s^2 = c^2 t_{\text{imag}}^2 + x^2$$

The difference now between this version of the Euclidean metric and the previous one is that the time unit is still positive but Wick's transformation causes imaginary time implications.

Dividing $s^2 = c^2\, t^2{}_{\text{imag}} + x^2$ by $c^2 t^2{}_{\text{imag}}$ gives: $t_r = t_{\text{imag}}\,[1 + (v/c)^2]^{1/2}$, where $t_r = s/c$, and $v = x/t_{\text{imag}}$. It should be noted that for $v = 0$, $t_r = t_{\text{imag}}$. But for $v = c$, $t_r = 1.141\ t_{\text{imag}}$. This last result not only allows $v \geq c$, but also indicates that relativistic time ($t_r$) can be greater than non-relativistic imaginary time ($t_{\text{imag}}$). Under this scheme a new concept emerges; namely, that of time elongation: the opposite of the time dilation, which in Standard Special Relativity would be given as: $t_r = t[1 - (v/c)^2]^{1/2}$ where $t_r = 0$ when $v = c$.

Not only would time elongation seem to contradict the concept of time dilation in Standard Special Relativity, but would also demand length elongation rather than length contraction, and mass decrease rather than mass increase to occur!

The notion of imaginary time used in the No-Boundary proposal makes predictions about time and length that are at odds with the demands of Standard Special Relativity, at least in respect to time and length at the dawn of the universe. If Standard Special Relativity did not apply at the dawn of the light cone, then there may be a reason to doubt its application in today's space–time.

## 4 Conclusions

The use of the Lorentzian metric and its conversion in the Euclidean metric to construct a light cone leads to conflicting results: a time dilation in the Lorentzian metric, but a time elongation in Hartle and Hawking's Euclidean metrics. Alternative metrics must maintain a relativistic frame of reference that is capable of maintaining causal structure of space–time and result in a better theory of relativity.

## References

1. Minkowski, H.: Space and time. In: The Principle of Relativity (1908)
2. Hartle, J., Hawking, S.: Phys. Rev. D **28**, 2960 (1983)
3. Tolman, R.: The Theory of the Relativity of Motion. University of California Press, Berkeley (1917)
4. Forshaw, J., Smith, A.G.: Dynamics and Relativity. Wiley, New York (2009)
5. Wesley, C.: Four Decades of Scientific Explanation. University of Pittsburgh (2006)

# Exploratory Data Analysis for Predicting Student's Grades

**Kailash Kumar**

**Abstract** Over the period, the academic level of the students has improved. However, it is found in many research studies that students find it difficult in getting success in some core courses like Mathematics and Reasoning. The recent development in machine learning techniques and various data mining tools has made it possible to extract useful information from the available raw data. This research paper analyzes the student performance dataset available on the University of California, Irvine (UCI) Machine Learning Repository. The student's grades are predicted using various machine learning techniques using Python programming in Jupyter Notebook.

**Keywords** Student performance · Educational data analysis · Grade prediction · Exploratory data analysis · Machine learning algorithms · Deep learning

## 1 Introduction

Students are supposed to undergo their studies online as educational institutions are closed due to safety measures in the current pandemic situation [1]. The students must grasp the subjects with clear understanding, which is possible by enhancing the quality of education. The fundamental reason for the decline in the quality of education is that students cannot comprehend and understand the subject [2].

The use and significance of data mining, machine learning and deep learning are extensively increasing in our day-to-day life, and so is with the educational learning management and assessment. The educational institutes and organizations are concerned with developing methods which can discover new knowledge from existing information. In this competitive world, educational organizations are mainly facing challenges in analyzing their performance and establish a proper plan to act upon it. Most educational institutes are focused on the performance of their students.

K. Kumar (✉)
College of Computing and Informatics, Saudi Electronic University, Riyadh 11673, Kingdom of Saudi Arabia
e-mail: k.kumar@seu.edu.sa



T. K. Sharma et al. (eds.), *Soft Computing: Theories and Applications*,
Advances in Intelligent Systems and Computing 1381,
https://doi.org/10.1007/978-981-16-1696-9_33

If somehow the performance of the student can be predicted, it would help the student, instructor and educational institute and parents to take necessary steps and timely decision to improve the performance of the student.

Measuring a student's performance is an essential part of any educational system. Predicting the grades of the students helps the students in preparing for the examinations.

In this paper, an open dataset available on the UCI Machine Learning Repository [3] has been taken for our research purpose. This dataset contains 33 columns, but for our research purpose, only half of the relevant and key columns are analyzed in this paper.

This paper is organized as follows. Section 2 presents a literature review about the related works. Section 3 outlines the research methodology and the proposed model for predicting the grades of the students. Section 4 deals with implementation, results and findings of the proposed research work. Last but not the least, the conclusion and future work are presented in Sect. 5.

## 2 Literature Review

Several research studies have been conducted to explore and predict the academic grades of the students. These studies facilitate the students in planning their educational activities and to take appropriate measures to improve the grades of the students.

A factorization model for predicting the students' performance was proposed [4]. The model was applied to some algebra courses. The factorization method works where the students' background knowledge is not available, and the data matrix is sparse. The dataset was split into a training set and test set data. In their comprehensive research, the tensor-based factorization was used to predict the success of the students. The problem was formulated as a recommender-based system. The proposed method was based on tensor factorization, and the temporal effect of the students' performance was added. The output of the model saved the logs of success or failure of the students.

The matrix factorization method does not work well when dealing with a small sample size, and hence, the performance and accuracy of the system degrade. A new recommender system was explored and investigated [5] to overcome this problem. The system used to accurately predict the class assessment performance of Stanford University (SU), University of Minnesota (UMN) and George Mason University (GMU) and next term grades of the students. The findings of the study reveal that the error rate is low when personalized multi-linear regression is used in conjunction with advance matrix factorization.

During the initial semesters, the students perform well due to high motivation, but over the period, their academic performance might decrease. A new model was proposed [6], which can be used in both regressions as well as in classification to predict the final grades of the students. The students were classified into two groups, namely performing well and performing poorly. They found in their study that in-class exams are better predictors than homework assessment. It also concluded that prediction allows the timely intervention of the course instructor to improve the academics of the students.

A regression model, along with variable selection and aggregation method, was used [7] to predict the performance of the graduate students. A dataset of 171 students from Eidgen ossische Technische Hochschule (ETH), Switzerland, was used, and they found in their research that the 54% of the variance could be explained for performance of the undergraduate students. Their findings can be used as a guiding principle for the admission committee.

The Dropout Early Warning System (DEWS) model focusing on early warning for dropout students was proposed by authors [8]. This model can assist the educational institutions in identifying the students at risk as well as identifying proper help and guidance to their students. This model is flexible, adaptable and series of several modules which used hidden Markov model to predict the dropout risks.

Authors in their research study [9] applied the algorithm ID3 to a small set of only 50 students. The study focused on identifying the attributes which affected the performance of the students. The students are classified into poor, good and very good categories. The results obtained were not accurate and biased as they used the same dataset for training and testing purpose.

Cognitive, demographic and personality traits are used [10] to predict the success at the higher education level. They used a mixed sample of 200 boys and 200 girls, making a total of 400 samples. They found in their research that there were significant difference between higher academic achievement for girls with high socio-economic status and boys with low socio-economic status in science stream. The academic performance was measured using behaviors, body mass index and self-esteem of more than 6000 youngsters [11] using machine learning algorithms. It was found in their research that students with low BMI, physical activities and good dietary habits perform well as compared to others.

The social and demographic variables are examined to predict the grades of the students [12]. They used the sample size of 300 students from the University of Punjab that was analyzed using linear regression. It was found that there is a positive correlation between the mother's income and academic performance of the students. Similarly, family income is found positively correlated with a student's academic performance.

An unsupervised machine learning algorithm like association rule mining, outlier detection and clustering was used by Ramaswami and Rathinasabapathy [13] to predict the performance of around 3000 students. The students were classified in poor, fair, very good and excellent categories at the time of admission.

Authors used the Bayesian networks on more than 5500 students and CHAID algorithm on 772 students [14]. In their research, they used 35 attributes such as body mass index, family size, Internet accessibility, family status and personal, demographic and social characteristics. The class attributes are classified into the pass and fail; poor, good and very good, etc. Four different machine learning algorithms were used on the 778 Portuguese students [15] and concluded that decision tree algorithm outperforms. A technique using Bayes net is used by Yu [16] to predict the grades for school going students of secondary standard. They used the dataset having 300 records and predicted the grades for the students.

## 3 Research Methodology

In this research paper, the dataset is analyzed for the exploratory data analysis.

The quality of input decides the quality of output. So, once the business hypothesis is ready, it makes sense to spend a lot of time and efforts. The data have to be cleaned and prepared to build a predictive model [16]. In this research paper, the following steps are used exploratory data analysis.

In this research paper, the factors are identified which affect the grades/performance of the students positively or negatively. Following factors are analyzed using machine learning techniques.

1. Family educational background (parent's education)
2. The distance of school from home
3. Students having more weekly time (more than 10 h)
4. Students who do not have school support
5. Students paying for extra classes
6. Students who went for nursery school
7. Students willing to pursue higher education
8. Good Internet connectivity
9. Romantic relations
10. Alcohol consumption.

## 4 Implementation and Results

The proposed model is implemented in Python using Jupyter Notebook. The exploratory data analysis is presented in this section.

It is found that there are no null values in the dataset. The next task is to compare the two schools GP and MS with the score G3. The output is shown in Fig. 1.

From the above line plot, it is found that school GP is outperforming school MS. Next, we compare age and sex with G3 as depicted in Fig. 2.

This analysis does not give much information from age and sex. Then, a parent's education is compared with G3. A swarm plot is drawn in Fig. 3 to predict the impact of parent's education on student's grades.

From the above graph, we can conclude that students having good family education background (parent's education) are performing better.

The next task in exploratory data analysis is to compare travel and study time with G3. This is shown in Fig. 4.

**Fig. 1** Graph between performance versus schools GP and MS

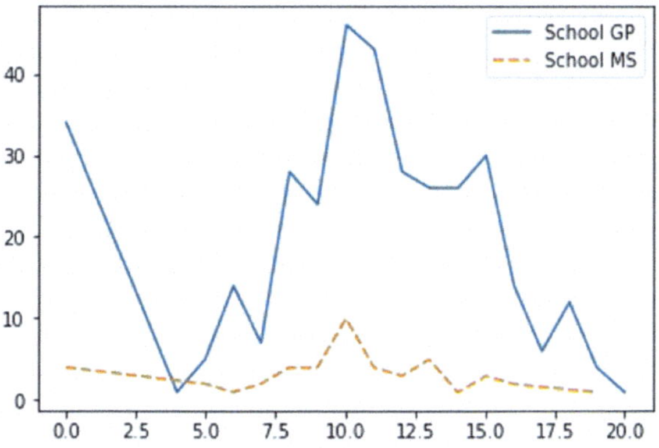


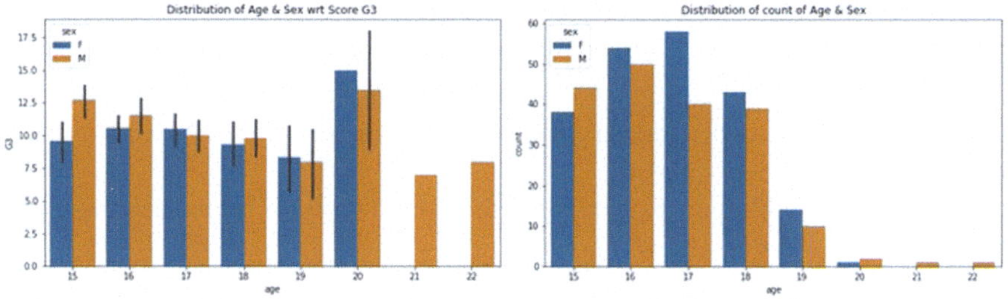


**Fig. 2** Graph between age and sex versus G3

**Fig. 3** Graph between parent's education versus G3

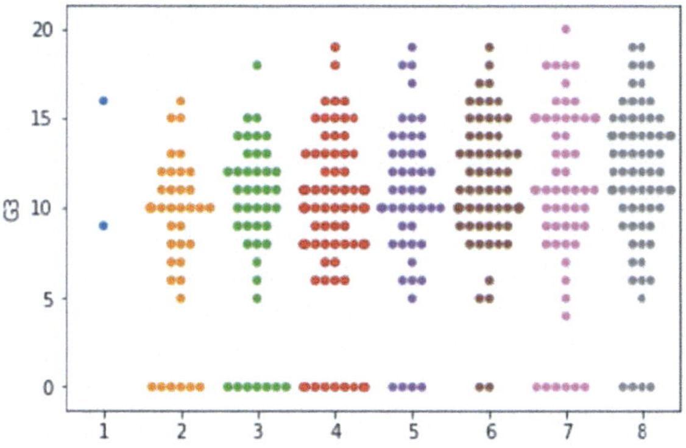


From the graph, it is apparent that students living near the school are scoring better than the far students. Also, it can be concluded that students having more weekly study time (over 10 h) are getting a better score.

Similarly, other features like school support, family support, paying for extra classes, devoting time for extracurricular activities, attended nursery classes, willing

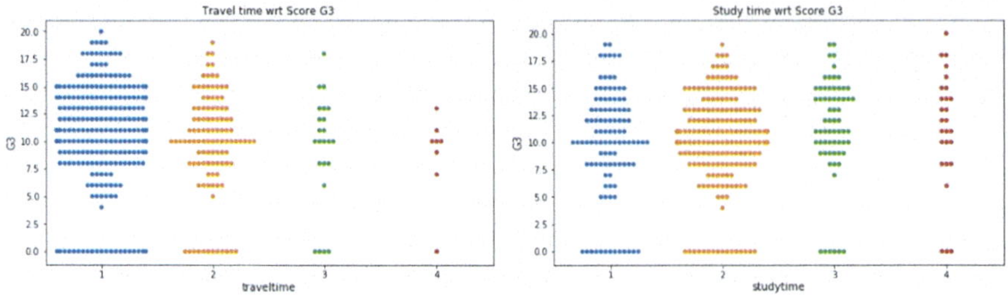


**Fig. 4** Graph between travel time and study time versus G3

to join higher education, Internet accessibility and romantic relations are explored for G3. These graphs are shown in Fig. 5.

From the above graphs, we can conclude the following observations.

1. The students who do not have support from the school do not perform well.
2. The students paying extra fees show negative trends.
3. The students attended nursery classes and outperform others who did not participate in nursery classes.
4. The students willing to pursue higher education perform better than those not willing to pursue higher education.

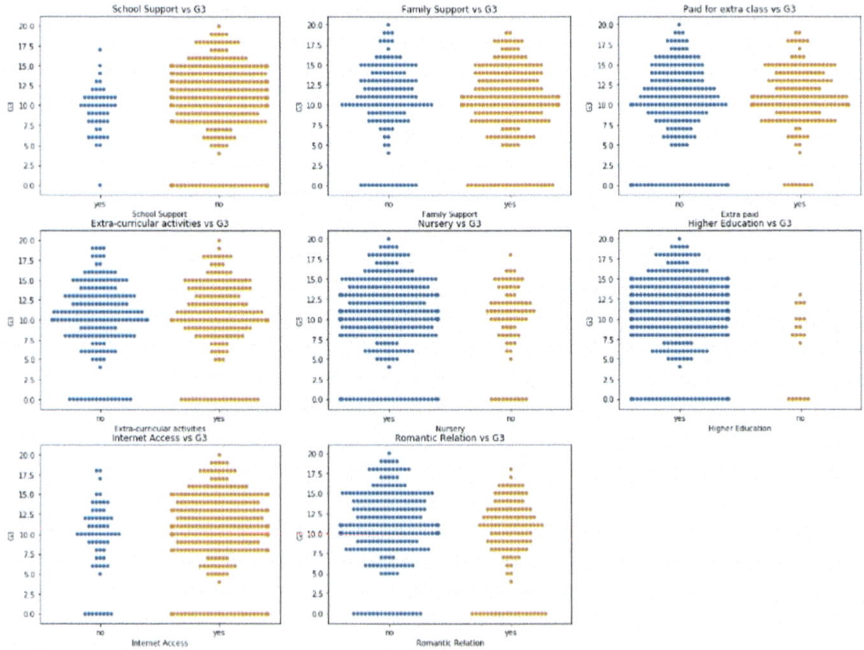


**Fig. 5** Other attributes versus G3

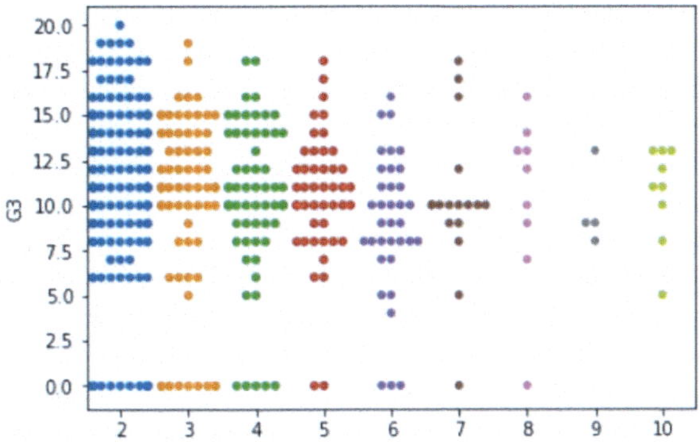


**Fig. 6** Graph between alcohol consumption versus grades

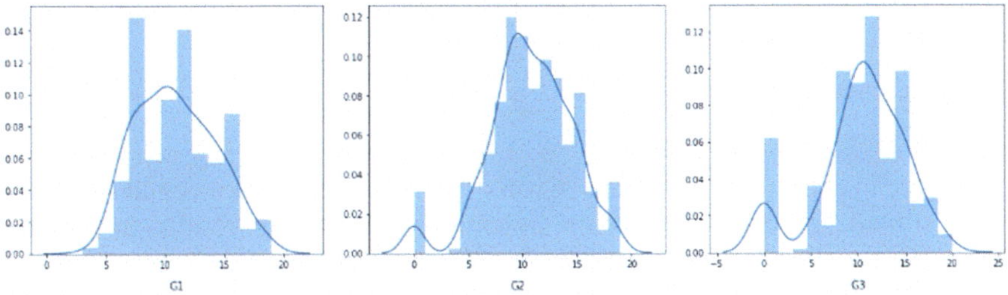


**Fig. 7** Probability distribution of grades for G1, G2 and G3

5. The students having Internet access perform better than others.
6. The students having no romantic relations perform better than those who have romantic relationships.

Next, we identify the effect of alcohol consumption on the grades of the students. This is shown in Fig. 6.

The above graph shows that students consuming more alcohol perform very poorly.

The next step is to find out the probability distribution of grades as shown in Fig. 7.

The last step in exploratory data analysis is to compare the grades G1 and G2 with G3.

It is clear from Fig. 8 that there is a linear relation between G1 and G3 as well as between G2 and G3.

## 5 Conclusion and Future Work

In this paper, the open dataset from UCI is taken for the experimental purpose to predict the grades of the students. The research paper started with a brief introduction, followed by an overview of the dataset used. In the next section, the literature review

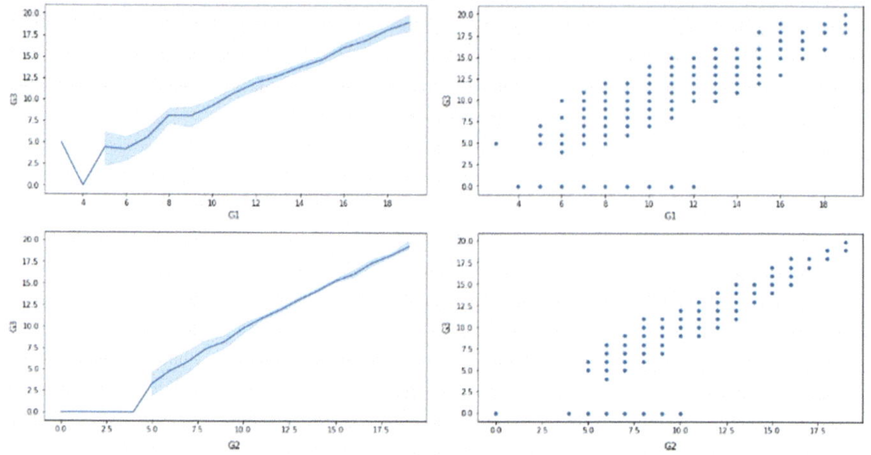


**Fig. 8** Compare grades G1 and G2 with G3

is presented in detail to show the related works. The literature review contains various proposed models, experimental setup, hypothesis and research results. In the next section, the detailed research methodology is presented and used to predict the grades of the students. Next section deals with the implementation of the proposed problem. The project is implemented in Python using Jupyter. The exploratory data analysis helps to predict the grades which in turn can help students to make an extra effort or make him drop a course to avoid going on probation.

**Acknowledgements** I wish to record our deep sense of gratitude and profound thanks to Saudi Electronic University, Riyadh, Kingdom of Saudi Arabia for inspiring guidance and constant encouragement with the work during all stages to bring this research paper into fruition.

## References

1. Goldberg, L., Parham, D., Coufal, K., Maeda, M., Scudder, R., Sechtem, P.: Peer review: The importance of education for best practice. J. College Teach. Learn. **7**, 71–84 (2010). https://doi.org/10.19030/tlc.v7i2.91
2. Cestone, C., Levine, R., Lane, D.: Peer assessment and evaluation in team-based learning. New Direct. Teach. Learn. 69–78. https://doi.org/10.1002/tl.334 (2008)
3. https://archive.ics.uci.edu/ml/machine-learning-databases/00320/
4. Thai-Nghe, N., Horvath, T., Schmidt-Thieme, L.: Factorization Models for Forecasting Student Performance, pp. 11–20 (2011)
5. Elbadrawy, A., Polyzou, A., Ren, Z., Sweeney, M., Karypis, G., Rangwala, H.: Predicting student performance using personalized analytics. Computer **49**, 61–69. https://doi.org/10.1109/MC.2016.119 (2016)
6. Meier, Y., et al.: Predicting grades. IEEE Trans. Signal Process. **64**(4), 959–972 (2016)

7. Zimmermann, J., Brodersen, K.H., Heinimann, H.R., Buhmann, J.M.: A modelbased approach to predicting graduate-level performance using indicators of undergraduate-level performance. JEDM-J. Educ. Data Mining **7**(3), 151–176 (2015)
8. Miranda, M., Parrini, F., Dalerum, F.: A categorization of recent network approaches to analyze trophic interactions. Methods Ecol. Evol. **4**, 897–905 (2013). https://doi.org/10.1111/2041-210X.12092
9. Iqbal, Z., Qadir, J., Mian, A., Kamiran, F.: Machine Learning Based Student Grade Prediction: A Case Study (2017)
10. Elbadrawy, A., Andkarypis, G.: Domain-aware grade prediction and top-n course recommendation. Boston, MA, Sept (2016)
11. Hijazi, S.T., Naqvi, S.: Factors affecting students' performance: a case of private colleges. Bangladesh e-J. Soc. **3**, 1–10 (2006)
12. Abuteir, M., El-Halees, A.: Mining educational data to improve students' performance: a case study. Int. J. Inf. Commun. Technol. Res. **2**, 140–146 (2012)
13. Ramaswami, M., Rathinasabapathy, R.: Student performance prediction modeling: a Bayesian network approach. Int. J. Comput. Intell. Inf. 1(4) (2012)
14. Cortez, P., Silva, A.: Using data mining to predict secondary school student performance. EUROSIS (2008)
15. Vandamme, J., Meskens, N., Superby, J.: Predicting academic performance by data mining methods. Educ. Econ. **15**, 405–419. https://doi.org/10.1080/09645290701409939 (2007)
16. Yu, C.H.: Exploratory data analysis in the context of data mining and resampling. Int. J. Psychol. Res. **3**, 9–22 (2010)

# The Misleading Implications of the Premise '*0 = 0*' in the Standard Special Relativity

**Talal Al-Ameen and Imad Muhi**

**Abstract** Mathematically speaking, it is justifiable to state that $1 = 1$. If $1 = 1$, then $1 = 1 \times A$ is also correct if and only if $A = +1$. But can the same be said of $0 = 0$? It is important to bear in mind that while 1 is a natural number, zero is not! There are difficulties associated with certain mathematical operations involving zero, such as dividing zero by zero, for they result in indeterminate values. Hence, unlike $1 = 1 \times A$, the premise $0 = 0 \times A$ would be correct for any value of $A$. Thus, any theory founded upon $0 = 0 \times A$, or any premise that permits more than one mathematical solution must be considered unsound for it would be liable to costly errors in its power to interpret the physical world accurately. In the course of the present work, it will be shown that standard special relativity is apparently tainted with a mathematical premise that is more inclined towards accepting $0 = 0 \times A$.

**Key words** Lorentz factor · Time dilation · Length contraction

## 1 Introduction

Although written in 1905, Einstein's Special Theory of Relativity, published in his Zur Elektrodynamik bewegter Körper (On the Electrodynamics of Moving Bodies) in 1905 [1], is still considered the standard theoretical approach in relativistic physics to measure and explain the relationship between space (and/or length in the direction of motion) and time within inertial frames of reference [1]. Time and length are fundamental concepts in physics. For inertial observers in relative motion, these fundamental concepts can be either absolute or relative, and they can have the same or different magnitudes, depending on 'Lorentzian', 'Minkowskian', 'Machian', 'Leibnizian', 'Maxwellian', 'Galilean', 'Newtonian' and 'Aristotelian' space times. Unlike

T. Al-Ameen · I. Muhi (✉)
Department of Mathematics and Natural Sciences, Prince Mohammad Bin Fahd University (PMU), Al Khobar, Kingdom of Saudi Arabia
e-mail: imuhi@pmu.edu.sa

T. Al-Ameen
e-mail: tameen@pmu.edu.sa

 
T. K. Sharma et al. (eds.), *Soft Computing: Theories and Applications*,
Advances in Intelligent Systems and Computing 1381,
https://doi.org/10.1007/978-981-16-1696-9_34

classical physics, the physics of special relativity rejects the concept of absolute time and length. Instead, special relativity theory considers that the same time and length spans have different magnitudes for different inertial observers in relative motion. We will assess in the present work some of the implications of these relativistic concepts according to Albert Einstein's theory of special relativity and show that using premise $0 = 0$ to build a theory of special relativity would lead to an incorrect conclusion on length contraction. However, the same premise surprisingly leads to a conclusion on time dilation. This paradox can be explained mathematically as one of the misleading implications of the premise $0 = 0$. For example, the conclusion $2 = 3$ can result if we consider $2 \times 0 = 3 \times 0$; then, by cancelling zero on both sides of the equation would result in a '$2 = 3$' fallacy.

## 2 The Mathematical Premise of the Standard Special Relativity

Science generally relies upon mathematical justifications to support premises about the physical world. In the absence of direct experimental evidence, the mathematics used to support a theory needs to be even more rigorous. Albert Einstein's theory of special relativity was presented to the world in 1905, and for more than a century, the verdict of mathematicians has accepted its mathematical premises as valid. At first, it was judged on the basis of its mathematical formulations (supported by thought experiments). In the course of time, Einstein's equations and thought experiments seemed to receive experimental confirmation. As long as the theory continues to receive experimental support, even to the present day, then why question the mathematics? But things might not be as they seem!

As far as we are aware, there has never been a peer-reviewed paper that aimed to evaluate a main mathematical premise from which the equations of Einstein's special relativity can surface, i.e. $0 = 0$. We believe that this starting point needs to be assessed, and Einstein's standard model of special relativity is inclined to accept the premise $0 = 0$, which is undoubtedly a mathematically justifiable premise. Our alternative special relativity (to be published) begins with another mathematically justifiable premise, $1 = 1$. Choosing to establish equations for special relativity on either $0 = 0$ or $1 = 1$ will have different implications. The question then becomes whether the implications of the two approaches are mathematically consistent.

Let us first consider the mathematical veracity of the premise $1 = 1$. If $1 = 1$, then $1 = 1 \times A$ is also correct if and only if $A = +1$. This is due to the possibility of $A \neq 1$; for example, if $A = 0$, then $1 = 0$, and if $A = -1$, then $1 = -1$, respectively! It is mathematically unacceptable for $1 \neq 1$. As only one mathematical value is accepted by '$A$', namely $+1$, $1 = 1 \times A$ is a sound mathematical premise. Any proposition that permits or gives the possibility for more than one mathematical solution would be considered an unsound premise in mathematics.

Evidently, the premise $1 = 1$ produces clear mathematical results and has to be accepted as a sound mathematical premise upon which a theory of special relativity can be established. Can the same be said of the premise $0 = 0$?

It is important to bear in mind that while 1 is a natural number, zero is not. Mathematically speaking, there are simply unavoidable difficulties associated with certain operations involving zero. These include dividing zero by zero (0/0) and zero to the power of zero (00), which both result in indeterminate values. Hence, unlike $1 = 1 \times A$, the premise $0 = 0 \times A$ would be correct for any value of $A$. Thus, any theory founded upon $0 = 0 \times A$, or any other premise that permits more than one mathematical solution must be considered unsound for it would be liable to costly errors in its power to interpret the physical world accurately.

Einstein's special relativity is apparently tainted with a mathematical premise that is more inclined towards accepting $0 = 0 \times A$! We can demonstrate how this premise $(0 = 0)$ is inherent in the standard model of special relativity by considering the expression $x = ct$ or $x - ct = 0$ where $x$ is the distance travelled by the photon in $t$ seconds of time, and $c$ is the speed of light. Such a premise can lead to peculiar mathematical implications such as 0/0. Although the premise '$0 = 0$' is not explicitly declared by the mathematics of the standard approach to special relativity, in respect to issues related to the light cone, it is nevertheless one of the overlooked premises in the theory:

$$\begin{aligned} &0 = 0 && \text{(mathematical premise)} \\ &0 = A \times 0 && \text{(where } A \text{ is any quantity)} \\ &0 = B \times 0 && \text{(where } B \text{ is any quantity).} \end{aligned}$$

$x^2 - c^2t^2 = (x - ct)(x + ct) = x'^2 - c^2t'^2 = (x' - ct')(x' + ct') = 0$ (a mathematical premise in the standard approach to special relativity) where $x - ct = 0$ (in the system $s$) and $x' - ct' = 0$ (in the system $s'$) [1].

Standard special relativity expresses the kinematics of light pulses propagating in parallel in system $s$ and $s'$ as follows:

$x' - ct' = A(x - ct)$. This, we believe, is indisputably '$0 = A \times 0$' in disguise.

$x' + ct' = B(x + ct)$. This, we believe, is indisputably '$0 = B \times 0$' in disguise.

Any theoretical approach founded upon such a mathematical stance is bound to be fraught with problems as we shall show.

## 3 Conundrums of Accepting the Premise '0 = 0' in the Standard Special Relativity

Solving the last two equations for $x'$ and $ct'$ yields

$$x' = x(A + B)/2 - ct(A - B)/2$$

$$ct' = ct(A + B)/2 - x(A - B)/2$$

or

$$x' = xR_1 - ctR_2 \quad \text{where} \quad R_1 = (A + B)/2, \quad \text{and} \quad R_2 = (A - B)/2.$$

$$ct' = ctR_1 - xR_2$$

Here, for $x' = 0$, we get $x = ctR_2/R_1$ or $x/t = v = cR_2/R_1$. From the standard approach to special relativity point of view, $v$ here is the velocity with which the origin of $s'$ is moving relative to $s$. Notice now that $x = vt$ (or put another way, $x\text{-}vt = 0$). This is a fundamental problem that is at odds with the initial premise of the standard approach to special relativity: $x - ct = 0$! Notice if we, under this oddity, use $v = cR_2/R_1$ in the relation $x' = xR_1 - ctR_2$, we arrive at $x' = R_1(x - vt)$. Now comparing $x' = R_1(x - vt)$ with Lorentz equation for length contraction: $x' = (x - vt)/\left[1 - (v/c)^2\right]^{1/2}$, we find that $R_1$ = Lorentz factor $= \gamma = 1/[1 - (v/c)^2]^{½}$!

If $x = ct$ and $x = vt$, then, simply put, this scenario tells us that $v = c$! If $v = c$, then there can be no way of developing a theory of special relativity because a moving system ($s'$) will be moving at the speed of light. For relativity to be a valid notion, $v \neq c$ or according to the standard approach to special relativity, the velocity of any massive object must be less than the speed of light ($v < c$).

In the above approach, '$0 = 0$' is an inherent premise that comes at a high cost. It makes the approach impotent. It only takes a few mathematical steps, from the above, to obtain the Lorentz transformation equations [3–6]:

$$x' = (x - vt)/\left[1 - (v/c)^2\right]^{1/2}$$

$$t' = \left[t - (vx)/c^2\right]/\left[1 - (v/c)^2\right]^{1/2}$$

(Here if $x' = 0$ corresponds to $x - vt = 0$, then $0 = \gamma \times 0$ or $\gamma = 1/[1 - (v/c)^2]^{½} = 0/0$!? Dividing zero by zero is considered mathematically indeterminate and could be the reason why, as shall be argued, the standard approach to special relativity lacks consistency in respect to length contraction.).

The first equation: $x' = (x - vt)/\left[1 - (v/c)^2\right]^{1/2}$ does not, however, lead to length contraction, as has been incorrectly assumed. The claim that the equation provides a proof for length contraction simply does not withstand mathematical scrutiny. If $x/t$ is considered greater than $v$ (or alternatively, $x > vt$), then a case for length contraction could be made: $x' = X/\left(1 - (v/c)^2\right)^{1/2}$ where $X = x - vt$ is the Galilean transformation: $X = ct - vt$, if $x = ct$. Unfortunately, such a distinction cannot be made between the two variables ($x$ and $vt$) as $v$ here is a quantity by which $x/t$ is expressed: What is $v$ if not $x/t$?

It should be noted that in the equation $x' = X/\left(1-(v/c)^2\right)^{1/2}$, only $X$ will contract, whereas $x'$ elongates (quite the opposite of what the standard approach to special relativity demands)! We need to be clear about what $X$ represents: So what is $X$? $X$ should not be confused with $x$, nor with $x'$. As such, it should be clear that $X$ is definitely not the $x$ found in standard special relativity's equation for length (and/or space) contraction:

$$x' = X/\left(1-(v/c)^2\right)^{1/2}$$

Even if we assume, for the sake of argument, that this, now controversial, length contraction equation $x' = X/\left(1-(v/c)^2\right)^{1/2}$ is correct, there should be absolutely no doubt that it cannot be claimed as a proof for length contraction and, therefore, should be questioned as a valid template for the length contraction equation in standard special relativity.

Surprisingly, in regard to the second equation for time dilation: $t' = \left[t-(vx)/c^2\right]/\left[1-(v/c)^2\right]^{1/2}$, no similar controversial implications exist. The equation does seem to be mathematically sound, thus leading to the time dilation equation in standard special relativity:

$t' = t(1-(v/c)^2)^{½}$ but only on the condition that $v = x/t$. This can be demonstrated as follows:

$$t' = \left[t-(vx)/c^2\right]/\left[1-(v/c)^2\right]^{1/2}$$

$t' = \left[t-(x^2)/tc^2\right]/\left[1-(v/c)^2\right]^{1/2}$ (substituting $v = x/t$ in the term: $t-(vx)$)
$t' = \left[t-(x^2t)/t^2c^2\right]/\left[1-(v/c)^2\right]^{1/2}$ (multiplying the term $x^2/tc^2$ by $t/t$)
$t' = \left[t-(v^2t)/c^2\right]/\left[1-(v/c)^2\right]^{1/2}$ (substituting $x^2/t^2 = v^2$)
$t' = t\left[1-(v^2/c^2)\right]/\left[1-(v/c)^2\right]^{1/2}$ ($t$ being the common factor).

This is Einstein's well-known time dilation equation:

$$t' = t\left[1-(v^2/c^2)\right]^{1/2}$$

It should once again be emphasized that compatibility between Lorentz and Einstein here can only occur if, and only if, $v = x/t$.

Interestingly, while $v = x/t$ jeopardizes the first equation, it validates the second equation. For a simple expression to be sound in one instant but not in one directly related to it, it is surely a worry for the mathematical reliability of the standard approach to special relativity.

## 4 Conclusions

In co-local events (such as $s$ and $s'$ moving simultaneously with respect to each other), light speed constancy demands $x = ct$ ($x^2 = c^2t^2$ or $x^2 - c^2t^2 = 0$) in $s$-frame, and $x' = ct'(x'^2 = c^2t'^2$ or $x'^2 - c^2t'^2 = 0)$ in $s'$-frame. If so, then $x^2 - c^2t^2 = 0 = x'^2 - c^2t'^2$. The arbitrary application of this premise under the conditions: $x = 0$ for $t \neq 0$ in the $s$-frame, and $x' = 0$ for $t' \neq 0$ in the $s'$-frame; or alternatively $t = 0$ for $x \neq 0$ in $s$-frame, and $t' = 0$ for $x' \neq 0$ in the $s'$-frame, is shown here to result in different conclusions for length contradiction and time dilation according to the standard approach to special relativity as theorized by Einstein.

## References

1. Einstein, A.: Zur Elektrodynamik bewegter Körper (On the Electrodynamics of Moving Bodies). Annalen der Physik (Berlin) (in German) (1905)
2. Collier, P.: A Most Incomprehensible Thing: Notes Towards a Very Gentle Introduction to the Mathematics of Relativity, 3rd edn. (2017)
3. David, M.: Introduction to Classical Mechanics. Cambridge University Press, Cambridge (2007)
4. Edwin, F., John, A.: Spacetime Physics, 1st edn. W. H. Freeman and Company, San Francisco (1966)
5. Tolman, R.: The Theory of the Relativity of Motion. University of California Press, Berkeley (1917)
6. Rindler, W.: Essential Relativity, 2nd edn. Springer, New York (1977)

# Shape-Based Video Watermarking Robust Against Affine Attacks

**Imad Muhi and Talal Al-Ameen**

**Abstract** The abstract should summarize the contents of the paper in short terms. With the increasing use of the Internet and other digital data, there is a strong need to protect the illegal duplication, editing, or modifying of these data. One such approach to meet this demand is to use watermarking. Such mark must be transparent to legitimate users of the media yet robust against geometric attacks or any image processing. This paper presents a new model, based on shape, for rendering video watermarking scheme resistant to geometric attacks and image processing. This is achieved by means of a new image transform to rotation, scaling, and translation (RST) based on ideas from shape theory and the utilization of the scale-invariant feature transform (SIFT) detector for locating robust feature points. Indeed any detector that produces such robust feature points can be used. This model was first developed and applied to standard image processing based on a new image transform into such RST invariant domain (Morris and Muhi El-Ddin in Proceedings of IPV08, 2008 [1]; El-Ddin in Watermarking schemes robust against affine attacks. Ph. D. Thesis, Claremont Graduate University, Los Angeles, 2008 [2]; Muhi El-Ddin et al. in Proceedings of the Fifth International Conference on Information Technology: New Generations (ITNG 2008), pp. 795–800, 2008 [3]; Muhi El-Ddin in Proceedings of the 41st Annual Asilomar Conference on Signals, Systems, and Computers, pp. 565–569, 2007 [4]). In this work, we utilize the shape theory and the curvelet transform to introduce an RST invariant map from a video scene to a set of *universal video watermarking prisms*. In this invariant domain, we have embedded a watermark into the low frequency component of the curvelet transforms of each of the three universal videos and assured RST invariance.

**Keywords** Curvelet transform · Geometric attacks · Invariant domains · Shape theory · SIFT · Video watermarking prisms

---

I. Muhi (✉) · T. Al-Ameen
Department of Mathematics and Natural Sciences, Prince Mohammad Bin Fahd University (PMU), Al Khobar, Kingdom of Saudi Arabia
e-mail: imuhi@pmu.edu.sa

T. Al-Ameen
e-mail: tameen@pmu.edu.sa

 
T. K. Sharma et al. (eds.), *Soft Computing: Theories and Applications*,
Advances in Intelligent Systems and Computing 1381,
https://doi.org/10.1007/978-981-16-1696-9_35

## 1 Introduction

In [1–4], a feature-based image watermarking method that is resistant to geometrical attacks RST was presented. The approach is based on regions rather than points. In [1], a new paradigm for video watermarking based on feature points was presented. This paper shows how this method can be applied using the *wrapped transform* implementation of the fast discrete curvelet transform (FDCT) [5, 6]—that is obtained from the curvelet toolbox and that can be downloaded from the Internet [7]. For the sake of completeness and to establish notation, we present the basics of the approach.

In Sect. 2, Affine Projective Area-Invariant Transformation by a distinguished triangle in the image is defined. And some examples are given. Section 3 shows how such a triangle may be found using the SIFT feature point algorithm [2] (scale-invariant feature transform). The image of the distinguished triangle must be found in an image that has been subjected to an affine attack. This work proposes a new method based on shape theory [2, 3]. In Sect. 4, some fundamental results from shape theory [2] are presented, and in Sect. 5, we talk about the curvelet transform, that was introduced in Candes et al. [5, 6, 8]. In Sect. 6, an outline in how this approach can be extended to video watermarking algorithm. In Sect. 7, an example of the proposed algorithm applied to video images is used. Finally, in Sect. 8 we summarize the results.

## 2 Affine Projective Area-Invariant Transformation

The key element of our construction, for regular and video images, is a triangle. To achieve the robustness required, the SIFT method [9] is chosen to choose stable feature points as the vertices of the triangle. This is essential as it is necessary to be able to identify and locate the equivalent triangles in the geometrically attacked image. However, our method is independent of the SIFT algorithm. The proposed method generates three of RST invariant image regions that are called watermarking zones [2]. A Projective Area-Invariants Transformation by defining a new local coordinate system for an image based on a distinguished triangle was developed [2]. The function relating the regular Cartesian coordinates of a point in the image to the new coordinates can be viewed as an image transformation. This is called this the Projective Area-Invariants Transformation.

Let $R_1$, $R_2$ and $R_3$ be any three non-collinear points and $A$ is any fourth point in the plane, and let the point $X$ be the intersection of the lines $R_1R_2$ and $R_3A$ as shown in Fig. 1.

From the triangles formed in Fig. 1, two affine invariants can be constructed [10] as follows:

$$r_1 = \frac{\text{area}(R_1R_2R_3)}{\text{area}(XR_2R_3)} \quad \text{and} \quad r_2 = \frac{\text{area}(XR_2R_3)}{\text{area}(AR_2R_3)} \tag{1}$$

**Fig. 1** Basic triangle

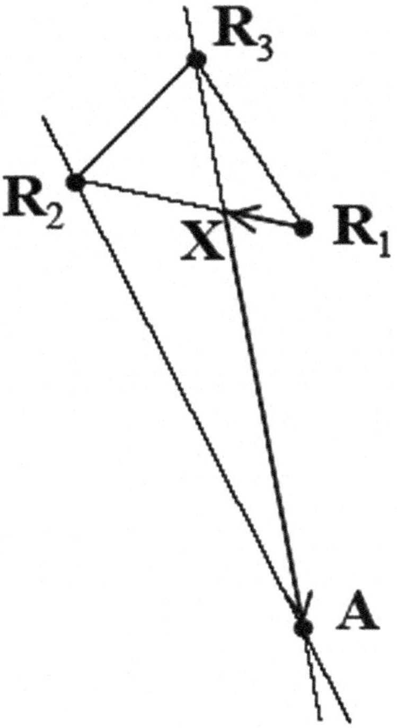


Each point $A = (x, y)$ is associated with a unique point $(r_1, r_2)$. Therefore, if $A$ and $X$ are located, then $r_1$ and $r_2$ can be found and vice versa. Furthermore, the interior of the triad $\Delta$ maps to the rectangle $[1, \infty) \times [1, \infty)$.

$$X = R_2 + \frac{1}{r_1}(R_1 - R_2) \quad \text{and} \quad A = R_3 + \frac{1}{r_2}(X - R_3) \tag{2}$$

The mapping $\phi$ is a $1 - 1$ transformation with inverse

$$\phi^{-1} : [1, \infty) \times [1, \infty) \to \Delta \quad \text{with} \quad (r_1, r_2) \to A. \tag{3}$$

This inverse map has the explicit form:

$$r_1 = \frac{\det(A - R_3, R_2 - R_1)}{\det(A - R_3, R_2 - R_3)} \quad \text{and} \quad r_2 = \frac{\det(R_2 - R_3, R_1 - R_2)}{\det(A - R_3, R_1 - R_2)} \tag{4}$$

This mapping depends upon the choice of vertex $R_3$ and so, in order to avoid this restriction, the triangle is divided into three quadrilateral regions $Q_i$, $i = 1, 2, 3$ by adding the centroid of the triangle as an additional vertex; see Fig. 6. In this way, three watermarking zones are obtained, one for each vertex, $R_1$, $R_2$ and $R_3$, of the triangle $\Delta$. Then for each $i$, the map

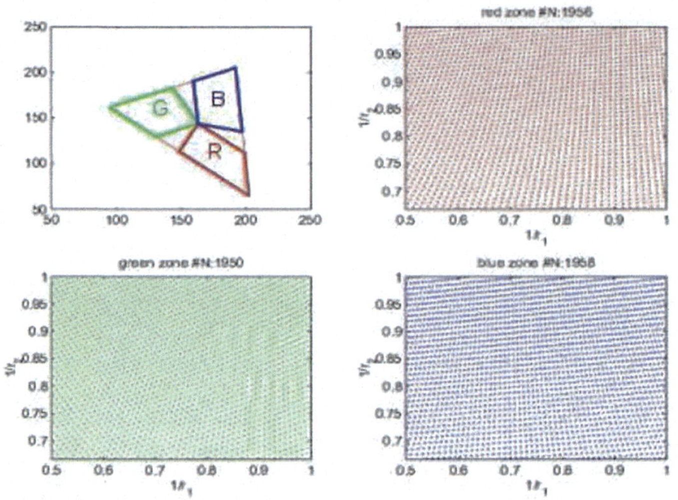


**Fig. 2** Three quadrilateral regions and the mapping zones

$$\phi_i : Q_i \rightarrow U_i = [1, 2] \times \left[1, \frac{3}{2}\right], \quad i = 1, 2, 3 \tag{5}$$

is defined. This map is affine invariant in the sense of Eq. (7).

These three watermarking zones are shown in Fig. 2. Each rectangle corresponds to a vertex, and each zone $Q_i$ is the pre-image of the same rectangle $U$.

In Fig. 2, the upper left image shows the three watermarking zones, one for each vertex $R_1$, $R_2$ and $R_3$, of the triangle $\Delta$. The rest three images represent the three rectangles $U_i$ , $i = 1, 2, 3$, (blue, green, and red), each one corresponds to one of the three vertices.

An $n \times m$ image $I(i, j)$ is defined on a rectangular grid. The set of 3-tuples is given by

$$\{i, j, I(i, j) | 1 \leq i \leq m, \quad 1 \leq j \leq n\} \tag{6}$$

Each of the mappings $\phi_i$ maps the image points in the $i$th zone $Q_i$ to an unstructured set of points in the rectangle $U_i$. An image on $U_i$ is defined [2] by introducing a grid on $U_i$ and then interpolating the unstructured data to that grid. In this work, a $64 \times 64$ grid is used. Thus, whatever the size of the original image, it is mapped to three square images. Each of these three resulting images is called a universal image (UI). The inverse of the mapping is implemented in a similar way. These are RST invariant domains. Having three universal images gives us choices of implanting the same watermark in each image.

These universal images are affine invariant in the sense that, for any affine mapping $T$, the following relation holds

$$\phi_i(TI, T\Delta) = \phi_i(I, \Delta), \quad i = 1, 2, 3 \tag{7}$$

In the original image of Lena, three quadrilateral regions have been selected that map to three square universal images. Figures 3 and 4 illustrate this.

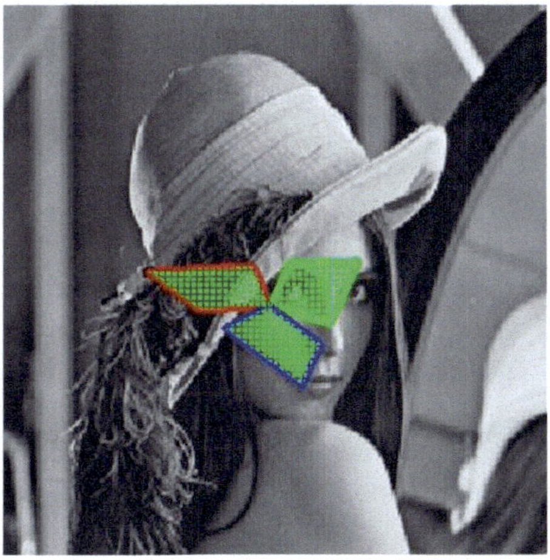

**Fig. 3** Three watermarking zones in the Lena image

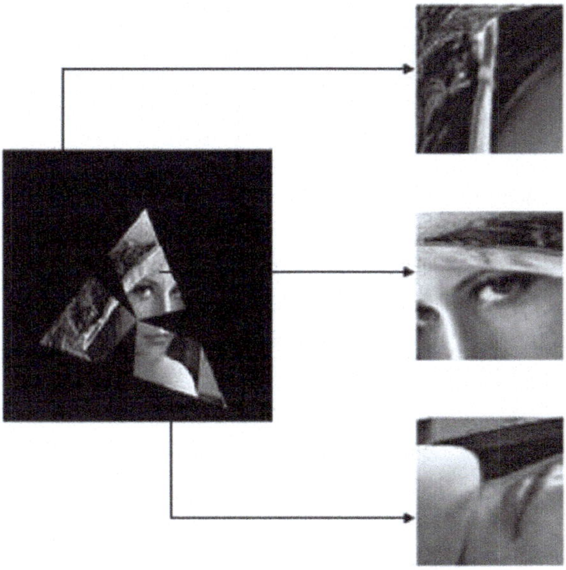

**Fig. 4** Mapping from the watermarking zones to the universal images

## 3 Generating Feature Points Using SIFT

SIFT was developed by Lowe [9] for image feature generation in object recognition. Using SIFT, feature points are detected by applying various size circular windows to an image. This can be done by directly applying them to the original image or by applying them to the transformed image. The original image $I(x, y)$ is repeatedly convolved with Gaussian

$$G(x, y, \sigma) = \frac{1}{2\pi\sigma^2} e^{-\left(\frac{x^2+y^2}{\sigma^2}\right)}$$

A sub-pixel image location, scale and orientation are associated with each SIFT feature. This additional information allows a classification of feature points according to their robustness across different scales. Candidate features that have a low contrast or are poorly localized along edges are removed by using a robustness measure.

Figure 1 shows typical output from SIFT. Each row gives the location of the feature together with a classification, 0 for 'edge' type features, 1 otherwise. Also, in the final column (prominence), it provides a measure of the robustness of the feature. Only type 1 points is taken and ordered by the absolute value of the prominence. The top eighty most prominent features are retained as candidates for vertices of an image triangle (Table 1).

Figure 5 shows the SIFT method applied to both Lena image and a rotated version of the image. In the figure, the 'edge' type features are shown as line segments and 'non-edge' by boxes. This method is restricted to non-edge features with the greatest significance. Each feature is given a measure of its robustness. Note that the features detected by SIFT still show in the rotated image.

**Table 1** Typical output from SIFT

| *x*loc | *y*loc | Scale | Size | Edge tag | Edge orientation | Prominence |
|---|---|---|---|---|---|---|
| 115.5243 | 210.3435 | 1.7346 | 2.0204 | 0 | 4.7124 | −1.4349 |
| 124.9223 | 196.7486 | 2.1036 | 2.3465 | 1 | 0 | −0.8935 |
| 156.2825 | 137.4315 | 2.1793 | 2.4197 | 0 | 1.5708 | −1.1225 |
| 158.9022 | 158.6591 | 1.8683 | 2.133 | 0 | 4.7124 | −0.9805 |
| 159.4298 | 179.2016 | 1.97 | 2.2228 | 0 | 5.4978 | −1.8961 |
| 187.7796 | 159.7414 | 2.242 | 2.482 | 0 | 1.5708 | −3.0167 |
| 213.5307 | 243.945 | 2.2539 | 2.494 | 1 | 4.3197 | −2.5103 |
| 216.9852 | 223.9251 | 2.0416 | 2.2882 | 1 | 4.7124 | −1.7238 |
| 217.7257 | 218.2623 | 2.1411 | 2.3824 | 0 | 1.5708 | −1.5793 |
| 226.8712 | 197.5393 | 2.0801 | 2.3242 | 0 | 1.5708 | −1.4637 |

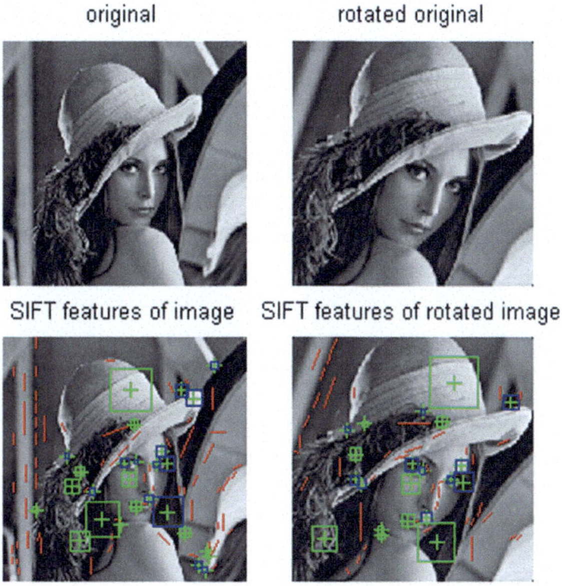


**Fig. 5** In the first row: original Lena, and its rotated version. In the second row: feature points of original Lena and its rotated version

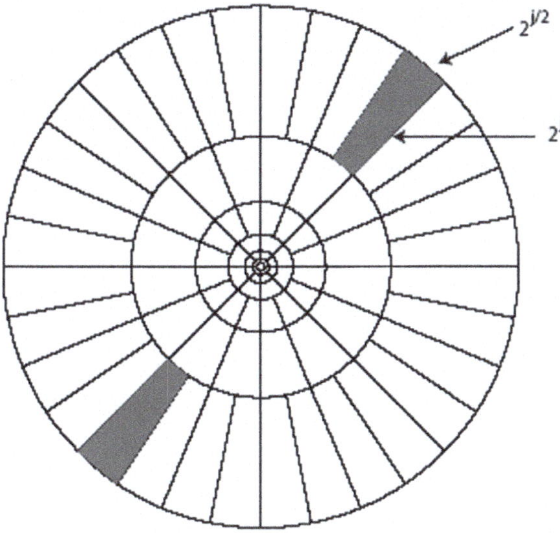


**Fig. 6** Curvelet tiling of the frequency plane

## 4 Shape Space

Consider a set of $k$ $(k > 2)$ points in the plane. For example, these could be $k$ feature points in an image. Such a set is called a $k$-ad. Each point with coordinates $(x, y)$ can be defines a complex number with real part $x$ and complex part $y$ and a $k$-ad is then defined by a complex $k$-tuple $z = (z_1, ..., z_k)$. The shape of the $k$-ad $z$ is defined to be the orbit of $z$ under translation, rotation and scaling. To remove the translation, subtract the barycenter $\overline{z} = \frac{1}{k}\sum_{i=1}^{k} z_i$ to get $z - \overline{z}$. Rotation of the $k$-ad by an angle $\theta$ and scaling by a factor $\lambda$ corresponds to multiplying $z - \overline{z}$ by the complex number $\lambda e^{i\theta}$. This provides a very efficient way to estimate the geometric distortions an image has undergone. If the original triangle is known, both the rotation angle and scale factor can be determined. Once these are known the effects can be reversed and have a better image from which to extract a watermark. The quantity $u = \frac{z-\overline{z}}{\|z-\overline{z}\|}$ is called the *preshape* of the shape of $z$. The *shape* of $z$ is the orbit $[z] = \{e^{i\theta}u \mid \theta \in [0, 2\pi)\}$. Two $k$-ads $w_1$ and $w_2$ that are related by an affine transformation have $[w_1] = [w_2]$. A distance function can be defined between two $k$-ads by means of the *Veronese–Whitney Map* $\psi : \Sigma_2^k \rightarrow H$ defined by $\psi([z]) = z z^*$ with $\|z\| = 1$. This maps the space of $k$-ads $\Sigma_2^k$ to the space of Hermitian matrices. The distance between $[u]$ and $[v]$ is then given by $d([u], [v]) = \|uu^* - vv^*\|_H$ where $\|\cdot\|_H$ is the Frobenius norm $\|M\|_H^2 = tr(MM^*)$ on the space of Hermitian matrices. To find equivalent triangles, the minimum distance pairs should be found.

## 5 The Curvelet Transform

The continuous curvelet transform (CCT) is a multi-scale transform with frame elements $\gamma_\mu(x)$ indexed by a location, scale, and orientation parameter $\mu$. The transform combines the time–frequency properties of wavelets with high directionality and anisotropy. Curvelets are constructed by tiling the frequency plane with the polar grid.

In Candes et al. [5, 6], the fast discrete curvelet transform (FDCT) is introduced. This discrete analog is faithful to the continuous version whilst giving results on the time scale of an FFT. This bears the same relationship to the CCT as the discrete wavelet transform (DWT) does to the CWT. In the case of the FDCT, the CCT tiling of the plane shown in Fig. 6 is replaced by the rectangular analog shown in Fig. 7.

## 6 Video Watermarking

Video watermarking involves embedding cryptographic information derived from frames of digital video into the video itself. Ideally, a user viewing the video cannot perceive a difference between the original, unmarked video and the marked video, but a watermark extraction application can read the watermark and obtain the embedded information. Because the watermark is part of the video, this technology works independently of the video file format or codec. This type of watermarking algorithm optimizes for three separate factors robustness, security, and perceptual fidelity. Attackers commonly use geometric attacks against watermarked videos. Rotation, scaling, translation, random bending, and cropping are common geometric attack.

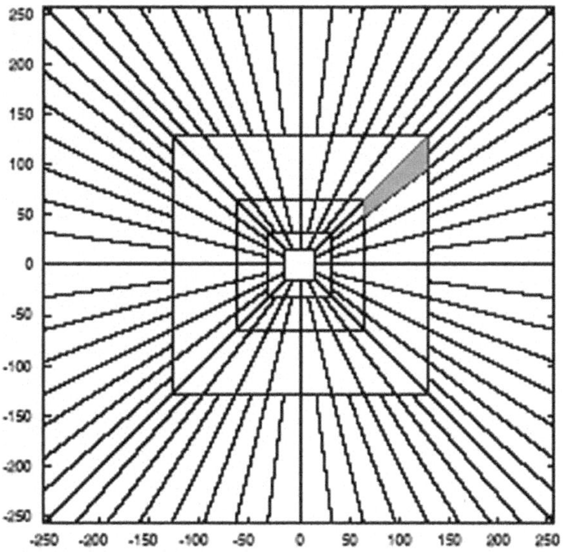


**Fig. 7** FDCT analog of Fig. 6

## 7 Example

It is important to make the video watermark RST invariant. See, for example, [10]. In this work, a way is proposed in which the developed image approach can be extended to the video realm. Video sequences are composed of consecutive still images, which can be independently processed by different image watermarking algorithms. The three-dimensional (3D) wavelet transforms [11] can be used, where the new technique is applied:

- The video is a segmented into scenes. This is done with a scene-change detection algorithm [8].

Figure 8 shows some scene changing frames from the video sample of 'Foreman.avi'.

Here, if the bottom left frame is taken, the watermark zones and universal images for it can determined. These are shown in Fig. 9.

By running through the frames in the scene, automatically a set of base triangles one for each frame can be generated. This is done by a modification of the triangle location algorithm. In each frame, the feature point triangle nearest in size to the triangle in the previous frame can be selected. Figure 10 shows an example of this process.

By selecting $N$ frames from a single scene, $N$ triples of universal images can be constructed. This gives us three image rectangles. Each is $L \times L \times N$ where $L \times L$ is the size of each universal image. These three image blocks are defined by an $N$-tuple of triangles $\{\Delta_i\}_{i=1}^{N}$ The dimensions $L$ and $N$ can be chosen to be dyadic powers. As

**Fig. 8** Scene breaks from 'Foreman.avi'

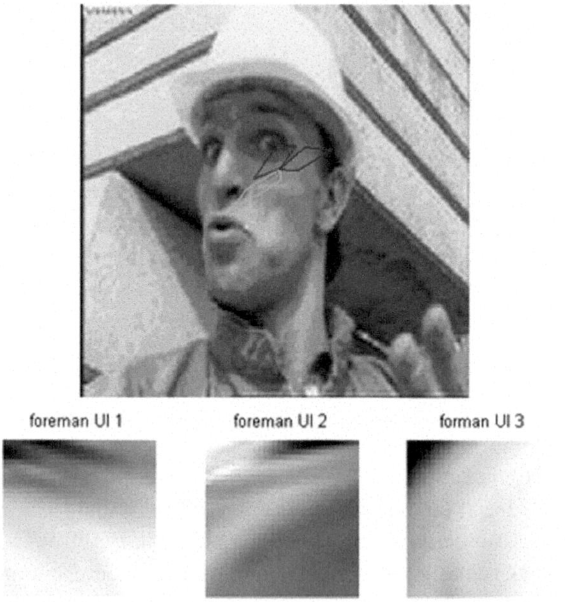


**Fig. 9** Watermarking zones and universal images of a single frame from Forman.avi

each triangle is made from feature points of a particular frame our ability to locate the triangle in that frame is excellent (Fig. 11).

The following transformation is constructed

$$\Phi_\alpha(S, \Delta) = (\varphi_\alpha(S_1, \Delta_1), ..., \varphi_\alpha(S_N, \Delta_N))$$

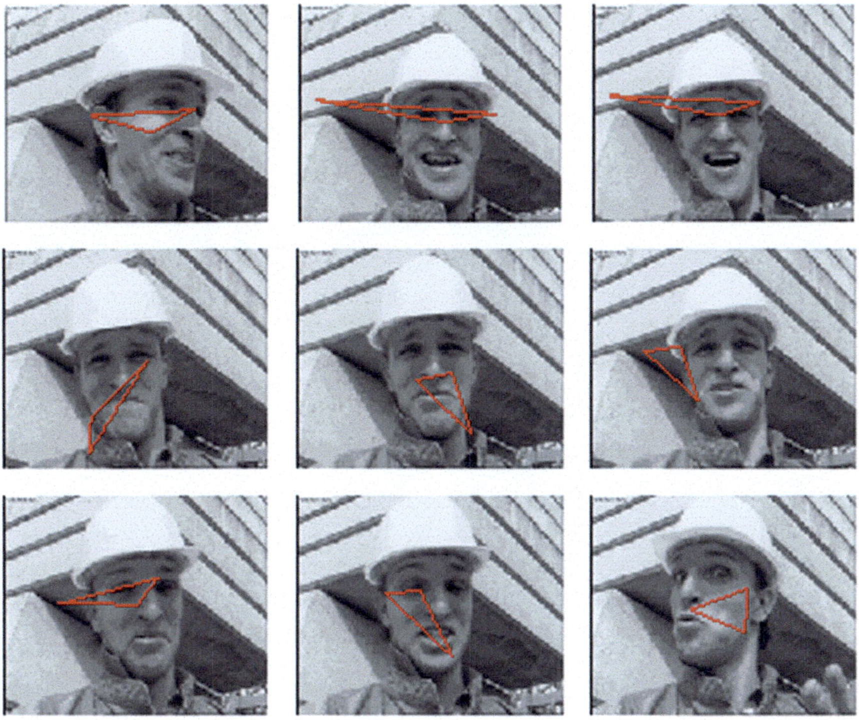

**Fig. 10** An automated triangle allocation for a set of frames from 'Foreman.avi'

This transformation maps a video scene $S$ onto three rectangular prisms. These are called *universal video watermarking prisms.* These prisms are then assembled into videos. Thus, each video is mapped to a set of three small videos. These *universal videos* are what watermarked. Each frame in these *universal videos* is RST invariant. They are a direct generalization of RST invariant universal images. The mapping from $S$ to a triple of watermarking prisms is invertible as each of the maps $\varphi_a$ is invertible.

- Embed watermarks into these universal *videos* in a way which is robust against imperfect triangle location. The *wrapped transform* of the fast discrete curvelet transform (FDCT) of [12] al is used.
- The watermark is embedded into the low-frequency component of the curvelet transforms of each of the three universal videos and map them back to the original image, adding the low-frequency component of each together in a Cox type way, and reconstructing the image. The same watermark is placed in each universal videos.
- To recover the watermark is to project out the low-frequency component of a universal *videos* and reverse the Cox steps. The position of the watermark is

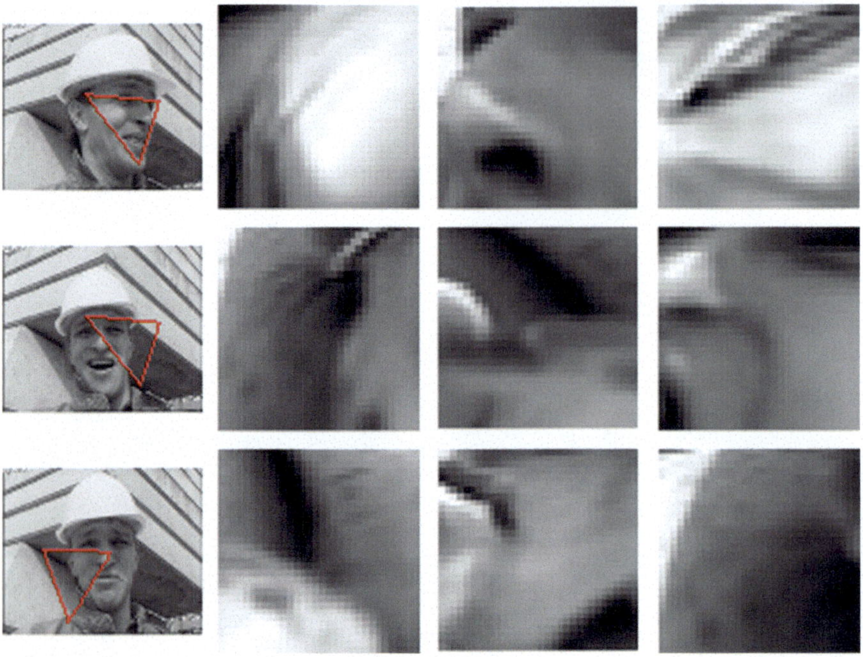

**Fig. 11** Universal images of three frames from 'Foreman.avi'

approximately constant when there is little change in the video frames. Fig. 12 shows some watermarked frames.

- Each watermarking region can be given a separate watermark, and each frame can be given a different set of watermarked if required. Each and every watermark can be extracted by means of the standard similarity technique [13].

## 8 Conclusion

This paper presents a new watermarking model based on shape theory and curvelet transform for rendering video watermarking scheme resistant to geometric attacks, such as rotation, scaling, and translation, RST. Indeed, by utilizing the shape theory and the curvelet transform, we have introduced an RST invariant map from a video scene to a set of universal video watermarking prisms. In this invariant domain, we have embedded a watermark into the low-frequency component of the curvelet transforms of each of the three universal videos and assured RST invariance. This method has some limitations for addressing some other geometric attacks, such as severe cropping. This issue together with other image processing could be addressed in the future work.

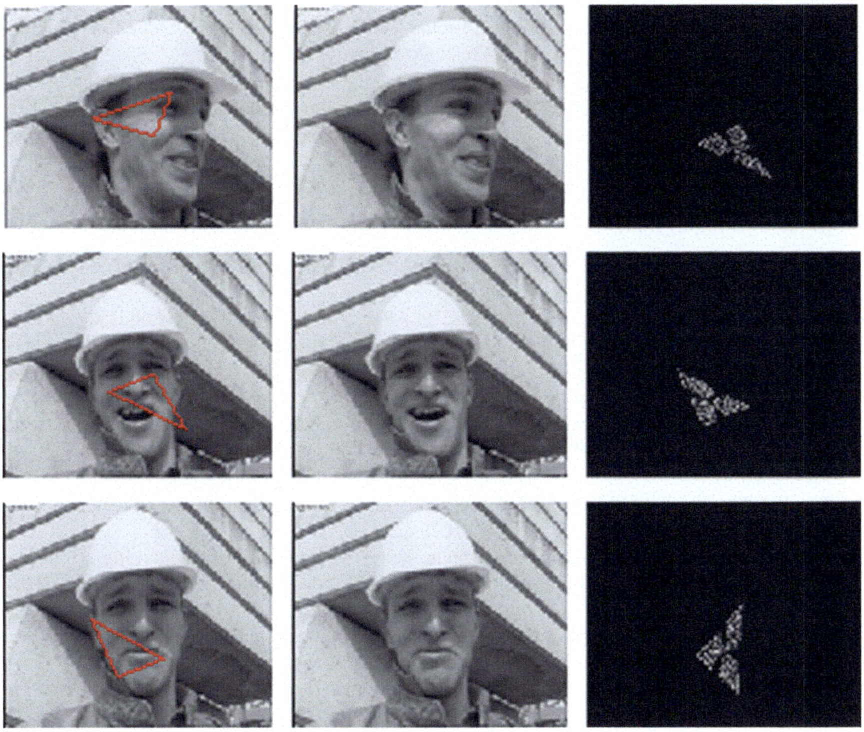

**Fig. 12** The frame, the watermarked frame, and the watermark

# References


1. Morris, H., Muhi El-Ddin, I.: A new image transform. In: Proceedings of IPV08 (2008)
2. El-Ddin, M.: Watermarking schemes robust against affine attacks. Ph. D. Thesis, Claremont Graduate University, Los Angeles, California (2008)
3. Muhi El-Ddin, I., Morris, H., Eyadat, M.: Watermarking: a new approach. In: Proceedings of the Fifth International Conference on Information Technology: New Generations (ITNG 2008), pp. 795–800 (2008)
4. Morris, H., Muhi El-Ddin, I.: Feature frame watermarking. In: Proceedings of the 41st Annual Asilomar Conference on Signals, Systems, and Computers, pp. 565–569 (2007)
5. Candès, E.J., Donoho, D.L.: Continuous curvelet transform: I. Resolution of the wave front set. Appl. Comput. Harmon. Anal. **19**(2), 162–197 (2005). Candès, E.J., Donoho, D.L.: Continuous curvelet transform: II. Discretization and frames, Applied and Computational Harmonic Analysis, vol. 19, Issue 2, pp 198–222 (2005)
6. Candès, E., Demanet, L.: Communications on pure and applied mathematics. In: The Curvelet Representation of Wave Propagators is Optimally Sparse, vol. 58, Issue 11, 1472–1528 (2005)
7. www.curvelet.org.
8. El'arbi, M., Ben Amar, C., Nicolas, H.: Video watermarking scheme resistant to geometric transformations. In: ICIP 2007. IEEE International Conference on Image Processing, vol. 5, pp. V-481–V-484, 16 Sept 2007–19 Oct 2007
9. Lowe, D.G.: Distinctive image features from scale-invariant keypoints. Int. J. Comput. Vis. **60**(2), 91–110 (2004)

10. Sparr, G.: Proceedings of the 13th International Conference on Pattern Recognition (Vienna), pp. 328–333. IEEE Computer Society Press, Washington, DC (1996)
11. Zou, F., Lu, Z., Ling, H.: A multiple watermarking algorithm based on CDMA technique. In: Proceedings of the 12th Annual ACM International Conference on Multimedia, pp. 424–427. ACM (2004)
12. Lee, C.-H., Lee, H.-K., Guseong-Dong, Yuseong-Gu, D.: Geometric attack resistant watermarking in wavelet transform domain. Opt. Exp. **13**(4), 1321 (2005)
13. Cox, I., Miller, M., Bloom, J.: Digital Watermarking. Morgan Kaufmann Publishers (2002). ISBN: 1-55860-714-5
14. Ettinger, S.: SIFT MATLAB Implementation. Intel (2002). https://robots.stanford.edu/cs223b/MatlabSIFT.zip
15. Doulaverakis, C., Vagionitis, S., Zervakis, M., Petrakis, E.: Adaptive methods for motion characterization and segmentation of MPEG compressed frame sequence. In: 1st International Conference on Image Analysis and Recognition (ICIAR'2004), Porto, Portugal, Sept./Oct. 2004, Proceedings. Part I, Springer Verlag (LNCS 3211), pp. 310–317
16. Wang, B., Wang, Y., Selesnick, I., Vetro, A.: Video coding using 3D dual-tree wavelet transform. EURASIP J. Image Video Process. **2007**, Article ID 42761, 15 p (2007). https://doi.org/10.1155/2007/42761
17. Bas, P., Chassery, J.-M., Macq, B.: Geometrically invariant watermarking using feature points. IEEE Trans. Image Process. **11**(9), 1014–1029 (2002)
18. Kankanhalli, M.S., Rajmohan, Ramakrishnan, K.R.: Content-based watermarking of images. In: Proceedings of 6th ACM International. Multimedia Conference, Bristol, U.K., Sept 1998, pp. 61–70
19. Kothari, A.M., Dwivedi, V., Thanki, R.M.: Watermarking techniques for copyright protection of videos. In: Proceedings of SoCTA '19. Book. ISBN: 978-3-319-92836-4
20. Celebi, E., Chen, J., Gopi, E.S., Neustein, A., Poor, H.V.: Signals and Communication Technology. In: Proceedings of SoCTA '19 (2019) ISSN: 1860-4862
21. Yadav, B., Kumar, A., Kumar, Y.: A robust digital image watermarking algorithm using DWT and SVD. In: Soft Computing: Theories and Applications. Springer (2018)
22. Cox, I.J., Doërr, G., Furon, T.: Watermarking is not cryptography. In: Shi, Y.Q., Jeon, B. (eds.) Digital Watermarking. IWDW 2006. Lecture Notes in Computer Science, vol. 4283. Springer, Berlin. https://doi.org/10.1007/11922841_1

# An Implementation of Single and Multiple Thresholds for Efficient Image Segmentation

**K. Deeparani and P. Sudhakar**

**Abstract** The image segmentation is the process of finding and grouping the correlated pixels in the particular image. There are different types that are available to find the correlated pixels in the image. In this paper, the single and multiple thresholding are used for segmentation to observe the different image objects in an image. At first, the input samples are converted into grayscale images, after that those images are reshaped and processed by single and multiple thresholding to produce the segmented image output. The single and multiple thresholding are based on calculating the mean value of an image and fix the threshold value for object discriminations. Out of several algorithm for image segmentation, the conventional method of region-based segmentation will provide good results. The objects discrimination is purely based on the level set of pixels available in the image. Before processing, the image reshaping is also performed for better visualization of the segmented image.

**Keywords** Image processing · Image segmentation · Thresholding · Multiple thresholds · Reshaping · Multiple object discrimination

## 1 Introduction

In the field of image processing [1], there is a similarity between the object tracking [2], object detection, and object discrimination [3]. The image segmentation concept is mostly related with the concept of object discrimination in an image [4, 5]. The image detail will be preserved in every concept of image processing [6, 7], which is also applicable for the segmentation [8, 9]. The thresholding is based on the observation of the mean value of the pixels in the particular frame [10, 11] and deciding the reference value [12]. The mean value and the reference value [13], which is also to be decided by the algorithm [14], used to observe the difference

K. Deeparani (✉)
Department of Computer Science, Bharathiar University, Coimbatore, Tamil Nadu 641046, India

P. Sudhakar
Department of Computer Science, Annamalai University, Chidambaram, Tamil Nadu, India



T. K. Sharma et al. (eds.), *Soft Computing: Theories and Applications*,
Advances in Intelligent Systems and Computing 1381,
https://doi.org/10.1007/978-981-16-1696-9_36

between the information pixels [15] and the non-information pixels for different object detection [16].

## 2 Related Work

Zhang et al. [3] implement the edge-based and region-based segmentation with different level sets with high synergy data points. The technique is derived from the conventional method, but the modified version of the conventional modal gives better results for image segmentation. Zhu et al. [4] concentrate on pixel resolution improvement using the super resolution techniques. Based on that concept, the super pixel is used in all dimensions to get the different object discrimination. Li et al. [1] used the machine learning for image segmentation with single-level set in an image. The supervised learning technique is used for the image segmentation with semantic data points. Huang et al. [6] find the discrimination in the video frames with the popularly established neural elements. The algorithm evaluates the similarities in the frames as well as the similarities between the frames and provides harmonic values to the neural networks for image segmentation. Jing et al. [2] also deal with the level sets of the images for the semantic segmentation. The algorithm tunes the processing area from coarse to fine for adjusting and keeping the exact values for processing, and it labels the threshold levels. Zhou et al. [16] used the unsupervised learning technique for the video frame processing which understands the changing property of the video frames.

## 3 Proposed Methodology

In the beginning, the input image will be converted into the grayscale image [17], and the image reshaping [18] is done for the betterment of further processing like thresholding [19]. The value of the threshold is 0.5 to discriminate the foreground information [7, 20] and the background information [21, 22]. If the value of the particular pixel is greater than 0.5, then it is considered as the foreground image information [23, 24], and if the value of the particular pixel is less than 0.5, then it is considered as the background image information in the single value thresholding which is represented in Fig. 1a.

If there is a need to find different objects in a single frame, then the multiple thresholding concept is used [25]. The mean value of the image is calculated, and the threshold value is fixed as less than 0.25, between 0.25 and 0.5, between 0.5 and 0.75, and between 0.75 and 1.0; based on this value, there are four different objects [26] can be distinguished in a single frame. The sequence diagram of this concept is also shown in Fig. 1b.

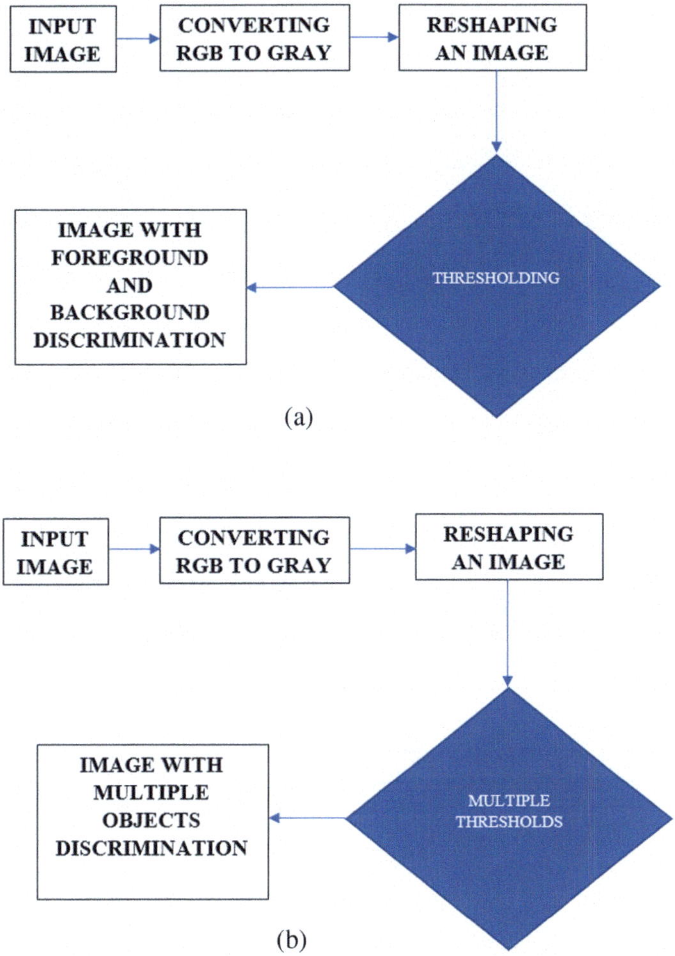


**Fig. 1 a, b** Flow sequence of proposed implementation

## 4 Results and Discussions

See Figs. 2, 3, 4, and 5.

## 5 Conclusion

There is a clear discrimination between foreground and background of an image while implementing the thresholding technique. If the single value threshold is used, then there are only two different objects will be available in the image. The multiple

Fig. 2 **a–f** Sample input images

thresholds are also implemented in this article to find out different objects in an image. The efficiency of the multiple threshold implementation is observed in this implementation by trying out with six different sample images which are shown in Fig. 2. In that, the foreground and background discrimination are clearly distinguishable based on the grouping of data points. The discrimination in the output is based on the illumination in an image which is clearly observed at the output images in

**Fig. 3** **a–f** Images after RGB to gray conversion

comparison with input images which are shown in Fig. 4. The image reshaping is also done before thresholding for the betterment of processing.

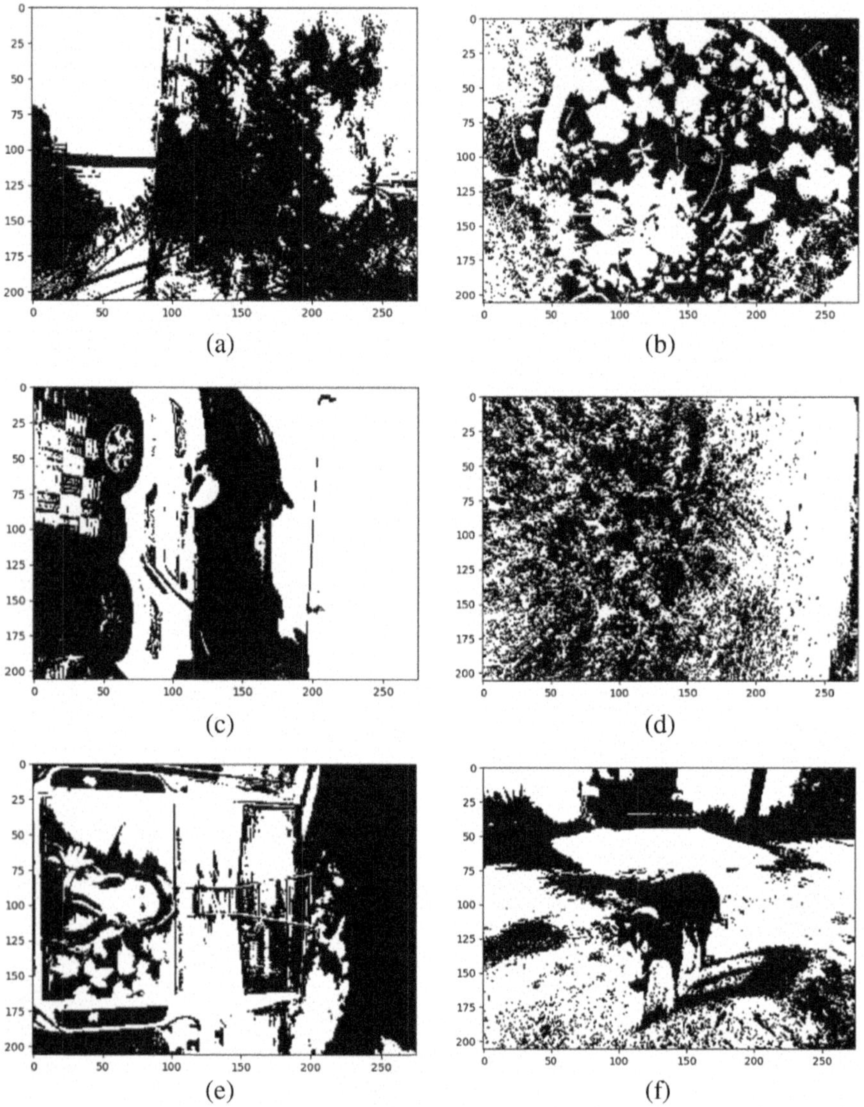

Fig. 4 a–f Images after single value threshold

**Fig. 5** **a–f** Images after multiple thresholds

# References


1. Li, X., Ma, H., Luo, X.: Weaklier supervised semantic segmentation with only one image level annotation per category. IEEE Trans. Image Process. **29**, 128–141 (2020). https://doi.org/10.1109/TIP.2019.2930874
2. Jing, L., Chen, Y., Tian, Y.: Coarse-to-fine semantic segmentation from image-level labels. IEEE Trans. Image Process. **29**, 225–236 (2020). https://doi.org/10.1109/TIP.2019.2926748

3. Zhang, W., Wang, X., You, W., Chen, J., Dai, P., Zhang, P.: RESLS: region and edge synergetic level set framework for image segmentation. IEEE Trans. Image Process. **29**, 57–71 (2020). https://doi.org/10.1109/TIP.2019.2928134
4. Zhu, H., Zhang, Q., Wang, Q., Li, H.: 4D light field superpixel and segmentation. IEEE Trans. Image Process. **29**, 85–99 (2020). https://doi.org/10.1109/TIP.2019.2927330
5. Yadav, B., Kumar, A., Kumar, Y.: A robust digital image watermarking algorithm using DWT and SVD. Soft Comput. Theories Appl. **583**(1), 25–36 (2018)
6. Huang, H., Xu, S., Cai, J., Liu, W., Hu, S.: Temporally coherent video harmonization using adversarial networks. IEEE Trans. Image Process. **29**, 214–224 (2020). https://doi.org/10.1109/TIP.2019.2925550
7. Mohan, E., Sugumar, R., Venkatachalam, K.: Automatic brain and tumor segmentation in MRI using fuzzy classification with integrated Bayesian. Int. J. Appl. Eng. Res. **9**(24), 25859–25870 (2014). ISSN: 25859–25870
8. Zhuo, T., Cheng, Z., Zhang, P., Wong, Y., Kankanhalli, M.: Unsupervised online video object segmentation with motion property understanding. IEEE Trans. Image Process. **29**, 237–249 (2020). https://doi.org/10.1109/TIP.2019.2930152
9. Saurin Sheth, C.K., Ajmera, A., Sharma, A., Patel, S.: Design and development of intelligent AGV using computer vision and artificial intelligence. Soft Comput. Theor. Appl. **583**(1), 337–349
10. Sindagi, V.A., Patel, V.M.: HA-CCN: hierarchical attention-based crowd counting network. IEEE Trans. Image Process. **29**, 323–335 (2020). https://doi.org/10.1109/TIP.2019.2928634
11. Somwanshi, D., Asthana, M., Agarwal, A.: An OTSU-discrete wavelet transform based secure watermark scheme. Soft Comput. Theories Appl. **583**(1), 273–283 (2018)
12. Ngo, L., Cha, J., Han, J.: Deep neural network regression for automated retinal layer segmentation in optical coherence tomography images. IEEE Trans. Image Process. **29**, 303–312 (2020). https://doi.org/10.1109/TIP.2019.2931461
13. Yu, J., et al.: Exemplar-based recursive instance segmentation with application to plant image analysis. IEEE Trans. Image Process. **29**, 389–404 (2020). https://doi.org/10.1109/TIP.2019.2923571
14. Abu, A., Diamant, R.: Enhanced fuzzy-based local information algorithm for sonar image segmentation. IEEE Trans. Image Process. **29**, 445–460 (2020). https://doi.org/10.1109/TIP.2019.2930148
15. Treece, G.: Morphology-based noise reduction: structural variation and thresholding in the bitonic filter. IEEE Trans. Image Process. **29**, 336–350 (2020). https://doi.org/10.1109/TIP.2019.2932572
16. Zhou, S., Nie, D., Adeli, E., Yin, J., Lian, J., Shen, D.: High-resolution encoder–decoder networks for low-contrast medical image segmentation. IEEE Trans. Image Process. **29**, 461–475 (2020). https://doi.org/10.1109/TIP.2019.2919937
17. Rizkallah, M., Su, X., Maugey, T., Guillemot, C.: Geometry-aware graph transforms for light field compact representation. IEEE Trans. Image Process. **29**, 602–616 (2020). https://doi.org/10.1109/TIP.2019.2928873
18. Dinh, T.H., Phung, M.D., Ha, Q.P.: Summit navigator: a novel approach for local maxima extraction. IEEE Trans. Image Process. **29**, 551–564 (2020). https://doi.org/10.1109/TIP.2019.2932501
19. Jiang, X., Ma, J., Jiang, J., Guo, X.: Robust feature matching using spatial clustering with heavy outliers. IEEE Trans. Image Process. **29**, 736–746 (2020). https://doi.org/10.1109/TIP.2019.2934572
20. Liao, T., Li, N.: Single-perspective warps in natural image stitching. IEEE Trans. Image Process. **29**, 724–735 (2020). https://doi.org/10.1109/TIP.2019.2934344
21. Rong, X., Yi, C., Tian, Y.: Unambiguous scene text segmentation with referring expression comprehension. IEEE Trans. Image Process. **29**, 591–601 (2020). https://doi.org/10.1109/TIP.2019.2930176
22. Pant, M., Ray, K., Sharma, T.K., Rawat, S., Bandyopadhyay, A.: Soft Computing: Theories and Applications, vol. 583, XXII, p. 865. Springer, Singapore

23. Touati, R., Mignotte, M., Dahmane, M.: Multimodal change detection in remote sensing images using an unsupervised pixel pairwise-based Markov random field model. IEEE Trans. Image Process. **29**, 757–767 (2020). https://doi.org/10.1109/TIP.2019.2933747
24. Pant, M., Ray, K., Sharma, T.K., Rawat, S., Bandyopadhyay, A.: Soft Computing: Theories and Applications, vol. 584, XXII, p. 783. Springer, Singapore
25. Lu, C., Xia, S., Shao, M., Fu, Y.: Arc-support line segments revisited: an efficient high-quality ellipse detection. IEEE Trans. Image Process. **29**, 768–781 (2020). https://doi.org/10.1109/TIP.2019.2934352
26. Guo, D., Pei, Y., Zheng, K., Yu, H., Lu, Y., Wang, S.: Degraded image semantic segmentation with dense-gram networks. IEEE Trans. Image Process. **29**, 782–795 (2020). https://doi.org/10.1109/TIP.2019.2936111

# Anisotrophic Filter-Based Detection of Brain Tumor

**N. Rupavathy and M. J. Carmel Mary Belinda**

**Abstract** Brain tumor is a destructive sickness which cannot be clearly recognized without MRI. In this undertaking, we stressed to see whether patient's brain has tumor or not from MRI picture utilizing MATLAB recreation. Tumor is a pre-phase of disease which has built up a significant issue in this period. Analysts are irritated to create strategies and medicines to adjust it. Mind tumor is an amazing cell improvement in the brain tissue and may not consistently be found in imaging stunts. Attractive resonance imaging (MRI) is a method which is reasonable to show the point by point picture of the scrutinized brain area. To clear the methods for morphological procedure on MRI picture, the picture was first sifted utilizing anisotropic diffusion filter to diminish contrast between continuous pixels. After that the picture was resized, and using a limit, worth picture was changed to a high contrast picture naturally. This essential channel is the conceivable spot for tumor nearness. A MRI anomalous mind pictures as exertion in the introduced strategy, anisotropic filtering focused on commotion evacuation, SVM classifier for division, and morphological tasks for removal of the overstated region from ordinary exceptional are the key stages of the introduced technique. Achieving clear MRI pictures of the mind and the tumor are the offensive of this strategy. The arrangement of the powers of the pixels on the sifted picture groups the tumor. Trial result introduced that the SVM has acquired 83 percent rightness in division. At long last, the portioned district of the tumor is put on the one of a kind picture for an unmistakable recognizable proof.

**Keywords** Anisotropic filtering · Morphological operations

## 1 Introduction

Magnetic resonance imaging (MRI) is viewed as now as a noteworthy apparatus for specialists. It conveys high greatness pictures within the human body. One should be cautious when managing fragile organs like the mind. A brain tumor

N. Rupavathy (✉) · M. J. Carmel Mary Belinda
Department of Computer Science and Engineering, Vel Tech Rangarajan Dr. Sagunthala R&D Institute of Science and Technology, Chennai, Tamil Nadu 600062, India

 
T. K. Sharma et al. (eds.), *Soft Computing: Theories and Applications*, Advances in Intelligent Systems and Computing 1381,
https://doi.org/10.1007/978-981-16-1696-9_37

is somewhat intracranial mass molded by unusual and uncontrolled cell division. Tumors can complete synapses or harm them by implication by starting aggravation, compacting different pieces of the brain, instigating cerebral edema, or by applying inner weight as they produce. Mind tumors are characterized into primary brain tumor and secondary (metastatic) brain tumor. Essential mind tumors, regardless of whether favorable (noncancerous) or dangerous (malignant), might be confined or broadened, while minor tumors could be in various areas. Tumor is generally characterized as the irregular development of the tissues. Mind tumor is an irregular type of tissue where cells develop and duplicate wildly; in brain, this system is unrestricted by the control typical cells. Brain malignant growth can be totaled among the most fatal and difficult ailments. Mind tumors can be essential or metastatic and either dangerous or considerate. A metastatic brain tumor is a disease that has feast from somewhere else in the structure to the mind. Clinical imaging modalities as in MRI, CT dissect generally rely on modernized improvement to make or show advanced photographs of within organs of the human body which benefits the specialists to envision the internal parts of the body. CT scanner, ultrasound, and magnetic resonance imaging seized over conventional x-bar imaging, by allowing the specialists to see the body's third assessment.

## 2 Related Work

Build up a gamma photographic camera reliant on a multi-wire relative chamber outfitted with a high rate, ordinal electronic read-out system for imaging applications in nuclear prescription. Benedicte et al. [1] report portrays early use of a hoarding sound information base right now including 50 subjects developed 20–72. Bricq [2] presents a uniting format for solo division of multimodal mind MR pictures with partial volume sway, inclination field cure, and information expected by a probabilistic guide book. Chan et al. [3] present a two-advance methodology, which coops locale and structure distortion, to find the restriction of an article from a picked early breaking point plan. Chunyan et al. [1] present deformable model-based system which is changed in the structure. Moreover, by the practical UI, the division can be mediated by head brilliantly at progressing. Corina et al. [4] base on the robotized extraction of the cerebrospinal fluid tissue limit, particularly around the ventricular surface, from progressive fundamental MRI of the brain picked up in imaging examinations of developing and dementia.

Various methodologies have been portrayed for gathering of brain tumors in MRI pictures, most particularly, support vector machine (SVM) [1], artificial neural network (ANN) [5], data-based methods [3], expectation–maximization (EM) computations, and fuzzy C-means (FCM) bundling. Gering and accomplices [6] applied the EM computations in the affirmation of oddities. This requires high computational exertion. The data-based procedures permitted to make progressively successful results for the division and request errands anyway these systems requires concentrated getting ready. In prosperity picture examination, the confirmation of tissue type (normal or psychotic) and tissue pathology plan is done by using surface. MR picture surface is exhibited to be significant for choosing the sort of the tumor [5] and to recognize Alzheimer's affliction [7]. Dana et al. [3] propose a variational mind tumor division count that grows present systems from surface division by using a high-dimensional rundown of capacities anticipated from MRI data and enrolled map books. Dimitri's et al. [6] present different blend deformable systems; we have been ascending for division and enlistment. These methodologies fuse metamorphs, a novel shape and feel blend deformable model structure, and the blend of deformable copies with graphical models and learning strategies. Elizabeth et al. [4] report to distinguish and count convolution variations from the norm on high goals MRI pictures offers a novel way to manage the noninvasive assurance of damage. Erik et al. arrange programmed division subject to controlled learning with a passing on multi-scale watershed division strategy.

Different strategies have been the future to manage this issue. A fractional outline incorporates surface model, deformable and dynamic contour model, and iterative making model. Gering et al. [7] who used another, exhaustively significant division device that depends upon intuitive deformable models executed as level sets. The instinctual rates for illuminating the level-set PDE give the client brief reaction on the breaking point settings, and thusly, clients can tune three separate cutoff points and switch the state of the ideal unendingly. Ahmed et al. [5] show genuinely and quantitatively the physiologically based assessment routs two standard division techniques.

## 3 Proposed Work

The mind is the foremost noteworthy aspect of the focal sensory framework. Mind tumor is an intracranial solid neoplasm. Tumors are shaped by an odd and uncontrolled cell division in the mind. Brain tumor is a destructive sickness which cannot be clearly recognized without MRI. To notice if there is a tumor in a given MRI Image. On this semi prepared picture, morphological tasks have been down to earth, and information on strength and territories of the imaginable areas was gotten. The division, location, and extraction of contaminated tumor, region from attractive resonance (MR) pictures territory fundamental concern yet a monotonous and time taking errand did by radiologists or clinical specialists, and their precision be dependent upon their experience as it were. Data is passed on completed pictures. Picture handling is where information picture is prepared to become yield additionally as a picture. To make visually, under consideration, object or image to know those processing techniques is the main objective in easier manner. The area of the tumor in the brain, likewise, quite modifies the danger of neurological poison levels that change the patient's greatness of life. At present, mind tumors are recognized by imaging just a while later the beginning of neurological manifestations. No early discovery systems are being used, even in people known to be at peril for explicit sorts of brain tumors by ethicalness of their hereditary cosmetics. Current histopathological grouping frameworks, which depend on the tumor's alleged cell of inception, have been set up for almost a century and were effective by the World Health Organization in 1999. Albeit palatable in numerous regards, they do not permit exact expectation of tumor conduct in the individual patient, nor do they control helpful dynamic as accurately as patients and doctors would certainly need. Different methodologies have been completed in the field of mind tumor identification. In existing work, screen tumor volume on T1-GAD follows change in tumor volume after some time. With productive treatment, tumor should get smaller or stopped creating. T1-GAD pictures are presumably going to be gotten at all associations. The improving edge of a tumor gives something reasonably objective to trail some time. It is a reasonable doubt that a working tumor will create. In this proposed system, here segment presents the materials, the wellspring of mind MR picture dataset, and the calculation used to perform brain MR tissue division. It is the initial phase in our proposed method. The reason for these methods is essentially preparing incorporates emptying low-repeat establishment uproar, normalizing the power of the individual particles pictures, removing reflections, and veiling portions of pictures.

# 4 Module Description

## *4.1 Anisotropic Filter*

Anisotropic filter is used to oust the establishment racket and thus in this manner shielding the edge focuses in the picture. In anisotropic channel, dissemination is consistently identified with the clamor angle, and smoothing the foundation commotion by sifting a fitting limit esteem is chosen. For this reason higher dispersion steady worth is picked contrast and the total estimation of the commotion slope in its edge. Head cover was built by thresholding the sifted picture. Coordinating power extends in all the pictures, the most noteworthy, and least powers are restricted to the span [0, 255].

## *4.2 Morphological Operations*

Morphological operations are easy to utilize and take a shot at the premise of set hypothesis. The vast majority of the tasks utilized here are blended of two procedures, enlargement and disintegration. The activity utilizes a little lattice structure called as organizing component. The shape and size of the organizing component have noteworthy effect on the conclusive outcome.

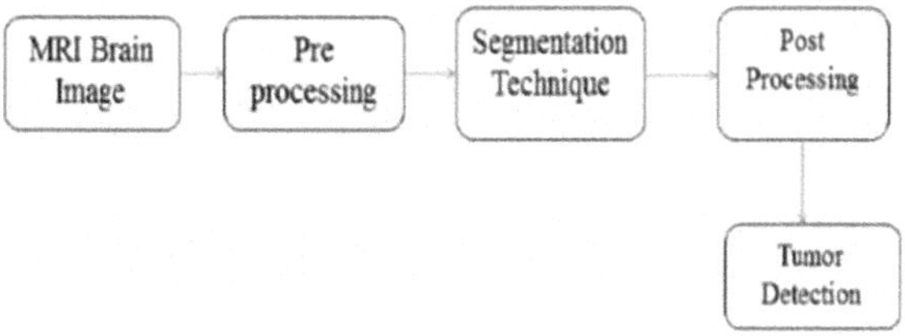


**Fig. 1** Architecture diagram

## 4.3 Segmentation

In PC vision, picture division is the route toward dividing progressed picture into different sections (sets of pixels, regardless called super-pixels). Picture division is normally used to find things and cutoff focuses (lines, turns, and so forth) in pictures. Significantly, more absolutely, picture division is the course toward giving out an engraving to each pixel in a picture, so that pixels with a similar name share certain characteristics. The delayed consequence of picture division is a great deal of areas that all.

## 4.4 General Architecture

In Fig. 1, it shows that the fundamental objective is to eliminate commotion from the pictures. In any case, the commotion cancelation should not obliterate the edges of the picture and diminishing the lucidity and nature of it. There are a few strategies for eliminating clamor, including Gaussian channel, contourlet change approach and wavelet thresholding approach, middle channel, and anisotropic dispersion channel. So, in our proposed strategy, the mind tumor fragments the loud MRI pictures utilizing anisotropic dispersion. Anisotropic dissemination channel is a technique for eliminating commotion which is proposed by Perona and Malik. This strategy is for smoothing the picture by safeguarding required edges and structures. Homogenous areas are profoundly smoothed, and solid edge locales are scarcely smoothed (to safeguard the structure).

## 4.5 Design Phase

### 4.5.1 Data Flow Diagram

In Fig. 2, it shows that the strategy incorporates non-straight anisotropic diffusion filter and morphological administrators to section the tumor. In picture preparing and PC vision, anisotropic scattering, furthermore called Perona–Malik scattering, is a strategy focusing on diminishing picture clatter without taking out basic bits

**Fig. 2** Data flow diagram

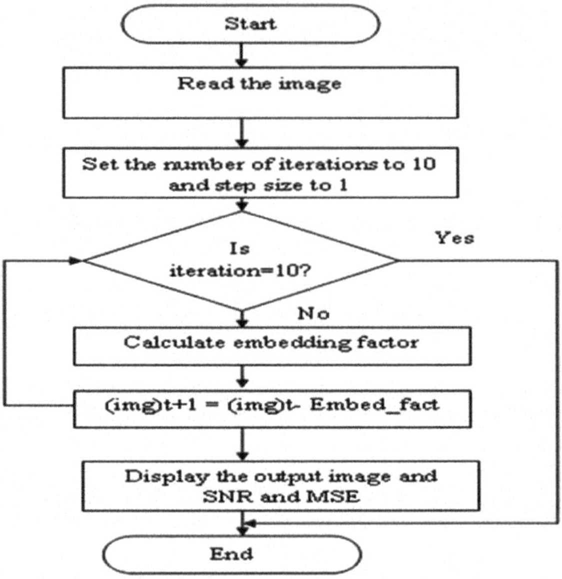


of the image content, normally edges, lines or various nuances that are critical for the comprehension of the image. Anisotropic scattering resembles the cycle that makes a scale space where an image makes a characterized gathering of dynamically and consistently expanding number of darkened pictures reliant on spread measure. This scattering cycle is a direct and space-invariant difference in the main picture. Anisotropic scattering is a hypothesis of this dispersal measure: It conveys a gathering of characterized pictures; anyway, each ensuing picture is a blend between the principal picture and a channel that depends upon the local substance of the primary picture. As a result, anisotropic scattering is a non-straight and space-variety change of the original image.

## 4.6 *Module Description*

### 4.6.1 Image Taken from MRI Scan

In the underlying stage, a picture of X-ray output of brain is taken. The article is characterized in the various pixels. Library capacities for morphological activity were utilized to segment the two picture and notice the tumor zone. The binary picture was disintegrated; i.e., all the sentiments in the area of a dark point were turned dark. At long last, we get the outskirt by taking away the disintegrated picture from unique paired picture. Anisotropic Diffusion Filter Shape-adjusted smoothing. It crops striking picture and a channel that depends upon the local substance of the unique picture. Accordingly, anisotropic spread is a non-direct and space-variety change of the main picture. Anisotropic filter obscures the pictures without evacuating any edge.

#### 4.6.2 Preprocessing

The fundamental endeavor of preprocessing is to improve the idea of the MR pictures and make it in a structure suitable for extra dealing with by human or machine vision system. Also, pre-handling helps with improving certain limits of MR pictures, for instance, improving the sign to-clatter extent, redesigning the visual appearance of MR picture, clearing the irrelevant uproar and undesired parts far out, smoothing the internal bit of the zone, and defending its edges. To improve the sign to clamor proportion and, therefore, the lucidity of the crude MR pictures, we applied versatile difference upgrade dependent on altered sigmoid capacity.

#### 4.6.3 Segmentation Technique

The division of the debased brain MR locale is refined through the going with ventures: In the fundamental development, the pre-taken care of mind MR picture is changed over into a consolidated picture with a limit for the cutoff of 128 being picked. The pixel respects more significant than the picked edge are wanted white, while others are isolated as dull; by virtue of this two, exceptional districts are acclimated to the ruined tumor tissues, which is overseen out. In the subsequent development, to slaughter white pixel, a separating activity of morphology is utilized.

#### 4.6.4 Post-processing

An edge happens when there is an abrupt and surprising force change of the picture. At whatever point, it is recognized a sudden alteration or a channel in the force of a specific picture, and the related pixel would be treated as an edge pixel. The calculation that has been advanced for the recognition of edge pixel underpins in recognizing the nature of the edge.

## 5 Implementation and Testing

### *5.1 Input*

This project is loaded in MATLAB; when we run it, a front end user interface is displayed which displays an option to select and upload the desired image and has an option to detect the tumor (Fig. 3).

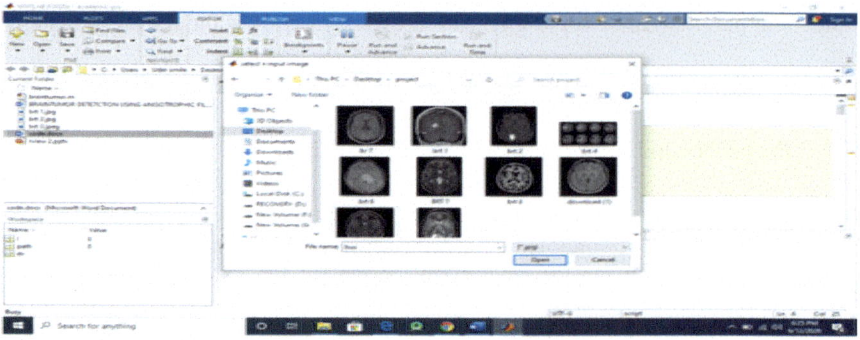

Fig. 3 Datasets

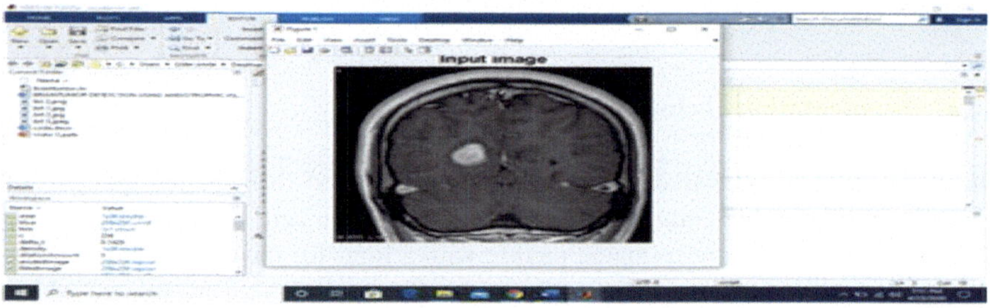


Fig. 4 Output of an input image

## 5.2 *Output*

The user interface consists of multiple options such as select the image, preprocessing, segmentation technique, and post-processing which are used to perform its morphological operations, and the screen not only shows output, but also input image and processed image (Figs. 4, 5, 6, 7, 8 and 9).

# 6 Results and Discussions

## 6.1 *Efficiency of the Proposed System*

The proposed system is efficient as it indicates a tumor region using anisotropic filtering, and the erode image is obtained through morphological operations.

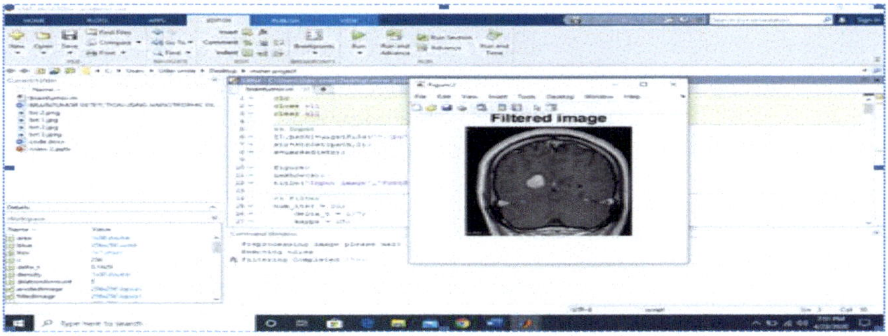


**Fig. 5** Output of filtered image

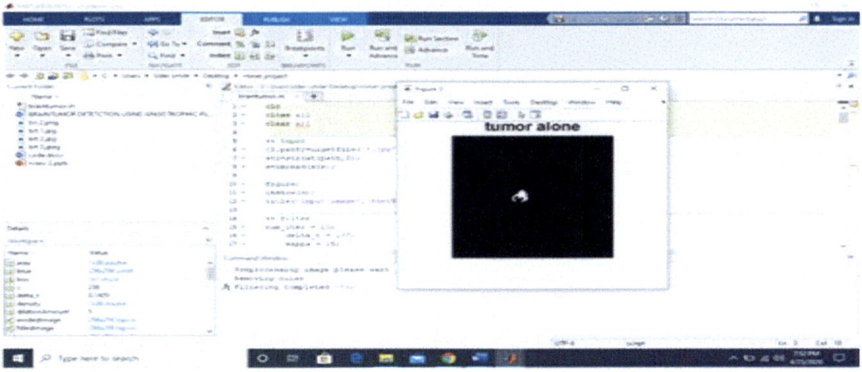


**Fig. 6** Output of tumor alone

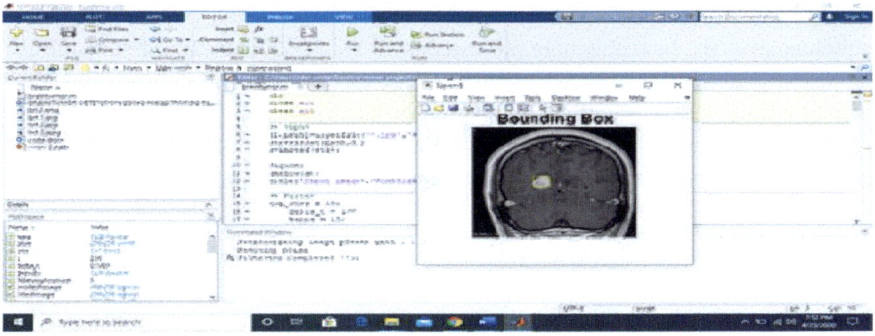


**Fig. 7** Output of bonding box

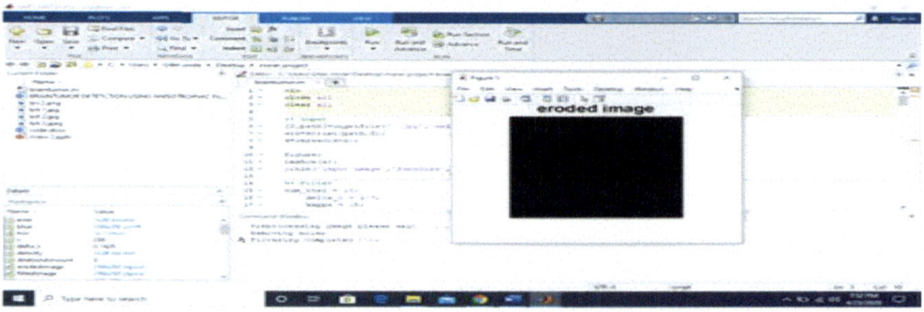


Fig. 8 Output of eroded image

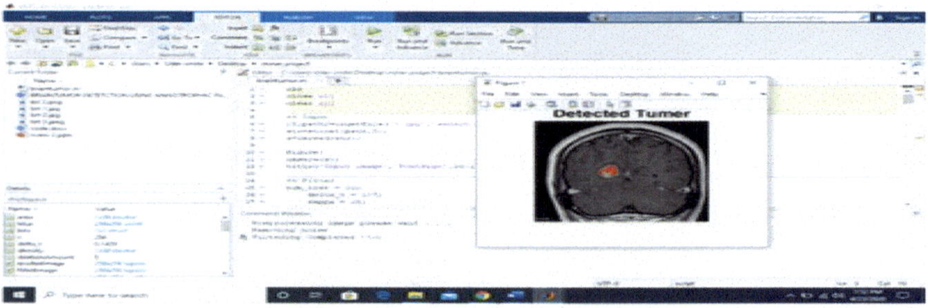


Fig. 9 Output of detected tumor

### 6.2 Results of the Proposed System

By applying different tests to the dull picture, we at last settle that anisotropic dispersion channel is the best channel contrasted with wiener and middle. In anisotropic dispersion channel, the extent of commotion expelled was extra; subsequently, it gives best outcome. At long last, the region of the tumor is determined. The norms are recorded beneath. The outcomes got are organized in underneath Table, which shows that the future technique is increasingly proficient contrasted with different channels.

## 7 Conclusion and Future Enhancements

### 7.1 Conclusion

It is seen that the proposed strategies have demonstrated higher vigor in separation of areas as a result of the low sign/clamor proportion portraying a large portion of clinical pictures data. The preprocessing stages are initially done through following calculations. The next procedure brain MRI is portioned utilizing anisotropic sifting, morphological operation, and segmentation. Clinical picture preparing is the most testing and rising field nowadays. Handling of MRI pictures is one of the piece of this field. This code makes MRI picture investigation open to more individuals who need not need to be expert on MRI imaging.

X-ray imaging is less destructive than X-beam. It is less lessened by bones. This undertaking can make MRI picture handling and tumor identification process quicker and less expensive. It very well may be noticed that anisotropic dissemination channel outflanks other sifting procedure in denoising in clinical pictures. Further, denoising execution can be improved by altering a few boundaries of separating technique. It can be stretched out to shading pictures.

## *7.2 Future Enhancements*

The procedure can be stretched out to 3D picture. The best possible anatomical position might be recognized fixed edges were utilized. AI might be executed to prepare the framework progressively change the limits. This is especially significant as MR imaging which is turning out to be standard demonstrative technique in clinical practice. It is likewise significant that any commonsense division calculation should manage 3D volume division rather than 2D cut by cut division since MRI subtleties are 3D in nature.

# References


1. Mortamet, B., Zeng, D., Gerig, G., Prastawa, M., Bullitt, E.: Effects of healthy aging measured by intracranial compartment volumes using designed MR brain database. In: International Conference on Medical Image Computing and Computer-Assisted Intervention, pp. 383–391 (2005)
2. Devos, A., Simonetti, A.W., VandGraaf, M., Lukas, L., Suykens, J.A.K., Vanhamme, L., Buydens, L.M.C., Heerschap, A., VanHuffel, S.: The use of multivariate MR imaging intensities versus metabolic data from MR spectroscopic imaging for brain tumour classification. J. Magnet. Reson. **173**, 218–228 (2005)
3. Corina, S., Drapaca, V.C., Colin, S.: Segmentation of tissue boundary evolution from brain MR Image sequences using multi-phase level sets. Comput. Vis. Image Understand. **100**(3), 312–329 (2005)
4. Yazdan-Shahmorad, A., Soltanian-Zadeh, H., Zoroofi, R.A.: MRSI brain tumor characterization using wavelet and wavelet packets feature spaces and artificial neural networks. In: 26th Annual International Conference IEEE Transactions on Engineering in Medicine and Biology Society, vol. 1, pp. 1810–1813. IEMBS '04, Sept 2004
5. Lefohn, A.E., Cates, J.E., Whitaker, R.T.: Interactive, GPU-based level sets for 3D segmentation. In: Conference on Medical Image Computing and Computer Assisted Intervention (MICCAI), pp. 564–572, April 2003

6. Amini, L., Soltanian-Zadeh, H., Lucas, C.: Automated segmentation of brain structure from MRI. In: Proceedings of the International Society for Magnetic Resonance in Medicine, p. 11 (2003)
7. Gering, D.T., Eric, W., Grimson, L., Kikinis, R.: Recognition Deviations from Normalcy for Braintumor Segmentation. MICCAI, LNCS vol. 2488, pp. 388–395. Springer, Berlin (2002)
8. Devos, A., Lukas, L., Simonetti, A.W., Suykens, J.A.K., Vanhamme, L., van der Graaf, M., Buydens, L.M.C., Heerschap, A., Van Huffel, S.: Does the combination of magnetic resonance imaging and spectroscopic imaging improves the classification of brain tumours? In: Proceedings of the 26th Annual International Conference of the IEEE MBS, 1–5 Sept 2004

# Health Monitoring System for Diabetic Patients

**K. Kishore Kumar, D. Manivannan, and N. K. Manikandan**

**Abstract** Diabetes is one of the main universal health problems. Consistent with the World Health Organization report, around 346 million people worldwide are affected by DM. DM may be a metabolic disease where the inappropriate controlling of blood sugar levels causes the danger of many diseases like attack, renal disorder, and kidney failure. In DM, the body does not properly use the insulin hormone secreted by the pancreas gland. There are numerous computerized methods for the diagnosis of DM, but the most drawback of those methods is that the patient has got to undergo several medical tests to give the input values to the computerized recognizing system which proves to be high cost and time taking process. With the rapid advancement within the field of AI, there are numerous techniques and algorithms during AI mostly used for the prediction and identification of various diseases. These algorithms in AI convince to be cost effective and time saving for diabetic patients and doctors. In the proposed model, we are recognizing diabetes using K-nearest neighbor algorithm which is most dominant model of artificial intelligence.

**Keywords** Artificial intelligence · K-nearest neighbor · Diabetes mellitus

## 1 Introduction

Nowadays, DM ought to become a typical international pathological state which can result in many health complications like vas diseases, kidney failure, disability, etc., hypoglycemic agent could be a natural endocrine that is secreted by the duct gland within the figure. Hypoglycemic agent turns the sugar in easier molecules that square measure used by the body cells to get energy. This conversion is affected because of the necessity for hypoglycemic agent, and also, the sugar starts obtaining accumulated within the blood. As a result, the blood glucose level starts increasing, and so, the person develops DM. It is divided into two types; first type is polygenic disorder that is often diagnosed in youngsters, and here, the body does not turn out

K. Kishore Kumar (✉) · D. Manivannan · N. K. Manikandan
Department of Computer Science and Engineering, Vel Tech Rangarajan Dr. Sagunthala R D Institute of Science and Technology, Chennai, Tamil Nadu, India



T. K. Sharma et al. (eds.), *Soft Computing: Theories and Applications*, Advances in Intelligent Systems and Computing 1381,
https://doi.org/10.1007/978-981-16-1696-9_38

hypoglycemic agent. Second is mostly diagnosed in adults. Here, either the body does not turn out enough hypoglycemic agent or cells ignore the hypoglycemic agent. Uncontrolled polygenic disorder could cause hyperglycemia or raised blood sugar. It is going to result in serious harm to the nerves and blood vessels. These days DM has become a typical international pathological state which can result in many health complications like vas diseases, kidney failure, disability, etc., consistent with the W.H.O report, and a complete three hundred million of the planet population are going to be laid low with polygenic disorder by 2025. Hypoglycemic agent could be a natural endocrine that is secreted by the duct gland within the figure. Hypoglycemic agent converts sugar into easier molecules that square measure used by the body cells to get energy. This conversion is affected because of the dearth of hypoglycemic agent, and also, the sugar starts obtaining accumulated within the blood.

## 2 Related Works

This chapter gives the overview of literature survey. This chapter represents some of the relevant work done by the researchers. Many existing techniques have been studied by the researchers on search system for proper diabetic prediction; by using python and spider, we review some of them below.

The author [1] discussed the use of AI in helper recognizable proof sicknesses have become a flow research hotspot. The ordinary strategy of conclusion polygenic sickness circulatory intricacy, diabetic fringe neuropathy lipoidemia, DM fringe pathology, and along these lines, the brilliant infections are to recognize Associate in Nursing audit report by a proficient specialist. It is the usage of clinical call uphold calculation for clinical content information which faces a test with the vanity level and exactness. We tend to arrange Associate in Nursing growing and learning framework to locate infections on the head of in our clinical content information that cowl a few kinds of physiological boundaries of people, as hematological boundaries, wee-wee boundaries, and natural science identification. To begin with, the data was swollen and revised. Second, the prepared information was taken care of into a 1D convolution neural organization with dropout and pooling. The arranged helper indicative frameworks have a profoundly right and strong exhibition square measure commonly utilized for starter ID and reference; in this way, it is not exclusively sparing a few HR anyway conjointly bringing about improved clinical symptomatic power.

The author in paper [2] says to improve the presentation and affectability of ceaseless perception frameworks for the location of persistent sicknesses, and the choice of ideal AI calculations is pivotal. As of now, the regularly utilized calculations face requirements, for example, high machine worth and absence of ideal element decision on application to time span flags along these lines lessening the power of such examination. Profound learning draws near, much the same as the convolution neural organization, and conquered these disadvantages by scheming robotized alternatives from a crude sign and ordering the inferred highlights. This plan shows savvy merits.

In any case, the utilization of completely associated multilayer recognition calculations has indicated low-order execution. This paper proposes to build up a changed profound learning convolution neural organization equation incorporated with help vector machines to manage the disadvantages blessing in multilayer perception and consequently rising the general presentation of time-frame discovery applications. The framework is substantial on time-frame breath signals for non-intrusive recognition of polygenic illness. The exhibition of this extended recipe is assessed and contrasted and the predominant strategy.

The author in [3] presents the non-obtrusive polygenic sickness expectation that has been picking up conspicuousness throughout the most recent decade. Among a few human serums assessed, human breath develops as a promising decision with dimethyl ketone levels in breath displaying Associate in Nursing as fair connection to glucose levels. Such a relationship sets up dimethyl ketone as a reasonable biomarker for the polygenic infection. The premier basic information examination approaches to research the biomarkers in breath for unknown location abuse include extraction and order calculations. Notwithstanding, obstacles like cycle worth and absence of a best component decision on the application to time-frame signals downsize the intensity of such examination.

The author in paper [4] presents the performance analysis of the competitive learning models on mathematician knowledge for automatic cluster choice and additionally studied and analyzed the performance of these algorithms, and randomized results have been analyzed on 2D mathematician knowledge with the training rate parameter unbroken easy for all algorithms. Algorithms utilized in their work embody bunch formula, competitive learning formula, and frequency sensitive competitive learning formula. Supervised learning machine algorithms square measure used for classification of the mathematician knowledge.

The author in paper [5] proposed singular sickness hazard forecast dependent on clinical history. This framework likewise predicts each danger dependent on their own clinical history information. Dataset is utilized for clinical coding and collaborative assessment and recommendation engine (CARE) data procedure; from this writing, it is seen that the AI calculations place a critical part in information revelation and structure the datasets particularly in clinical distinguishing proof with the clinical information.

The author in paper [6] presents diabetes as a non-contagious illness that is prompting the never-ending issue and significant medical issues. World Health Organization reports diabetes, and it is difficult that impacts an individual genuinely, monetarily, and financially over the families. The reports state that more than ten lakh deaths were caused because of the unmanageable phase of well-being leading to death. In excess of twenty lakhs deaths caused because of the danger variables of diabetes like heart and different sicknesses.

The author in paper [7] presents diabetes and is an illness caused because of the all-encompassing degree of sugar fixation in the blood. In this proposed model, they examined different classifiers, and choice support framework is proposed which uses the AdaBoost model with decision stump as a base classifier for classification. The accuracy gain for AdaBoost calculation with decision stump as a classifier is 80.72, which is more vital appeared differently in relation to that of support vector machine.

## 3 Proposed Work

Figure 1 represents the proposed framework for health monitoring system, which deals with clinical presentation of polygenic disease in a very patient is that the symptomatic options bestowed by the patients. This feature is an indication of the malady cause and has direct impact in guiding clinicians concerning the choice to require. In case of classifying positive and negative polygenic disease, the subsequent parameters we have taken are age, weight, insulin, smoke, and age first smoked, and wherever the survey was taking.

### *3.1 Data Preprocessing*

Information pre-handling significantly affects the presentation of regulated learning models on the grounds that problematic examples likely lead to wrong yields. This examination is additionally used to watch the changing pattern in style deals. Information preprocessing is a significant advance to set up the information to shape a QSPR

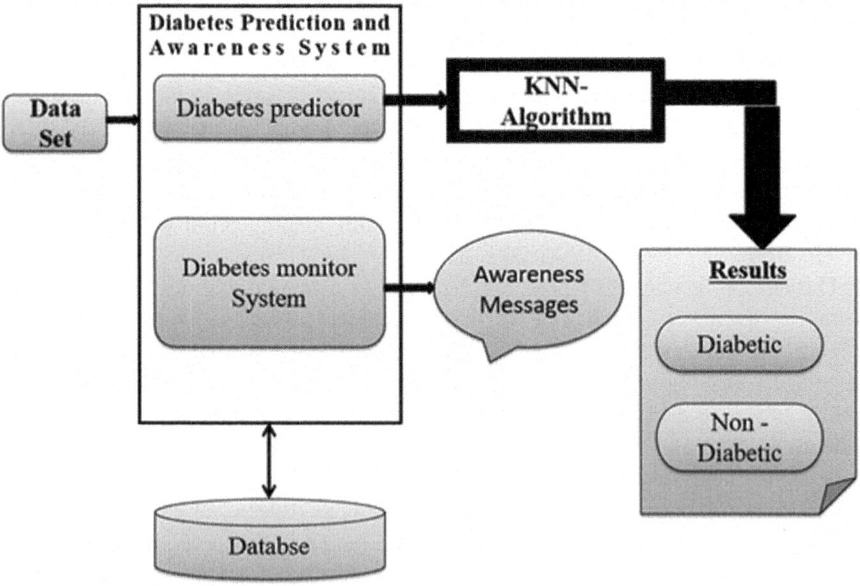


**Fig. 1** Proposed framework for health monitoring system

model. There are numerous ground-breaking steps in information preprocessing, for example, information cleaning, information extraction, and quality determination.

### *3.2 Data Splitting*

The basic objective of machine learning is to sum up on such a great amount of part of the data examples used to prepare models. We may want to check the model to appraise the nature of its example speculation for data the model has not been prepared on. Nonetheless, because of future examples which have obscure objective qualities, we tend to not check the exactness of our forecasts for future occurrences by and by, and we might want to utilize an assortment of the information that we tend to, in general as of now, see the response for as an intermediary for future information. Assessing the model with consistent information that was utilized for work is not helpful, because it rewards models which will "recollect" the work data, as threatening. A typical system needs all offered marked data and splits it into work and examination subset.

### *3.3 Training Data*

Training data is labeled information wont to train your machine learning algorithms and increase accuracy. Machine learning models do not seem to be too totally different from a person's kid. Once a toddler observes a brand new object, say for instance a dog, and receives constant feedback from its atmosphere, the kid is ready to find out that this new machine can also learn once they see enough relevant information. Victimization is that you will be able to model algorithms to search out relationships, sight patterns, perceive advanced issues, and create choices. Eventually, the standard, variety, and amount of your coaching information verify the success of your machine learning models.

### *3.4 Test Data*

The test set might be a lot of perceptions used to evaluate the exhibition of the model utilizing some presentation metric. It is compulsory that no perceptions from the preparation set square measure encased inside the test set. In the event that the test set will contain models from the preparation set, it will be inconvenient to evaluate whether the algorithmic principle has figured out how to sum up from the preparation set or has just remembered it (Fig. 2).

```
 1.59008473]
 [-0.22446668 -1.32874488 -1.06099801 ... -0.65073254 -0.60203068
   0.74654563]
 [-0.22446668 -0.45661411  0.13570575 ... -0.77166568 -0.51505559
  -1.32985833]]
{'n_neighbors': 19}
0.8053819444444444
Accuracy of K-NN classifier on training set: 0.79
Accuracy of K-NN classifier on test set: 0.72
TP - True Positive 109
FP - False Positive 14
FN - False Negative 39
TN - True Negative 30
Accuracy Rate: 0.7239583333333334

In [14]:
```

**Fig. 2** Classification result of diabetic disease prediction

## 4 Conclusion

Diabetes mellitus is a sickness, which offers ascend to numerous issues. The forecasting and deciding this infection by utilizing an AI algorithm predict the yield in less time. There are many existing frameworks; we found that the precision is not acceptable, and the aftereffects of utilizing the KNN algorithm have better outcomes. The outcome, which just utilized the KNN algorithm, has a superior presentation in the given patient dataset. Here, we have utilized the K-nearest neighbor system for the ID of diabetes mellitus. In this, we determined exactness, mistake rates, and misclassification rate. The exactness rate is demonstrating that various yields of the information of the test dataset are equivalent to the yield of the information of various qualities of the preparation dataset. The mistake rate is showing that various yields of the information of the test dataset are not equivalent to the yield of the information of various qualities of the preparation dataset. The yield is showing that as the estimation of k expands, the exactness rate and mistake rate additionally increment. KNN is one of the most proficient artificial insight algorithms that are generally utilized for medical care or recognizable proof purposes. Precise and efficient outcomes can be gained through this algorithm.

## References

1. Sahai, A., Waters, B.: Fuzzy Identity-Based Encryption, vol. 3494, pp. 457–473 . Springer, Berlin (2005)
2. Uygu, C., Esin, D.: An automatic diabetes diagnosis system based on LDA wavelet support vector machine classifier. Expert Syst. Appl. **38**, 8311–8315 (2011)
3. Robertson, G., Lehmann, E.D., Sandham, W., Hamilton, D.: Blood glucose prediction using artificial neural networks trained with the AIDA diabetes simulator: a proof-of-concept pilot study. J. Electr. Comput. Eng. **2011**, 681786 (2011). https://doi.org/10.1155/2011/681786

4. Lewko, A., Waters, B.: Decentralizing Attribute-Based Encryption, vol. 6632, pp. 568–588. Springer, Berlin (2011)
5. Jegan, C.: Classification of diabetes disease using support vector machine. Microcomput. Dev. **3**, 1797–1801 (2014)
6. Wang, A., Chen, G., Yang, J., Zhao, S., Chang, C.Y.: A comparative study on human activity recognition using inertial sensors in a smartphone. IEEE Sens. J. **16**(11), 4566–4578 (2016)

# Comparative Study of Machine Learning with Novel Feature Extraction and Transfer Learning to Perform Detection of Glaucoma in Fundus Retinal Images

**Deepti Ravishyam and Dhanalakshmi Samiappan**

**Abstract** Glaucoma detection from fundus retinal images is a difficult task which requires years of practice and expertise in the domain. In this paper, we propose a novel method of statistical feature extraction from fundus retinal images and test its feasibility in glaucoma detection with machine learning algorithms. We combine the results of these algorithms by applying ensemble learning to create one optimized predicted output. We also apply transfer learning and compare the results obtained. The random forest model and ensemble learning model performed better than other conventional models. Using the conventional machine learning algorithms, the highest accuracy, sensitivity and AUC of 83.42%, 74.62% and 0.82, respectively, were obtained by ensemble learning, and the highest specificity of 94.74% was observed in the random forest algorithm. Using the transfer learning algorithms, the best performance was obtained with VGG19 model with AUC 0.8919 followed by GoogLeNet with an AUC of 0.8872.

**Keywords** Glaucoma · Machine learning · Transfer learning · Fundus · Template matching · Statistics

## 1 Introduction

Glaucoma is a complex neuro-retinal disease, which causes progressive damage to the optic nerve head [1]. It is the second most cause for loss of vision after cataract [1]. It is not completely curable and causes irreversible loss of vision which makes it important to detect the disease in its early stages, to reduce its progression and hence prevent from complete blindness [2]. Thus, automation of glaucoma diagnosis is very essential and pertinent. In a healthy eye, the neuro-retinal rim follows a configuration

D. Ravishyam (✉) · D. Samiappan
Department of Electronics and Communication Engineering, SRM Institute of Science and Technology, Kattankulathur 603203, India

D. Samiappan
e-mail: dhanalas@srmist.edu.in

 
T. K. Sharma et al. (eds.), *Soft Computing: Theories and Applications*,
Advances in Intelligent Systems and Computing 1381,
https://doi.org/10.1007/978-981-16-1696-9_40

**Fig. 1** ISNT rule [3]

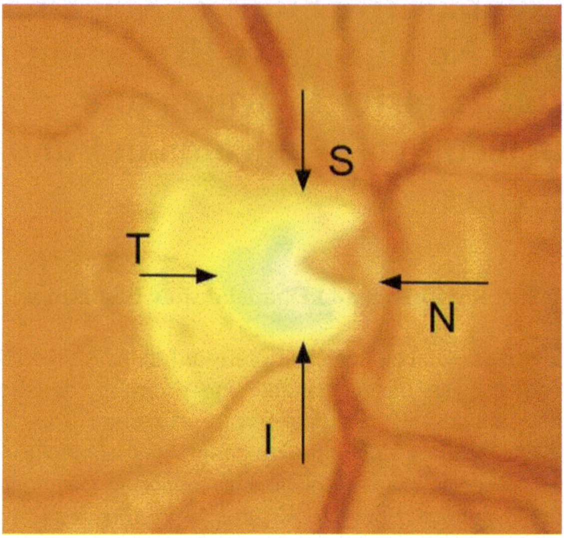


known as the ISNT rule as shown in Fig. 1. It states that rim thickness follows the order Inferior > Superior > Nasal > Temporal. When this criterion fails, it is a suspect of glaucoma [3].

Some of the previous works involving machine learning involved application of empirical wavelet transform followed by correntropy-based feature extraction and a support vector machine algorithm. Textons [4] and filters are applied followed by machine learning algorithms [5]. Another research work classifies glaucoma with fundus images using optic-disc localization and segmentation [6]. Mean, median [7] and other image processing techniques [8] as we know it are used.

Most techniques make use of various filters and transforms for image processing and feature extraction. We extract statistical features in terms of distributions to obtain 12 different features while keeping the essence of the ISNT rule. We then apply machine learning and transfer learning models to perform diagnosis of glaucoma. The block diagram in Fig. 2 gives an overview of the proposed method.

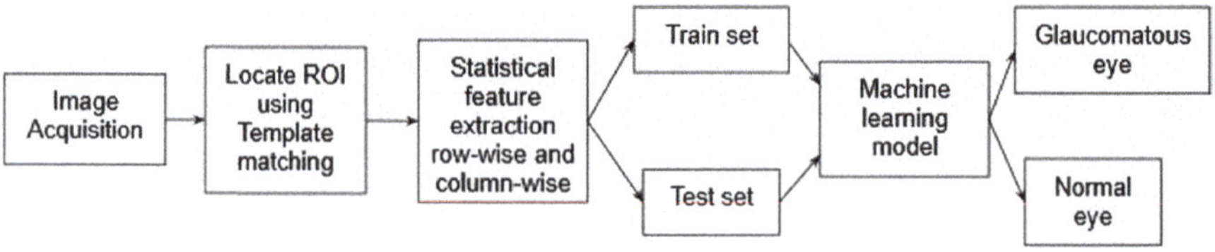


**Fig. 2** Block diagram for glaucoma diagnosis

## 2 Image Acquisition

RIM-ONE is an open-source fundus retinal coloured image dataset which is accurate in the section of optic nerve head (ONH) [9]. We use RIM-ONE dataset which has three releases comprising of 766 images having 311 glaucomatous and 455 healthy retinal fundus images.

## 3 Image Processing

In order to automate the diagnosis of glaucoma, we identify the region of interest (ROI) and then extract features to apply conventional machine learning algorithms. The following subsections explain the same in detail.

**Location of Region of Interest with Template Matching**

The region of interest in the retinal fundus image of human eye is the optic nerve head [10]. Therefore, the part of the image other than the optic nerve head is discarded, which in turn reduces the computational power required. Template matching was performed which effectively matched the region of interest [11]. The three-layered RGB images obtained are converted to a single-layer greyscale image (Fig. 3b) to reduce computational complexity. Through experimentation, it was concluded that the aspect ratio of $385 \times 385$ was suitable for further processing without compromising on the quality for further image processing and feature extraction.

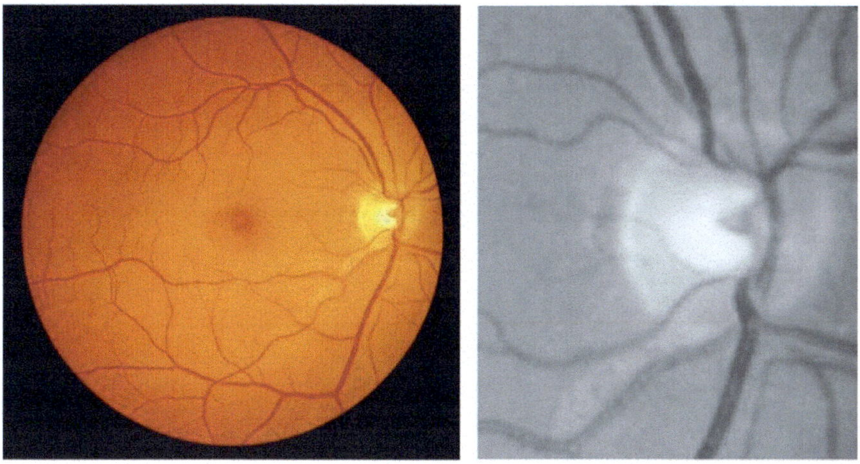

(a) (b)

**Fig. 3** **a** Sample image, **b** grayscale RoI image

### Feature Extraction for Conventional Machine Learning Algorithms

We experiment the feasibility of using row-wise and column-wise statistical properties of the image over the kernel by using filtering or transform methods [12] used in conventional image processing techniques [8]. We use more of an intuitive approach with six pairs of features, for row-wise and column-wise distribution, namely mean, median, geometric mean, kurtosis, standard deviation and skewness.

In Fig. 4, mean of each row of pixels is computed from row 1 to row 385 and plotted. As shown in the graph, the mean of the 218th row has the maximum mean value of 119 due to more brightness in those set of pixels. The figure also showcases the difference between the row-wise mean of a glaucomatous image and a healthy image. It is clear that the peak in healthy fundus image is narrower than the glaucomatous fundus image. Similarly, we extract the mean for each column of pixels.

Fundus eye images are quite symmetric as the neuro-retinal optic cup and disc are almost concentric circles. Skewness [13] was chosen, as it measures lack of symmetry. Since the kurtosis vary significantly for normal eye and glaucomatous eye

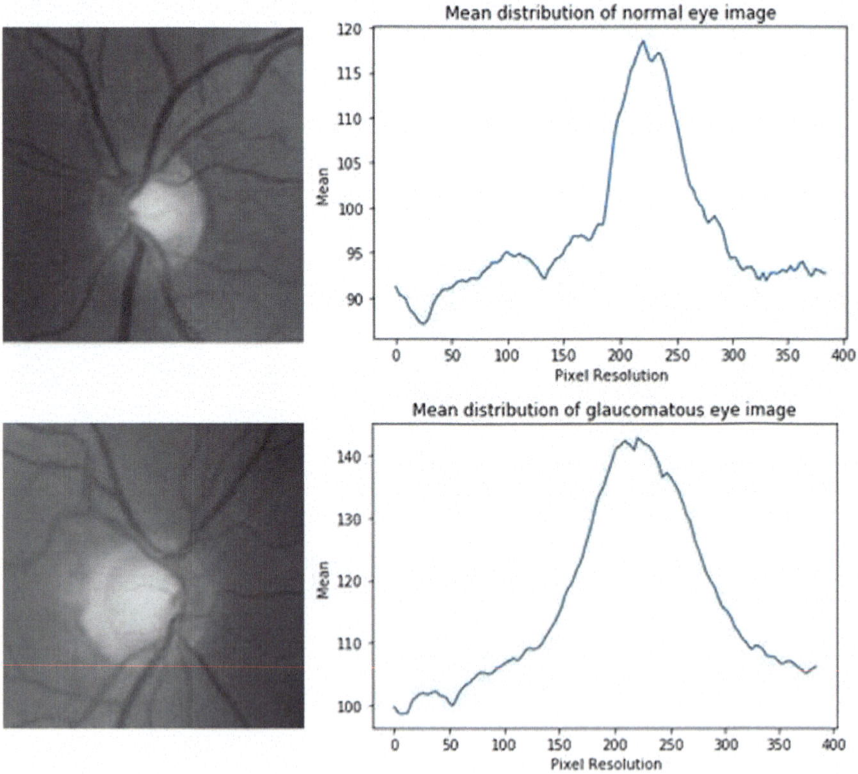


**Fig. 4** Row-wise mean values between glaucomatous and normal fundus images

images, it can easily classify both classes of images accurately [14]. Geometric mean [15] was considered as it showed significant differences between both eye images. We then compute, both row-wise and column-wise, the average median, average standard deviation, average skewness, average kurtosis and average geometric mean for each pixel-row and each pixel-column of glaucomatous eye images and normal eye images. It was found that these features show significant difference between the two classes (Fig. 5).

## 4 Conventional Machine Learning Algorithms

The features obtained after extracting the statistical properties constitutes to about 4 million instances. The data was then split into a training sample of 585 images and test sample of 181 images. The below machine learning algorithms were applied on the extracted features, and the models are then evaluated on new data on various metrics.

**Support Vector Machine (SVM)**

SVM is a supervised learning algorithm comprising of kernel functions which can be used for classification. If the data is linearly separable, it is termed as a linear SVM classifier. If not, then kernel functions are used to map or transform the linearly inseparable data to higher dimensions where the data is presumed to be linearly separable [16–18]. Support vector machines scale relatively well to high-dimensional data, and the risk of overfitting is less. A radial basis function kernel was applied to the SVM classifier, and an accuracy of 79.00% was obtained.

**Random Forest**

Random forest classifier is an ensemble method consisting of several decision trees that collectively decide on the end result, i.e. if an image is that of a glaucomatous eye or normal eye. A majority vote is taken from several trees to obtain the predicted output. A random forest model of 400 decision trees with a minimum sample split of 2 was applied to obtain an accuracy of 79.55% which was highest among all conventional algorithms used.

**K-Nearest Neighbours (KNN)**

KNN is a supervised machine learning algorithm that is robust to noisy data [5, 19]. We use a KNN classifier with a minkowski distance metric of 2 (Euclidean distance) and $k$ being 3 with a leaf size of 30. The model was then tested on the test images, and an accuracy of 76.24% was obtained.

**Logistic Regression**

Logistic regression is a binary probabilistic classifier algorithm that calculates the probability of a data being in class 0 or class 1. A penalty of $L2$ was applied, and an intercept was fit with a scaling of 1 to obtain an accuracy of 75.13%.

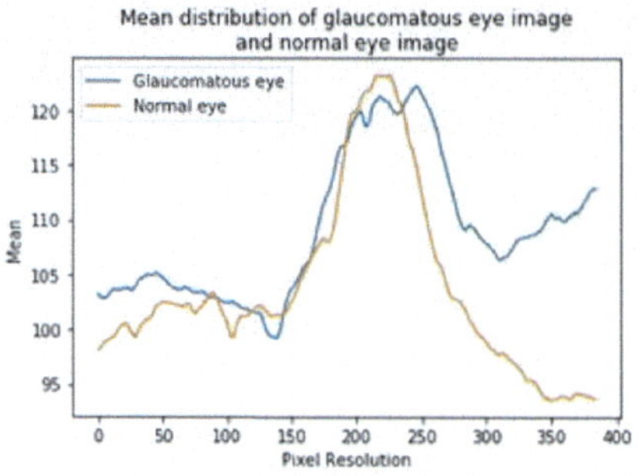


(a) **Mean of the pixels in ROI**

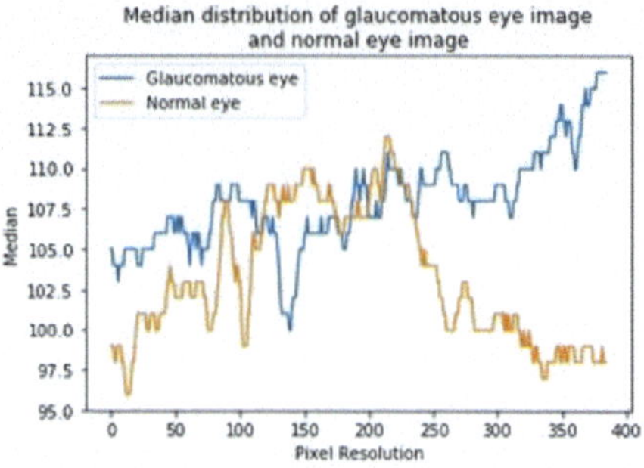


(b) **Median of pixels in ROI**

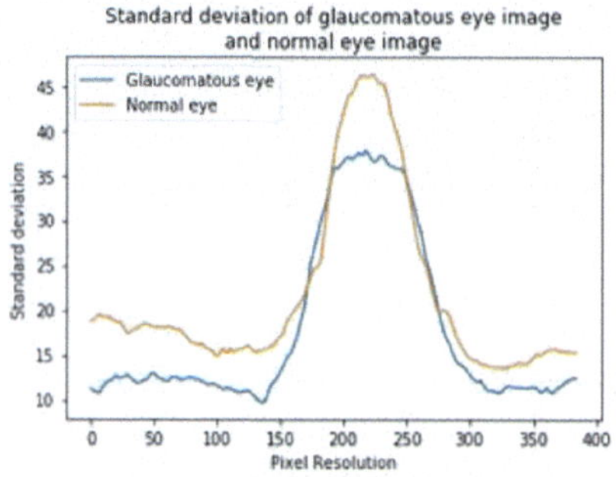


(c) **Standard deviation of the pixels in ROI**

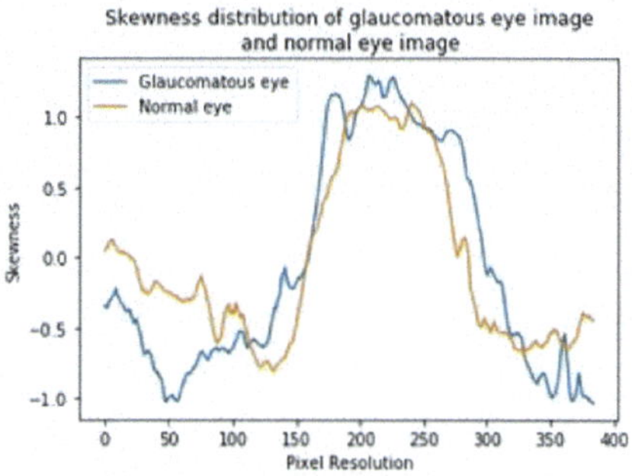


(d) **Skewness of the pixels in ROI**

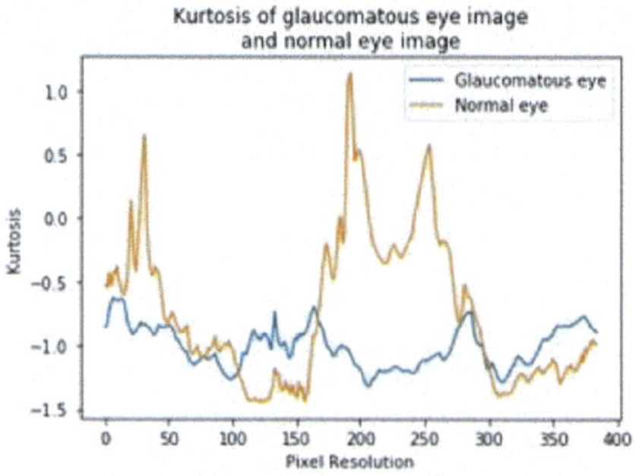


(e) **Kurtosis of the pixels in ROI**

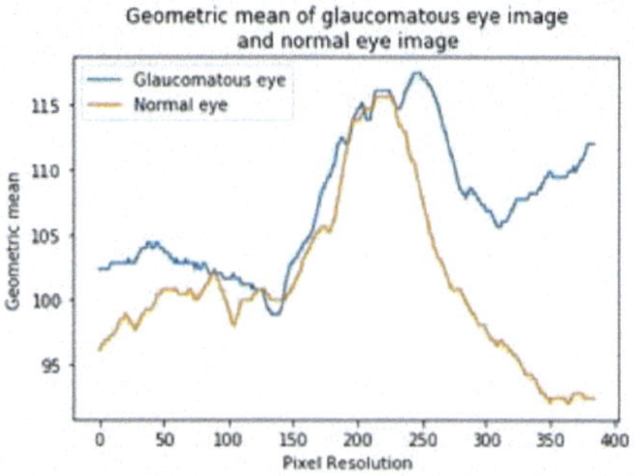


(f) **G-Mean of the pixels in ROI**

**Fig. 5** Visualizing the distribution of features extracted from pixel intensity. **a** Mean of pixels in ROI, **b** median of pixels in ROI, **c** standard deviation of pixels in ROI, **d** skewness of pixels in ROI, **e** kurtosis of pixels in ROI, **f** g-mean of pixels in ROI

## Ensemble Learning

Ensemble learning is a method of combining individual model predictions together to improve the prediction probability and reduce variance and bias in the predictions. It uses multiple algorithms to obtain better predictions than constituent stand-alone

**Table 1** Description of the pre-trained models used for transfer learning

| Architecture | Input size of the image (pixels) | Number of convolutional layers | Number of fully connected layers | Activation function/special function used |
|---|---|---|---|---|
| AlexNet | 227 × 227 × 3 | 5 | 3 | ReLU |
| VGG16 | 224 × 224 × 3 | 13 | 3 | ReLU |
| VGG19 | 224 × 224 × 3 | 16 | 3 | ReLU |
| GoogLeNet | 224 × 224 × 3 | 22 | 0 | Inception module and bottleneck layer |
| ResNet-50 | 224 × 224 × 3 | 50 | 0 | Residual block or skip connection |

algorithms [20]. The majority voting technique of ensemble learning was used [21]. Each of the four models, namely logistic regression, KNN, SVM and random forest were trained on the entire dataset, and the majority of the final outputs was obtained as the ensemble output.

## 5 Transfer Learning Algorithms

Transfer learning makes use of the knowledge gained while solving one set of images and apply it to a different set of images. The new dataset can be used to fine-tune the weights of the existing pre-trained model to suit itself, thus reducing training time and resulting in lower generalization error compared to training a CNN from scratch [22]. Five architectures were used as shown in Table 1. The primary difference between the models is the depth, building block, convolutional layers and fully connected layers. The data was split into a training sample of 585 images and test sample of 181 images. Adam optimization with a batch size of 32, learning rate of 0.001 and a loss function as categorical cross-entropy was applied to the following models.

## 6 Results and Discussion

Conventional machine learning and transfer learning were implemented on the RIM-ONE dataset. The following are the results obtained for the conventional machine learning algorithms (Tables 2 and 3).

The random forest model and ensemble learning model seem to perform better than all other conventional models. The highest sensitivity, accuracy and F1 score of 74.62%, 83.42% and 0.82, respectively, were seen in ensemble learning. However, the highest specificity of 94.74% was observed in the random forest algorithm.

**Table 2** Results obtained with conventional machine learning algorithms

| Model | Sensitivity (%) | Specificity (%) | Accuracy (%) | *F*1 score | AUC in ROC |
|---|---|---|---|---|---|
| SVM | 58.20 | 91.22 | 79.00 | 0.76 | 0.75 |
| Random Forest | 53.73 | **94.74** | 79.55 | 0.75 | 0.78 |
| KNN | 53.73 | 89.47 | 76.24 | 0.73 | 0.66 |
| Logistic Regression | 70.14 | 78.07 | 75.13 | 0.74 | 0.74 |
| Ensemble learning | **74.62** | 88.59 | **83.42** | **0.82** | **0.82** |

**Table 3** Results obtained with pre-trained convolutional neural networks

| Architecture | Sensitivity (%) | Specificity (%) | AUC in ROC | Accuracy (%) |
|---|---|---|---|---|
| AlexNet | 79.82 | 80.23 | 0.8102 | 78.82 |
| VGG16 | 82.54 | 81.33 | 0.8290 | 79.26 |
| VGG19 | 88.91 | 85.72 | **0.8919** | 84.91 |
| GoogLeNet | 86.31 | **87.23** | 0.8872 | 81.63 |
| ResNet-50 | **89.80** | 86.25 | 0.8831 | **85.92** |

Considering only conventional models, they have a maximum AUC of 0.78 obtained from random forest algorithm. It is interesting to note that random forest classifies true positives slightly better than the ensemble model. However, the ensemble model is the best at classifying false positives. The confusion matrix in Fig. 6b implies that the ensemble model classified the images with 50 true positives and 101 true negatives.

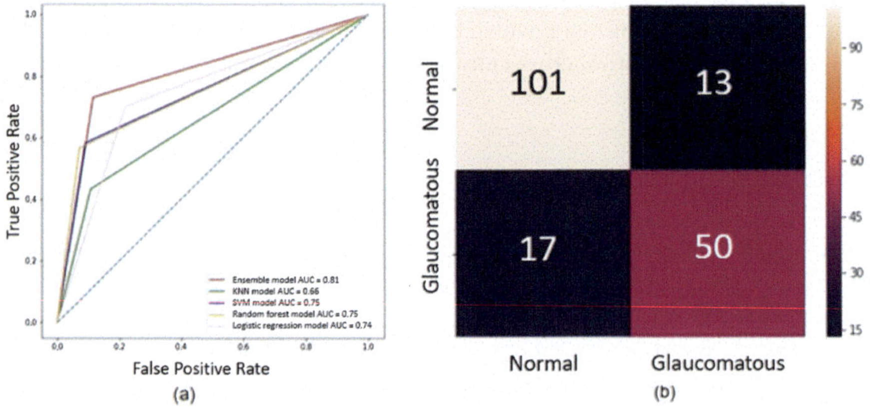


**Fig. 6** **a** Comparative ROC curve for all conventional machine learning models, **b** confusion matrix for ensemble model

**Table 4** Confusion matrix for the test dataset with 181 images with VGG19 model

| | Normal | Glaucomatous |
|---|---|---|
| Normal | 64 | 15 |
| Glaucomatous | 8 | 94 |

ResNet with 50 layers had the highest sensitivity of 89.80%. VGG19 model performed the best, exhibiting the highest AUC of 0.8919. It is evident that the best pre-trained CNN models are VGG19 and GoogLeNet. Plotting the confusion matrix (Table 4) for the test images assured that VGG19 indeed performed well for our data. It seems that the transfer learning model performed slightly better at detecting glaucomatous eyes, while the machine learning models performed better at detecting normal eyes.

Singh [6] uses a private dataset of 63 images with wavelet feature extraction along with PCA obtained an accuracy of 94.7% with both KNN and SVM algorithms. Our dataset comprised of 766 images (455 healthy eye and 311 glaucomatous eye images) which is a more balanced, less biased dataset. Our work gives an accuracy of 83.42% with ensemble learning and 85.92% with ResNet-50 using a larger, more balanced dataset.

## 7 Conclusion

Feature extraction is a vital part which greatly affects any model's performance. We can see that our novel method of feature extraction for the machine learning models follows an intuitive approach by computing the six features row-wise and six features column-wise [11, 13–15] to capture the essence of the ISNT rule [3] which is used by ophthalmologists and worked successfully in the classification of glaucoma.

It is seen that among machine learning models, the ensemble model [20, 21] gave highest accuracy of 83.42%, highest sensitivity of 74.62%, highest F1 score of 0.82 and highest AUC of 0.82. However, the highest specificity of 94.74% was obtained in the random forest algorithm compared to 88.59% obtained in ensemble learning.

VGG19 and GoogLeNet models with AUC of 0.8919 and 0.8872, respectively, performed better compared to other architectures. This can be due to their high complexity and large number of layers in the CNN.

The feature extraction performed by transfer learning uses convolution and pooling in order to extract features from the image [23]. However, they take hours to train. The conventional machine learning models, on the other hand, were trained within seconds. Also, the feature extraction was performed manually prior to training. One major difference between conventional models and transfer learning is that the latter learns the feature extraction during training [24]; while in the former, the features are manually extracted and given as input to the algorithm [18], which contributes to difference in performance.

There are some notable advantages associated with the current method. Our approach is user friendly, cost effective, and automated. The machine learning algorithms do not require much computational power. Thus, our method can be easily used at medical centres and in regions with limited access to ophthalmologists.

**Acknowledgements** We sincerely thank N.M.S. Chaitanya, S.R. Shruthi Vidya and S. Chaitanya for their assistance in obtaining the fundus image database, preprocessing of the images.

## References

1. Cook, C., Foster, P.: Epidemiology of glaucoma: what's new? Can. J. Ophthalmol. **47**(3), 223–226 (2012)
2. Weinreb, R.N., Aung, T., Medeiros, F.A.: The pathophysiology and treatment of glaucoma: a review. JAMA **311**(18), 1901–1911 (2014). https://doi.org/10.1001/jama.2014.3192
3. Ruengkitpinyo, W., Kongprawechnon, W., Kondo, T., Bunnun, P., Kaneko, H.: Glaucoma screening using rim width based on ISNT rule. In: 6th International Conference of Information and Communication Technology for Embedded Systems (IC-ICTES) IEEE, pp. 1–5 (2015). https://doi.org/10.1109/ictemsys.2015.7110827
4. Dana, K.J.: Computational texture and patterns: from textons to deep learning. Synth. Lect. Comput. Vis. **8**(3), 1–113 (2018)
5. Acharya, U.R., Bhat, S., Koh, J.E., Bhandary, S.V., Adeli, H.: A novel algorithm to detect glaucoma risk using texton and local configuration pattern features extracted from fundus images. Comput. Biol. Med. **88**, 72–83 (2017)
6. Singh, A., Dutta, M.K., ParthaSarathi, M., Uher, V., Burget, R.: Image processing based automatic diagnosis of glaucoma using wavelet features of segmented optic disc from fundus image. Comput. Methods Programs Biomed. **124**, 108–120 (2016)
7. Gupta, G.: Algorithm for image processing using improved median filter and comparison of mean, median and improved median filter. Int. J. Soft Comput. Eng. (IJSCE) **1**(5), 304–311 (2011)
8. Kumar, V., Gupta, P.: Importance of statistical measures in digital image processing. Int. J. Emerg. Technol. Adv. Eng. **2**(8), 56–62 (2012)
9. Fumero, F., Alayón, S., Sanchez, J.L., Sigut, J., Gonzalez-Hernandez, M.: RIM-ONE: an open retinal image database for optic nerve evaluation. In: 2011 24th International Symposium on Computer-Based Medical Systems (CBMS), pp. 1–6. IEEE (2011)
10. Lee, E.J., Han, J.C., Park, D.Y., & Kee, C.: A neuroglia-based interpretation of glaucomatous neuroretinal rim thinning in the optic nerve head. Prog. Retinal Eye Res. 100840 (2020). https://doi.org/10.1016/j.preteyeres.2020.100840
11. Wankhede, P.R., Khanchandani, K.B.: Optic disc detection using histogram based template matching. In: International Conference on Signal Processing, Communication, Power and Embedded System (SCOPES) (2016). https://doi.org/10.1109/scopes.2016.7955765
12. Maheshwari, S., Pachori, R.B., Acharya, U.R.: Automated diagnosis of glaucoma using empirical wavelet transform and correntropy features extracted from fundus images. IEEE J. Biomed. Health Inform. **21**(3), 803–813 (2016)
13. Patil, P.N., Bagkavos, D., Wood, A.T.A.: A measure of asymmetry based on a new necessary and sufficient condition for symmetry. Sankhya A **76**(1), 123–145 (2013). https://doi.org/10.1007/s13171-013-0034-z
14. Elangovan, P., Nath, M.K., Mishra, M.: Statistical parameters for glaucoma detection from color fundus images. Procedia Comput. Sci. **171**, 2675–2683 (2020). https://doi.org/10.1016/j.procs.2020.04.290

15. Senthil Kumar, T., Helen Prabha, K.: Geometric mean filter with grayscale morphological method to enhance the RNFL thickness in the SD-OCT images. Multimedia Tools Appl. **77**(8) (2018)
16. David, D.S., Jayachandran, A.: A new expert system based on hybrid colour and structure descriptor and machine learning algorithms for early glaucoma diagnosis. Multimedia Tools Appl. **79**(7), 5213–5224 (2020)
17. Dey, A., Dey, K.N.: Automated glaucoma detection from fundus images of eye using statistical feature extraction methods and support vector machine classification. Ind. Interact. Innov. Sci. Eng. Technol. 511–521 (2017). https://doi.org/10.1007/978-981-10-3953-9_49
18. Dey, A., Bandyopadhyay, S.: Automated glaucoma detection using support vector machine classification method. J. Adv. Med. Med. Res. **11**(12), 1–12 (2015). https://doi.org/10.9734/BJMMR/2016/19617
19. Hagiwara, Y., Koh, J.E.W., Tan, J.H., Bhandary, S.V., Laude, A., Ciaccio, E.J., Acharya, U.R.: Computer-aided diagnosis of glaucoma using fundus images: a review. Comput. Methods Progr. Biomed. **165**, 1–12 (2018). https://doi.org/10.1016/j.cmpb.2018.07.012
20. Sagi, O., Rokach, L.: Ensemble learning: A survey. Wiley Interdisc. Rev. Data Mining Knowl. Discov. **8**(4), e1249 (2018). https://doi.org/10.1002/widm.1249
21. Zilly, J., Buhmann, J.M., Mahapatra, D.: Glaucoma detection using entropy sampling and ensemble learning for automatic optic cup and disc segmentation. Comput. Med. Imaging Graph. **55**, 28–41 (2017)
22. Kim, M., Zuallaert, J., De Neve, W.: Few-shot learning using a small-sized dataset of high-resolution FUNDUS images for glaucoma diagnosis. In: Proceedings of the 2nd International Workshop on Multimedia for Personal Health and Health Care. Association for Computing Machinery, New York, MMHealth, pp. 89–92 (2017)
23. Krizhevsky, A., Sutskever, I., Hinton, G.E.: Imagenet classification with deep convolutional neural networks. In: Advances in Neural Information Processing Systems, pp. 1097–1105 (2012)
24. Diaz-Pinto, A., Morales, S., Naranjo, V., et al.: CNNs for automatic glaucoma assessment using fundus images: an extensive validation. BioMed. Eng. OnLine **18**, 29 (2019). https://doi.org/10.1186/s12938-019-0649-y

# Detection of Jaundice in Neonates Using Artificial Intelligence

**S. Nihila, T. Rajalakshmi, Shradha Suman Panda, Nyelham Lhazay, and Ganga Devi Giri**

**Abstract** Neonatal Jaundice is a condition that occurs in infants in the first few weeks of their birth. Jaundice tends to appear when the bilirubin level surpasses 5 mg/dl in the infants during their birth in nearly sixty percent of the full-term babies and eighty percent of the preterm babies. Techniques for automatic Jaundice detection have been developed, but the outcome is not accurate enough to attract the doctors for their usage. The aim of this project is for the early detection of Jaundice in neonates non-invasively and with greater accuracy. This study involved 37 normal infants and 22 Jaundice prone infants. The images of the infants need to be captured with a smartphone camera. Algorithms like face detection, skin detection, colour map transformation and white balancing were applied to the specific region of interest (ROI), and eight quantitative features were extracted from the processed image. Principle component analysis (PCA) was used in this study to minimize data redundancy. PCA indicated that features like skewness, entropy, standard deviation and mean possess more information for further classification of normal and Jaundice affected neonates. Extracted statistical features are fed as an input to the machine learners such as SVM and ensemble regression for further classification of infants as Jaundice affected or normal. The analysis result indicates that out of 37 neonates, 22 were classified as Jaundice affected babies, and 15 were normal. The machine learning approach showed an accuracy of 81.1%

S. Nihila · T. Rajalakshmi (✉) · S. S. Panda · N. Lhazay · G. D. Giri
Department of Biomedical Engineering, SRMIST, Kattankulathur, India
e-mail: rajalakt@srmist.edu.in

S. Nihila
e-mail: nihila_sathish@srmuniv.edu.in

S. S. Panda
e-mail: shradha_kumar@srmuniv.edu.in

N. Lhazay
e-mail: nyelhamlhazay_@srmuniv.edu.in

G. D. Giri
e-mail: gangadevigiri_@srmuniv.edu.in



T. K. Sharma et al. (eds.), *Soft Computing: Theories and Applications*,
Advances in Intelligent Systems and Computing 1381,
https://doi.org/10.1007/978-981-16-1696-9_41

**Keywords** Neonatal Jaundice · Image processing · Skin detection · Face detection · PCA · Feature extraction · Machine learning

## 1 Introduction

Jaundice or hyperbilirubinemia is a symptom of elevated bilirubin level but not a disease if it is not taken care of at an early stage. Hyperbilirubinaemia can be caused by the lack of water in the body, lack of nutrients and blood exudation from blood vessels into the tissues [1]. Bilirubin level can be determined using a conventional method, that is, transcutaneous bilirubin meter. The methodology includes the measurement of serum bilirubin. The yellow coloration of skin is visible after 24 h after birth. Jaundice is observed in neonates if the bilirubin concentration crosses 2 mg/dl and below that could be considered normal [2]. Phototherapy is needed when the bilirubin exceeds 5 mg/Dl, depending on the neonates' weight. Jaundice can be divided into three types, i.e. physiological Jaundice, pathological Jaundice and breast-milk Jaundice. Bilirubin production in the newborn is two to three times per kilogram body weight than in adults [3].

Aydin et al. [4] suggested detecting Neonatal Jaundice using an advanced image processing technique. For analysing the captured images taken by smartphones, eight colour calibration cards were used. The colour calibration card can be used once and had to be disposed of after use and replaced with a new one. This process is followed to avoid the risk of infection in neonates. Moreover, unwanted improvement has been made by the removal of unwanted background in the image processing part. Taylor [5] et al. proposed a method for analysing digital images taken by smartphones to estimate the bilirubin level concentration in newborn infants having Jaundice. In this study, the colour calibration card is placed on the neonate's sternum, and a set of standardized images are taken for further analysis. Two separate decision rules were considered for determining the level of the newborn's bilirubin.

Miah et al. [6] proposed a method that utilizes only the eye's sclera region and is applicable only for adults. Since it does not apply to neonates, the proposed study will not focus on sclera detection. It is not possible to take sclera images from neonates since it can cause discomfort to their eyes by the flashing camera. Hence, image processing is done based on facial pigmentation. Falk and Jensen [7] encountered image inconsistencies, which led them to develop a novel colour correction solution, using Gaussian process regression (GPR), a machine learning approach that adapts exceptionally well to unseen colours. The colour correction solution is used to correct the ambient lighting conditions. Still, it is quite expensive, and so in this proposed study, we are using a white balancing algorithm to correct the same. Saini and Kumar [8] published a comparative study on three non-invasive Jaundice detection techniques—JM-103, BilliCheck and BilliCam. JM-103 uses two wavelengths and a dual optical path system and can detect bilirubin concentration up to 14 mg/dl. BilliCheck uses multiple wavelengths and spectrometers to measure bilirubin and detect bilirubin concentration up to 14 mg/dl. BilliCam uses a paper-based colour calibration card and an embedded camera to precisely detect bilirubin concentration up to 17 mg/dl.

This paper aims to develop a method for detecting Jaundice in infants using PCA to obtain statistically significant features and machine learners for classifying Jaundice and non-Jaundice babies. The study focuses on the face for statistical feature extraction instead of the whole-body region or sternum, as mentioned in other related studies. In this study, facial skin, as it is considered to be the most sensitive and thinnest region of our body, shows the symptom of Jaundice (yellow to orange pigmentation) in the initial stage and gradually spreads down to lower parts of the body in later stages, thereby ensuring early detection of Jaundice accurately. No colour checker or colour calibration card is used for that matter in this study; instead, a direct colour feature extraction is done for determining normal and neonates with Jaundice, thereby saving the cost. Any android or iOS camera of any resolution can be used to capture the baby's image. Research today is limited for Jaundice detection for neonates. The technique involves image processing and machine learning to detect Jaundice, which showed its robustness with a range of lighting conditions and varying skin tones. Hence, the proposed study focuses on the non-invasive Jaundice detection approach for neonates, thereby classifying the babies as normal or prone to Jaundice.

## 2 Materials and Method

In this proposed study, around 22 normal neonates and 15 Jaundice affected neonates were taken for the analysis. Pictures of the babies were taken using a smartphone, and these data were loaded to MATLAB R2018a® environment for processing. Secondly, these images were analysed using image processing techniques to detect the face. Face was considered for detection since it is the thinnest and sensitive part of the skin compared to other parts. After detection of face, a skin detection algorithm based on pixel similarity was done. Viola–Jones algorithm can be used to detect people's faces, noses, eyes, mouth or upper body face detection. Skin detection can be done using colour map transformation where the pixels are compared with that of the actual skin colour [8]. It is then followed by colour map transformation to convert images to various colour spaces. Finally, features like standard deviation, skewness, kurtosis, energy, entropy, homogeneity and correlation were extracted for distinguishing between Jaundice affected and normal babies.

Figure 1 shows the various steps and procedures involved to detect the presence of Jaundice non-invasively. The methodology for the detection of Jaundice is elaborated in detail in the below sections. A smartphone camera was used as the primary tool for data collection. Neonatal Jaundice was evaluated by a direct method; unlike the conventional way, blood samples are required for analysis. The camera of the smartphone was used to capture the picture of the neonate focussing on the face. These data (Photos) were loaded into the MATLAB environment for processing. The smartphone was placed in a downward-facing direction. The images were taken

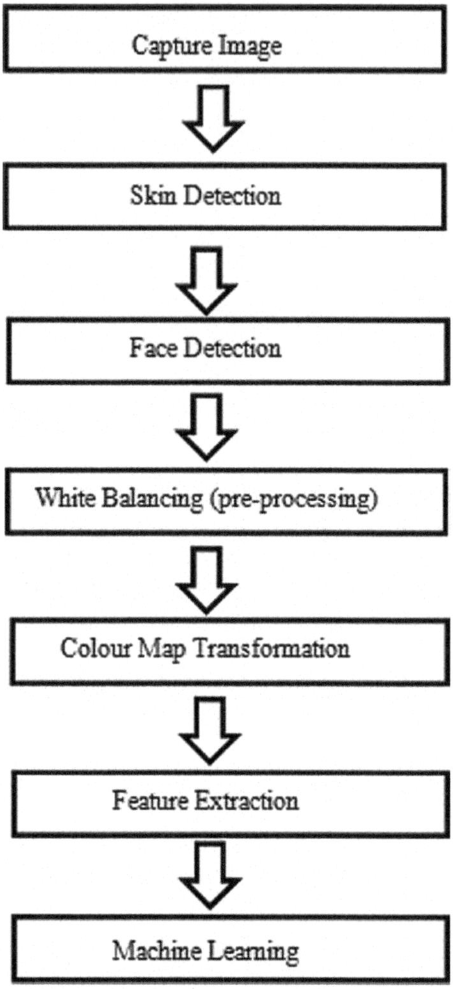


**Fig. 1** Block diagram for detection of Jaundice

after ensuring that the babies are in a calm posture. The skin colour thresholding method was implemented for skin detection where skin pixel is differentiated from non-skin pixels. Skin detection is performed by obtaining YCbCr colour spaces from the image. After the image's colour space transformation, the skin region was brought based on specific intervals of colour space components. The resultant image undergoes decision function based on the Chroma components, i.e. Cb and Cr from YCbCbr and Hue from HSV. It produces a skin map and accordingly has the output image [9]. In this study, skin detection is implemented, followed by a face detection algorithm to extract various statistical features from the processed images. Face detection identifies the human face(s) in images or videos by analysing the images and comparing the patterns. In this proposed study, the Viola–Jones face detection algorithm is considered.

White balancing (pre-processing) is done to remove unrealistic colours and adjust the intensities of the picture. After performing skin detection and face detection algorithms on the image, the detected face region is pre-processed by white balancing. After the white balancing colour map transformation is applied, colour map transformation is the process of converting an image in one colour space to another colour space. The transformed image looks like the original image to preserve the information. The image obtained after skin detection is in the RGB colour model and was converted to YCbCr and lab colour spaces. Hence, three colour channels were obtained for analysis. Furthermore, individual colour channels like R, G, B; L, a,b; Y, Cb, Cr were separated from their respective colour spaces and were used for feature calculations. The feature is the relevant information of an image that can be used for analysis. In this step, the images processed in YCbCr and lab colour spaces, and images in RGB colour space were taken, and for each colour channel, eight features were extracted, namely mean, standard deviation, Kurtosis, skewness, energy level, entropy, homogeneity and correlation. Principle component analysis was done to reduce complexity in high-dimensional data. Data are summarized, and dimensionality is reduced. PCA is implemented in the proposed study to identify features that are more statically significant. In this proposed study, out of the eight significant features extracted from the colour space models, only four features were considered more statistically significant features such as mean, entropy, standard deviation and skewness and were obtained after the application of PCA on it and further considered as an input to the machine learning and regression analysis. They act as a classifier in classifying the newborn infants as normal or affected with Jaundice and further aids in estimating the bilirubin level in neonates. For estimating the bilirubin level in newborn infants, two regression models were considered, i.e. the SVM regression model and ensemble regression model. In this study, one of the popular machine learning tools for regression, i.e. SVM regression analysis, is used. Training an SVM model is considered for obtaining greater accuracy on extracted data sets. It consists of all the main features which aid in characterizing the maximum margin algorithm. Proper selection of a particular kernel type and kernel function parameters is needed. Further, they are based on application-domain knowledge and should reflect the distribution of training data's input values. The advantage of using this model is that they can avoid the difficulties of using a linear function in high-dimensional feature space. In this proposed work, an improved machine learning algorithm, i.e. classification ensemble, is used for classifying the neonates as normal or affected by Jaundice and further aids in estimation of the bilirubin level. Measurement of the bilirubin level is important for obtaining the level of Jaundice in newborns. They are more effective when used with decision trees. Classification ensemble stores the data used for training and can resume training if required. The most significant advantage of using this model is that the classification rate and accuracy for Jaundice detection are raised compared to other classifiers. For enhancing the generalization performance and reducing the model uncertainty, ensemble classifiers have been extensively used.

## 3 Results and Discussion

In this proposed study, images from 37 subjects, of which 22 were normal, and the remaining 15 showed the symptoms of Jaundice were considered. These images are taken from Scientech and Pubmed databases. They are of unknown age and gender in specific. Figure 2 depicts the original image of the baby before image processing. Figure 3 shows the skin detected image after processing. Figure 4 shows the detected face region obtained by using Viola–Jones algorithm after skin detection. Figure 5 shows the white balanced image (pre-processed) after the removal of unwanted unnatural colours.

Figure 2 depicts the original image of the baby before image processing.

Figure 6a depicts the red channel extracted from the pre-processed image; the red color's brightness is obtained since green and blue appear lighter. Figure 6b illustrates the green channel extracted from the pre-processed image brightness of the green colour which is obtained since red and blue appear lighter. Figure 6c depicts the blue channel extracted from the pre-processed image. The brightness of the blue colour is obtained since green and red appear lighter. The RGB colour channels preserve the natural texture of the images, i.e. the information is preserved.

Figure 7a depicts the Y (luma) channel extracted from the pre-processed image. Y component filters the luminance (brightness) from the image and is more sensitive to human eyes and needs to be more accurate. Figure 7b depicts the chrominance blue (Cb) channel extracted from the pre-processed image, and Fig. 7c illustrates the chrominance red (Cr) channel extracted from the pre-processed image. Cb and Cr subtract red and blue components. However, it is not sensitive to human eyes and need not be accurate. This colour space is used to remove false colours. For YCbCr, they are linearity of transformation and stability of calculations.

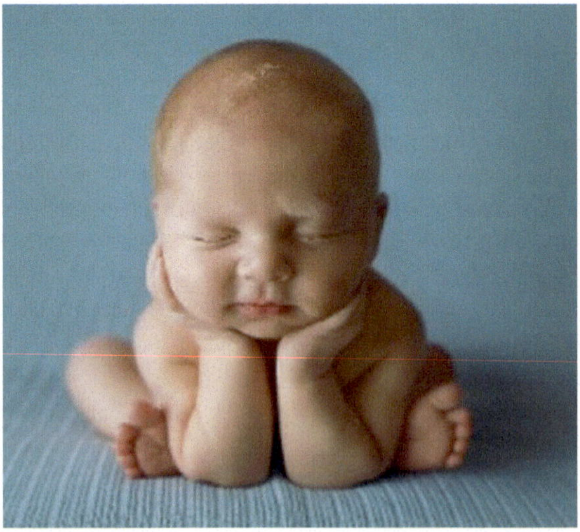

**Fig. 2** Original image

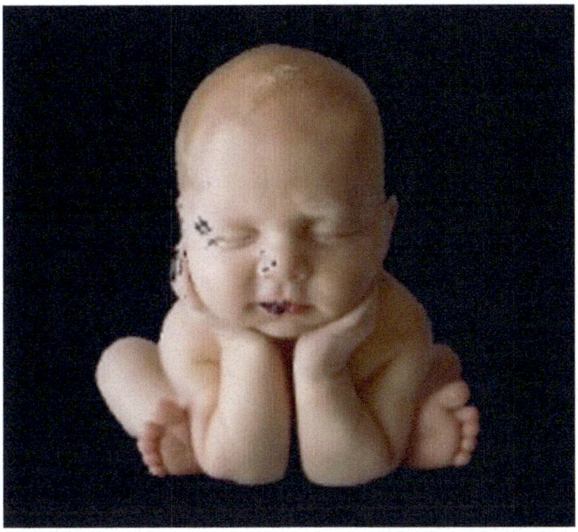

**Fig. 3** Skin detected image

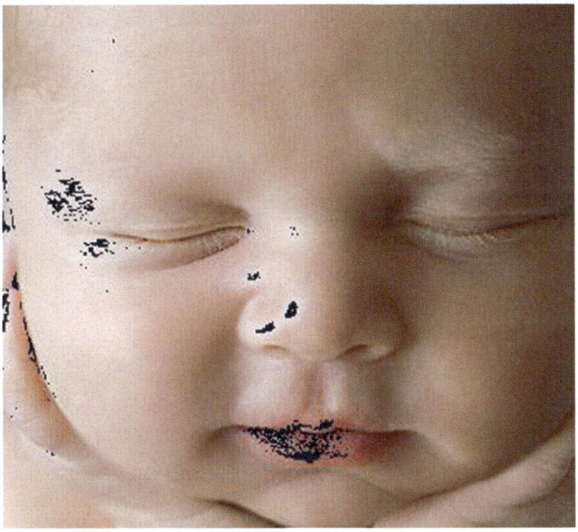

**Fig. 4** Face detected image

Figure 8a depicts the light (L) channel extracted from the pre-processed image. L gives information about how darker or lighter than the human eye sees colour. Figure 8b depicts the (ranges from red to green) channel extracted from the pre-processed image, and Fig. 8c depicts the b (ranges from blue to yellow) channel extracted from the pre-processed image. Colour channels a and b represent neutral colours unaffected by external factors, unlike RGB colour space, thereby providing precise colour value by finding a colour's attributes. For the lab channel, there is

**Fig. 5** Pre-processed image

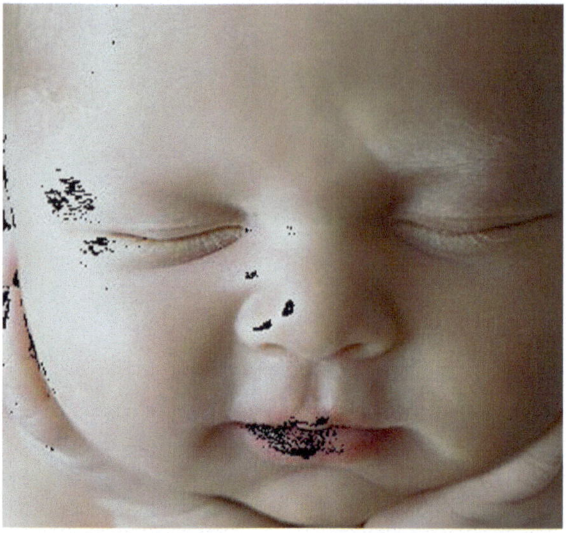

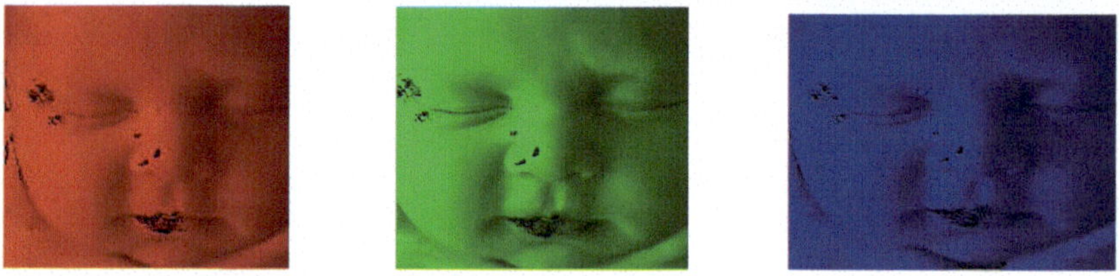

(a) Red Channel Image (b) Green Channel Image (c) Blue Channel Image

**Fig. 6** RGB colour space images

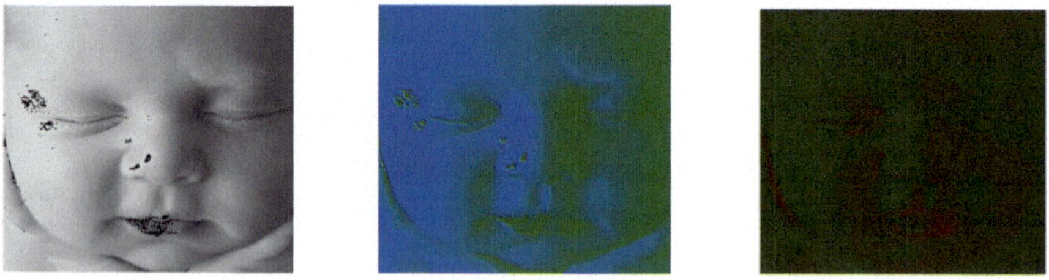

(a) Y Channel Image (b) Cb Channel Image (c) Cr Channel Image

**Fig. 7** YCbCr colour space images

(a) L Channel Image (b) a Channel Image (c) b Channel Image

**Fig. 8** Lab colour space images

**Table 1** Features extracted from colour spaces

| Features | Mean | Standard deviation | Entropy | Skewness | Correlation | Energy | Kurtosis | Homogeneity |
|---|---|---|---|---|---|---|---|---|
| Normal | 24.89 | 2.29 | 0.93 | 0.07 | 0.88 | 769.6 | 3.79 | 0.93 |
| Jaundice affected neonates | 37.4 | 5.49 | 0.82 | −0.25 | 0.90 | 849.0 | 3.22 | 0.94 |

the stability of calculation and perceptual uniformity. Table 1 depicts the value of statistical features extracted from color space.

Figure 9 depicts the results after the application of PCA. The prediction model is plotted against response (Condition of the neonate) and recorded number (number of subjects) where yellow dots represent the predicted value. The blue dots represent the true value. Variation and RMS errors between them were analysed. When all the features were taken as input to the machine learner, the RMS error was more, and

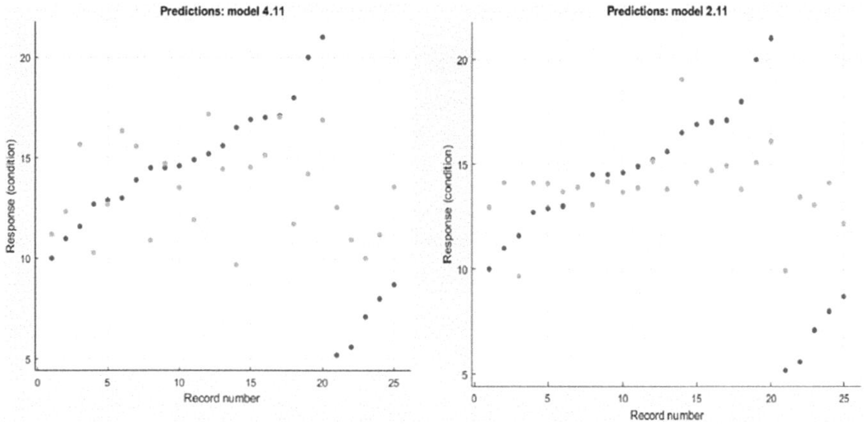


**Fig. 9** Output of principal component analysis

**Table 2** Statistical features extracted to detect the presence of Jaundice

| PCA selected features | Mean | Standard deviation | Entropy | Skewness |
|---|---|---|---|---|
| Normal neonates | 24.89 | 2.29 | 0.93 | 0.07 |
| Jaundice affected neonates | 37.40 | 5.49 | 0.82 | −0.25 |

variance was less, as shown by model 2.11. When features like energy, skewness, mean and standard deviation and the variance were maximum, and the RMS error was minimal.

Table 2 depicts the statistical features extracted after performing the principal component analysis (reduces data redundancy). Out of the eight features extracted from the colour space model, after the application of PCA, only four statistical features like mean, entropy, standard deviation and skewness were considered to be strongly significant. It is observed that for mean, the normal range is 17–30 approximately and in the case of babies affected by Jaundice ranges from 30 to 51 approximately; hence, it can be concluded that mean and standard deviation provide a better accurate range for analysis. It is observed that for standard deviation, the normal range is 1–3 approximately and in the case of babies affected by Jaundice ranges from 3 to 11 approximately. Entropy showed a slight variation among normal and Jaundice affected babies (For normal babies, it is slightly higher). However, skewness did not exhibit any specific pattern, and no accurate result can be inferred from this data.

Confusion matrix aids in the comparison of a classified image, i.e. the comparison between normal babies and Jaundice affected babies. It contains information regarding actual and predicted classifications done by a classification system. Figure 10 shows the SVM model's confusion matrix, where the x-axis represents the predicted class, and the y-axis represents true class, and the outcome shows the true positive rate and false-negative rate. The figure depicts that 20 out of 22 babies

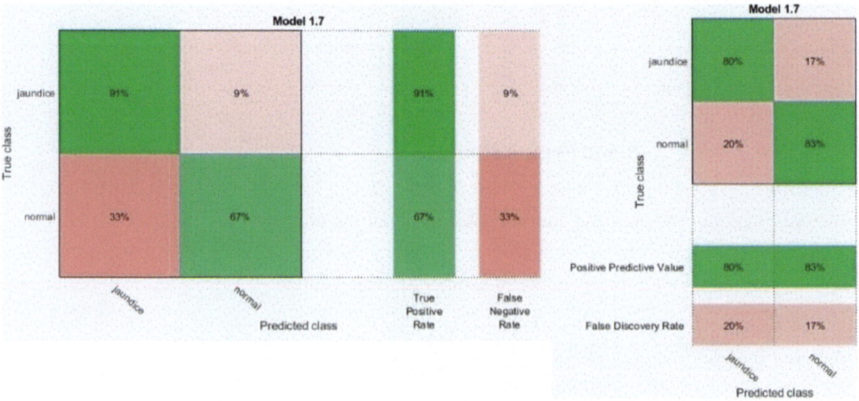


**Fig. 10** Confusion matrix for support vector machine classifier

were classified as Jaundice affected babies, 10 out of 15 babies were classified as normal, 2 babies who were affected by Jaundice were classified as normal, and 5 normal babies were classified as Jaundice affected babies. With the help of data in the matrix, the performance of such a system is classified.

Figure 10 shows the confusion matrix for the SVM model depicted that 20 out of 22 babies were classified as Jaundice affected babies, 10 out of 15 babies were classified as normal, 2 babies who were affected by Jaundice were classified as normal, and 5 normal babies were classified as Jaundice affected babies. Figure 11 shows the scatter plot that depicts the performance of the support vector machine with 81.1% accuracy graphically.

The ensemble regression model act as a classifier in classifying the babies as normal or Jaundice affected. The classification rate and accuracy for detecting Jaundice in neonates are analysed using this model. For reducing the model uncertainty, ensemble classifiers have been extensively used. Figure 12 shows the ensemble regression model's confusion matrix, where the x-axis represents the predicted class, and the y-axis represents the true class. The outcome shows the true positive rate and false-negative rate. Figure 12 further depicts that 16 out of 22 babies were classified as Jaundice affected babies, 11 out of 15 babies were classified as normal, 6 babies who were Jaundice affected were classified as normal, and 4 normal babies were classified as Jaundice. Figure 12 shows the confusion matrix for the ensemble regression model. It is observed that 16 out of 22 babies were classified as Jaundice affected babies, 11 out of 15 babies were classified as normal, 6 babies who were Jaundice affected were classified as normal, and 4 normal babies were classified as Jaundice. Figure 13 shows scatter plot that depicts the performance of the ensemble regression learner with 73% accuracy.

## 4 Conclusion

The proposed study is to detect Neonatal Jaundice at an early stage. The developed technique can be considered as a simple and accessible tool for self-assessing Neonatal Jaundice. The proposed system is accurate in diagnosing Neonatal Jaundice and can reduce the need for blood sampling. Of all the methods tested during this project, facial detection seemed more promising. This method could be used for all skin types and populations of varying skin pigmentation. Hyperbilirubinemia causes severe damage to the brain, and non-invasive monitoring of Jaundice using a smartphone can pick up such cases early. Using the extracted features, normal and Jaundice affected babies can be distinguished. Future work includes the extraction of more features from images and using more classifier models (machine learners) to determine the bilirubin level.

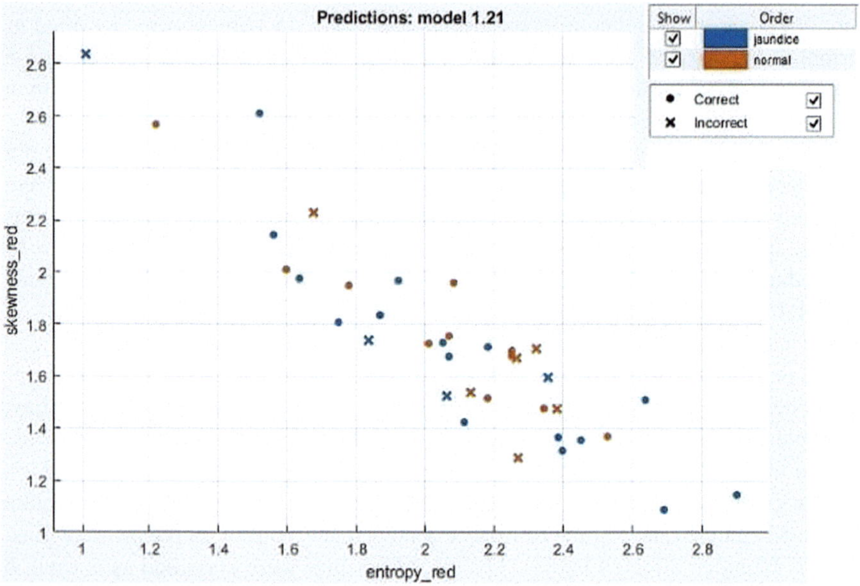


**Fig. 11** Scatter plot (Support vector machine)

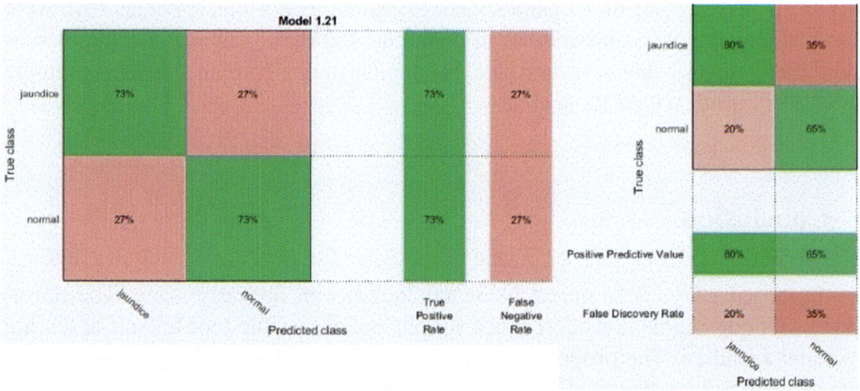


**Fig. 12** Confusion matrix (Shows the performance of the ensemble regression learner in percentage)

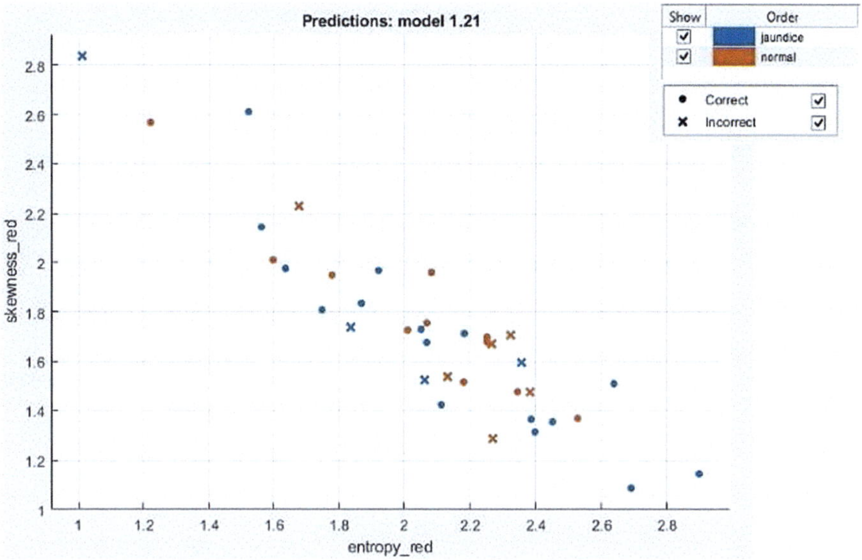


**Fig. 13** Scatter plot (Ensemble regression learner)

## References


1. Daunhawer, I., Kasser, S., Koch, G., et al.: Enhanced early prediction of clinically relevant neonatal hyperbilirubinemia with machine learning. Pediatr. Res. **86**, 122–127 (2019). https://doi.org/10.1038/s41390-019-0384-x
2. Abbas, M.W., Shamshad, T., Ashraf, M.A., Javaid, R.: Jaundice: a basic review. Int. J. Res. Med. Sci. **4**, 1313–1319 (2016)
3. Werwath, S.:. Automatic coding for neonatal jaundice from free text data using ensemble methods (2018). arXiv:1805.01054v1[cs.CL]
4. Aydin, M., Hardalaç, F., Ural, B., Karap, S.: Neonatal jaundice detection system. J. Med. Syst. **40** (2016) https://doi.org/10.1007/s10916-016-0523-4
5. Taylor, J.A., Stout, J.W., de Greef, L., Goel, M., Patel, S., Chung, E.K., et al.: Use of a smartphone app to assess neonatal Jaundice. Pediatr. e20170312 (2017)
6. Miah, M.M.M., et al.: Non-invasive bilirubin level quantification and jaundice detection by sclera image processing. In: IEEE Global Humanitarian Technology Conference (GHTC) (pp. 1–7), Seattle, WA, USA (2019). https://doi.org/https://doi.org/10.1109/GHTC46095.2019.9033059
7. Falk, H.H., Jensen, O.D.: A machine learning approach for jaundice detection using color corrected smartphone images. MS thesis. NTNU (2018)
8. Kumar, A., Saini, N.: Comparison of non-invasive Bilirubin detection techniques for Jaundice prediction. Int. J. Innov. Res. Electr. Electron. Instrument. Control Eng. **4**(5), 151–153 (2016)
9. Kolkur, S., Kalbande, D., et al.: Human Skin Detection using RGB, HSV, and YCbCr colour models. Adv. Intell. Syst. Res. **137**, 324–332 (2017)

# An Automatic Cardio Pulmonary Resuscitator—A Device for People Affected with Cardiac Arrest

**Sakthivel Sankaran, M. Pradeep Kumar, and C. Geethaanandhi**

**Abstract** Cardiac arrest is a sudden loss of electrical conduction to the heart that ceases the pumping action of the heart. This in turn may be due to some coronary artery disease. When a person with cardiac arrest has not been treated or given first aid within a few minutes, then it may lead to brain death followed by fatality. The common first aid given to people who have cardiac arrest is the cardio pulmonary resuscitation (CPR). CPR is a technique where the rescuer will take the chance to do the CPR on the victim. CPR techniques involve constant chest compressions followed by respiratory support periodically. In certain cases, CPR is required for almost an hour or more. During such cases, the rescuer cannot perform CPR for more than an hour. To overcome this problem, our team ended up with the design of an automated device for chest compressions and respiratory support and finally ensures the exact operation of CPR.

**Keywords** CPR · Cardiac arrest · Chest compression · Coronary artery disease

## 1 Introduction

In today's modern world, people are running behind technology and failed to take care of their health. Due to food habits and stressful life, people tend to have many diseases and disorders. Cardiac diseases are one such familiarized disease among people of all ages. Most of the cardiac diseases are mainly due to dietary, health imbalance, and mental imbalance. People nowadays do not bother about their health parameters which in turn caused cardiac diseases. Most cardiac diseases are related to the blockage of the vessels to the heart. The blockage of blood vessels may be due to the clogging of cholesterol on the walls of the blood vessel supplying the heart.

On analyzing the literature and many articles from renowned researchers worldwide, the CPR device is one of the most needed devices for emergency services. Like fire extinguishers and other emergency tools, CPR devices should be available

S. Sankaran (✉) · M. Pradeep Kumar · C. Geethaanandhi
Department of Biomedical Engineering, Kalasalingam Academy of Research and Education, Krishnankoil, India

 
T. K. Sharma et al. (eds.), *Soft Computing: Theories and Applications*,
Advances in Intelligent Systems and Computing 1381,
https://doi.org/10.1007/978-981-16-1696-9_42

everywhere to help people who have cardiac arrest all of a sudden. Researchers and developers across the world developed a prototype in such a way it satisfies the manual CPR method. In most of the articles when analyzed, we can realize that all these devices are developed to do chest compressions alone [1]. CPR is meant to provide chest compressions and ventilation support. If the device only provides chest compression, then the device is not suited for CPR. While using these devices, the ventilation is provided separately with the help of other devices. We as a team interfered with this problem and framed a prototype that ensures chest compression and ventilation support as well. This completes the method of CPR to its fullest. The flow of the article is in such a way that it introduces the term cardiac arrest and the sudden cardiac death along with the death toll due to cardiac arrest in Sect. 2. It is followed by the treatment for cardiac arrest and the technology comparison in Sect. 3. Section 4 followed the proposed methodology; in Sect. 5, the system description has been elaborated, and in Sect. 6, the working description has been stated. In Sect. 7, the importance and impact of the device have been briefed and have been concluded in Sect. 8.

## 2 Cardiac Arrest

### *2.1 Sudden Cardiac Arrest*

Cardiac arrest is a sudden cessation of the heart's activity due to loss of electrical conduction to the heart. In such a case, the heart will fail to do the pumping action. Once the heart does not pump, the blood supply to the brain will be stopped all of a sudden and the victim will be unconscious in less time. This condition is called sudden cardiac arrest. The causative factor of cardiac arrest may be due to the victim's health ailments viz. coronary heart disease, dysrhythmia, respiratory arrest, electrocution, drowning, choking, etc. A person will be given CPR when the victim is unconscious or pulse is weak or no pulse, no breathing, or difficulty in breathing.

### *2.2 Death Rate*

Cardiac arrest condition is also a heart ailment which is a rapid action that in turn may lead to brain death sometimes fatal. In the USA, the death rate due to CVD is declined by 41%; whereas in India, it rose around 34%. A recent review by Hygriv Rao collected data through various resources that evaluated the death toll in India in a particular state [2]. A total of 22,724 people were evaluated using medical students of the state, and the verbal autopsy was conducted. As a result, the mortality rate due to sudden cardiac death is about 10%. The patient's data has been analyzed from the physicians and surgeons of various hospitals and finally provided the statistics for the year 2012. When comparing the statistics with developed countries, it is very clear that the mortality rate in India due to sudden cardiac death is high. It has also been reported that the injury due to CPR is equally imposed on manual as well as automated procedures [3].

## 3 CPR Procedure

Cardiac arrest condition can occur to anyone anytime irrespective of their health condition. Cardio pulmonary resuscitation (CPR) is the first aid given to a victim's encountered cardiac arrest. CPR will be provided to those who are unconscious, and the pulse is weak or no pulse has been detected, when the victim has a lack of respiration or breathless condition, drowning, and choking [4]. The right procedures to perform the CPR are as follows: Place the palm of one hand on the center of the chest and place another hand above the first hand and club it together. Keep the arms straight then push hard and fast at the rate of a minimum of 100 pushes per minute and to a depth of 2–2.5 in. [5]. After every 30 compressions of the chest, keep the victim head upright and hold it to perform ventilation support. The ventilation support can be provided by mouth-to-mouth resuscitation or by keeping the patient on a ventilator. If the CPR has not been provided in 0–4 min of the patient's collapse, there may be brain damage unlikely. When it crosses 4–6 min, there may be possible chances for brain damage. If CPR has not been provided after 6–10 min of the patient's collapse, then there will be more probability of brain damage. If the timeline crosses above 10 min, then there is more probability of brain death [6].

### *3.1 Automated Chest Compression Device (ACCD)*

Automated chest compression devices are one of the advancements in providing the major part of CPR, i.e., chest compression. Many researchers and health care providers developed models and products for automatic chest compressions. One such device that is currently available in the market for performing chest compression is LUCAS™ CPR [7]. This automated mechanical chest compression device has piston-like support that hits the chest and does the process of chest compression as per the fixed rate. Apart from LUCAS, there are other devices like auto pulse by ZOLL circulation Inc. that is also available that provide high-quality chest compressions to the patients and is widely been used in many hospitals [8].

One of the major limitations is that these devices are designed to provide only the chest compression part of CPR. CPR is satisfied with chest compression followed by pulmonary support. These devices, hence, do not complete the functionality of CPR [1].

## 4 Proposed Methodology

Our team has decided to lift the limitations of CPR devices and designed a novel device that can provide continuous chest compression followed by pulmonary support as well. This novel device, thus, satisfies the procedure of CPR completely and will act as an emergent device to serve the people affected with cardiac arrest. The device is designed in such a way that for every 15 s of chest compression will be interrupted by pulmonary support for 5 s. The proposed method involves the usage of one motor for chest compression; whereas for pulmonary support, another motor has been used. The chest compression part of the device uses six linked gears with the motor. The clockwise and counter-clockwise directional movement of the motor ensures the movement of the band attached to it on the ends. Figure 1 explains the complete block diagram of the device.

For pulmonary support, we took the concept of an air blower. We designed a small blower like device using a motor fixed with blades at the end which acts as a fan. Further, it has been enclosed in a closed chamber where the tip is provided with a hose for connection with the ventilation tube. This, in turn, helps the victim to have direct support from the device rather than mouth to mouth resuscitation. These two motors are controlled parallel using an Arduino microcontroller. The proper delay

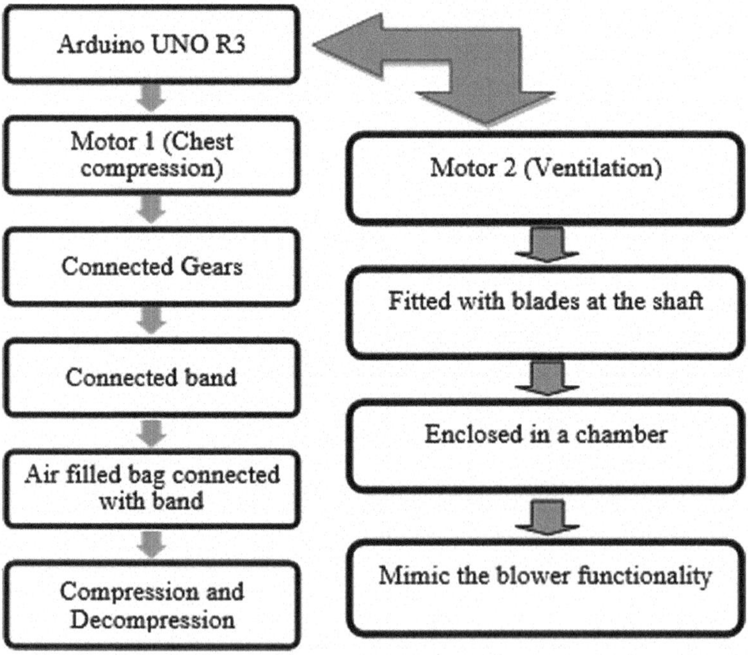


**Fig. 1** Block diagram of automated CPR device

has been provided to the controller to control these two motors which in turn satisfies the procedural operation of cardiopulmonary resuscitation.

## 5 System Description

The automated cardiopulmonary resuscitator designed by our team has used a 12V DC motor to drive the main frame of the work. As we have designed a prototype to show the proof of concept, we used a 12 V DC motor. In the future, if there is a need to give more pressure, we, in turn, will switch to some higher capacity motor. The design of the chest compression device and ventilation support device is simulated with help of CAD software (Figs. 2 and 3). The work plane of the chest compressing unit comprises two gears and four pulleys inter-connected with one another. The two gears play a major role in directing the functionality of the chest compression unit. A microcontroller has been used to control the motor to rotate clockwise and counter clockwise as well. The ventilation part has been designed in such a way that it acts as a blower. The design involves a closed chamber with a hole at the bottom to draw air from the environment. A motor is fixed inside the chamber in such a way that the shaft of the motor is inside the chamber. The shaft is fitted with small blades that when driving acts as a blower and are controlled by the microcontroller.

**Fig. 2** Design of chest compression device

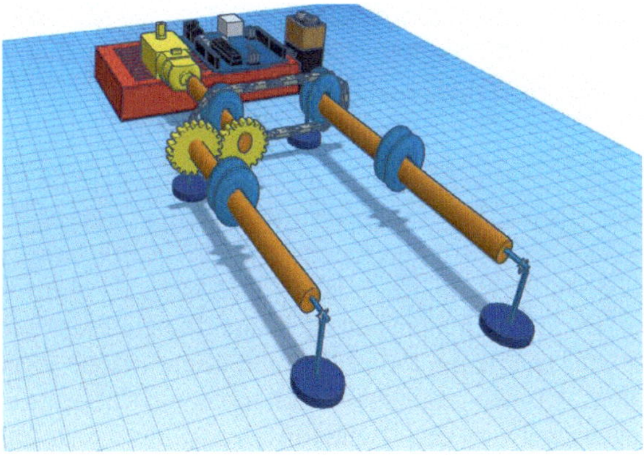

**Fig. 3** Design of ventilation support device

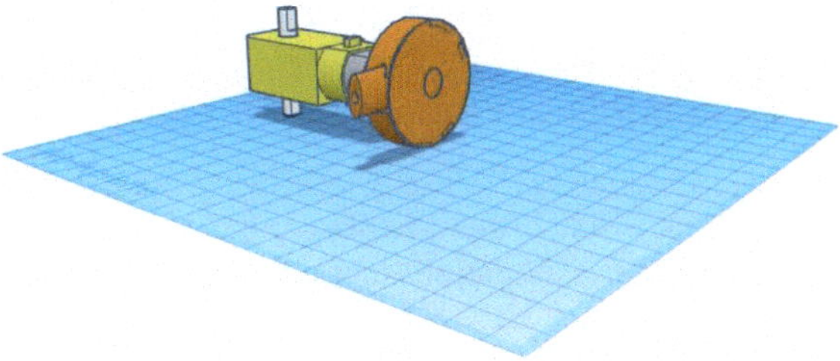

### 5.1 DC Motor

A DC motor is an electrical device that uses direct current to convert electrical energy to mechanical energy. The energy conversion has been achieved with the help of electromagnetism where two permanent magnets have been placed on the sides that act as repulsing energy. The DC motor is mainly of two types brushed and brushless motors. The brushed DC motor uses a small unit between the two poles of the magnet, whereas the brushless motor does not use it. Mechanical energy generated by the motor is used to drive the belt attached to the pulley part.

### 5.2 Arduino UNO R3

Arduino UNO R3 is an open-source microcontroller based on a microchip ATmega328P microcontroller developed by Arduino cc. This microcontroller board can be interfaced with any of the external modules, and there are specific expansion boards and circuits. It has integrated I/O pins that include 14 digital I/O pins and 6 analog I/O pins. It is programmed using Arduino integrated development environment (IDE) using a type B USB cable. It is powered using the type B USB cable or 7–20 V power supply. The ATmega328 comes with a boot loader which allows the user to upload a new code to the board without using an external programmer.

### 5.3 L293D Motor Driver

The motor driver is an electronic integrated circuit chip that is used to control the motor as per the code or instruction from the user. It acts as an interface between the motor and Arduino. The L293D driver chip consists of two H-bridges. H-bridge is one of the simplest circuits which are used to drive low current rated motors. It has 16 pins with 8 pins on each side. It includes two input pins and two output pins and one enable pin for each motor. There will be separate pins for Vcc and ground. The IC has two grounds for each motor since it deals with heavy currents. So, the IC gets heated very easily, and thus, it requires a heat sink to release the heat. The two ground pins are at a distance where the heat will be released through the gap.

## 6 Working Description

As we already stated that the device has two units one for chest compression and another one for ventilation support. The working descriptions of each unit are discussed below.

### 6.1 Chest Compression Device

An automated chest compression device uses the motor which is programmed in such a way that it rotates clockwise and counter clockwise direction. Two gears have been fixed parallel to ensure the movement concurrently. One gear is fixed directly to the motor shaft where the other gear is allowed to touch the sides of the first gear. When the motor rotates clockwise, the first gear, in turn, rotates the second gear counter clockwise. Pulley has been fixed with a shaft from the gears, and in the end, the band has been fixed. The other end of the bands will be fixed to the air-filled bag which will be placed over the chest at the center. The air-filled bag is used to ensure the axial compression of the chest. If it has not been used, then the bands will do lateral compression which is not needed for CPR.

### 6.2 Ventilation Support Device

The ventilation support device's design is inspired by air blowers. The design includes a closed chamber with an opening for a 5 V DC motor shaft to enter inside. The other end of the chamber has a hose to allow the external air. The shaft of the motor is affixed with blades that allow the setup to act like a fan. The border of the chamber has a small opening for a pipeline to the face mask which will be connected to the patient at the end. This setup entirely supports ventilation and is controlled by the Arduino as per the timing prescribed for the device.

Arduino is programmed in such a way that it operates the chest compression device's motor for 15 s and then drives the ventilation support motor for 5 s after the driving of the chest compression motor. This completes the functionality of the CPR procedure, and it will also act as a life-supporting device.

## 7 Importance and Impact

The automated CPR device can be useful as an emergency unit that serves people during cardiac arrest. CPR procedure is a life-saving process; if it has not been done on time and perfectly following the procedures, then it may lead to fatality. To overcome these complications, this device can serve well as a life savior which repeats the procedure of chest compression followed by ventilation support. People who assist during the CPR procedure may be hesitant to do the mouth-to-mouth resuscitation, whereas AMBU bags need the right positioning of the victim's head which cannot be done by a common man. The ventilation support device is designed in such a way that it provides continuous ventilation support to the victim without any time delay. This will reduce the amount of death toll due to cardiac arrest and has a huge impact in developing countries like India.

## 8 Conclusion

Cardio pulmonary resuscitation is the emergency life-saving procedure applied to people who encountered sudden cardiac arrest. If the CPR has not been done on time to the victim correctly, then in less than 10 minutes, there may be a higher risk of brain death which may lead to fatality. To enhance the perfection of CPR, the device designed will ensure the right CPR procedure, provide maximum support, and also act as a life savior at the right time. It serves as emergency equipment at public places with an easy user interface, anyone can operate it perfectly, thus, ensures CPR has been given at the right time. Comparing with the other chest compression devices which only provide chest compression for CPR, this device provides complete functionality of CPR.

## 9 Future Work

To enhance the perfection and accuracy of the device, the programming for the same can be done using LabVIEW technology. Using this type of programming will also provide the maximum support for the interfacing which has been proven by the previous applications [9–11].

**Acknowledgements** The authors would like to thank the management of Kalasalingam Academy of Research and Education. The authors also extend their regards to the department of Biomedical Engineering for their support. The authors extend their sincere thanks to KLU NI- Technology Innovation Centre for providing computational and hardware testing facilities.

## References

1. Remino, C., Baronio, M., Pellegrini, N.: Automatic and manual devices for cardiopulmonary resuscitation: a review. Adv. Mech. Eng. **10**(1), 1–14 (2018). https://doi.org/10.1177/1687814017748749
2. Hygriv Rao, B.: Global burden of sudden cardiac death and insights from India. Indian Heart J. **66**(Suppl1), S18–S23 (2014). https://doi.org/10.1016/j.ihj.2013.11.009
3. Ondruschka, B., Baier, C., Bayer, R., Hammer, N., Drebler, J., Bernhard, M.: Chest compression-associated injuries in cardiac arrest patients treatedwith manual chest compressionsversus automated chest compression devices (LUCAS II)—a forensic autopsy-based comparison. Forens. Sci. Med. Pathol. J. **14**(4), 515–525 (2018). https://doi.org/10.1007/s12024-018-0024-5
4. Poole, K., Couper, K., Smyth, M.A., Yeung, J., Perkins, G.D.: Mechanical CPR: Who? When? How? Crit. Care **22**, 140 (2018). https://doi.org/10.1186/s13054-018-2059-0
5. Yeung, J., Meeks, R., Edelson, D., Gao, F., Soar, J., Perkins, G.D.: The use of CPR feedback/prompt devices during training and CPR performance: a systematic review. Resuscitation J. **80** (7), 743–751 (2009). https://doi.org/10.1016/j.resuscitation.2009.04.012
6. National Healthcare Provider Solutions, https://nhcps.com/world-first- aid-day-a-day-dedicated-to-saving-lives/

7. RuberttsonS, E.L., Smekal, D.: Mechanical chest compressions and simultaneous defibrillation vs. conventional cardiopulmonary resuscitation in out-of-hospital cardiac arrest the LINC Randomized Trial-JAMA. **311**(1), 53–61 (2014). https://doi.org/10.1001/jama.2013.282538
8. Couper, K., Smyth, M., Perkins, G.D.: Mechanical devices for chest com- pression: to use or not to use? Curr. Opin. Crit. Care **21**(3), 188–194 (2015). https://doi.org/10.1097/MCC.0000000000000200
9. Sankaran, S., et al.: A survey report on the emerging technologies on assistive device for visually challenged people for analyzing traffic rules. In: 2020 International Conference on Communication and Signal Processing (ICCSP) (pp. 0582–0587), Chennai, India (2020). https://doi.org/10.1109/ICCSP48568.2020.9182335
10. Sankaran, S., Pallikonda Rajasekaran, M., Govindaraj, V., Raja Mahaif, R., Chinnaiah, G., Kamarajan, R., et al.: Assistive device for visually challenged people to read books using lab view. Int. J. Innov. Technol. Explor. Eng. (IJITEE) **9**(2S2) (2019). ISSN: 2278-3075
11. Kumar, M.P., Thilagaraj, M., Sakthivel, S., Maduraiveeran, C., Rajasekaran, M.P., Rama, S.: Sign language translator using LabVIEW enabled with internet of things. In: Smart Intelligent Computing and Applications, pp. 603–612 (2019)

# Dysphagia and Silent Aspiration Monitoring for Patients with Neurological Disorder Using Pharyngeal Deglutitive Biomechanics

Dhivya Balaji, G. Niranjana, Usha Dalvi, and Varshini Karthik

**Abstract** Deglutition is the act of engulfing food or drink to pass down the throat into the digestive tract. It utilizes tongue and several smooth muscles of the pharynx and oesophagus. The initiation of swallowing is voluntary, but once the deglutition process starts, it becomes involuntary. Swallowing happens in four phases. An issue in any of the phases of swallowing leads to many swallowing disorders, dysphagia being one among them. Neurological degenerative conditions like Parkinson's disease and stroke tend to affect the various phases of swallowing in the long run. Our study focuses on the pharyngeal phase. The most important aspect of the pharyngeal phase is the superior and anterior movement of the hyoid bone and larynx, to keep the air ducts out of the pathway of the bolus. Conditions leading to the impairment of the muscles and nerves related to the pharyngeal phase lead to aspirations and swallowing troubles. This work aims at the development of a simple and cost-effective means for real-time assessment of deglutition problems. It was observed that the movements of the hyoid bone can give signals that hint difficulty in swallowing. The waveforms obtained show distinctive features for various consistencies (solid, semi-solid, liquid) of bolus. The swallowing signals are found to be different between healthy and neurological disorder subjects. Hence, the prototype designed can be used as a candidate to pick up and analyse deglutition sounds, and on miniaturization, the prototype can be used for clinical correlation.

**Keywords** Swallowing · Dysphagia · Aspirations · Neurological · Bolus · Pharyngeal

---

D. Balaji · G. Niranjana · U. Dalvi · V. Karthik (✉)
SRM Institute of Science and Technology, Kattankulathur, India
e-mail: varshink@srmist.edu.in

D. Balaji
e-mail: dhivya_balaji@srmist.edu.in

G. Niranjana
e-mail: niranjana_gangaprasad@srmist.edu.in

U. Dalvi
e-mail: ushaa@srmist.edu.in

 
T. K. Sharma et al. (eds.), *Soft Computing: Theories and Applications*,
Advances in Intelligent Systems and Computing 1381,
https://doi.org/10.1007/978-981-16-1696-9_43

## 1 Introduction

Deglutition (swallowing) is a process by which the food in any form such as solid or liquid passes from the mouth through the throat, the pharynx and the oesophagus with the epiglottis remaining closed [1, 2] (Fig. 1). Bolus is a mixture of the food entering the oral cavity and the saliva. This occurs during the oral phase of swallowing. If the food enters through the trachea, then various abnormalities arise. The process by which food passes through trachea is called pulmonary aspiration. It results in the bolus entering the lower respiratory tracts. After the oral phase, the food passes through the back of the mouth by the help of the tongue, thereby entering the pharyngeal phase of swallowing. In this phase, the bolus passes through the folds of the cavity such that it does not enter the airway. Furthermore, airway protection is also provided by the raising of the larynx and the covering by the epiglottis. The bolus then passes through the final stage of swallowing which is the oesophageal phase where the bolus enters the oesophagus, a muscular tube, from where it is transferred through other parts of the alimentary canal through peristaltic motions. If the bolus enters, the airway aspirations occur.

Dysphagia [3] is the medical term used to denote any difficulty in swallowing and is common in patients with neuro-motor impairment [4]. It can be of two types based on the cause of the difficulty. If the difficulty to swallow is because of the bolus being stuck in the neck or chest, then the type of dysphagia is oesophageal. If the difficulty is found in initiating a swallow, then it is termed as oropharyngeal dysphagia. The fibre-optic techniques and the video-fluoroscopic techniques that are being followed currently are invasive examinations for dysphagia and require

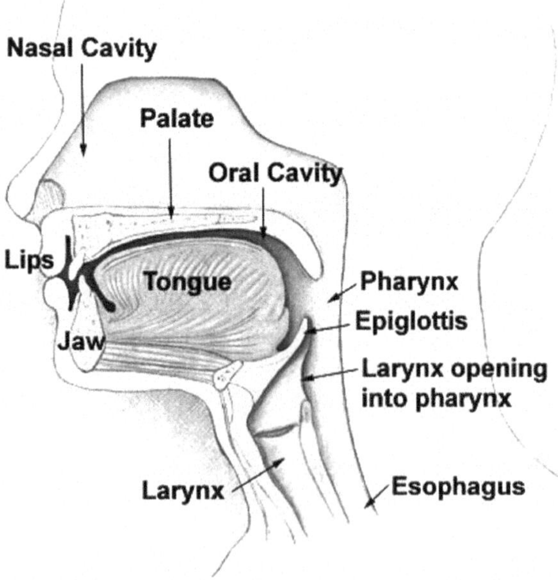


Fig. 1 Anatomy of the mouth and the parts of the oral cavity involved in swallowing [2]

sophisticated biomedical equipments. There are also several latest and non-invasive procedures in the literature [5–16] that have been used to obtain swallowing signals and detect aspirations.

This work aims at presenting a device that can monitor aspirations in real time and is also cost effective when compared to existing standard fibre optic and fluoroscopic techniques. The biomechanics of the hyoid bone are recorded using a piezoelectric sensor that is placed on the hyoid bone externally. These signals are then represented as waveforms after being amplified using an amplifying and filtering circuit. Signal analysis is performed thereafter to detect swallowing problems and silent aspirations in comparison with normal swallowing signals.

# 2 Materials and Methods

## *2.1 Sensors and Circuitry*

Piezoelectric sensors are used to measure physiological displacements. Piezoelectric materials generate an electric potential when mechanically strained, and conversely, an electric potential can cause physical deformation of the material. The principle of operation is that, when an asymmetrical crystal lattice is distorted, a charge reorientation takes place, causing a relative displacement of negative and positive charges. The displaced internal charges induce surfaces of opposite polarity on opposite sides of the crystal. Surface charge can be determined by measuring the difference in voltage between electrodes attached to the surfaces. The total induced charge is directly proportional to the applied force [17]. Here, we have used the piezoelectric phenomenon of generation of electric potential by the mechanical strain produced due to the movement of the hyoid bone during swallowing. The sensor is made up of ceramic material, whereas the metal plate covering is made up of nickel–brass alloy. The sensor is of 4000 Hz oscillating frequency type. It was observed that the normal frequency for swallowing recorded by bend sensors was of the frequency of 1000 Hz [18]. Also, there are chances of the sensor electrode getting corroded if the signal to be picked up is of a frequency that is very much greater than the sensor frequency type [19] with no prior knowledge about the exact range of abnormal frequency produced during swallowing. Considering all these factors, a 4000Hz sensor was used.

The signals picked up are of very low amplitude. So, an amplification circuit was used to amplify the signals to a considerable amount. The circuitry used in the study has two piezoelectric sensors that pick up movement signals non-invasively from the throat at the region of the hyoid bone. Each of the two sensors are then connected to an instrumentation amplifier that amplifies the electrically modified mechanical signals as picked up by the sensor. The outputs from the amplifiers are then allowed to pass through a low pass filter from where the output is taken to a digital storage oscilloscope for recording and monitoring the signals on a real-time basis. Also, there are two 9V batteries connected to the circuit. Instrumentation amplifier is a type of

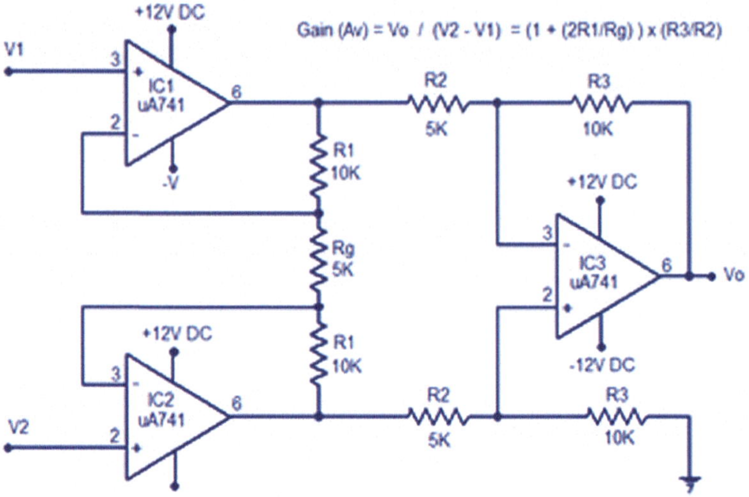


**Fig. 2** Instrumentation amplifier circuit

differential amplifier that has additional input buffer stages. This helps in impedance matching. These amplifiers are commonly used in industrial test and measurement application. The instrumentation amplifier also has some useful features like low offset voltage, high input resistance, high gain, etc.

The voltage gain of the instrumentation amplifier can be expressed by using the equation below where $R1$, $R2$, $R3$, Rg are resistors, Vo, $V1$, $V2$ are voltages, and Av is the voltage gain.

$$\text{Av} = \text{Vo}/(V2 - V1) = (1 + 2R1/\text{Rg}) \times R3/R2 \tag{1}$$

Practical instrumentation amplifier circuit shown in (Fig. 2.) has been used in this study. The amplifier operates using +/−12V DC. Resistor values of 10k, 5k, 10K and 5K were used for $R1$, $R2$, $R3$ and Rg, respectively, and the voltage gain (Av) produced at the output would be around 10 as obtained from Eq. (1). Also, a potentiometer with an adjustable gain of up to 5K is used to have a variable gain. The circuit board developed as a prototype is shown in Fig. 3.

## 2.2 *Experimental Procedure*

The two electrodes, i.e. the sensors are placed horizontally on the either side of the neck region where the hyoid bone movements are felt for the subject as shown in Fig. 4. The exact positioning can vary slightly from one person to another. Once the position is finalized, the sensors are adhered to the patient's skin using a medical tape which in future can be incorporated in a collar band. The batteries are connected to

**Fig. 3** Sensors and circuit board used in the prototype

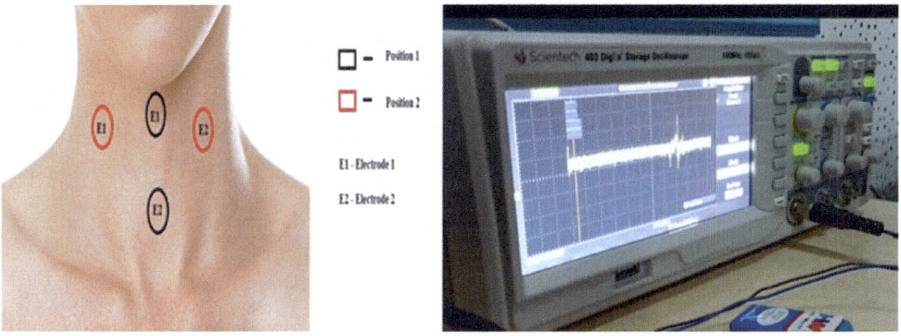


**Fig. 4** Electrode placement and sample signal displayed on an oscilloscope

the respective slots, and the signal acquisition is performed using a digital storage oscilloscope.

The real-time signals obtained from the person under study can be observed on the screen of the oscilloscope. As the person swallows, distinct waveforms appear on the screen of the oscilloscope. The signals were taken for the swallow of various types of foods such as solids and liquids and also with only saliva for the purpose of study. Also, for the positioning of the electrodes, various placements were experimented surrounding the throat. The best and similar results were observed in two of those positions. One such position is being vertically placed above and below the hyoid bone as seen as position 1 in Fig. 4 or being placed on either sides of the hyoid bone as seen as position 2 in Fig. 4. The waveforms produced can be easily observed on the oscilloscope screen as shown in Fig. 4. Such variations can be useful for nurses feeding food to patients suffering from any neural disorder or swallowing trouble, to monitor any abnormality such as silent aspirations, while the patient is swallowing. Also, the undesired muscle noise and other artefacts are removed using low pass filter.

## 3 Results and Discussion

The signals were acquired for 35 people of different age groups spanning from 10–70 years of age. Institute ethics clearance has been obtained (ethics clearance number-1544/IEC/2018) for the same. The peak values obtained from the signals will show the variation in the amplitudes between them. No similar patterns between the signals were observed. The sample swallowing waveforms are shown in Fig. 5. The waveforms obtained have been analysed under two categories. Category 1: Various bolus consistencies and Category 2: Healthy versus neurological disorder subjects.

### *3.1 Category 1: Various Bolus Consistencies*

Subjects were asked to swallow different consistencies of food to check the bolus swallowing for variation in signal waveform. The different consistencies considered were solid, semi-solid and liquid. Sample waveforms are shown in Fig. 5. Distinctive difference was observed between the three bolus consistencies which can be used to detect silent aspirations even for bolus type variations.

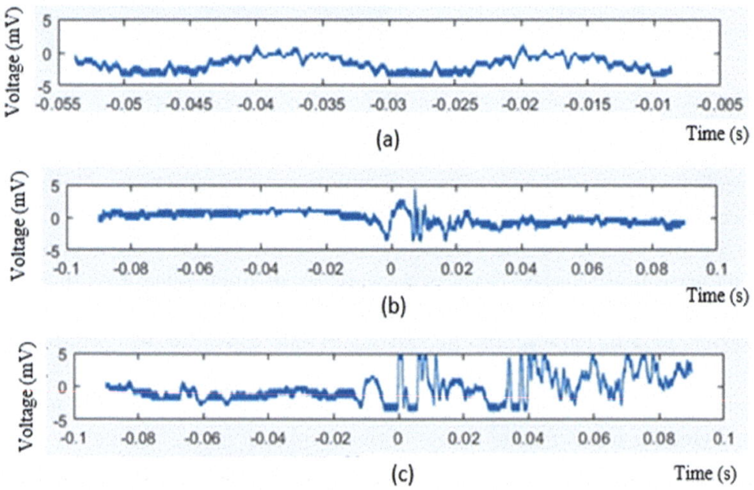


**Fig. 5** Sample waveforms showing swallowing signals for different bolus (food) consistencies **a** solid **b** semi-solid **c** liquid.

### 3.2 *Category 2: Healthy Versus Neurological disorder subjects*

A comparative study was done between healthy subjects and neurological disorder subjects. Fig. 6 shows the comparison (different bolus consistencies) for the two kinds of subjects. It is clearly observed that signals obtained for neurological disorder (stroke) subjects are different from normal subjects. It was observed that there was an increase in voltage values when the difficulty of swallowing increased.

From the study using the prototype developed, it is concluded that the signals from the movement of hyoid bone showed almost constant voltage whereas for coughing (due to silent aspirations) of the patient the voltages decreased considerably. Any difficulty in swallowing had an increase in voltage values. So, the future plan of work would be to develop an alarm that could give notifications or ring on a mobile phone by detecting any such fluctuations of the voltage than a preset threshold value. This set threshold value has to be manually arrived at after a preliminary test as the value varies from one person to another and also with age.

## 4 Conclusion and Future Work

The sensor of the prototype developed is connected to a simple circuit consisting of an instrumentation amplifier and low pass filters. The movements of deglutition were recorded by the sensor from subjects of different age groups. The threshold for normal and abnormal swallowing for individuals will have to be customized. The prototype could be miniaturized further, and an additional arrangement to pick up deglutition sounds for clinical correlation could be developed in the future. The prototype developed could be of help to the clinicians and also the nurses to monitor the patient condition on a real-time basis while feeding food and also at rest.

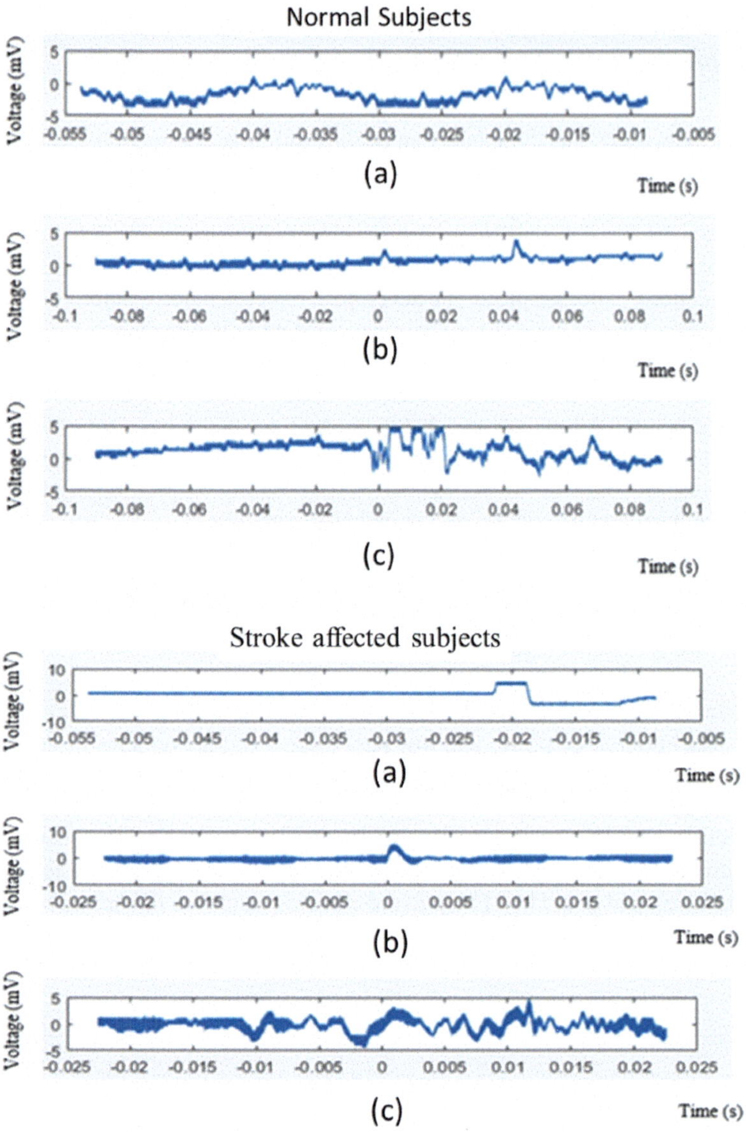


**Fig. 6** Sample waveform showing swallowing signals for normal healthy subject (top) and stroke affected subject (bottom) for various bolus (food) consistencies **a** solid **b** semi-solid **c** liquid

**Acknowledgements** We acknowledge the Audiology and Speech Pathology Department, SRM Medical College and Hospital, for the support in testing the device

## References


1. Shaker, R., Belafsky, P.C., Postma, G.N., Easterling, C. (eds.): Principles of deglutition: a multidisciplinary text for swallowing and its disorders. Springer Science & Business Media (2012)
2. Short, M.J., Levin-Goldstein, D.: Head. Cengage Learning, Neck and Dental Anatomy (2016)
3. Groher, M.E., Crary, M.A.: Dysphagia-E-Book: clinical management in adults and children. Elsevier Health Sciences (2020)
4. Krishnan, G., Prasad, T.: Compensatory strategies and feedback of hyolaryngeal excursion for swallow in treatment of swallowing difficulties in parkinsons disease: a case report. J. All India Inst. Speech Hearing., **32** (2013)
5. Sarraf-Shirazi, S., Baril, J.F., Moussavi, Z.: Characteristics of the swallowing sounds recorded in the ear, nose and on trachea. Med. Biol. Eng. Compu. **50**(8), 885–890 (2012)
6. Lazareck, L.J., Moussavi, Z.M.: Classification of normal and dysphagic swallows by acoustical means. IEEE Trans. Biomed. Eng. **51**(12), 2103–2112 (2004)
7. Spadotto, A.A., Pereira, J.C., Guido, R.C., Papa, J.P., Falcao, A.X., Gatto, A.R., Schelp, A.O. et al.: Oropharyngeal dysphagia identification using wavelets and optimum path forest. In: 2008 3rd International Symposium on Communications, Control and Signal Processing, pp. 735–740. IEEE (Mar 2008)
8. Sazonov, E.S., Makeyev, O., Schuckers, S., Lopez-Meyer, P., Melanson, E.L., Neuman, M.R.: Automatic detection of swallowing events by acoustical means for applications of monitoring of ingestive behavior. IEEE Trans. Biomed. Eng. **57**(3), 626–633 (2009)
9. Kandori, A., Yamamoto, T., Sano, Y., Oonuma, M., Miyashita, T., Murata, M., Sakoda, S.: Simple magnetic swallowing detection system. IEEE Sens. J. **12**(4), 805–811 (2011)
10. Jayatilake, D., Suzuki, K., Teramoto, Y., Ueno, T., Nakai, K., Hidaka, K., Matsumura, A. et al.: Swallowscope: a smartphone based device for the assessment of swallowing ability. In: IEEE-EMBS International Conference on Biomedical and Health Informatics (BHI), pp. 697–700. IEEE, (Jun 2014)
11. Marik, P.E., Kaplan, D.: Aspiration pneumonia and dysphagia in the elderly. Chest **124**(1), 328–336 (2003)
12. Jones, B. (ed.): Normal and abnormal swallowing: imaging in diagnosis and therapy. Springer Science & Business Media (2012)
13. Khalifa, Y., Coyle, J.L., Sejdić, E.: Non-invasive identification of swallows via deep learning in high resolution cervical auscultation recordings. Sci. Rep. **10**(1), 1–13 (2020)
14. Perry, B.J., Stipancic, K.L., Martino, R., Plowman, E.K., Green, J.R.: Biomechanical biomarkers of tongue impairment during swallowing in persons diagnosed with amyotrophic lateral sclerosis. Dysphagia (2020)
15. Taveira, K.V.M., Santos, R.S., Leão, B.L.C.D., Stechman Neto, J., Pernambuco, L., Silva, L.K.D., Porporatti, A.L.: Diagnostic validity of methods for assessment of swallowing sounds: a systematic review. Braz. J. Otorhinolaryngol. **84**(5), 638–652 (2018)
16. Wang, C.M., Shieh, W.Y., Weng, Y.H., Hsu, Y.H., Wu, Y.R.: Non-invasive assessment determine the swallowing and respiration dysfunction in early Parkinson's disease. Parkinsonism Relat. Disord. **42**, 22–27 (2017)
17. Rahman, Md. S.: Bio-signals and transducers. Proceedings of the Short Course on Biomedical Instrumentation. 1. 5.1–5.19 & 5A.1–5A.4 (2008)
18. Li, Q., Hori, K., Minagi, Y., Ono, T., Chen, Y.J., Kondo, J., Maeda, Y., et al.: Development of a system to monitor laryngeal movement during swallowing using a bend sensor. PLoS One **8**(8), e70850 (2013)
19. Retrieved from https://www.sparkfun.com/datasheets/Sensors/Flex/MSI-techman.pdf.

# Infrared Thermographic Analysis of Interrelation Between Blood Pressure and Temperature of Forearm for Normal, High and Low BP Subjects

**S. P. Angeline Kirubha, J. Akshaya, N. Dhesigan, and R. Yamini**

**Abstract** The thermal variation in body has been detected, monitored, analysed using IR thermography which is a non-invasive clinical imaging method. In the precursory work, the empirical relationship and interrelationship between temperature (°C) and blood pressure (mmHg) are derived at three different regions (common interosseous artery, radial artery and palmar arch regions) in forearm for normal, low and high BP subjects. It is done for the study setup, consisting 40 subjects (age group of 20–50 years), where 15 high blood pressure subjects, 15 normal blood pressure subjects and 10 low blood pressure subjects are chosen. The future work is the development of an IR-based sensor which would assist in measuring the blood pressure of a body from the surface of the skin. This IR sensor can be used by diabetic patient, sports persons to know their blood pressure instantly as IR sensor can fit in smart watch.

**Keywords** Infrared thermography · Empirical relationship · Surface temperature · BP subjects

## 1 Introduction

Surface temperature changes can be quantitatively measured by portable IR thermographic technology, which have been used in diagnosis and monitoring of diabetic neuropathy, peripheral vascular disease [10, 16]. Tendinitis and carpel tunnel syndrome are caused due to reduced blood flow. IR thermography investigate surface skin temperature which is function of subcutaneous perfusion [2]. PT100 sensor and thermal image recognition can be used for real-time human surface skin temperature analysis. Human thermal comfort can be evaluated by thermographic imaging technology which is non-invasive, non-destructive for measuring temperature [9]. Thermal response can be studied using IR thermography [1, 5, 6, 8, 17]. Blood vessel diameter, blood flow direction, velocity in target blood vessel can be quantitatively

S. P. A. Kirubha (✉) · J. Akshaya · N. Dhesigan · R. Yamini
Biomedical Engineering Department, SRM Institute of Science and Technology, Chennai, India



T. K. Sharma et al. (eds.), *Soft Computing: Theories and Applications*,
Advances in Intelligent Systems and Computing 1381,
https://doi.org/10.1007/978-981-16-1696-9_44

and qualitatively detected using thermographic imaging technique [3]. The detailed study of relationship between blood pressure and environmental temperature was carried in previous study which has significant impact because thermos sensitivity appears to be related to hypertension [7,11–14]. During vasodilation, more heat is lost as blood flow to surface tissue increases. During vasoconstriction, heat is lost, as the blood flow to surface tissue decreases [5]. The objective of this study is to derive the empirical relationship between blood pressure and temperature for normal, high and low BP subjects at three regions, namely common interosseous artery, radial artery and palmar arch.

## 2 Materials

### *2.1 Experimental Data Collection*

The experiment was conducted at the campus of SRM University, India, from the month of July to October 2018. The subjects were clinically studied and excluded if they were aware of any kind of disease. A total number of 40 subjects of age varying from 20–50 years were analysed for predicting the interrelationship between temperature and blood pressure. Out of these 40 subjects, 15 have high blood pressure, 15 have normal blood pressure and 10 have low blood pressure.

In thermography procedure, FLIR A305SC is used for capturing thermal images with resolution 320 × 240 pixels and accommodates manual and autofocus standard. They are mainly equipped with RJ-45 Gigabit Ethernet connection that produces images at rate of 60 Hz. It has focal length of 18 mm and the image frequency of 9 Hz. Manual sphygmomanometer and stethoscope are used to measure the blood pressure during inflation and deflation. The thermal images of forearm obtained were analysed using FLIR tools.

## 3 Methodology

### *3.1 Standard Thermal Image Acquisition Protocol*

The thermal images are captured in closed air-conditioned dark room. A proper air ventilation should be maintained around 18–22 degrees °C. Thermal camera should be placed at 15 m from subject forearm to thermal camera as shown in Fig. 1a, for correct focusing. While capturing images at area of interest, the hand should be placed in a dry place over a black background to avoid the shadow effect in images. The thermal images are captured by focusing the camera vision on subject's forearm at three different regions with varying pressure ranging from 0 to 140 mmHg during inflation and from 140 to 0 mmHg during deflation. Then, the acquired images

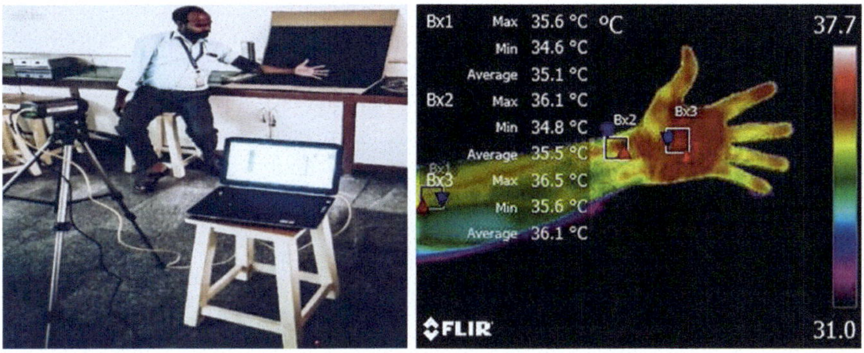


(a) (b)

**Fig. 1** **a** Experimental setup for thermal image acquisition; **b** region of interest where temperature variation is measured—Thermal view [Bx1- Common interosseous artery region, Bx2—Radial artery region, Bx3—Palmer arch region]

are studied and analysed using the software FLIR tool. The average temperature is measured at these three different regions in forearm as shown in Fig. 1b.

# 4 Results and Discussion

## *4.1 Temperature Variation in Normal BP, High BP and Low BP subjects at Common Interosseous Artery, Radial Artery and Palmar Arch Regions*

From Table 1, in normal BP subjects, when the blood pressure increases from 0 to 140 mmHg (at the interval of 20 mmHg), the temperature increases and when the blood pressure decreases from 140 to 0 mmHg, the temperature decreases simultaneously. But temperature at palmar arch is high, at radial artery is low, whereas temperature at common interosseous artery is moderate in comparison with other two regions. In high BP subjects, the temperature increases during inflation, at all three regions. During deflation, temperature continues to increase at all three regions. In low BP subjects, the temperature decreases during inflation at all three regions. It also shows that during deflation temperature continues to decrease at all three regions. From Table 2, in common interosseous artery and radial artery region the temperature in low BP subjects is low, for normal BP subject's temperature is medium and in high BP subjects' temperature is high. At palmar arch, the temperature in low BP subjects is low but in normal BP subjects the temperature is high, and in high BP subjects, the temperature will be medium.

**Table 1** Temperature variation among normal, high and low BP subjects at common interosseous, radial and palmar arteries

| Blood Pressure (mmHg) | Average surface temperature during inflation(°C) | | | | | | | | | Average surface temperature during deflation(°C) | | | | | | | | |
|---|---|---|---|---|---|---|---|---|---|---|---|---|---|---|---|---|---|---|
| | Normal BP subjects | | | High BP subjects | | | Low BP subjects | | | Normal BP subjects | | | High BP subjects | | | Low BP subjects | | |
| | Common Interosseous Artery | Radial Artery | Palmar Arch | Common Interosseous Artery | Radial Artery | Palmar Arch | Common Interosseous Artery | Radial Artery | Palmar Arch | Common Interosseous Artery | Radial Artery | Palmar Arch | Common Interosseous Artery | Radial Artery | Palmar Arch | Common Interosseous Artery | Radial Artery | Palmar Arch |
| **0** | 34.04 | 33.94 | 34.82 | 34.47 | 34.63 | 34.64 | 32.93 | 33.82 | 33.96 | 34.07 | 33.98 | 34.66 | 34.77 | 34.79 | 35.00 | 32.41 | 3.39 | 33.41 |
| **20** | 34.05 | 33.94 | 34.82 | 34.47 | 34.63 | 34.64 | 32.91 | 33.70 | 33.81 | 34.07 | 33.99 | 34.67 | 34.77 | 34.79 | 35.00 | 32.41 | 33.39 | 33.41 |
| **40** | 34.08 | 33.96 | 35.16 | 34.49 | 34.66 | 34.69 | 32.90 | 33.69 | 33.81 | 34.09 | 34.02 | 34.80 | 34.76 | 34.79 | 34.97 | 32.41 | 33.49 | 33.58 |
| **60** | 34.11 | 33.99 | 35.19 | 34.50 | 34.66 | 34.69 | 32.77 | 33.65 | 33.76 | 34.13 | 34.03 | 34.89 | 34.76 | 34.79 | 34.97 | 32.41 | 33.49 | 33.58 |
| **80** | 34.13 | 34.02 | 35.01 | 34.56 | 34.69 | 34.71 | 32.61 | 33.65 | 33.75 | 34.16 | 34.07 | 34.90 | 34.74 | 34.76 | 34.97 | 32.41 | 33.50 | 33.61 |
| **100** | 34.18 | 34.07 | 35.16 | 34.57 | 34.73 | 34.87 | 32.60 | 33.61 | 33.70 | 34.19 | 34.10 | 34.93 | 34.71 | 34.76 | 34.96 | 32.41 | 33.51 | 33.61 |
| **120** | 34.21 | 34.11 | 35.20 | 34.64 | 34.74 | 34.87 | 32.51 | 33.59 | 33.69 | 34.20 | 34.12 | 34.98 | 34.71 | 34.76 | 34.94 | 32.46 | 33.51 | 33.64 |
| **140** | 34.23 | 34.13 | 35.26 | 34.67 | 34.76 | 34.90 | 32.50 | 33.51 | 33.64 | 34.23 | 34.13 | 35.26 | 34.70 | 34.76 | 34.90 | 32.50 | 3.51 | 33.64 |

**Table 2** Temperature variation during inflation and deflation in normal, high and low BP subjects at common interosseous, radial and palmar arteries

| Region of interest | Blood pressure (mmHg) | Temperature for normal blood pressure subjects (°C) | | Temperature for high blood pressure subjects (°C) | | Temperature for low blood pressure subjects (°C) | |
|---|---|---|---|---|---|---|---|
| | | Inflation | Deflation | Inflation | Deflation | Inflation | Deflation |
| Common interosseous artery | 0 | 34.04 | 34.07 | 34.47 | 34.77 | 32.93 | 32.41 |
| | 20 | 34.04 | 34.07 | 34.47 | 34.77 | 32.91 | 32.41 |
| | 40 | 34.08 | 34.09 | 34.49 | 34.76 | 32.90 | 32.41 |
| | 60 | 34.11 | 34.13 | 34.50 | 34.76 | 32.77 | 32.41 |
| | 80 | 34.13 | 34.16 | 34.56 | 34.74 | 32.61 | 32.41 |
| | 100 | 34.18 | 34.19 | 34.57 | 34.71 | 32.60 | 32.41 |
| | 120 | 34.21 | 34.20 | 34.64 | 34.71 | 32.51 | 32.46 |
| | 140 | 34.23 | 34.23 | 34.67 | 34.70 | 32.50 | 32.50 |
| Radial artery | 0 | 33.94 | 33.98 | 34.63 | 34.79 | 33.82 | 33.39 |
| | 20 | 33.94 | 33.99 | 34.63 | 34.79 | 33.70 | 33.39 |
| | 40 | 33.96 | 34.02 | 34.66 | 34.79 | 33.69 | 33.49 |
| | 60 | 33.99 | 34.03 | 34.66 | 34.79 | 33.65 | 33.49 |
| | 80 | 34.02 | 34.07 | 34.69 | 34.76 | 33.65 | 33.50 |
| | 100 | 34.07 | 34.10 | 34.73 | 34.76 | 33.61 | 33.51 |
| | 120 | 34.11 | 34.12 | 34.74 | 34.76 | 33.59 | 33.51 |
| | 140 | 34.13 | 34.13 | 34.76 | 34.76 | 33.51 | 33.51 |
| Palmar arch | 0 | 34.82 | 34.66 | 34.64 | 35.00 | 33.96 | 33.41 |
| | 20 | 34.82 | 34.67 | 34.64 | 35.00 | 33.81 | 33.41 |
| | 40 | 35.16 | 34.80 | 34.69 | 34.97 | 33.81 | 33.58 |
| | 60 | 35.19 | 34.89 | 34.69 | 34.97 | 33.76 | 33.58 |
| | 80 | 35.01 | 34.90 | 34.71 | 34.97 | 33.75 | 33.61 |
| | 100 | 35.16 | 34.93 | 34.87 | 34.96 | 33.70 | 33.61 |
| | 120 | 35.20 | 34.98 | 34.87 | 34.94 | 33.69 | 33.64 |
| | 140 | 35.26 | 35.26 | 34.90 | 34.90 | 33.64 | 33.64 |

### *4.2 Analysing the Empirical Relationship of Temperature at Common Interosseous Artery, Radial Artery and Palmar Arch Regions for Normal, High and Low Blood Pressure Subjects*

From Fig. 2 , the linear line of normal and high BP subjects intersects at pressure 100 mmHg. The temperature difference between the normal and high BP subjects' increases from 100 to 0 mmHg (temperature of high BP subjects' increases and that of normal decreases). By comparing the inflation BP and deflation BP from Figs. 3, 4 and 5, it is shown that the temperature difference between inflation and deflation

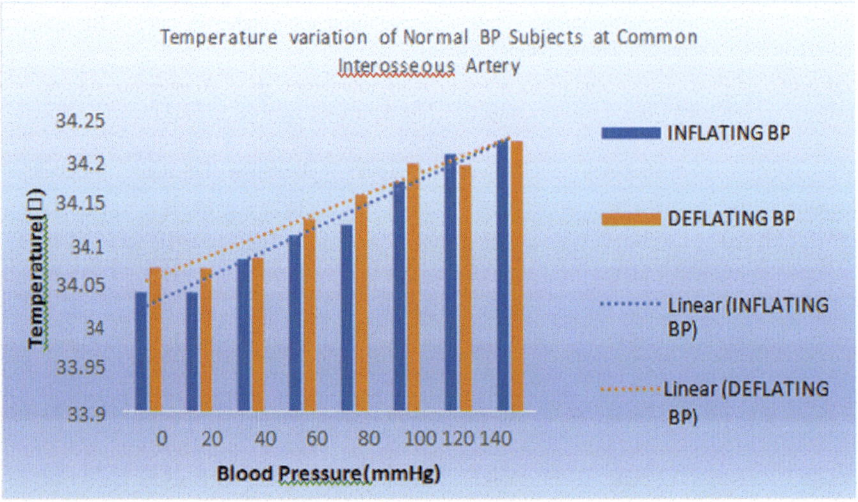


**Fig. 2** Temperature variation of normal BP subjects at common interosseous artery

decreases with increase in blood pressure from 0 to 140 mmHg, but becomes equal at blood pressure 140 mmHg in Figs. 8 and 9. By comparing the temperature during the inflation and deflation of Figs. 6 and 7, it depicts that the difference in temperature gradually decreases from 0 to 140 mmHg and becomes equal at blood pressure of 140 mmHg, but becomes equal at blood pressure 140 mmHg. In Fig. 8, temperature difference between inflation and deflation is high at 0 mmHg compared to Fig. 6. By comparing the temperature during inflation and deflation of Fig. 8, it depicts that there is a huge difference between the temperatures of inflation and deflation which is higher than the difference between them at the other two regions (common interosseous and radial artery). By comparing the temperature during inflation BP and deflation BP of Figs. 8, 9 and 10, it depicts that there is difference between the temperatures of inflation and deflation which gradually decreases but becomes equal at blood pressure 140 mmHg at common interosseous (Fig. 9) and radial artery (Fig. 10). The difference in temperature is higher in radial artery compared to common interosseous artery and that of palmar arch is higher than radial artery. From Figs. 6, 7 and 8, in comparison with the temperature peaks for both inflating and deflating for all three regions common interosseous artery, radial artery and palmar arch, it depicts that the linear lines of both inflating and deflating do not meet each other at common interosseous artery region. In radial artery region, the linear lines of both inflating and deflating appear to meet, whereas in palmar arch region, the linear lines of both inflating and deflating cross each other at pressure 120 mmHg. Table 3 infers that during inflation, the empirical formula at all three regions for normal and high BP subject is positive; this remarks that there is an increase in temperature, whereas for low BP subjects, the empirical relationship is negative at all three regions and this depicts that there is a decrease in temperature. During deflation observed in Table 3,

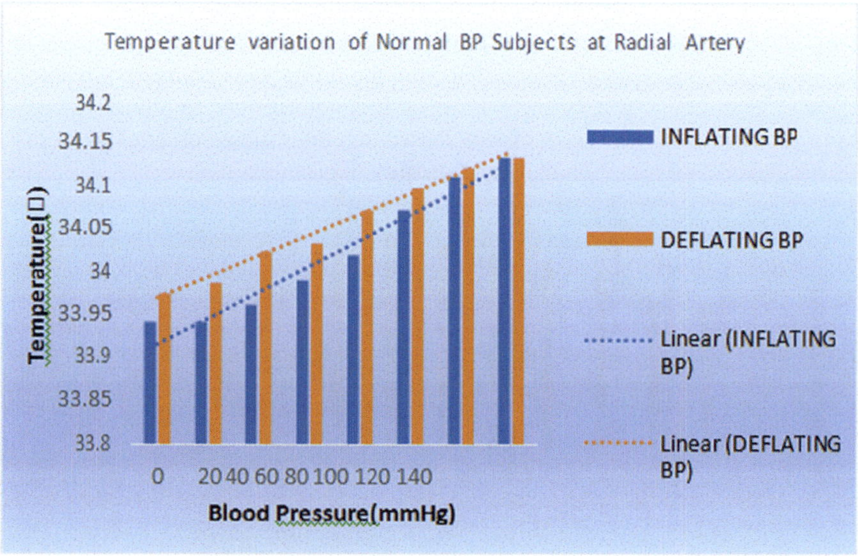


**Fig. 3** Temperature variation of normal BP subjects at radial artery

the empirical formula is negative for all the three regions of normal and low BP, and this infers that there is a decrease in temperature, whereas for high BP subjects, the empirical formula is positive at all three regions which infer that there is an increase in temperature.

The characterization of blood vessel diameter, blood flow direction and velocity in target blood vessel is carried out by thermography using vasoconstriction and vasodilation concept [10]. Therefore, the surface skin temperature was quantitatively measured for diagnosis of diabetic neuropathy, ulcer patients and peripheral vascular disease. Whereas in this work, the interrelation between blood pressure and temperature of forearm for normal, high and low BP subjects at three regions is examined. In the anatomical study of blood flow and temperature at this region, it is explained that blood flow at radial artery is high; therefore, when the blood reaches the palmar arch, temperature tends to increase compared to that of region at radial artery. It is also noticed that the temperature at common interosseous artery region is low. The empirical equation is also established. This is achieved by noting the temperatures at three different regions, namely common interosseous artery, radial artery and palmar arch, for normal, high and low BP subjects and by studying the relationship between the temperature and pressure. Deficiency like hypoxia causes decrease in body temperature; this is recorded using IR thermograph [15]. Whereas in this work, we have analysed the temperature variation for normal, low and high BP subjects. The pre-typing and post-typing lead to variation in temperature of hand (when it becomes cold). This is due to reduction in blood flow [3]. Whereas in this work, we have analysed for low, normal and high BP subjects where the blood flow is less and high which lead to decrease and increase in temperature, respectively.

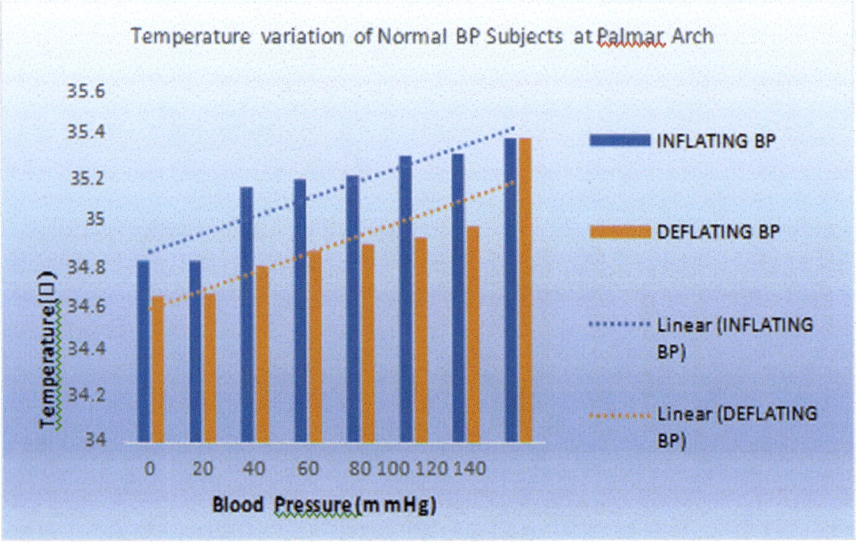


**Fig. 4** Temperature variation of normal BP subjects at palmar arch

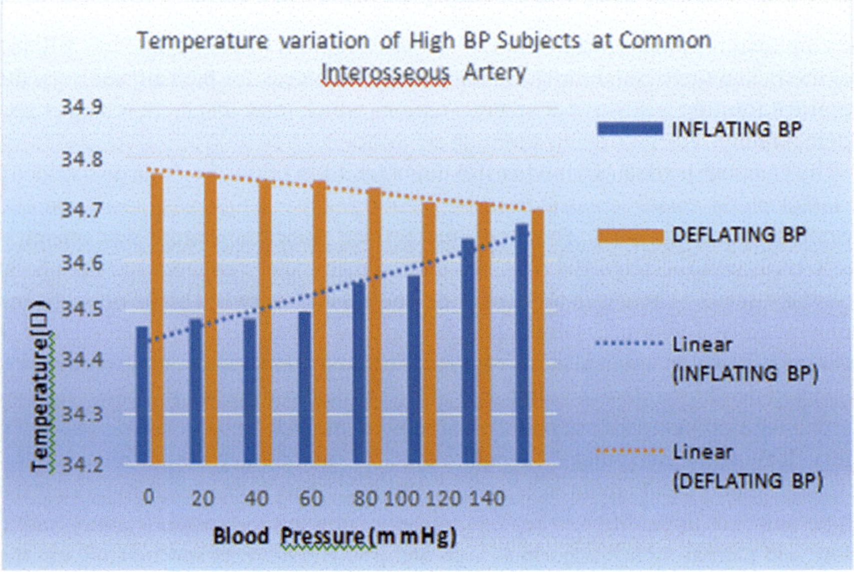


**Fig. 5** Temperature variation of high BP subjects at common interosseous artery

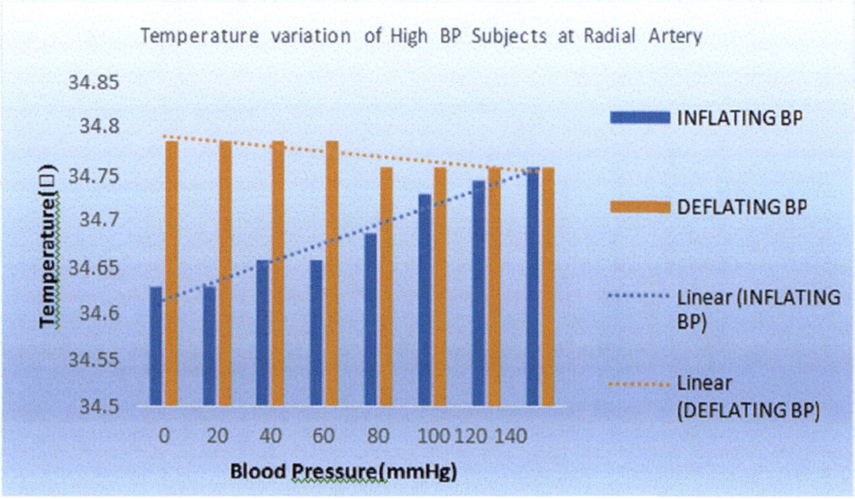


**Fig. 6** Temperature variation of high BP subjects at radial artery

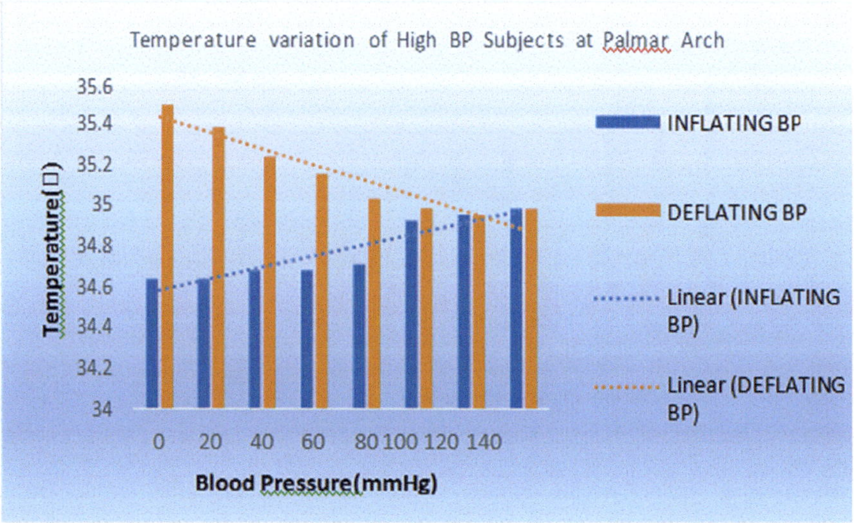


**Fig. 7** Temperature variation of high BP subjects at palmar arch

Therefore, pt100 sensor is biocompatible and suitable for all subjects [4]. It can improvise the level of accuracy to determine the relationship between temperature and blood pressure.

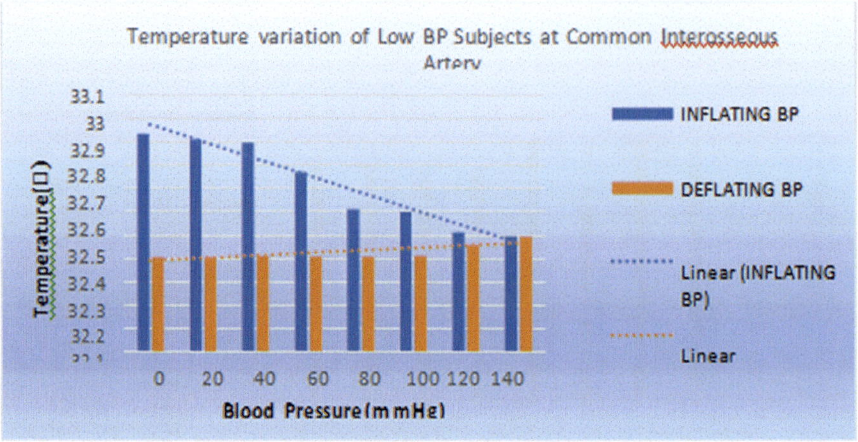


**Fig. 8** Temperature variation of low BP subjects at common interosseous artery

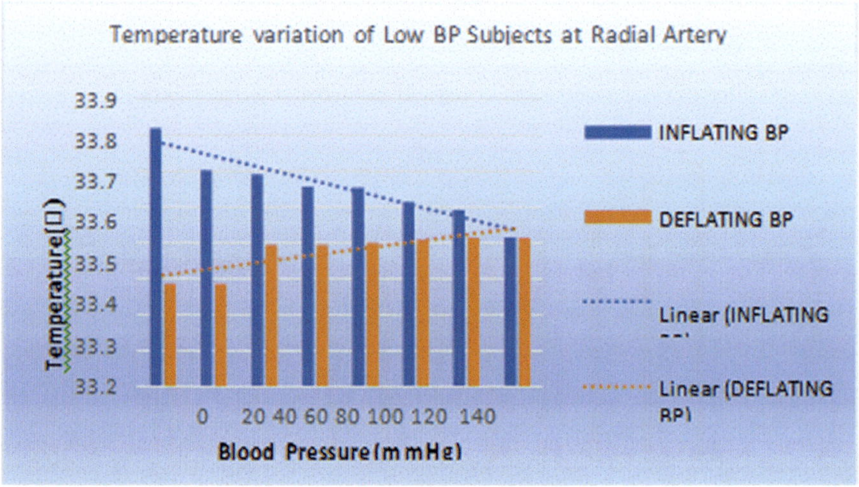


**Fig. 9** Temperature variation of low bp subjects at radial artery

## 5 Conclusion

In this study, we have analysed that variation in blood pressure will lead to a change in temperature of a particular region of our body. From the present study, it is examined that subjects having normal blood pressure, the temperature at three different regions, i.e. common interosseous artery, radial artery and palmar arch, increases during inflation (0–140 mmHg) but at the time of deflation (140–0 mmHg), it had a downfall. In case of high BP subjects, the temperature at three different regions

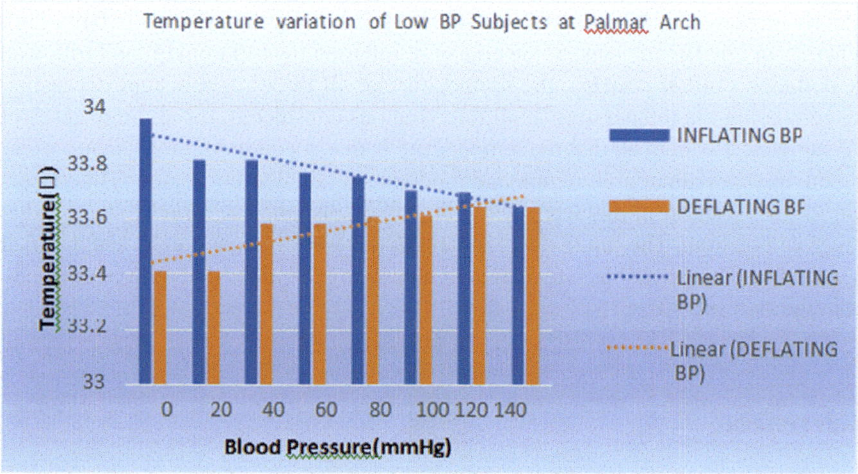


**Fig. 10** Temperature variation of low BP subjects at palmar arch

**Table 3** Empirical relationship between temperature and blood pressure during inflation and deflation

| | BP subjects | Name of the arteries | Empirical relationship | ↑/↓ |
|---|---|---|---|---|
| Inflating | Normal BP | Common Interosseous artery | $y = 0.0286x + 33.999$ | ↑ |
| | | Radial Artery | $y = 0.0303x + 33.886$ | ↑ |
| | | Palmar Arch | $y = 0.0807x + 34.785$ | ↑ |
| | High BP | Common Interosseous Artery | $y = 0.0298x + 34.414$ | ↑ |
| | | Radial Artery | $y = 0.0204x + 34.594$ | ↑ |
| | | Palmar Arch | $y = 0.0563x + 34.527$ | ↑ |
| | Low BP | Common Interosseous Artery | $y = -0.0726x + 33.043$ | ↓ |
| | | Radial Artery | $y = -0.0346x + 33.809$ | ↓ |
| | | Palmar Arch | $y = -0.0371x + 33.933$ | ↓ |
| Deflation | Normal BP | Common Interosseous Artery | $y = -0.0245x + 34.254$ | ↓ |
| | | Radial Artery | $y = -0.0051x + 34.745$ | ↓ |
| | | Palmar Arch | $y = -0.0827x + 35.271$ | ↓ |
| | High BP | Common Interosseous Artery | $y = 0.261x + 33.941$ | ↑ |
| | | Radial Artery | $y = 0.0051x + 34.745$ | ↑ |
| | | Palmar Arch | $y = 0.0772x + 34.814$ | ↑ |
| | Low BP | Common Interosseous Artery | $y = -0.0108x + 32.475$ | ↓ |
| | | Radial Artery | $y = -0.0187x + 33.558$ | ↓ |
| | | Palmar Arch | $y = -0.0347x + 33.717$ | ↓ |

linearly increases for inflation (0–140 mmHg) followed by deflation (140–0 mmHg). Whereas in low BP subjects, at the course of inflation (0–140 mmHg) and followed by deflation (140–0 mmHg), the temperature at all three regions kept deceasing. By comparing the temperature at all three regions considered, for all three cases of BP subjects, it is noticed that the temperature is low at common interosseous artery region, high at palmar arch region and moderate at radial artery region. This occurs due to the variation in diameter of the blood vessels, and the blood flow at radial artery will be high. The future development of this work includes an IR senor which is wearable. This wearable sensor aids in measuring the blood pressure of our body from the skin surface.

## References

1. Baldinelli, G., Bianchi, F., Rotili, A., Costarelli, D., Seracini, M., GianlucaVinti, F.A., Evangelisti, L.: A model for the improvement of thermal bridges quantitative assessment by infrared thermography. Appl. Energy **211**, 854–864 (2018). https://doi.org/10.1016/j.apenergy.2017.11.091
2. Gold, J.E., Cherniack, M., Buchholz, B.: Infrared thermography for examination of skin temperature in the dorsal hand of office workers. Eur. J. Appl. Physiol. **93**(1–2), 245–251 (2004). https://doi.org/10.1007/s00421-004-1210-6
3. Ashish Saxena, E.Y.K., Ng, VigneshRaman: Thermographic venous blood flow characterization with external cooling stimulation. Infrared Phys. Technol. **90**, 8–19 (2018). https://doi.org/10.1016/j.infrared.2018.02.001
4. Silva, N.C.M., Castro, H.A., Carvalho, L.C., Chaves, É.C.L., Ruela, L.O., Iunes, D.H.: Reliability of infrared thermography images in the analysis of the plantar surface temperature in diabetes mellitus. J. Chiropr. Med. **17**, 30–35 (2018). https://doi.org/10.1016/j.jcm.2017.10.006
5. Tanda, G.: The use of infrared thermography to detect the skin temperature response to physical activity. J. Phys.: Conf. Ser., IOP Publishing Ltd 655 (2015)
6. Becerik-Gerber, B., Castro, G., Ghahramani, A.: IGlass: infrared thermography for learning thermoregulation performance, US patent US201702675755A (2017)
7. Kunes, J., Tremblay, J., Bellavance, F., Hamet, P.: Influence of environmental temperature on the blood pressure of hypertensive patients in Montréal. Am J Hypertens **4**, 422–426 (1991). https://doi.org/10.1093/ajh/4.5.422
8. Ramesh, J., Thiruvengadam, J.: Application of Dynamic Thermogram. Comput. Sig. Process. Anal. **490**, 129–137 (2018)
9. Metzmacher, H., Wölki, D., Schmidt, C., Treeck, C.: Real-time human skin temperature analysis using thermal image recognition for thermal comfort assessment. Energy Buildings **158**, 1079–1086 (2018). https://doi.org/10.1016/j.enbuild.2017.09.032
10. Gatt, A., Formosa, C., Cassar, K., Camilleri, K.P., De Raffaele, C., Mizzi, A., Azzopardi, C., Mizzi, S., Falzon, O., Cristina, S., Chockalingam, N.: Thermographic patterns of the upper and lower limbs: baseline data. Int. J. Vasc. Med. **2015**, 1–9 (2015)
11. An, B.-Y., Wang, D., Liu, X.-J., Guan, H.-M., Wei, H.-X., Ren, Z.-B.: The effect of environmental factors in urban forests on blood pressure and heart rate in university students. J. For. Res. **24**, 27–34 (2018)
12. Yang, L., Li, L., Lewington, S., et al.: Outdoor temperature, blood pressure, and cardiovascular mortality. Eur. Heart J. **36**, 1178–1185 (2015)
13. Paolo, G., Melvyn, R., Ritabrata, D., Gracik, T.W., Morishita, Lu., Bard, M., Robert, L., Jackson Elizabeth, A., Fitzner Craig, A., Claudio, F., Brook Robert, D.: Particulate matter air pollution

and ambient temperature, opposing effects on blood pressure in high-risk cardiac patients. J. Hypertens. **33**, 2032–2038 (2015). https://doi.org/10.1097/HJH.0000000000000663
14. Ooi, W.L., Barrett, S., Hossain, M., et al.: Patterns of orthostatic blood pressure change and their clinical correlates in a frail elderly population. JAMA **277**(16), 1299–1304 (1997). https://doi.org/10.1001/jama.1997.03540400049030
15. Jones, D., Covins, S.F., Miller, G.E., Morrison, K.I., Clark, A.G., Calcott, S.D., Anderson, A.M., Lucas, S.J.E., Imray, C.H.E.: Infrared thermographic analysis of surface temperature of hands during exposure to Norobaric hypoxia. High Altitude Med. Biol. **19**(4) (2018). https://doi.org/10.1089/ham.2018.0008
16. Jorge, J., Harford, M., Villarroel, M., Chaichulee, S., Davidson, S., Finnegan, E., Clark, S.H., Young, J.D., Watkinson, P.J., Tarassenko, L.: Non-contact assessment of peripheral artery haemodynamics using infrared video thermography. IEEE Trans. Biomed. Eng. X, pp 1–12 (2020). https://doi.org/10.1109/TBME.2020.2999539
17. MacRae, B.A., Annaheim, S., Spengler, C.M., Rossi, R.M.: Skin temperature measurement using contact thermometry: a systematic review of setup variables and their effects on measured values. Front Physiol **9**, 29 (2018). https://doi.org/10.3389/fphys.2018.00029

# WSN Routing Based on Optimal and Energy Efficient Using Hybrid Antlion and K-Means Optimization

**K. Murugan, G. Harikrishnan, R. Mothi, T. Venkatesh, T. Jagadesh, and M. Supriya**

**Abstract** Energy saving is a major optimization challenge for wireless sensor networks (WSNs). Clustering methods are used in order to make an energy efficient network. Cluster head (CH) selection plays a significant role in hierarchical-based WSNs. CH needs to gather data from its participants and it is important to compile and forward the information to BS. It takes extra energy, thus. Due to the incorrect range of CH, the lifespan of a network is impacted. In order to render a network that is energy efficient, appropriate optimization algorithms must then be used to choose the CH. K-means and antlion optimization algorithms are merged in this research to form a hybrid clustering strategy in an optimized way to do cluster analysis. The antlion optimization (ALO) model is a global stochastic optimization model and is referred to as HKALO. The suggested clustering-based routing protocol K-means is an energy efficient and optimum packet size is considered based on radio parameters and channel state. This approach decreases the energy consumption of each individual node and extends the lifespan of the network. The transition of data between the head of the cluster to the member of the cluster and the head of the cluster to the base station is accomplished by assuming different degrees of control. In terms of network life and throughput, as demonstrated by simulation results, the proposed algorithm showed better efficiency compared to the traditional clustering algorithm.

**Keywords** Wireless sensor networks · Clustering · Optimization

---

K. Murugan (✉) · T. Venkatesh · T. Jagadesh · M. Supriya
KPR Institute of Engineering and Technology, Coimbatore, India

G. Harikrishnan
HolyGrace Academy of Engineering, Mala, Thirussur, India

R. Mothi
Mahendra Engineering College, Namakkal, India

 
T. K. Sharma et al. (eds.), *Soft Computing: Theories and Applications*,
Advances in Intelligent Systems and Computing 1381,
https://doi.org/10.1007/978-981-16-1696-9_45

## 1 Introduction

A distributed network is a wireless sensor network. It has nodes for sensors (SNs). To detect the data and to upgrade the base station BS with the sensed data, SNs are used [1]. This WSN technology is used for application fields such as military, medical, offshore industries and processes because of its management and simple implementation. WSN's key challenges are node distribution, energy-aware clustering and location recognition. In WSN, by using clustering technology, the energy of sensor nodes can be well maintained. Clustering is the grouping process for sensor nodes. In order to create the clusters, the different groups are created by splitting the network. There is going to be a cluster head (CH) for each cluster. In the cluster, data sensed by participants of the cluster is gathered by the head of the cluster and the collected data is aggregated to be sent to BS. Routing protocols focused on clusters demonstrated their connectivity efficiency and scalability [2].

In WSNs, the resources of the network can be saved easily by using the routing protocol based on different clusters. Sensor nodes, using data aggregation, are organized as a cluster. Fusion approaches are used on the head of the cluster to ensure the efficient transmission to the base station of sensed data and are often used for energy storage. Intracluster connectivity reduces the distance from the sensor node and energy usage. In recent days, naturally motivated algorithms have been able to provide the latest optimization methods for WSN applications. Figure 1 demonstrates a routing protocol based on clustering.

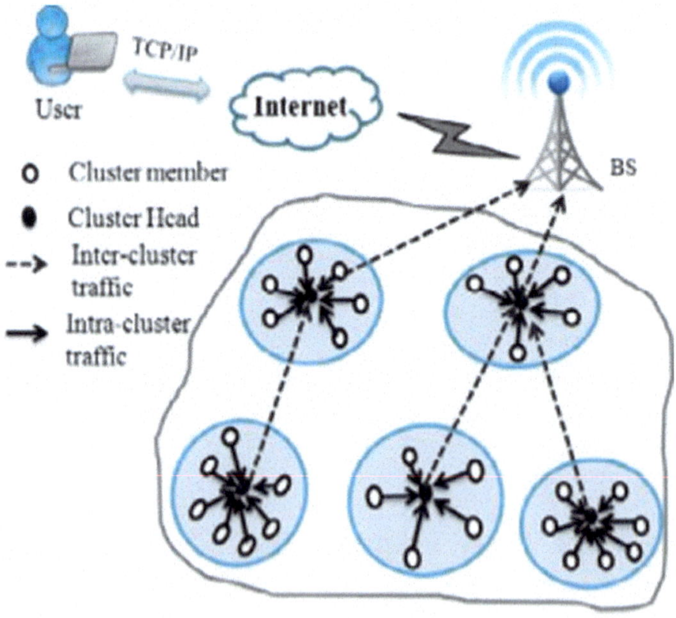


**Fig. 1** Cluster-based protocol for WSNs

Various clustering algorithms have been developed for WSNs. The analysis of these algorithms is presented on the basis of heuristic and bio-inspired techniques. On bio-inspired algorithms, more attention is seen. In [3–6], separate clustering algorithms based on heuristic techniques [3–6] are introduced for WSNs. LEACH is the most popular algorithm for clustering. In this algorithm, CH is chosen by SNs depending on a certain probability value. As compared to the minimal transmission energy (MTE) protocol, LEACH saves large quantities of energy and increases the lifespan of the network. But SN with very little energy can be chosen as CH in this algorithm. Such an SN will quickly die and will shorten a network's lifespan. To solve the problems with LEACH, algorithms are used in [7] HEED and [8] PEGASIS.

In PEGASIS, the chain is generated by SNs, such that data transfer takes place only between adjacent nodes. In each round, CH is picked from the SN chain. PEGASIS has good energy efficiency as compared to LEACH, but it should not be found in a large-sized network. Various WSN clustering protocol algorithms are introduced in [9–12] to increase the lifespan of the network. Different levels of hierarchy are added in [13] to form a TL-LEACH to increase the lifespan of a LEACH network. The secondary CH and non-CH are chosen using intracluster distance. Because of this, the network creates an energy surplus. The sensed data is transmitted to BS by means of the M-LEACH algorithm in [14]. It is an alternative to direct contact that, relative to TL-LEACH and LEACH, has a high benefit for energy consumption.

In this, cluster forming process and data transfer between CHs through multihop transmission are not taken into account. In [15], the VLEACH protocol is introduced by choosing trim CHs in addition to main CHs to increase the longevity of a network. When compared with LEACH, the experimental results showed good efficiency. Additional power is required by SN to pick the trim CH. Protocol E-LEACH is suggested in [16]. In order to increase the network's lifespan, this protocol is used to minimize energy consumption during the CH selection process. A cluster-based framework to estimate the energy in WSN was proposed in [17]. To boost the life of the WSN, numerous clustering algorithms are generated that are influenced by biological behaviour. In unified LEACH [18], simulated annealing is applied. In LEACH-C, BS performs the CH selection to boost efficiency.

LEACH-C considers the energy and intracluster gap of nodes to efficiently pick the CH in order to increase the lifespan of the network. Energy inefficiency can be caused by the avoidance of the formation of clusters. A PSO technique chooses the optimal position of the CHs in [19]. The key focus of this work is intracluster distance reduction. In this method, the sink distance is not taken into consideration and energy efficiency gain relies on this distance in direct network-to-sink contact. The energy disparity can be caused by the distance dependent assignment of non-CH nodes to CH [20–25].

In this work, by incorporating the reformed new bio-inspired antlion optimization (ALO) algorithm with clusters based on routing protocol, network structure is rendered in a better way. In simulation, output is measured and the experimental hardware test bed is used to verify it.

## 2 System Model

Many sensor nodes and the base station (BS) are included in this device model around the sensing region. By classifying the nodes, cluster members (CM) and cluster heads (CH) are created. After sensing the area, the data is transmitted to CH by CM. Information obtained is aggregated and given by CH to BS. It is a protocol for unified clustering. Selection of CH happens at the base station and all CMs of the cluster are told regarding the CH selected. Consumed energy is decreased in sensor nodes due to reduced overhead control.

In the data transmission process, the WSN absorbs a large amount of resources. The optimum fixed packet size can be used to reduce this energy consumption [3] since,

$$E_{\text{opt}} = \frac{\sqrt{C_o^2 \frac{4C_o}{\text{In}(1-p)}} C_o}{2} \tag{1}$$

where $C_o = \propto + \frac{K_2}{K_1}$ and header bits in packet is given by$\alpha$. Amount of energy used in communication of payload is given by$K_1$, start-up energy consumption of transceiver is given by $K_2$ and channel bit error rate (BER) is given by *p*.

$$E_{\text{TX}}(L_{\text{opt}}, x) = L_{\text{opt}} * E_{\text{ele}} + L_{\text{opt}} * \varepsilon_{\text{amp}} \tag{2}$$

$$E_{\text{RX}}(L_{\text{opt}}) = L_{\text{opt}} * E_{\text{ele}} \tag{3}$$

where distance between sensor nodes is represented as x, $E_{\text{TX}}$ gives energy consumption of transmitter and $E_{\text{RX}}$ gives energy consumption of receiver, energy consumed by transmitter and electronic is represented as$E_{\text{ele}}$, and amplifier energy consumption is represented as $\varepsilon_{\text{amp}}$ which is in transmitter sensor nodes and it can be computed as,

$$\varepsilon_{\text{amp}} = \varepsilon_{\text{fs}} * x^2 \text{ for } x \le x_{\text{th}} \tag{4}$$

$$= \varepsilon_{\text{mp}} * x^4 \text{ for } x \ge x_{\text{th}} \tag{5}$$

where threshold value is represented by $x_{\text{th}}$, amplification energy parameters are represented by $\varepsilon_{\text{fs}}$ and$\varepsilon_{\text{mp}}$.

The multipath fading channel model is utilized if $x_{\text{th}}$ is less than the distance x. Free space propagation model is used in other cases. Total energy needed by an optimum packet is computed as,

$$E_{\text{total}} = E_{\text{TX}} + E_{\text{RX}} + E_{\text{DA}} \tag{6}$$

$$E_{\text{total}} = \left(L_{\text{opt}} * E_{\text{ele}} + L_{\text{opt}} * \varepsilon_{\text{amp}}\right) + \left(L_{\text{opt}} * E_{\text{ele}}\right) + E_{\text{DA}} \quad (7)$$

$$E_{\text{total}} = L_{\text{opt}}\left(2 * E_{\text{ele}} + \varepsilon_{\text{amp}}\right) + E_{\text{DA}} \quad (8)$$

In a network with $N$ number of sensor nodes, distance between CH and CM is given by $x_{\text{CH}}$ and distance between BS and CH is given by $x_{\text{BS}}$. The energy required for a sensor node to transmit L bits to its CH is computed as,

$$E_{\text{CM}} = E_{\text{ini}} - E_{\text{TX}}\left(L_{\text{opt},x}\right) \quad (9)$$

The energy needed to aggregate and transmit a data to BS by a CH is computed as,

$$E_{\text{CH}} = E_{\text{ini}} - E_{\text{RX}}\left(L_{\text{opt}}\right) - E_{\text{DA}} - E_{\text{TX}}\left(L_{\text{opt},x}\right) \quad (10)$$

where energy consumed by CM is represented by $E_{\text{CM}}$, energy consumed by CH is represented by $E_{\text{CH}}$ and each node's initial energy is given by $E_{\text{ini}}$.

## 3 Proposed Methodology

In this analysis, the centralized clustering approach based on K-means clustering is suggested to form a cluster and CH selection is made using the optimal size of a decision variable packet. High overhead benefit is imposed on short packets, whereas long packets produce a high error rate. The packet length can be varied according to the channel conditions in order to maximize its throughput [20]. In a resource constrained and autonomous network, due to the increased overhead and expense of resource control, this form of variable packet size is preferable. Thus, energy usage is minimized in the proposed work by providing an optimal packet size that increases the lifespan of the sensor nodes in WSN [20]. In the suggested scheme, there are three stages:

A. **Initialization Phase**

In the network area, the initialization request (IRQ) is transmitted by BS in the first step to all sensor nodes. The initialization reply (IRP) response is sent to BS as a reply after receiving the IRQ response. The energy of the node and its location are included in the IRP message [8].

B. **Cluster Formation Phase**

Proposed hybrid K-means and ALO clustering algorithm.

K-means clustering is a clustering method of hard partitioning and it is an unsupervised method. K clusters are computed from data on the basis of the objective function J and the objective function is expressed as,

$$J = \sum_{i=1}^{k} \sum_{j=1}^{N} d^2(C_i - X_j) \tag{11}$$

where the square Euclidean distance between the data point of the ith cluster centroid and jth is defined as d^2 (C-i-X-j), N is assigned the total number of data points. Points are allocated to the clusters with the least distance from the centroid depending on the measured distance. For all points of the cluster, the mean value is computed. The new centroid cluster is created for the next iteration by assigning the mean value. This mechanism continues until centroid convergence happens. A big purpose of the K-means algorithm is to minimize objective functions.

The strategy of antlion optimization is an algorithm inspired by nature. In this algorithm [19], the hunting behaviour of antlion larvae is followed. In sand, the antlion larva runs in a circular direction and uses its jaw to create a conical hole to throw the sand. The bottom of the cone is covered by the larva after trap excavation, and it waits to trap the ants in its hole. After discovering that it has a prey on its trap and prey is slipped into the hole, the sand is tossed outside by antlion.

If jaw captures the prey, prey is dragged towards antlion by antlion and it eats the prey. This method is structured mathematically in order to perform optimization. There are five key stages in this strategy, namely random walking ants, constructing traps, trapping ants in traps, capturing prey and restoring traps.

Ants pass by a random walk in the quest space. This random walk could be influenced by traps of antlions. Adjust the location of the ant with a random walk in each iteration. Equation (12) is used to crate the random walk of the iteration t. To guarantee that the random walk does not surpass the search space limits, normalization is used. Using (13), the random walk normalization is achieved as,

$$x(t) = [0, \text{cumsum}(2r(t_1) - 1), \text{cumsum}(2r(t_2) - 1), \ldots \text{cumsum}(2r(t_n) - 1)] \tag{12}$$

where $r(t) = 1$ if $r$ and $> 0.5$ or 0 if $r$ and $\leq 0.5$.

$$X_i^t = \frac{(X_i^t - a_i) \times (b_i - c_i^t)}{(d_i^t - a_i)} + c_i \tag{13}$$

$a_i$ represents minimum value of a random walk and $b_i$ represents the maximum value of random walk in ith variable. Minimum of ith variable is represented as $c_i^t$ at iteration t. Maximum of ith variable is represented as $d_i^t$ at iteration t.

Antlions creates the traps. The ant's random walk is affected by this. Equations (14) and (15) explain this process mathematically.

$$c_i^t = \text{Antlion}_j^t + c^t \tag{14}$$

$$d_i^t = \text{Antlion}_j^t + d^t \tag{15}$$

where position of antlion is represented at iteration t. Minimum of all the variables is represented as $c^t$ at iteration t. Maximum of all the variables is represented as $d^t$ at iteration t. Minimum of ith variable is represented as $c_i^t$ at iteration t. Maximum of ith variable is represented as $d_i^t$ at iteration t.

For optimization, antlions are selected using roulette wheel selection approach and it is done based on their value of fitness. Elite is saved in every iteration by a fittest antlion find in that iteration. The ant's movement is affected by the elite. Selected antlion's random walk and elite are used to update the ants position. This is due to the fact that, ants are walking around selected elite and antlion. This is expressed in the following equation.

$$\text{Ant}_i^t = \frac{R_A^t + R_E^t}{2} \tag{16}$$

Random walk around antlion is given by $R_A^t$ and roulette wheel selects it in the tth iteration and random walk around elite is represented by $R_E^t$ at tth iteration. Position of ant i is given by $\text{Ant}_i^t$ at iteration t. All ants fitness value is computed. The antlion is replaced by an ant, if it has fitness value greater than the fitness value of antlion. The antlion is used to replace the elite, if it has fitness value greater than the elite.

The K-means clustering algorithm is combined with the antlion optimization algorithm in the proposed work to form a hybrid algorithm to increase the K-means algorithm clustering efficiency. In the first step, the number of clusters that must be formed is computed. Using the calculated minimum Euclidean distance, all data points form clusters. In the next step, the optimized cluster centroid must be computed for each cluster that is created. Each cluster is randomly initialized as an antlion and ant population to maximize it.

The objective function of the K-means clustering system is used to measure the fitness score of both ants and antlions. The antlion refers to an elite with a minimal health score. This is required to minimize the average number of distances within the cluster. The best location of the cluster centriode is determined by optimizing the antlion in each cluster. For the algorithm K-means, centroids are considered to be this returned elite.

## 4 Hybrid K-Means and Alo Clustering Algorithm

INPUT: data set D with N instances and M attributes, the number of iterations T, number of ants A, number of clusters K, total ant antlion population P, number of antlions L.

OUTPUT: Optimized cluster centroids.

BEGIN ALGORITHM:

Select K random points as cluster centres.

WHILE the end criterion is not satisfied.

FOR each point.

From cluster centres, Euclidean distance of each point is computed.
With minimum Euclidean distance, a point is assigned to a cluster.
END FOR.
In each cluster, mean of all points is computed.
Mean values are used to assign a new cluster centres.
END WHILE.
Return K clusters.
FOR each cluster.
First population of ants are initialized from data set.
First population of antlions initialized randomly.
Using objective function, fitness of ants and antlions are computed.
Antlion with minimum fitness value is selected as an elite.
WHILE the end criterion is not satisfied.
FOR every ant.
Roulette wheel is used to select antlion.
Minimum and maximum of all variables are updated.
Using Eq. (12), random walk is created.
Using Eq. (13), random walk is normalized.
Using Eq. (16), position of ant is updated.
END FOR.
Compute all ants fitness value.
Ant is used to replace an antlion ant if f(Anti) < f (Antlionj).
Update elite if f (Antlionj) < f (elite).
END WHILE.
Return elite K.
END FOR.
Elite is selected as cluster 1′s new centre.
END ALGORITHM.
Cluster Head Selection Phase.

Selection of CH is initiated after the formation of cluster by grouping the nodes. Two weight functions are computed for the selection of CH and they are given by,

$$W_n = c_1 * E_n + c_2 * D_{cc} \tag{17}$$

$$W_s = c_1 * E_{\text{total}} + c_2 * \text{Avg}(D_{cc}) \tag{18}$$

where i = 1, 2, 3, 4….N, constants are given by $c_1$ and $c_2$ [8], weight of each node in cluster is represented as $W_n$, residual energy of ith node is represented as $E_n$, distance of ith node to cluster centre is represented as $D_{cc}$, standard weight of a node is represented as $W_s$ and it may be considered for cluster head selection, amount of energy required to transceiver as well as to aggregate is given by $E_{\text{total}}$ and average distance of all nodes to cluster centre is represented as $\text{Avg}(D_{cc})$.

A distinction is made between these two weights in a single round. The node having the weight equal to standard weight is chosen as a CH. This knowledge about

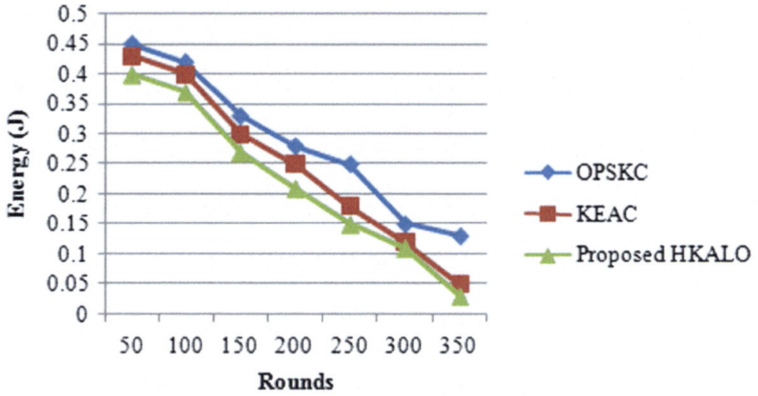


**Fig. 2** Network energy conservation

CH selection is conveyed by BS to all sensor nodes in the cluster. It changes the routing table of the sensor nodes. For transmission from CM to CH and CH to BS, various degrees of power are presumed in a scheme proposed.

## 5 Result and Discussion

In a $100 \times 100m^2$ field, this work assumes that the cluster is occupying an area of N/K. This model is implemented in MATLAB and following lists the assumptions made in the network. After deployment, BS and sensors nodes are static. In the far field of sensing region, there will be only one BS. In the field of a network, sensors are deployed randomly with equal amount of initial energy and they are homogeneous. Information is forwarded to BS by cluster head after aggregating all the gathered information. Use Rayleigh fading channel model [20]. Energy of decoding and length of the trailer $\tau$ are assigned with zero. Delay as well as packet re-transmissions are not taken into consideration.

The computed energy conservation of a network in every round is shown in Fig. 2. When compared to KEAC scheme and optimal packet size K-means clustering (OPSKC) methods, less energy is consumed by the proposed HKALO scheme as illustrated in the figure and it shows the efficiency of it.

## 6 Conclusion

Most of the energy is lost by the sensor nodes in the WSN during the data transmission stage. The proposed work uses the fixed size packets to reduce the energy consumption in data transmission process. The cumulative energy available for both sending and receiving these packets is determined. By using this measured energy value with the average node distance to the cluster, the standard weight is determined. The lifespan of the node and the throughput of the network are increased by this approach. The findings of the simulation demonstrate that by massively conserving electricity, this approach has extended the lifespan of a network. In the future, performance assessment measures such as reliability and latency can be used to measure performance, and by using multiple error correcting methods, reliability can also be increased.

## References

1. Akyildiz, I.F., Su, W., Sankarasubramaniam, Y., Cayirci, E.: A Survey on Sensor Networks. IEEE Commun Mag **40**(8), 102–114 (2002)
2. Akkaya, K., Younis, M.: A survey on routing protocols for wireless sensor networks. Ad. Hoc. Netw. **3**(3), 325–349 (2005)
3. W.B. Heinzelman et al., "Energy Efficient Communication Protocol for Wireless Micro Sensor Networks", Proceedings of 33rd Hawaii International Conference on System Sciences, pp. 112–117, 2000.
4. Liu, X.Y., et al.: CDC: compressive data collection for wireless sensor networks. IEEE Trans. Parallel Distrib. Syst. **26**(8), 2188–2197 (2015)
5. Xiang, L. et al.: "Compressed data aggregation for energy efficient wireless sensor networks", Proceedings of 8th IEEE Communications Society Conference on Sensor, Mesh and Adhoc Communications and Networks, pp. 46–54 (2011)
6. Xu, X., et al.: Hierarchical data aggregation using compressive sensing (HDACS) in wsns. ACM Trans. Sens. Netw. **11**(3), 45–59 (2015)
7. Younis, O., Fahmy, S.: HEED: hybrid energy efficient distributed clustering approach for ad hoc sensor networks. IEEE Trans. Mob. Comput. **3**(4), 366–379 (2004)
8. Lindsey, S., Raghavendra, C.S.: PEGASIS: power efficient gathering in sensor information systems. Proc. IEEE Int. Aerosp. Conf. **3**, 1125–1130 (2002)
9. Yanjun, Y., Cao, Q., Vasilakos, A.V.: EDAL: An energy-efficient, delay aware, and lifetime-balancing data collection protocol for heterogeneous wireless sensor networks. IEEE ACM Trans. Networking **23**(3), 810–823 (2015)
10. Yanjun, Y., Cao, Q., Vasilakos, A.V.: "EDAL: an energy efficient, delay-aware and lifetime-balancing data collection protocol for wireless sensor networks". Proc. 10th Int. Conf. Mobile Ad hoc Sens. Syst. pp. 182–190 (2013)
11. Wei, G., et al.: Prediction-based data aggregation in wireless sensor networks: combining grey model and kalman filter. Comput. Commun. **34**(6), 793–802 (2011)
12. Han, K.: Algorithm design for data communications in duty cycled wireless sensor networks: a survey. IEEE Commun. Mag. **51**(7), 107–113 (2013)
13. Loscri, V.: A two-level hierarchy for low-energy adaptive clustering hierarchy (TL-LEACH). Proc. IEEE Int. Conf. Veh. Technol. **62**(3), 1809–1815 (2005)
14. Xiaoyan, M.: "Study and design on clustering routing protocols of wireless sensor networks", Ph.D. Dissertation, Department of Control Science and Engineering, Zhejiang University (2006)
15. Yassein, M.B.: Improvement of LEACH protocol of wireless sensor networks (VLEACH). Int. J. Digit. Content Technol. Appl. **3**(2), 132–136 (2009)
16. Fan, X., Yulin, S.: "Improvement on LEACH protocol of wireless sensor network". Proc. Int. Conf. Sens. Technol. Appl., pp. 260–264 (2007)
17. Zhu, N., Vasilakos, A.V.: A generic framework for energy evaluation on wireless sensor networks. Wireless Netw. **22**(4), 1199–1220 (2015)

18. Heinzelman, W.B., Chandrakasan, A.P., Balakrishnan, H.: An application specific protocol architecture for wireless microsensor networks. IEEE Trans. Wireless Commun. **1**(4), 660–670 (2002)
19. Tillet, J.: "Cluster head identification in adhoc sensor networks using particle swarm optimization". Proceedings
20. Sankarasubramaniam, Y., Akyildiz, I. E., Mchughlin, S.W.: "Energy efficiency based packet size optimization in wireless sensor networks." pp. 1–8 (2003)
21. Swami, V., Kumar, S., Jain, S.: An improved spider monkey optimization algorithm. In: Pant, M., Ray, K., Sharma, T., Rawat, S., Bandyopadhyay, A. (eds.) Soft Computing: Theories and Applications. Advances in Intelligent Systems and Computing, vol. 583. Springer, Singapore (2018)
22. Singh, V., Saini, G.L.: DTN-enabled routing protocols and their potential influence on vehicular ad hoc networks. In: Pant, M., Ray, K., Sharma, T., Rawat, S., Bandyopadhyay, A. (eds.) Soft Computing: Theories and Applications. Advances in Intelligent Systems and Computing, vol. 583. Springer, Singapore (2018). https://doi.org/10.1007/978-981-10-5687-1_33
23. Dass, A., Srivastava, S.: On comparing performance of conventional fuzzy system with recurrent fuzzy system. In: Pant, M., Ray, K., Sharma, T., Rawat, S., Bandyopadhyay, A. (eds.) Soft Computing: Theories and Applications. Advances in Intelligent Systems and Computing, vol. 583. Springer, Singapore (2018)
24. Rajkumar, A., Helen, D.: New arithmetic operations in inverse of triskaidecagonal fuzzy number using alpha cut. In: Pant, M., Ray, K., Sharma, T., Rawat, S., Bandyopadhyay, A. (eds.) Soft Computing: Theories and Applications. Advances in Intelligent Systems and Computing, vol. 583. Springer, Singapore (2018)
25. Rajak, R., Shukla, D., Alim, A.: Modified critical path and top-level attributes (MCPTL)-based task scheduling algorithm in parallel computing. In: Pant, M., Ray, K., Sharma, T., Rawat, S., Bandyopadhyay, A. (eds.) Soft Computing: Theories and Applications. Advances in Intelligent Systems and Computing, vol. 583. Springer, Singapore (2018)

# Privacy Preserving System and Data Integrity Using Distributed Hash Table in Blockchain

**R. Angeline and R. Vani**

**Abstract** With an expanding number of information focused on assaults on cloud-based workers, the hope is to actualize a blockchain backend so as to build security and diminish weaknesses. A customary cloud worker is unified consequently powerless to being a solitary purpose of disappointment. Utilizing certificateless cryptography with blockchain as a stage, the backend becomes decentralized wiping out the previously mentioned issue. Information would be put away in distributed hash tables (DHT) while the pointers to the DHT would be put away in the blockchain. Utilizing this technique can fulfill the three fundamental boundaries of certifiability, transparency, and availability. Edge computing would be actualized to deal with the IoT information at the organization edge rather than a distant cloud. Clients at that point do not have to stress over unapproved endeavors to get to information as vindictive endeavors can undoubtedly be identified. The whole framework is then discernible and responsible.

**Keywords** Blockchain · Distributed hash tables (DHT) · Certificateless cryptography · Edge computing

## 1 Introduction

In this day and age, information is one of the most important resources an individual can have. Companies esteem information exceptionally and the information they have about their clients is an enormous generator of income. That being stated, our information is defenseless against countless penetrates and assaults by programmers

R. Angeline (✉)
Department of Computer Science Engineering, SRM Institute of Science and Technology, Ramapuram, Chennai 600089, India
e-mail: angelinr1@srmist.edu.in

R. Vani
Department of Electronics and Communication Engineering, SRM Institute of Science and Technology, Ramapuram, Chennai 600089, India
e-mail: vanir@srmist.edu.in



T. K. Sharma et al. (eds.), *Soft Computing: Theories and Applications*,
Advances in Intelligent Systems and Computing 1381,
https://doi.org/10.1007/978-981-16-1696-9_46

who need admittance to our information. It is significant accordingly to guarantee that our information is protected and secure. In this paper, a model that can enormously improve the security and honesty of our information has been actualized. In the proposed framework, a lot more secure and proficient option is to the customary stockpiling advancements accessible. The recommended elective is a system that executes the idea of blockchain and edge figuring to store the information [1]. Penetrating a blockchain model is exponentially harder when contrasted with conventional distributed storage advancements. In our framework, the information is dispersed more than a few squares which guarantees the security of the information. The usage of edge registering is to improve the data transmission and therefore the general calculation measure. This model can be applied in different fields and its application isn't restricted to the field of the insurance of client information. The idea of blockchain has a few use-cases; from being be utilized in a political race cycle to guarantee that the votes are not altered to the field of cash move and digital money.

## 2 Existing Framework

The current framework manages the information stockpiling in a cloud-based stage. A cloud-based stage is alluded to a method of putting away information in a unified way; for example, all the information in the dataset is kept to restricted capacity both concerning memory and geology. A unified information stockpiling will stay under the reconnaissance of the dataset [2]. This information must be changed with the consent of the referenced administrator. Thus, regardless, if any information is to be controlled the client more likely than not needed authorizations from the administrator. In the current situation, everything the information that requires to be taken care of to the information base goes legitimately from the source to the information base through the administrator. Despite the fact that it becomes advantageous that the information stays in a similar spot it carries with it various burdens. There are some key highlights that are absent from the framework that makes it out of date contrasted with the proposed framework. It isn't distributable which builds the heap on the sole worker of activity, and it isn't as responsible when contrasted with the proposed framework and isn't unchanging. The nonattendance of these highlights makes the new framework more proficient and pertinent. There have been numerous situations where the classification of the information base has been undermined by an aggressor. This absence of an additional layer of security to the information base likewise adds to the numerous detriments of the paper. There are no basic wellbeing calculations for this situation, which carries us to the substantially more helpful proposed framework. Inconveniences of existing framework:

- The existing framework requires significant level stockpiling and computational necessities on the cloud worker.
- It works on a conventional brought together worker. Subsequently, there is a high chance of a solitary purpose of disappointment.

- The existing framework expects ability to store the huge scope IoT information.
- Sensitive information from the worker can be effortlessly spilled.
- Traditional public key foundation (PKI) with acquaints a lot of excess due with the usage of authentications.
- Identity-based encryption (IBE): It is an elective that empowers a client's public key to be made utilizing his/her character, so different substances can utilize the public key to confirm the client's personality.
- However, IBE experiences the key Escrow issue; the client's private key is accessible to the Key Age Community (KGC). Because of this, it is extremely unlikely to confirm a client except if it is expected the KGC is a totally confided in specialist.

## 3 Proposed Framework

The proposed framework means to actualize a blockchain-based stage. Through blockchain, delicate information is appropriated and the degree of security intensifies complex. A framework offers another level of straightforwardness, security, and responsibility. It comprises of squares that are connected to each other to a public record that can confirm and record exchanges. Excavators are a gathering of clients who cooperate to make these squares. Excavators take care of issues and are compensated reasonably to include blocks where the exchanges are then recorded onto the blockchain. When these exchanges are recorded onto the blockchain, they are delivered changeless and are perpetual in nature. This undertaking is an endeavor at bringing blockchain and Web of Things together. Disseminated hash tables (DHTs) can be utilized to store information. The information completely isn't put away in the blockchain; however, the pointers to the DHT stockpiling locations can be put away in the blockchain.

The other angle to the proposed framework is the utilization of edge registering. Edge registering offers a few focal points to customary distributed computing as it measures information at the organization edge rather than the far off cloud. This aids in giving a higher scope of transfer speed and improves calculation. Consider the case of a clinical framework that utilizes inserts, an edge worker in closeness to a patient can gather the necessary information from the tangible gadgets present in these inserts. The edge worker winds up doing the accompanying:

1 Performs cryptographic calculations in the IoT gadgets
2. Collects information from the sensors present in the IoT gadgets and advances it to the DHT's.

More or less, the work process of the proposed framework would be characterized by the sensors present in the gadgets conveying ongoing information to the edge worker which oversees capacity of information in the DHT through the blockchain. The accompanying ways show how the blockchain is acting like a confided in outer gathering in this situation:

1. As the edge gadget sends information acquired from sensors to the DHT, it plays out an exchange onto the blockchain proclaiming the cycle of information stockpiling in the DHT. The blockchain then checks, records, and records the exchange.
2. During an information demand from DHT to IoT gadget, the blockchain goes about as an outside gathering and confirms the requestor. On the off-chance that the exchange is endorsed, the information is sent while keeping up a serious extent of security [3, 4].

In our endeavor to incorporate blockchain with IoT, an assortment of difficulties is normal. The greatest one is to deal with the security and personality of the different gadgets so validation and approval should be possible by means of the blockchain in a safe and idiot proof way. In the blockchain stage, since the excavators check and record for exchanges guarantee that they do not have any admittance to delicate information. Consequently, a cryptographic instrument or something to that affects and is expected to accomplish the ideal degree of security [5]. Different conventional components like public key foundation (PKI), character-based encryption (IBE), all have their impediments. The public key framework carries with itself a serious extent of excess while the personality-based encryption framework experiences the Escrow issue; the KGC (Key Age Place) knows the client's private key. There is no technique for confirming a client until and except if it is accepted that the KGC is a confided in source. Accordingly, to understand these restrictions, certificateless cryptography comes in. It is not quite the same as IBE as the client's character produces the client's public key and a mystery key that the KGC is uninformed of. Through this technique, an IoT gadget can append its public key to the information solicitation to the blockchain where the public key can be checked by the diggers of the blockchain [6].

## *3.1 System Architecture*

The architecture diagram gives the representation for data storage structure which is as shown in Fig. 1

### 3.1.1 Blockchain Agreement Instrument

A normal blockchain network fills in as a shared (P2P) network without an outer specialist to meddle with activities. Since blockchain is a stage that permits a chain of exchanges to be checked and included onto blocks, the agreement calculation is a significant factor with respect to the working of the blockchain [1]. The plan of the agreement system calculation is an essential point which decides the manner in which squares are included. The primary use-instance of the blockchain, for example, Bitcoin utilized the confirmation of work agreement calculation. The overall thought

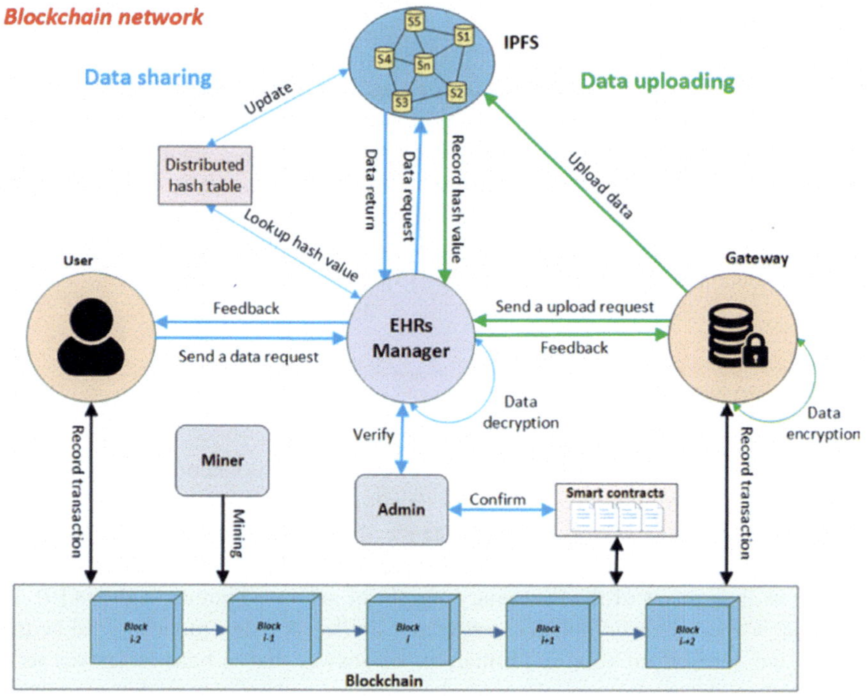


Fig. 1 Data storage structure

behind this calculation was that the exchanges to be performed and confirmed by a gathering of diggers. The disadvantage with this calculation is that it can handle just seven exchanges for every second (7 t/s), and it devours an excessive amount of power [7].

The issue with respect to the force utilization can be explained by changing the calculation to confirmation of helpful work. For this situation, minors give inalienably valuable and dependable reports that consider reasonable and plausible execution of the blockchain.

### 3.1.2 Blockchain Exchanges

Customarily, blockchain is utilized as a stage to mine digital currency. Nonetheless, it is being utilized as a stage to communicate with IoT information for capacity and assurance. Hence, when an IoT gadget sends a solicitation for information access or capacity administrations, it is supposed to be an exchange as to the proposed framework.

### 3.1.3 Edge Figuring

An edge gadget is a gadget that performs edge figuring. It makes it simpler to perform calculations and deals with the capacity of information. For IoT gadgets, it is utilized to transfer messages or exchanges. The personality of the IoT gadgets is overseen where the edge worker stores a duplicate of characters and aids in age of keys for validation by the KGC. The edge calculation worker helps in scrambling information that is delicate before sending for capacity. In conclusion, the edge worker diligently assembles data and advances it to the DHT [8].

## 4 Implementation

With regard to the validation cycle, there are different components that could have been utilized that would have made the execution simple; nonetheless, they do not fit the standards for ideal working. For instance, the elliptic-bend technique for cryptography (EEC) is an easy to actualize component of client confirmation. It is advantageous as it disposes of the customary weight of advanced authentications [9]. In any case, it does not accomplish client responsibility. Another model would be the case of the zero-information confirmation, an answer that is hard to convey as a framework with an all-out feeling of obscurity isn't plausible. In this manner, a framework is required where the IoT gadgets are doled out exceptional characters to keep up a capacity framework that is responsible and secure. Certificateless cryptography accomplishes both verification [10] and responsibility. As each IoT gadget is allotted a special personality, public keys can be created which are effectively unquestionable by different clients. Since the extraordinary ID of every gadget is the essential idea, regardless of whether a gadget disavows its present key, it can create another pair. A Key Age Place (KGC) utilizes the client's personality to make a fractional private key. This key is used alongside its mystery key to make the private key. As this private key is just accessible and known to the client, the KGC does not know about it, accordingly supporting the degree of security which implies the escrow issue is evaded [11].

To establish the keys, the following steps are, $PK_A$ and $SK_A$ for a user A.

1. The security parameters are given in the setup algorithm and get the output as system parameters K and a secret master key MSK.
2. The system parameter is given ad the input to partial private key generation algorithm and partial private key will be generated and the output will be given to the entity A.
3. The secret value generation algorithm runs by the user, and $X_A$ will be used to transform the partial to private key.
4. The private key $SK_A$ is generated using private key generation algorithm.
5. The public key generation algorithm generates $PK_A$ which will be broadcast to the public.

The above advances show age of public key sets using certificateless cryptography, which are to encode, decode, sign, and confirm capacities—[11].

The accompanying calculations are utilized for the IoT gadgets to connect themselves with the blockchain.

Step:1. Registration:

At the point when an IoT gadget demands access and to enlist itself on the blockchain, it contacts the KGC. The KGC created a fractional private key which it at that point transfers the message back to the gadget mentioned in the algorithm 1.

**Algorithm 1** Device Registration

**Input:** $ID_A$
**Output:** $PK_A$, $SK_A$
**procedure** KEYGEN($1^\lambda$, $ID_A$)
  $X_A \leftarrow SValGen(K, ID_A)$
SendRequest ($ID_A$)
  RecvReq($PSK_A$, $Sign_{SK_K}(ID_A)$)
  $V \leftarrow Ver(ID_A, Sign_{SK_K}(ID_A), SK_K)$
  if $V = Valid$ then
    $SK_A \leftarrow SKeyGen(K, PSK_A, X_A)$
    $PK_A \leftarrow PKeyGen(K, X_A)$
  **end if**
  **return**
**end procedure**

Step:2. Verify a transaction:

After effective enrollment, the IoT gadget would now be able to store information onto the blockchain. To confirm a transaction, diggers check for the accompanying boundaries mentioned in the algorithm 2:

**Algorithm 2** Verify A Transaction

**Input:** $T_A$, $\sigma_{T_A}$
**Output: a verified** $T_A$
**procedure** VERTRANS( $T_A$, $\sigma_{T_A}$)
  $s \leftarrow 0$
  $V_1 \leftarrow VerID(ID_A, PK_A, K)$
  **if** $V_1 = Valid$ **then**
    $V_2 \leftarrow Ver(T_A, \sigma_{T_A}, ID_A, PK_A)$
  elseAbort
    **if** $V_2 = Valid$ **then**
      $s \leftarrow 1$
    elseAbort
    **end if**
  **end if**
  **return**
**end procedure**

i. On the off-chance that the public key is gotten from the interesting character related with the gadget.
ii. The public key ought to be obvious with the exchange that is agreed upon.

Step:3. Data Insurance and Capacity:

Information taken from continuous items, for example, inserts and wearables are thought of. To get to the profoundly secret information, the clinical gadgets makes an entrance control list. This determines who has the correct degrees of approval to get to the information mentioned in the algorithm 3.

**Algorithm 3** Store Data

**Input:** $ID_A$, $ACL$

**Output: a verified** $T_A$

```
procedure SETACL(ACL)
    Create T_A = (PK_A, ID_A, ACL, Addr)
    Broadcast (T_A, σ_A)
     return
end procedure
procedure VERTRANS(T_A, σ_A)          D run by the miners
    s ← 0
    V_1 ← VerID(ID_A, PK_A, K)
    if V_1 = Valid then
        V_2 ← Ver(T_A, σ_A, ID_A, PK_A)
    elseAbort
        if V_2 = Valid then
            s ← 1
        elseAbort
        end if
    end if
     return
end procedure
```

Step:4. Accessing Informatio

As the diggers are answerable for including blocks containing exchanges, they additionally confirm access demands before allowing approval. So as to approve an exchange, it needs to pass a 2-venture approval measure that is featured in the calculation beneath mentioned in the algorithm 4 [12].

**Algorithm 4** Access Data

```
Input: IDA||Addr, IDB
Output: a verified TB
procedure REQUESTDATA(IDB, IDA||Addr)
    Create TB = (IDB, IDA||Addr)
    σTB ← Sign(K, TB, SkB)
    Broadcast (TB, σTB)
     return
end procedure
procedure VERTRANS(TB, σTB)        ▷ run by the miners
    s ← 0
    V1 ← VerID(IDB, PKB, K)
    if V1 = Valid then
        V2 ← Ver(TB, σTB, IDB, PKB)
    elseAbort
        if V2 = Valid then
            if IDB ∈ ACL then
                s ← 1
            elseAbort
            end if
        end if
    end if
     return
end procedure
```

## 4.1 Security

### 4.1.1 Convention Security

A. Algorithms
1–4 intend to show a safe confirmation convention utilizing certificateless cryptography. A potential programmer won't be ready to unravel the private key of a client as it gets hard to recognize plaintext and ciphertext. Likewise, producing an advanced mark for any exchange will be inconceivable as the confirmation capacity won't acknowledge the phony key.

B. Privacy
To look after protection, re-encryption is a cycle that empowers and permits encryption of information under a public key to be changed or adjusted into information under an alternate public key without the need to unscramble the message. This outcome in information is sharing to turn out to be truly plausible and simple with the utilization of re-encryption. The cycle of re-encryption is finished by the DHT hub. When the cycle is done, it presents the exchange on the square which must be included onto the blockchain ensuring that the information is encoded and mixed under the gadget's public key [13].

C. Accountability and Detectability
As the premise of blockchain is to incorporate the component of straightforwardness, responsibility and discernibility are accomplished. There is a ceaseless cycle of logging at whatever point a gadget demands admittance to information. Accordingly, any malevolent endeavor to get to information can be effectively recorded. The pernicious gadget can be identified and can likewise be hindered to forestall future assaults. This likewise decreases the odds of refusal-of-administrations (DOS) assaults from happening [14].

D. Blockchain Security
As blockchain turns out to be more standard and more IoT and different sorts of utilizations begin incorporating with blockchain, the fascination of building secure arrangements increments as the exchange expense increments bringing about higher prizes for the excavators.

## 5 Conclusion and Future Work

In the above paper, the proposed framework that actualizes the idea of blockchain just as edge processing to upgrade the security and uprightness of the client's information. The information is put away in disseminated hash tables (DHT). The decentralization of the information guarantees that the information turns out to be less defenseless to security penetrates. Edge figuring has been utilized as it gives various points of interest over the conventional distributed computing measure. Favorable circumstances remember an improvement for transmission capacity just as all the more registering power. Likewise utilized certificateless cryptography over conventional systems like public key foundation (PKI) and character-based encryption (IBE) as it helps in keeping up the obscurity of the client. These executions can extraordinarily improve the respectability of client information. In spite of the fact that the gave components and usage ensure better security for client information, upgrades can at present be made. In this setting take a gander at executions that could prepare for future work and exploration regarding the matter of information insurance.

One such improvement is the usage of recursive queries in the disseminated hash tables. There are two kinds of query tasks: iterative queries and recursive queries. In iterative queries, a hub will demand for a key from another neighboring hub. In the event that the said key isn't found in the hub, the hub will restore other close hubs. The mentioning hub at that point questions different hubs for the keys. In the recursive query technique, a hub will demand for a key from another hub which will at that point demand another hub till the key is found. The information is then passed along the chain back to the client. Utilizing recursive queries can improve the recovery season of the information. The quick admittance to the information can improve the perceivability and subsequently the general security and honesty of the information.

## References


1. Dutta, A., Mathew, R.V., Rajeev, S., Angeline, R.: Data integrity and privacy preserving framework using blockchain and edge computing. Int. J. Adv. Sci. Technol. **29**(10s), 7340–7348 (2020)
2. Yadav, S., Saroliya, A.: "Patient-specific modeling for pattern identification in clinical events using data mining models and evolutionary computing." In: Pant, M., Ray, K., Sharma, T., Rawat, S., Bandyopadhyay, A. (eds.) Soft Computing: Theories and Applications. Advances in Intelligent Systems and Computing, vol. 583. Springer (2018)
3. Xu, J., et al.: Health chain: a Blockchain-Based Privacy Preserving Scheme for Large-Scale Health Data. IEEE Internet Things J. **6**(5), 8770–8781 (2019)
4. Vani, R., Thendral, N., Kavitha, J.C., Bhavani, N.P.G.: "Intelligent traffic control system with priority to emergency vehicles". IOP Conf. Ser.: Mater. Sci. Eng. **455**(1) (2018)
5. Mishra, N., Sharma, T.K., Sharma, V., Vimal, V.: "Secure framework for data security in cloud computing." In: Pant, M., Ray, K., Sharma, T., Rawat, S., Bandyopadhyay, A.: (eds.) Soft Computing: Theories and Applications. Advances in Intelligent Systems and Computing, vol. 583. Springer (2018)
6. Dwivedi, A.D., Srivastava, G., Dhar, S., Singh, R.: "A decentralized privacy-preserving healthcare blockchain for IoT. Sensors **19**(2), 326 (2019)
7. Al Omar, Rahman, M.S., Basu, A., Kiyomoto, S.: "Medibchain: a blockchain based privacy preserving platform for healthcare data." International Conference on Security, Privacy and Anonymity in Computation, pp. 534–543. Springer (2017)
8. Kang, J., Yu, R., Huang, X., Wu, M., Maharjan, S., Xie, S., Zhang, Y.: "Blockchain for secure and efficient data sharing in vehicular edge computing and networks." IEEE Internet Things J. (2018)
9. Angeline, R., Dutta, R., Yadav, R.K., Sharma, M.: "Password authentication system using coordinates and location". Int. J. Innovative Technol. Exploring Eng. **9**(1) (2019)
10. Wang, H., Zhang, J.: Blockchain based data integrity verification for large-scale IoT data. IEEE Access **7**, 164996–165006 (2019)
11. Li, R., Song, T., Mei, B., Li, H., Cheng, X., Sun, L.: Blockchain for large-scale internet of things data storage and protection. IEEE Trans. Serv. Comput. **12**(5), 762–771 (2019)
12. Hylock, R.H., Zeng, X.: "A blockchain framework for patient-centered health records and exchange (health chain): evaluation and proof-of-concept study." J. Med. Internet Res. 21(8) (2019)
13. Kadam, S., Meshram, A., Suryawanshi, S.: "Blockchain for healthcare: privacy preserving medical record". Int. J. Comput. Appl. **975**, 8887
14. Nizamuddin, N., Hasan, H. R., Salah, K.: "IPFSblockchain-based authenticity of online publications." In International Conference on Blockchain. pp. 199–212, Springer (2018)

# Intelligent Tourniquet System for Emergency AID Using Wireless Network

**T. Venkatesh, M. Nisha Angeline, S. K. Manikandan, S. Priyadharshini, K. Murugan, K. Siddharthraju, and T. Shanmugaraja**

**Abstract** It is important to use a tourniquet in case of accidents during explosions and weapon injuries in order to overcome serious injuries and blood loss. When using a tourniquet, there are a couple of important things to remember. If it is not tightened well to avoid the loss of blood, it continues and the life of the wounded individual becomes very dangerous. If more pressure is given and if it is not relaxed and hardly squeezed, serious problems can be caused. The aim of this product is to create an intelligent tourniquet device that can successfully stop bleeding by automatically supplying the limb with the necessary pressure to prevent further bleeding from occurring in the wounds. The device built would be commonly used in vehicle first aid kits, ambulances, hospital emergency departments, fire vehicles, first aid agencies, hazardous workplaces,s and so many civilian areas.

**Keywords** Tourniquet · Blood loss

## 1 Introduction

The significant purpose behind the passing of individuals during mishaps is loss of blood. Blood assumes an essential function in human body, so it is valuable to spare each drop of them. All over world, individuals face this issue. For the individuals who work in ventures managing monster machines, emergency clinics, military and safeguard-related positions, and so forth to defeat this issue, we are thinking of this item. The item configuration incorporates emergency treatment, tourniquet, remote

T. Venkatesh (✉) · K. Murugan · K. Siddharthraju · T. Shanmugaraja
Department of ECE, KPR Institute of Engineering and Technology, Coimbatore, India

M. N. Angeline
Department of ECE, Velalar College of Engineering and Technology, Erode, India

S. K. Manikandan
Department of EEE, Velalar College of Engineering and Technology, Erode, India

S. Priyadharshini
Department of ECE, Tamilnadu College of Engineering, Coimbatore, India

 
T. K. Sharma et al. (eds.), *Soft Computing: Theories and Applications*,
Advances in Intelligent Systems and Computing 1381,
https://doi.org/10.1007/978-981-16-1696-9_47

organization, sun-based board. This sort of mishaps may happen in streets, emergency clinics, military, and so forth individuals who gets harmed with hefty blood loss. The created item will be broadly utilized in ambulances, medical aid packs for vehicles, and crisis administrations for emergency clinics [1–6].

## 2 Tourniquet

A tourniquet is a tight band which is utilized to capture or control the progression of blood during the mishaps. A straightforward tourniquet should be possible by a rope and a stick. The rope is changed into a circle which must be effectively positioned over the harmed parts, and the stick is surrendered to the circle. It is fixed firmly by curving the stick. This may capture draining, however, results or harm may happen. There are three sorts of tourniquets present they are crisis tourniquets, careful tourniquets, and restoration tourniquets. In our item, we are utilizing crisis tourniquet framework (Fig. 1).

**Types and Practical Use of Tourniquet**

**Types**

**Surgical Tourniquet**

We use elastic ring tourniquets or silicone ring tourniquets in this form of surgical tourniquet, which are automated mechanical devices that do not require electricity or wires in any way. After the sheets that are on the patient, this kind of tourniquet should be present.

**Fig. 1** Tourniquet

**Emergency Tourniquet**

Emergency tourniquets are sleeve like structure that is intended to capture serious blood misfortune. They are hung around the appendage and fixed bit by bit. The plan and structure of crisis tourniquets permit quick reaction to the harmed people.

**Pneumatic Tourniquets**

Pneumatic tourniquets use compacted gas to sleeve to capture blood misfortune. The managing gadget on this sort of tourniquet can control the specific measure of weight created in sleeve applied on the appendage. Presently, the weight is created by a focal compacted air gracefully or by an electrically determined siphon.

**Practical Use of Tourniquet**.

We first find where the blood originates from and give the injured injury hands with consistent weight. Spot a tourniquet 2–3 inches nearer to the wound if the blood misfortune is not forestalled. The tourniquet can presently be applied safely. We can control and screen blood pressure by the utilization of the tourniquet. In the event that the draining is not captured at this point, at that point, we put a second tourniquet closer to the injury.

**Application of Tourniquet**

1. When serious wounds occur and normal operated tourniquets are harmless to the patient, tourniquets must be used.
2. To regulate blood misfortune, tourniquets should be placed closer to the injury.
3. We should not remove this when a tourniquet is fixed in a position. Despite the fact that, because of its quality, the person feels uncomfortable.
4. Essential weight should be added to the tourniquet before it captures the drainage.
5. When the tourniquet is set, the circumstance must be noted down.

## 3 Working

This framework incorporates parts like Arduino, 7 section show, driver engine, pressure sleeve, pressure engine, 5 V power flexibly. Right off the bat, we give power flexibly to begin the gadget utilizing start button. When the gadget is turned on the driver engine will create 230 V, then the AC engine will stifle the air to fill the sleeve with the air. Here, we likewise use addition and decrement catches to arrangement the weight territory, and it will be shown on the 7 fragment show. Additionally, we can change the weight rate by utilizing those catches. Presently, we utilize a sensor to detect the scope of the weight and show the readings in the 3 digit 7 portion show. In the event that the weight territory is surpassed, we can lessen it by utilizing the decrement button. On the off chance that the reach is diminished, we utilize the addition button. We are utilizing Bluetooth as the remote medium to on and off the whole gadget. In the event that the Arduino produces 5 V ground information, 230 V will

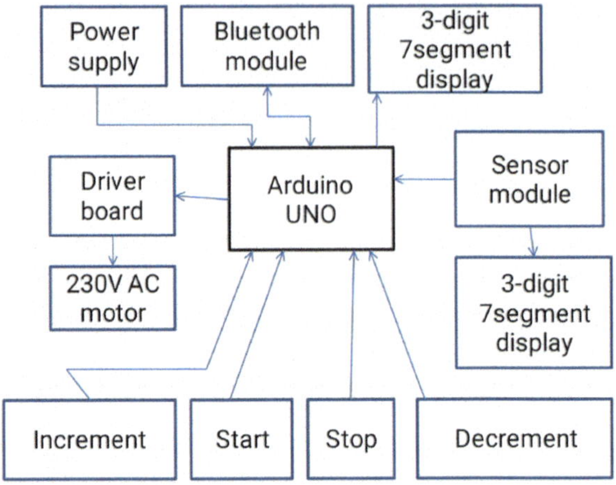


**Fig. 2** Block diagram of working module

jump on. On the off chance that there is no voltage present in the Arduino, at that point, the 230 V cannot be turned on. By utilizing this sort of system, the passing rate during mishaps is diminished because of the quick access gave by the organization. For building up this item, we use proteus 8 programming for animating the item. Inserted C program is fused in this item. On the off chance that there is no power present while getting to the gadget, we utilize sun-oriented boards to give the reinforcement to the gadget.

**Block Diagram**

See Fig. 2.

## 4 Utilization of Wireless Network

The remote neighborhood (WLAN) uses radio proliferations to interface frameworks, for example, workstations, play stations, and cell phones to get to the Internet and its applications. What is more, when you associate with a Wi-Fi at public organization area you are associated with that business kind of remote organization.

1. **Portability**
   In wired organization gadgets, individuals will in general adhere on to a spot for getting to the organization. Though in remote, we can access in from any spot.
2. **Comfort**
   In remote, we can get to the organization accessibility from any spot inside the specific inclusion or from any Wi-Fi.
3. **Simple Setup**
   Utilization of links are limited, so the speed for getting to the organization is expanded.

4. **Security**
   Remote gives high security, for example, powerful security assurance.
5. **Profitability**
   Utilizing remote organization, it is available for individuals to be more synergistic to one another when they are working in an industry or an association.
6. **Cost**

It is more financially savvy when contrasted with the wired organizations.

## 5 Solar Panel

The panel board is a plan of solar-based cells that can change light source into power. What is more, by joining the limit of different sun-based boards, a piece of a power prerequisite can be used. Right now, contingent upon the different sorts of board, 6–20% of the light source can be changed into power. This is called as "yield" of the sun-oriented board. The innovation is therefore being created, and the yield should improve further. By utilizing sun-powered boards, we can change daylight into power. The solar board delivers direct current. We can utilize this current in the home or office, and it must be changed to rotating power wellspring of 230 V. This should be possible by the inverter or converter.

**Features**

A. **Benefits**

1. Because of remote correspondence framework significant distance availability is conceivable.
2. This item is cost effective.
3. Demise rate during mishaps are diminished.
4. Speed is high.

B. **Demerits**

1. In the event that there is no mindfulness with respect to the useful use of tourniquet, at that point, the results will be most exceedingly terrible.

## 6 Conclusion

In the current item, they utilize wired framework. So, if there is any equipment deficiency or force misfortune, the absolute framework will get imploded. To stay away from such sort of occurrences, we utilize remote imparting organization, and furthermore, we are utilizing sun-based board to give power reinforcement to the gadget in the event of nonappearance of the power.

## References


1. Budak, E.İ: Development of Tourniquet System For Military Use (master thesis). TOBB University of Economics and Technology, Ankara, TURKEY (2017)
2. Unlu, A., Petrone, P., Guvenc, I., Kaymak, S., Arslan, G., Kaya, E., Yılmaz, S., Cetinkaya, R.A., Ege, T., Ozer, M.T., Kilic, S.: Combat application tourniquet (CAT) eradicates popliteal pulses effectively by correcting the windlass turn degrees: a trial on 145 participants. Eur. J. Trauma. Emerg. Surg. **43**(5), 605–609 (2017)
3. Ünlü, A., Cetinkaya, R.A., Ege, T., Ozmen, P., Hurmeric, V., Ozer, M.T., Petrone, P.: Role 2 military hospitals: results of a new trauma care concept on 170 casualties. Eur. J. Trauma. Emerg. Surg. **41**(2), 149–155 (2015)
4. Unlu, A., vd.: An evaluation of combat application tourniquets on training military personnel: changes in application times and success rates in three successive phases. J. Royal Army Med. Corps JRAMC-2014 (2014)
5. Kragh, Jr. J.F., vd.: Practical use of emergency tourniquets to stop bleeding in major limb trauma. J. Trauma Acute Care Surg. **64**(2), S38–S50 (2008)
6. Kragh, J.F., Jr., Walters, T.J., Baer, D.G., Fox, C.J., Wade, C.E., Salinas, J., Holcomb, J.B.: Practical use of emergency tourniquets to stop bleeding in major limb trauma. J. Trauma Acute Care Surg. **64**(2), S38–S50 (2008)

# Performance Comparison of TDS Controllers for Water Purification System with Dynamic Input

**Rajat Jaiswal and Irshad Ahmad Ansari**

**Abstract** Monitoring of the water total dissolved solids (TDS) has remarkable importance for industrial, agricultural, and domestic purposes. The controller designed and simulated in this paper controls and displays the TDS of water. This controller assembled with industrial and domestic purifiers has better control to frequent changing TDS of input water. Controller parameters in this paper are derived based on Ziegler Nicolas tuning method. Working of the controller will be explained here by discussion of several simulation models, and advantages over other controllers will also be studied. In order to have better and fast control over frequent change in TDS, a mathematical model for domestic water purification reverse osmosis (RO) system is taken from previous published paper. Then, the system model is used for testing and comparing P, PI, and PID controller; after the comparisons, a PID controller is selected because of its capability to settle fast and is then compared with constant and random TDS inputs. Task of controlling the TDS is done through real time and continuous controlling of a mixing valve.

**Keywords** Total dissolved solids (TDS) · Reverse osmosis (RO) · Ziegler Nicolas · P · PI · PID controller

## 1 Introduction

The entire water source available on the earth surface has some amount of solvents dissolved in it [1, 2]. Some sources have high concentration of certain salts like calcium, iron, etc., while others have some other salts present depending on the area where the source lie or from where the source flows. Water has high solubility characteristics and thus accumulates different solvents. It also dissolves carbon dioxide present in air and produces carbonic acid [3], which imparts corrosive behavior to the water. These solvents lead to hardening of the water. Hardness can be removed

R. Jaiswal · I. A. Ansari (✉)
Electronics and Communication Engineering, PDPM Indian Institute of Information Technology, Design and Manufacturing, Jabalpur, M.P 482005, India
e-mail: irshad@iiitdmj.ac.in

 
T. K. Sharma et al. (eds.), *Soft Computing: Theories and Applications*,
Advances in Intelligent Systems and Computing 1381,
https://doi.org/10.1007/978-981-16-1696-9_48

using different techniques [4]. Some hardness is temporary which can be removed by boiling, while ions of calcium, magnesium with chlorides, sulfates, and nitrates make water permanently hard. TDS is the measurement of dissolved solvents and is denoted in parts per million (ppm). As most of the equipments used in industries are made of metals, there is a high chance of corrosion if high TDS water is being fed. In case of food industries, it is crucial to maintain TDS level as even small change in TDS can change the taste and pH of the end product [5]. Similarly, for water used for domestic purposes like cooking, washing and drinking, issues like more cooking time, less lather with soap, health, and taste related problems may arise.

The RO system is used mostly by every desalination plant for reducing TDS. The mechanism of RO is discussed in [6]. In this paper, RO system's mathematical model is taken from [7], where transfer function is obtained using system identification tool of MATLAB [8].

Ziegler Nicolas tuning method [9] is the most popular method to find the controller parameters and is used in this paper to find out proportional, integral, and derivative gain of the PID controller.

The RO system model was tested well for constant TDS input in [7], but in practical situations TDS does not remain constant and vary in a certain range. This paper examines the model for randomly changing TDS of water.

The RO model transfer function has been tested with P, PI, and PID controller and has been further concluded with simulation results.

## 2 RO Model Estimation

RO system approximate transfer function has been derived from experiments in [7]. The open-loop transfer function of the RO system as shown in Eq. 1 has been adopted from [7], which has been derived from the output curve using the system identification tool of MATLAB. The data collected from the RO model was preprocessed, so that model estimated is more generalized.

$$G(S) = \frac{823.71}{0.7316s^2 + 1.2021s + 1} \times e^{-0.0070017S} \tag{1}$$

## 3 Controller Designing

Controllers are designed according to the plant to which they are to be connected. A unity feedback control system has been shown in Fig. 1.

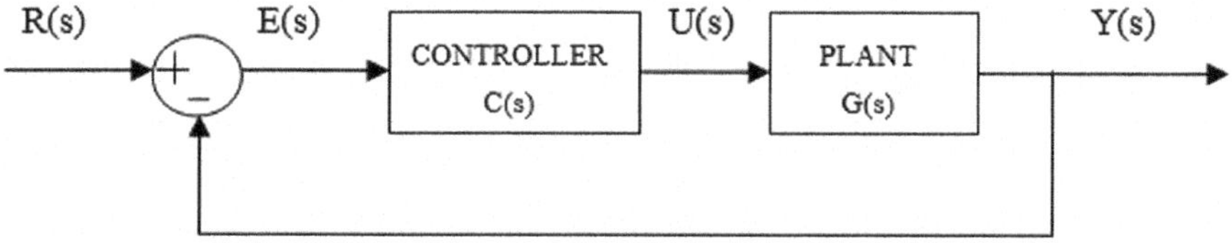


**Fig. 1** Feedback control system

### *3.1 Proportional Controller P*

Proportional is pure gain with value $K_P$. In this controller, transient and steady-state response is achieved by adjusting the $K_P$ value in the system. This controller depends on the error signal. This controller increases the systems undamped natural frequency, thus increasing the speed of the system with reduction in systems damping ratio leading to increase in system overshoot. This controller can reduce error, but cannot completely eliminate [10, 11]. This controller works well in a certain band of gain value. The steady-state offset to the reference input reduces with increase in gain value. But only till an upper gain value and then loses stability. This controller is best suited to those systems which are admissible to constant steady-state error [12]. The output from the proportional controller is given by Eq. 2.

$$U(s) = K_P\, E(s) \tag{2}$$

### *3.2 Proportional Integral PI*

Proportional integral has two gain parameters $K_P$ and Ki to determine the design of the controller to attain particular steady-state response [13]. This controller provides both past and present control. The integral expression responds to the past error by the elimination of the area under the error curve. The integral control places a pole at the origin which leads to zero steady-state error. Until the error becomes zero, the integral response will continuously increase, so called automatic reset controller [14]. With increase in Ki, damping ratio decreases and thus approaches overshoot and instability [10, 11]. This controller is slower than P and PID controller. This controller is best suited to system where zero steady-state error is essential for controlling. The output from the proportional integral controller is given by Eq. 3.

$$U(s) = E(s)\left(K_P + \frac{K_i}{s}\right) \tag{3}$$

### 3.3 Proportional Integral Derivative PID

Proportional integral derivative has three gain parameters $K_P$, $K_i$, and $K_d$ to refine both transient and steady-state response of the system. This controller provides past, present, and future control. The derivative expression responds to rate of change of error signal as input and not as error itself. Derivative control has the capability to control error before it arises. Usually, $K_d$ value is kept small as big values make controller prone to noise [14]. System approaches instability with slow control loops and noisy feedback. This controller places two zeros and one pole at the origin. The pole helps limit the high-frequency gain of the system [15–17]. Derivative control does not alter systems undamped natural frequency, so the speed is not changed, but it increases damping ratio, thus reducing the systems overshoot [15, 16]. The output from the PID controller is given by Eq. 4, and the transfer function of the PID controller in time constant form is given by Eq. 5.

$$U(s) = E(s)\left(K_P + sK_d + \frac{K_i}{s}\right) \tag{4}$$

$$C(s) = K_P\left(1 + \frac{1}{T_i s} + T_d s\right) \tag{5}$$

## 4 Controller Parameter Approximation

For finding out approximate values of $K_P$, $T_i$, and $T_d$, i.e., for properly tuning the PID controller, Ziegler Nicolas step response tuning method is used in this paper to achieve this objective.

This method of tuning is based on the open-loop step response of the stable system. In this, the step response of the system should look like $S$ shape, characterized by parameters $L$ (delay time) and $T$ (time constant), as shown in Fig. 2 [18].

The point of maximum slope is marked as base, and a tangent to the curve is drawn. The intersection of the tangent with the time axis is marked as $L$. The point where the tangent intersects the line $z(t) = F$ is marked, and a line perpendicular to the time axis is drawn. The distance between this point and $L$ gives us $T$. PID parameters can be found out using the relationship between $a = F \times L/T$ and $L$. Relationship table is as shown in Table 1 [9]. Figure 3 represents the open loop plant system. The delay is ignored here to check the performance of controller without incorporating the same.

Giving step input to the open-loop plant system.

Analyzing Fig. 4, following information is obtained F = 823.71, L = 0.23 (approx.), and T = 1.997 (approx.) Thus, a = 94.868 so Table 1 can be updated as shown in Table 2.

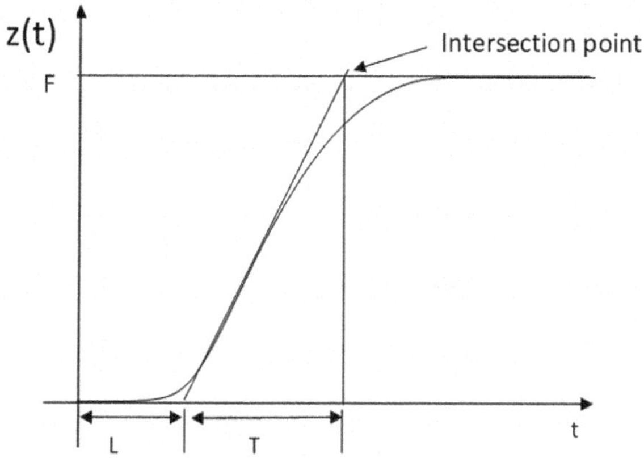


**Fig. 2** Open-loop system step response

**Table 1** Ziegler Nicolas tuning parameters

| Controller | $K_P$ | $T_i = K_P/K_i$ | $T_d = K_d/K_P$ |
|---|---|---|---|
| *P* | 1/a | – | – |
| *PI* | 0.9/a | 3L | – |
| *PID* | 1.2/a | 2L | L/2 |

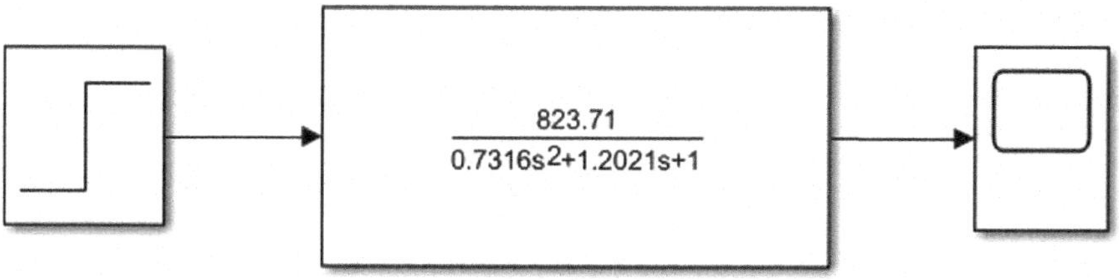


**Fig. 3** Open-loop plant system

In order to have a comparative analysis of all the proposed controllers, simulation is performed in MATLAB software (Fig. 5).

Input TDS level is shown by random input block with mean = 120 and variance = 38.1. Here, TDS level is found to be varying between (120 ± 10) (approx.) as shown in Fig. 6.

Using controller parameters from Table 3, tuning of P, PI, and PID controllers has been performed and is assembled with the RO plant. Simulating the above system, following results have been obtained as shown in Fig. 7.

From the above graph, it can be noted that the system with a PID controller has less settling time and also approaches mean TDS value of 120 early than P and PI.

As it can be seen that PID controller is controlling better, so it is simulated for constant input and random input to check whether the controller works in the same manner, as shown in Fig. 8.

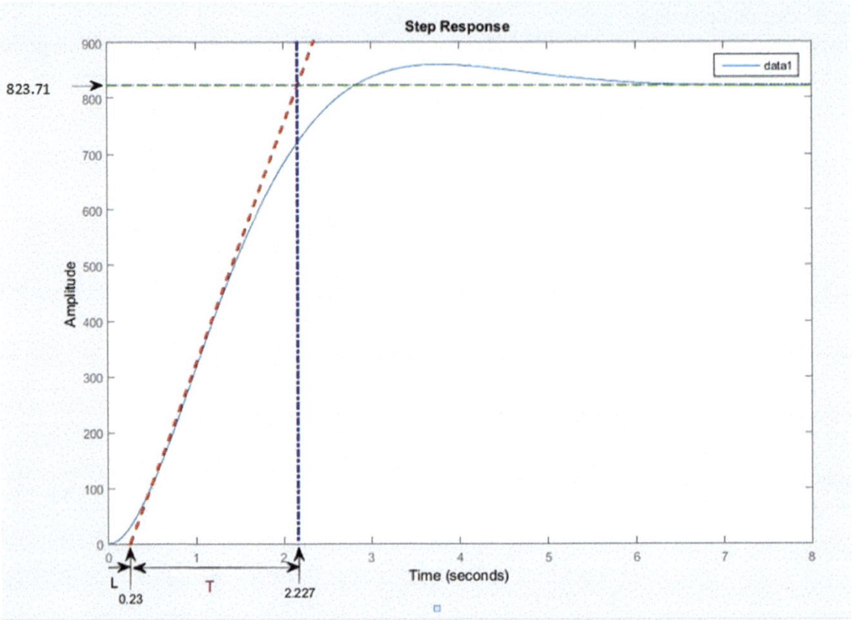


**Fig. 4** Open-loop plant system step response

**Table 2** Ziegler Nicolas tuning parameters

| Controller | $K_P$ | $T_i$ | $T_d$ |
|---|---|---|---|
| *P* | 0.0105 | – | – |
| *PI* | 0.0095 | 0.69 | – |
| *PID* | 0.0127 | 0.46 | 0.115 |

## 5 Conclusion

For effective control of frequent changing TDS of water, the PID controller designed in this paper can be employed. Based on above comparative study, it can be concluded that the designed controller works well with both constant and random input. Limitation of the design is that fixed TDS cannot be desired at the output as input keeps on varying within a certain range. The plant model has been obtained from [7] so delay is not considered during PID controller design for the sake of generalization. Further, controllers using fuzzy and ANFIS [19–22] will be designed for TDS control with dynamic input.

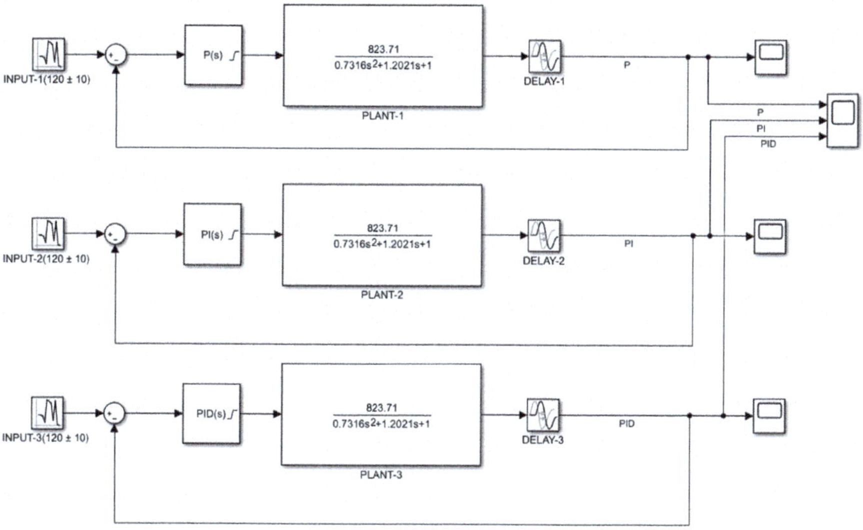


**Fig. 5** Simulation of the complete system in MATLAB environment

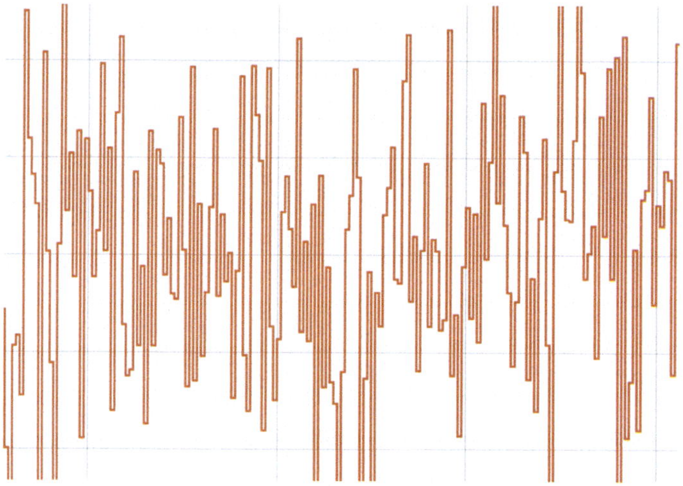

**Fig. 6** Random TDS input ranging (120 ± 10)

**Table 3** Controller parameters

| Controller | $K_P$ | $K_i$ | $K_d$ |
|---|---|---|---|
| *P* | 0.0105 | – | – |
| *PI* | 0.0095 | 0.0138 | – |
| *PID* | 0.0127 | 0.0275 | 0.0015 |

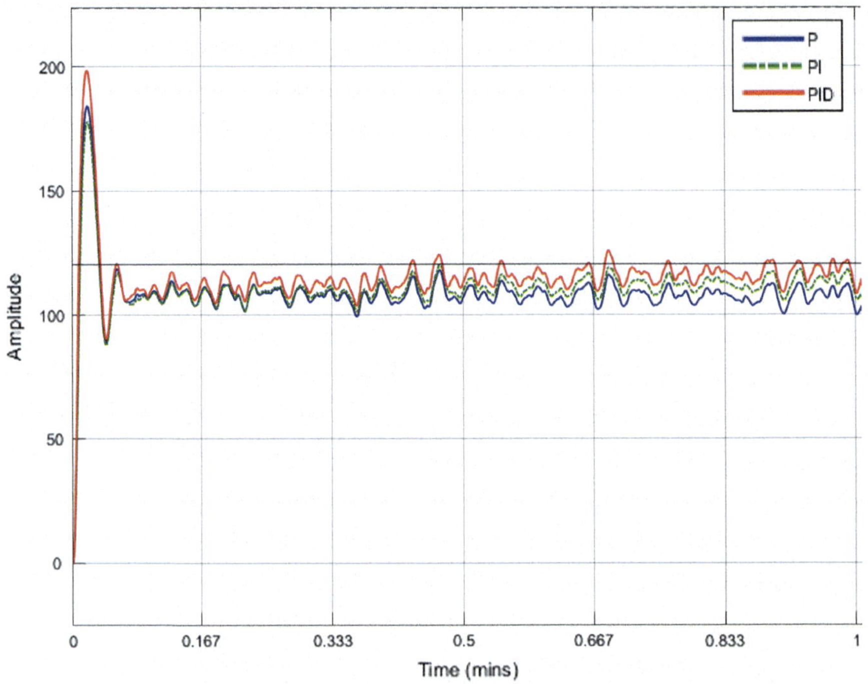


**Fig. 7** Comparative simulation graph

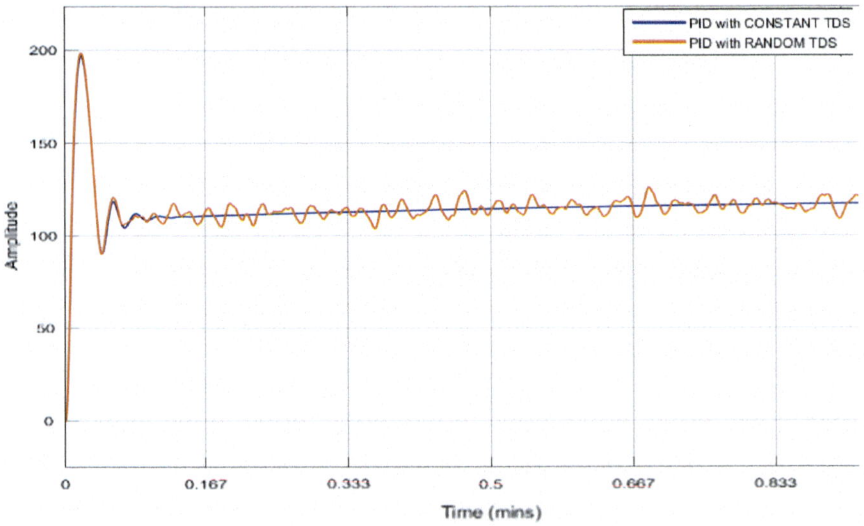


**Fig. 8** Comparative simulation graph for constant and random TDS input

## References


1. Shaikh, A.M., Mandre, P.N.: Seasonal study of physico-chemical parameters drinking water in Khed( Lote ) industrial area. Shodh, SamikshaaurMulyankan (Int. Res. J.)—ISSN-0974–2832 vol. II(7) (August 2009)
2. Jain, C.K., Bandyopadhyay, A., Bhadra, A.: Assessment of ground water quality for drinking purpose, District Nainital, Uttarakhand India. Environ Monit Assess **166**, 663–676 (2010)
3. Weiss, R.F.: Carbon dioxide in water and seawater: the solubility of a non-ideal gas. Elsevier Scientific Publishing Company, Amsterdam—Printed in The Netherlands Marine Chemistry, **2**, 203–215 (1974)
4. Younos, T., Tulou, K.E.: Overview of desalination techniques. J. Contemp. Water Res. J. Contemp. Water Res. Educ. **132**, 3–10 (2005)
5. Bavar, M., Sarrafzadeh, M.H., Asgharnejad, H., Firouz, H.N.: Water management methods in food industry: corn refinery as a case study. J. Food Eng. (2018). Doi: https://doi.org/10.1016/j.jfoodeng.2018.06.018
6. Tamura, M., Matsudo, Shinbo, A., Wako: reverse osmosis process and equipment. Patent No.: US 6,303,037 B1 (16 Oct 2001)
7. Mousavi Mashhadi, S.K., Yadollahi, H., Marvian Mashhad, A.: Design and manufacture of TDS measurement and control system for water purification in reverse osmosis by PID fuzzy logic controller with the ability tocompensate effects of temperature on measurement. Turk. J. Elec. Eng. Comp. Sci. **24**, 2589–2608 (2016). https://doi.org/10.3906/elk-1402-65
8. Ljung, L.: System identification toolbox for use with MATLAB. https://www.researchgate.net/publication/37405937
9. Copeland, B.R.: The design of PID controllers using Ziegler Nichols tuning, March 2008, https://educypedia.karadimov.info/library/Ziegler_Nichols.pdf
10. Bucek, V.J.: Control Systems Continuous and Discrete. Prentice-Hall, Inc., New Jersey (1989)
11. Li, S., Jiang, Q.: On pid parameters tuning method based on Matlab/Simulink, (2), 408–411 (2011)
12. Charles, L., Phillips, R.D.: Feedback control systems. Prentice -Hall, Inc., New Jersey (2002)
13. Chen, Y., Ma, Y., Yun, W.: Application of improved genetic algorithm in PID controller parameters optimization. TELKOMNIKA Indonesian J. Electr. Eng. **11**(3), 1524–1530 (2013)
14. Owen, F.: Designing and tuning PID controllers in Frank Owen. Editors. Control Systems Engineering A Practical Approach. California. 2012: 4–8
15. Michael, A., Johunson, M.H.: PID Control: New Identification and Design Method. Spring-Verlag London Limited, London (2005)
16. Xie, W., Duan, J.: The design and simulation of fuzzy PID parameter self-tuning controller. TELKOMNIKA Indonesian J. Electr. Eng. **14**(2), 293–297 (2015)
17. Omar, M., Ebrahim, M.A., Ghany, A.A., Bendary, F.: Tuning of PID controller for load frequency problem via harmony, search algorithm. Indonesian J. Electr. Eng. Comput. Sci. **1**(2), 255–263 (2016)
18. Astrom, K., Hagglund, T.: PID Controllers: Theory, Design, and Tuning. Instrument Society of America, USA (1995)
19. Dass, A., Srivastava, S.: On comparing performance of conventional fuzzy system with recurrent fuzzy system. In: Pant, M., Ray, K., Sharma, T., Rawat, S., Bandyopadhyay, A. (eds.) Soft Computing: Theories and Applications. Advances in Intelligent Systems and Computing, vol. 583. Springer, Singapore (2018). https://doi.org/https://doi.org/10.1007/978-981-10-5687-1_35
20. Giri, J.P., Giri, P.J., Chadge, R.: Neural network-based prediction of productivity parameters. In: Pant, M., Ray, K., Sharma, T., Rawat, S., Bandyopadhyay, A. (eds.) Soft Computing: Theories and Applications. Advances in Intelligent Systems and Computing, vol. 583. Springer, Singapore (2018). https://doi.org/https://doi.org/10.1007/978-981-10-5687-1_8

21. Irawan, A., Azahar, M.I.P.: Cascade control strategy on servo pneumatic system with fuzzy self-adaptive system. J Control Autom. Electr. Syst. **31**, 1412–1425 (2020). https://doi.org/10.1007/s40313-020-00642-4
22. Borase, R.P., Maghade, D.K., Sondkar, S.Y. et al.: A review of PID control, tuning methods and applications. Int. J. Dynam. Control (2020). https://doi.org/10.1007/s40435-020-00665-4

# On Generalized Matrix Summability $\phi - |U, \gamma; \delta|_r$ of Orthogonal Series

**Smita Sonker**

**Abstract** In the present paper, two theorems on $\phi - |U, \gamma; \delta|_r$ summability for $(1 \leq r \leq 2)$ of orthogonal series have been proved, and various known and new results have been deduced.

**Keywords** Infinite series · Orthogonal series · Hölder's inequality · Matrix summability · Abel's theorem

**Mathematics Subject Classification 2010 :** 40F05 · 40D15 · 42C15

## 1 Introduction

Let $\sum a_n$ be an infinite series whose partial sum's sequence is given by $\{s_n\}$ and $U = (u_{nv})$ denotes a normal matrix. Then $U$ defines a transformation which maps $s = \{s_n\}$ to $U_S = \{U_n(s)\}$, where

$$U_n(s) = \sum_{v=0}^{v} u_{nv} s_v, \qquad n = 0, 1, 2, \ldots [13]. \tag{1}$$

In 1957, Flett [1] defined that $\sum a_n$ is absolutely summable $|U|_r$; $r \geq 1$, if

$$\sum_{n=0}^{\infty} n^{r-1} |U_n(s) - U_{n-1}(s)|^r$$

converges and thus $\sum a_n \in |U|_r$.

In [2], Flett extended the definition of absolute summability by introducing a parameter $\delta$, i.e., $\sum a_n$ is absolutely summable $|U, \delta|_r$; $r \geq 1, \delta \geq 0$, if

S. Sonker (✉)
National Institute of Technology, Kurukshetra, Haryana 136119, India



T. K. Sharma et al. (eds.), *Soft Computing: Theories and Applications*,
Advances in Intelligent Systems and Computing 1381,
https://doi.org/10.1007/978-981-16-1696-9_49

$$\sum_{n=0}^{\infty} n^{\delta r+r-1}|U_n(s) - U_{n-1}(s)|^r$$

converges and thus $\sum a_n \in |U, \delta|_r$.

Also, $\sum a_n$ is absolutely summable $|U, \delta; \gamma|_r; r \geq 1, \delta \geq 0$ and $\gamma$ is a real number, if [3]

$$\sum_{n=0}^{\infty} n^{\gamma(\delta r+r-1)}|U_n(s) - U_{n-1}(s)|^r.$$

converges and thus $\sum a_n \in |U, \delta; \gamma|_r$.

For positive real numbers $\varphi = \{\varphi_n\}$, $\sum a_n$ is absolutely $\varphi - |U, \delta|_r$ summable; $r \geq 1$ and $\delta \geq 0$, if [4]

$$\sum_{n=0}^{\infty} \varphi_n^{\delta r+r-1}|U_n(s) - U_{n-1}(s)|^r$$

converges and thus $\sum a_n \in \varphi - |U; \delta|_r$.

The series is absolutely $\varphi - |U, \delta; \gamma|_r$ summable $r \geq 1$, if

$$\sum_{n=0}^{\infty} \varphi_n^{\gamma(\delta r+r-1)}|U_n(s) - U_{n-1}(s)|^r.$$

converge and thus $\sum a_n \in \varphi - |U, \delta; \gamma|_r$

If $\varphi_n$ = n ∀ n, then $\varphi - |U, \delta; \gamma|_r$ summability is reduced to summability $|U, \delta; \gamma|_r$.

The convolution $(p * q)_n$ of $p = \{p_n\}$ and $q = \{q_n\}$ is defined by

$$(p * q)_n = \sum_{s=0}^{n} p_s q_{n-s} = \sum_{s=0}^{n} p_{n-s} q_s.$$

The generalized Nörlund transform of $\{s_n\}$ is $\{t_n^{p,q}\}$ get by taking

$$t_n^{p,q} = \frac{1}{(p * q)_n} \sum_{s=0}^{n} p_{n-s} q_s.$$

where $(p * q)_n \neq 0$ ∀ n.

The $\sum a_n$ is $|N, p, q|$ summable, if $\sum_{n=0}^{\infty} |t_n^{p,q} - t_{n-1}^{p,q}|$converges and thus $\sum a_n \in |N, p, q|$.

Tanaka [5] introduced the idea of $|N, p, q|$ summability.

Let $\{\psi_j\}$ denote an orthonormal system which is defined on $(a, b)$. We suppose that $g \in L^2(a, b)$ and

$$g(x) \sim \sum_{j=0}^{\infty} d_j \psi_j(x), \tag{2}$$

where

$$d_j = \int_a^b g(x)\varphi_j(x) \qquad (j = 0, 1, 2, ......).$$

**Notations**: Assuming $R_n = (p * q)_n$, $\quad R_n^j = \sum_{m=j}^{n} p_{n-m}q_m$

$$\text{and} \qquad R_n^{n+1} = 0, \qquad R_n^0 = R_n.$$

In 1981, Okuyama and Tsuchikura [6] proved a theorem regarding Riesz summability of orthogonal series. Okuyama [7] established two theorems regarding $|N, p, q|$ summability. In 2012, Krasniqi [1] developed two results related to $|A, \delta|_k$ -summability of orthogonal series, and also Rajkumar and Helen [8, 9] gave results on new arithmetic operations. In this paper, work has been done to extend the results of Krasniqi [1] into more general matrix summability $\varphi - |U, \gamma; \delta|_r$ of orthogonal series.

## 2 Known Result

Krasniqi [1] demonstrated the following two results related to $|A, \delta|_k$ -summability of orthogonal series.

**Theorem 1** ([1]) *If the series*

$$\sum_{n=0}^{\infty} \left\{ n^{2(\delta+1-\frac{1}{k})} \sum_{j=0}^{n} |\widehat{a}_{n,j}|^2 |c_j|^2 \right\}^{\frac{k}{2}}$$

*converges for* $1 \leq k \leq 2$, *then the orthogonal series* $\sum_{j=0}^{\infty} c_j\varphi_j(x)$ *is* $|A, \delta|_k$ *-summable almost everywhere.*

**Theorem 2** ([1]) *Let* $1 \leq k \leq 2$ *and let* $\{\Omega(n)\}$ *be a positive sequence with* $\left\{\frac{\Omega(n)}{n}\right\}$ *is non-increasing and the series* $\sum_{n=1}^{\infty} \frac{1}{n\Omega(n)}$ *converges. If* $\sum_{n=1}^{\infty} |c_n|^2 \Omega^{\frac{2}{k}-1}(n)\omega^{(k)}(A, \delta; n)$ *converges, then the orthogonal series* $\sum_{j=0}^{\infty} c_j\varphi_j(x)$ *is* $|A, \delta|_k$ *-summable almost everywhere, where* $\omega^{(k)}(A, \delta; n)$ *is defined by* $\omega^{(k)}(A, \delta; j) = \frac{1}{j^{\frac{2}{k}-1}} \sum_{n=j}^{\infty} n^{2(\delta+\frac{1}{k})} |\widehat{a}_{n,j}|^2$. *The objective of this article is to generalize the known results for* $\varphi - |U, \gamma; \delta|_r$ *summability of the orthogonal series given by (2), where* $1 \leq r \leq 2$ . *With a normal matrix* $U = (u_{nv})$, *two associated lower matrices* $\overline{U} = (\overline{u}_{nv})$ *and* $\widehat{U} = (\widehat{u}_{nv})$ *are as follows:*

$$\overline{u}_{nv} = \sum_{i=v}^{n} u_{ni}, \qquad n, i = 0, 1, 2, \ldots$$

*and*

$$\widehat{u}_{00} = \overline{u}_{00} = u_{00}, \qquad \widehat{u}_{nv} = \overline{u}_{nv} - \overline{u}_{n-1,v}, \qquad n = 1, 2, .....$$

## 3 Lemmas

The lemma [10] given below is required for our result.

**Lemma 1** *If $V_n(t) \in L(E)$ are non-negative functions and*

$$\sum_{n=1}^{\infty} \int_E V_n(t) < \infty, \tag{3}$$

*then the series*

$$\sum_{n=1}^{\infty} V_n(t)$$

*converges almost everywhere on E to a function $V(t) \in L(E)$ over E. Also, the series (3) is also convergent to $V(t)$ in the norm of $L(E)$.*

## 4 Main Result

**Theorem 3** *If the series*

$$\sum_{n=0}^{\infty} \left\{ \varphi_n^{2\gamma(\delta+1-\frac{1}{r})} \sum_{j=0}^{n} |\widehat{u}_{n,j}|^2 |d_j|^2 \right\}^{\frac{r}{2}} \tag{4}$$

*converges for $1 \leq r \leq 2$, then the orthogonal series $\sum_{n=0}^{\infty} d_n \psi_n(x)$ is summable $\varphi - |U, \gamma; \delta|_r$ almost everywhere.*

***Proof*** Let $s_v(x) = \sum_{j=0}^{v} d_j \psi_j(x)$ denotes *vth* partial sums of (2) and $U_n(s)(x)$ be the matrix transform of $s_v(x)$.

So,

$$\begin{aligned} U_n(s)(x) &= \sum_{v=0}^{n} u_{nv} s_v(x) = \sum_{v=0}^{n} u_{nv} \sum_{j=0}^{v} d_j \psi_j(x) \\ &= \sum_{j=0}^{n} d_j \psi_j(x) \sum_{v=j}^{n} u_{nv} = \sum_{j=0}^{n} \overline{u}_{nj} d_j \psi_j(x) \end{aligned}$$

Thus,

$$\begin{aligned} U_n(s)(x) - U_{n-1}(s)(x) &= \sum_{j=0}^{n} \overline{u}_{nj} d_j \psi_j(x) - \sum_{j=0}^{n-1} \overline{u}_{nj} d_j \psi_j(x) \\ &= \overline{u}_{nn} d_n \psi_n(x) + \sum_{j=0}^{n-1} (\overline{u}_{n,j} - \overline{u}_{n-1,j}) d_j \psi_j(x) \\ &= \overline{u}_{nn} d_n \psi_n(x) + \sum_{j=0}^{n-1} \widehat{u}_{n,j} d_j \psi_j(x) \\ &= \sum_{j=0}^{n} \widehat{u}_{n,j} d_j \psi_j(x). \end{aligned}$$

There are following facts for using $1 \le r \le 2$:

1. We take $1 < r < 2$ because from the definition of $|U|_r$ summability, we have $r \ge 1$ and Hölder's inequality is applied for $p = \frac{2}{r} > 1$, $q = \frac{2}{2-r}$, so $r < 2$.
2. For $r = 1, 2$, we may apply Schwarz's inequality.

Hence, we may take $1 \le r \le 2$.

By means of Hölder's inequality and orthonormality, we get

$$\begin{aligned} \int_a^b |U_n(s)(x) - U_{n-1}(s)(x)|^r dx &\le (b-a)^{1-\frac{r}{2}} \left( \int_a^b |U_n(s)(x) - U_{n-1}(s)(x)|^2 \, \mathrm{d}x \right)^{\frac{r}{2}} \\ &= (b-a)^{1-\frac{r}{2}} \left( \int_a^b \left| \sum_{j=0}^{n} \widehat{u}_{n,j} d_j \psi_j(x) \right|^2 \mathrm{d}x \right)^{\frac{r}{2}} \\ &= (b-a)^{1-\frac{r}{2}} \left( \sum_{j=0}^{n} |\widehat{u}_{n,j}|^2 |d_j|^2 \right)^{\frac{r}{2}}. \end{aligned}$$

Thus, the series

$$\begin{aligned} &\sum_{n=1}^{\infty} \varphi_n^{\gamma(\delta r + r - 1)} \int_a^b |U_n(s)(x) - U_{n-1}(s)(x)|^r \mathrm{d}x \\ &\le C \sum_{n=1}^{\infty} \left\{ \varphi_n^{2\gamma\left(\delta + 1 - \frac{1}{r}\right)} \sum_{j=0}^{n} |\widehat{u}_{n,j}|^2 |d_j|^2 \right\}^{\frac{r}{2}}, \end{aligned}$$

(where $C$ represents a constant depending only on $r$ and taking different values in different situations).

Thus, using (4)

$$\sum_{n=1}^{\infty} \varphi_n^{\gamma(\delta r+r-1)} \int_a^b |U_n(s)(x) - U_{n-1}(s)(x)|^r dx < \infty.$$

Now, $|U_n(s)(x) - U_{n-1}(s)(x)|^r \geq 0$. So, by means of lemma (3), proof of theorem (3) is implied.

**Theorem 4** *Let $1 \leq r \leq 2$ and $\{\Omega(n)\}$ be a positive sequence with $\left\{\frac{\Omega(n)}{\varphi_n}\right\}$ is non-increasing and $\sum_{n=1}^{\infty} \frac{1}{\varphi_n \Omega(n)}$ converges. If*

$$\sum_{n=1}^{\infty} |d_n|^2 \Omega^{\frac{2}{r}-1}(n) \omega^{(k)}(A, \gamma, \delta; \varphi_n) \tag{5}$$

*converges, then the orthogonal series $\sum_{j=0}^{\infty} d_j \psi_j(x)$ is summable $\varphi - |U, \gamma; \delta|_r$ almost everywhere, where $\omega^{(k)}(U, \gamma, \delta; \varphi_n)$ is defined by*

$$\omega^{(k)}(U, \gamma, \delta; \varphi_j) = \frac{1}{[\varphi_j]^{\frac{2}{r}-1}} \sum_{n=j}^{\infty} \varphi_n^{2\gamma(\delta+1-\frac{1}{r})+2(\frac{2}{r}-1)} |\widehat{u}_{n,j}|^2. \tag{6}$$

***Proof*** By using Hölder's inequality to condition, we get

$$\begin{aligned}
&\sum_{n=1}^{\infty} \varphi_n^{\gamma(\delta r+r-1)} \int_a^b |U_n(s)(x) - U_{n-1}(s)(x)|^r \mathrm{d}x \\
&\leq C \sum_{n=1}^{\infty} \left\{ \varphi_n^{2\gamma(\delta+1-\frac{1}{r})} \sum_{j=0}^{n} |\widehat{u}_{n,j}|^2 |d_j|^2 \right\}^{\frac{r}{2}} \\
&= C \sum_{n=1}^{\infty} \frac{1}{(\varphi_n \Omega(n))^{1-\frac{r}{2}}} \\
&\times \left\{ \varphi_n^{2\gamma(\delta+1-\frac{1}{r})} (\varphi_n \Omega(n))^{\frac{2}{r}(1-\frac{r}{2})} \sum_{j=0}^{n} |\widehat{u}_{n,j}|^2 |d_j|^2 \right\}^{\frac{r}{2}} \\
&= C \sum_{n=1}^{\infty} \frac{1}{(\varphi_n \Omega(n))^{1-\frac{r}{2}}}
\end{aligned}$$

$$\times \left\{ \varphi_n^{2\gamma(\delta+1-\frac{1}{r})+\frac{2}{r}-1} (\Omega(n))^{\frac{2}{r}-1} \sum_{j=0}^{n} |\widehat{u}_{n,j}|^2 |d_j|^2 \right\}^{\frac{r}{2}}$$

$$\leq C \left( \sum_{n=1}^{\infty} (\varphi_n \Omega(n) \right)^{1-\frac{r}{2}}$$

$$\times \left( \sum_{n=1}^{\infty} \varphi_n^{2\gamma(\delta+1-\frac{1}{r})+\frac{2}{r}-1} (\Omega(n))^{\frac{2}{r}-1} \sum_{j=0}^{n} |\widehat{u}_{n,j}|^2 |d_j|^2 \right)^{\frac{r}{2}}$$

$$\leq C \left\{ \sum_{j=1}^{\infty} |d_j|^2 \sum_{n=j}^{\infty} \varphi_n^{2\gamma(\delta+1-\frac{1}{r})+\frac{2}{r}-1} (\Omega(n))^{\frac{2}{r}-1} |\widehat{u}_{n,j}|^2 \right\}^{\frac{r}{2}}$$

$$= C \left\{ \sum_{j=1}^{\infty} |d_j|^2 (\Omega(j))^{\frac{2}{r}-1} \frac{1}{[\varphi_j]^{\frac{2}{r}-1}} \right\}^{\frac{r}{2}}$$

$$\times \left\{ \sum_{n=j}^{\infty} \varphi_n^{2\gamma(\delta+1-\frac{1}{r})+2(\frac{2}{r}-1)} |\widehat{u}_{n,j}|^2 \right\}^{\frac{r}{2}}$$

$$= C \left\{ \sum_{j=1}^{\infty} |d_j|^2 (\Omega(j))^{\frac{2}{r}-1} \omega^{(k)}(U, \gamma, \delta; \varphi_j) \right\}^{\frac{r}{2}},$$

which is finite by the condition (5).

Now, $|U_n(s)(x) - U_{n-1}(s)(x)|^r \geq 0$. So, by means of lemma 3, proof of theorem (4) is implied.

Particular cases of the main results:

1. Taking $u_{n,v} = \frac{p_{n-v}q_v}{R_n}$, where $R_n = \sum_{v=0}^{n} p_v q_{n-v}$, we get the generalized Nörlund means $U_n(s) = \frac{1}{R_n} \sum_{v=0}^{n} p_{n-v} q_v s_v$ and summability $\varphi - |U, \gamma; \delta|_r$ reduces to $\varphi - |N, p_n, q_n, \gamma; \delta|_r$ summability.
2. Taking $u_{n,v} = \frac{1}{n+1}$, we get the Cesàro means $U_n(s) = \frac{1}{n+1} \sum_{v=0}^{n} s_v$ and summability $\varphi - |U, \gamma; \delta|_r$ reduces to $\varphi - |C, 1, \gamma; \delta|_r$ summability.
3. Taking $u_{n,v} = \frac{\binom{n-v+\alpha+1}{\alpha-1}}{\binom{n+\alpha}{\alpha}}$, $(0 \leq \alpha \leq 1)$, we get the Cesàro means (of order $\alpha$) $U_n(s) = \frac{1}{\binom{n+\alpha}{\alpha}} \sum_{v=0}^{n} \frac{\binom{n-v+\alpha+1}{\alpha-1}}{\binom{n+\alpha}{\alpha}} s_v$ and summability $\varphi - |U, \gamma; \delta|_r$ reduces to $\varphi - |C, \alpha, \gamma; \delta|_r$ summability.
4. Taking $u_{n,v} = \frac{p_{n-v}}{P_n}$, we get the Nörlund means $U_n(s) = \frac{1}{P_n} \sum_{v=0}^{n} p_{n-v} s_v$ and summability $\varphi - |U, \gamma; \delta|_r$ reduces to $\varphi - |N, p_n, \gamma; \delta|_r$ summability.

5. Taking $u_{n,v} = \frac{q_n}{Q_n}$, we get the Riesz means $U_n(s) = \frac{1}{Q_n}\sum_{v=0}^{n} q_v s_v$ and summability $\varphi - |U, \gamma; \delta|_r$ reduces to $\varphi - |\overline{N}, q_n, \gamma; \delta|_r$ summability.
6. Taking $u_{n,v} = \frac{1}{(n-v+1)\log n}$, we get the Harmonic means $U_n(s) = \frac{1}{\log n}\sum_{v=0}^{n} \frac{1}{(n-v+1)} s_v$ and summability $\varphi - |U, \gamma; \delta|_r$ reduces to $\varphi - |H, 1, \gamma; \delta|_r$ summability.
7. Taking $u_{n,v} = \frac{1}{n+1}\frac{1}{P_v}\sum_{k=0}^{v} p_{v-k}$, we get the $t_n^{CN}$ means [11] $U_n(s) = \frac{1}{n+1}\sum_{v=0}^{n} \frac{1}{P_v}\sum_{k=0}^{v} p_{v-k} s_k$ and summability $\varphi - |U, \gamma; \delta|_r$ reduces to $\varphi - |C^1.N_P, \gamma; \delta|_r$ summability.

## 5 Corollaries

**Corollary 1** *If for $1 \le r \le 2$ the series*

$$\sum_{n=0}^{\infty} \left( \frac{n^{\gamma(\delta+1-\frac{1}{r})} p_n}{P_n P_{n-1}} \right)^r \left\{ \sum_{j=1}^{n} p_{n-j}^2 \left( \frac{P_n}{p_n} - \frac{P_{n-j}}{p_{n-j}} \right)^2 |d_j|^2 \right\}^{\frac{r}{2}}$$

*converges, then orthogonal series $\sum_{j=0}^{\infty} d_j \psi_j(x)$ is $|N, p, \gamma; \delta|_r$- summable almost everywhere.*

***Proof*** Taking $u_{n,v} = \frac{p_{n-v}}{p_n}$ we get

$$\begin{aligned} \widehat{u}_{n,j} &= \overline{u}_{n,j} - \overline{u}_{n-1,j} \\ &= \frac{p_n}{P_n P_{n-1}} \left( \frac{P_n}{p_n} - \frac{P_{n-j}}{p_{n-j}} \right) p_{n-j}. \end{aligned}$$

Hence, on putting $\varphi_n = n$ in Theorem 3, it required result holds.

**Corollary 2** *If for $1 \le r \le 2$ the series*

$$\sum_{n=0}^{\infty} \left( \frac{n^{\gamma(\delta+1-\frac{1}{r})} q_n}{Q_n Q_{n-1}} \right)^r \left\{ \sum_{j=1}^{n} Q_{n-j}^2 |d_j|^2 \right\}^{\frac{r}{2}}$$

*converges, then orthogonal series $\sum_{j=0}^{\infty} d_j \psi_j(x)$ is $|\overline{N}, q, \gamma; \delta|_r$- summable almost everywhere.*

***Proof*** Taking $u_{n,v} = \frac{q_v}{Q_n}$ we find that

$$\widehat{u}_{n,j} = \overline{u}_{n,j} - \overline{u}_{n-1,j} = -\frac{q_n Q_{j-1}}{Q_n Q_{n-1}}$$

Hence, on putting $\varphi_n = n$ in Theorem 3, it required result holds.

**Corollary 3** ([12]) *If the series*

$$\sum_{n=0}^{\infty} \frac{p_n}{P_n P_{n-1}} \left\{ \sum_{j=1}^{n} p_{n-j}^2 \left( \frac{P_n}{p_n} - \frac{P_{n-j}}{p_{n-j}} \right)^2 |d_j|^2 \right\}^{\frac{1}{2}}$$

*converges, then orthogonal series* $\sum_{j=0}^{\infty} d_j \psi_j(x)$ *is* $|N, p|$*- summable almost everywhere.*

***Proof*** On putting $\gamma = 1, \delta = 0$ and $r = 1$ in Corollary 1, it required result holds.

**Corollary 4** ([6]) *If the series*

$$\sum_{n=0}^{\infty} \frac{q_n}{Q_n Q_{n-1}} \left\{ \sum_{j=1}^{n} Q_{n-j}^2 |d_j|^2 \right\}^{\frac{1}{2}}$$

*converges, then orthogonal series* $\sum_{j=0}^{\infty} d_j \psi_j(x)$ *is* $|\overline{N}, q|$*- summable almost everywhere.*

***Proof*** On putting $\gamma = 1, \delta = 0$ and $r = 1$ in Corollary 2, it required result holds.

**Acknowledgements** The author offers her true thanks to the Science and Engineering Research Board for giving financial support through Project No.: EEQ/2018/000393.

## References

1. Kransniqi, XhZ: On $|A, \delta|_k$- summability of orthogonal series. Mathematica Bohemica **137**(1), 17–25 (2012)
2. Flett, T.M.: On an extension of absolute summability and some theorems of Littlewood and Paley. Proc. Lond. Math. Soc. **7**, 113–141 (1957)
3. Tuncer, A.N.: On generalized absolute Cesàro summability factors. Ann. Polon. Math. **78**, 25–29 (2002)
4. Özarslan, H.S., Ari, T.: Absolute matrix summability methods. Appl. Math. Lett. **24**, 2102–2106 (2011)
5. Tanaka, M.: On generalized Nörlund methods of summability. Bull. Austral. Math. Soc. **19**, 381–402 (1978)
6. Okuyama, Y., Tsuchikura, T.: On the absolute Riesz summability of orthogonal series. Anal. Math. **7**, 199–208 (1981)
7. Okuyama, Y.: On the absolute generalized Nörlund summability of orthogonal series. Tamkang J. Math. **33**, 161–165 (2002)
8. Rajkumar, A., Helen, D.: New arithmetic operations of triskaidecagonal fuzzy number using alpha cut. Soft Comput.: Theories Appl. **583**(1), 125–135 (2017)
9. Rajkumar, A., Helen, D.: New arithmetic operations in inverse of triskaidecagonal fuzzy number using alpha cut. Soft Comput.: Theories Appl. **583**(1), 115–123 (2017)
10. Natanson, I.P.: Theory of Functions of a Real Variable, 2 vols. Frederick Ungar, New York (1955–1961). MR 16–804, 26 # 6309

11. Flett, T.M.: Some more theorems concerning the absolute summability of Fourier series and power series. Proc. Lond. Math. Soc. **8**, 357–387 (1958)
12. Okuyama, Y.: On the absolute Nörlund summability of orthogonal series. Proc. Jpn. Acad. **54**, 113–118 (1978)
13. Tanović-Miller, N.: On strong summability. Glas. Mat. **34**, 87–97 (1979)

# Security Analysis of SVD-Based Watermarking Schemes and Possible Solutions

**Siddharth Bhalerao, Irshad Ahmad Ansari, and Anil Kumar**

**Abstract** Copyright protection requires a robust image watermarking technique. With robustness, high imperceptibility is a much appreciated trait of a watermarked image. In past years, singular value decomposition (SVD) has emerged as a tool for robust image watermarking. SVD is frequently combined with transforms like discrete wavelet transform (DWT) and discrete cosine transform (DCT) to produce high-quality watermarked images. SVD-based watermarking can withstand large number of image processing attacks, but they are vulnerable to false positive (FP) problem. The false positive problem results in falsely establishing illegal user as the rightful owner of the image. In this paper, security analysis of SVD-based image watermarking techniques is provided. The inherent problems present in such methods are examined, and possible solutions are suggested.

**Keywords** Singular value decomposition · False positive · Image watermarking · Ownership · Check

## 1 Introduction

Image watermarking is done to establish integrity and authenticity of an image. In image watermarking, an image called as watermark is embedded in another image called as host. Watermarking serves different purposes [1]. When purpose of watermarking is copyright protection, it is expected that watermark is preserved even in case of attacks on image. In such case, watermarking is called as robust watermarking

S. Bhalerao · I. A. Ansari (✉) · A. Kumar
Electronics and Communication Engineering, PDPM Indian Institute of Information Technology, Design and Manufacturing, Jabalpur 482005, India
e-mail: irshad@iiitdmj.ac.in

S. Bhalerao
e-mail: siddharthbhalerao@gmail.com

A. Kumar
e-mail: anilkdee@gmail.com



T. K. Sharma et al. (eds.), *Soft Computing: Theories and Applications*,
Advances in Intelligent Systems and Computing 1381,
https://doi.org/10.1007/978-981-16-1696-9_50

[2, 3]; it is done to identify the rightful owner of the image. When watermarking is done to establish image authenticity, it is expected that watermark should get destroyed in case of attacks; such watermarking is called as fragile watermarking [3]. Watermarking can be done in spatial as well as transform domain. In transform domain, transforms like DWT and DCT have gained popularity owing to their ability to accurately model human visual system; they also provide high imperceptibility and robustness [4]. To improve robustness, combined DWT-SVD or DCT-SVD-based methods were reported [5–7]. These SVD-based methods improve robustness but suffer from problem of false positive detection [8]. In [9–12], various workarounds are suggested to eradicate the false positive problem. The method mentioned in [9] removes false positive problem by selectively embedding watermark bits in left singular matrix. Similarly, in [10] and [11], the false positive problem is removed by using principal components (PC) for embedding, instead of $S$ matrix.

Singular value decomposition (SVD) is an elegant approach toward matrix factorization. When singular value decomposition is performed on any rectangular matrix, it results in generation of three separate components (1). The three components are named $U$, $S$ and $V$. The matrix $U$ and $V$ are orthogonal in nature, and their columns are termed as left singular and right singular vectors, respectively. The diagonal values of matrix $S$ are called as singular values. Considering the matrix $X$ with dimensions $m \times n$, the matrix $S$ is of same size as $X$. The matrices $U$ and $V$ are of dimensions $m \times m$ and $n \times n$, respectively.

$$X = U \times S \times V^{\mathrm{T}} \tag{1}$$

In this paper, security analysis of two different SVD-based schemes [13, 14] is done. The method proposed by Zear et al. [13] was an SVD-based approach using both, DWT and DCT. The main aim of their work was to strengthen the security of medical images. The false positive problem faced by SVD-based approaches is discussed using their work. In the method proposed by Araghi and Manaf [14], the two-level authentication process is used to detect the false positive scenario. The security analysis finds out that, the level 1 authentication present in [14] is an unnecessary step and the security depends on level 2 process only. Secondly, it is also established that the second level of authentication can be easily deceived, thereby authenticating the forged image.

## 2 Security Analysis of Zear et al. [13]

In [13], a secure and robust multiple image watermarking was proposed for telemedicine applications. It was a non-blind technique, and three different watermarks were embedded in a cover medical image. The three watermarks included image watermark in form of lump image, and two text watermarks. One of the text watermarks was patient symptoms data, and other was doctor's signature. The image watermark was scrambled using Arnold transform before embedding it. The host image is initially decomposed by applying three-level DWT. The text watermarks were embedded in LH2 and LL3 sub-bands. The image watermark was embedded in LH1 sub-band. DCT was applied on LH1 sub-band and on scrambled image watermark. The SVD was computed for DCT coefficients of LHI sub-band and image watermark coefficients. The $S$ component of host image was updated using $S$ component of watermark. The image was repacked to get the watermarked image. The process is reversed during extraction to get the embedded watermarks. The extraction side requires cover image as well as watermark image. When the watermarked image is attacked knowingly or unknowingly, the lump image watermark gets distorted. Therefore, a backpropagation neural network was proposed to improve the watermark image, and this adds to robustness of the method.

## *2.1 Analysis and Discussion*

In the method proposed by Zear et al. [13], security was greatly increased by introduction of multiple watermarks. The robustness of their method was also enhanced by introduction of a backpropagation neural network. In the midst of all the improvements, the method was prone to the FP attack, and this was also acknowledged by the authors. During embedding stage, the singular values of watermark image (W) were introduced in singular values of host image (H) to get the watermarked image (WI). During extraction, singular values of watermarked image ($S_{\mathrm{wat}}$) and host image ($S_{\mathrm{h}}$) are used to get the singular values of embedded watermark ($S_{\mathrm{wext}}$). The watermark image is decomposed using SVD and obtained $U_{\mathrm{w}}$ and $V_{\mathrm{w}}$ components are combined with $S_{\mathrm{wext}}$ to get the embedded watermark (2).

$$W_{\mathrm{ext}} = U_{\mathrm{w}} \times S_{\mathrm{wext}} \times V_{\mathrm{w}}^{\mathrm{T}} \tag{2}$$

In false positive attack, an illegal user may claim ownership of the image by extracting a different watermark from the watermarked image. Let us assume, that the attacker decomposes an illegal watermark image ($W_{\mathrm{iw}}$) to get 3 SVD components $U_{\mathrm{iw}}$, $S_{\mathrm{iw}}$ and $V_{\mathrm{iw}}$. The singular vectors contain major share of information, therefore, when they are combined with $S_{wext}$, the recovered image will be similar to $W_{\mathrm{iw}}$. The resultant image will be distorted, but still be largely intact, the attacker may stake an ambiguous claim of ownership of image. Thus, false positive attack is nullifying the robustness of watermarked image.

**Possible Solution** The solution to FP problem is suggested by authors themselves, which is the use of shuffled SVD (SSVD) in place of SVD. Another possible solution to solve false positive problem is embedding principal components in place to singular values (3). The principal component is product of singular matrix $U$ with $S$ matrix.

$$PC = U \times S \tag{3}$$

# 3 Security Analysis of Araghi and Manaf [14]

In [14], a robust image watermarking technique is proposed by Araghi and Manaf, which is based on DWT and two-level SVD. The main offering of their method is that, in their algorithm, there is no dependence on size of host and watermark images. The watermark image is repeated to form an image same size as of host image. Thus, a method, independent of image size, is obtained. The watermark is embedded in HH sub-band using 2D-SVD. Another major contribution of their method is two-level authentication process. The two-level authentication ensures immunity of their scheme against false positive and false negative problems. First level of security is embedding of 8-bit digital signature post watermark embedding. The second level of authentication is based on DWT-SVD hashing described in [15]. During extraction, digital signature is extracted first, intact digital signature implies no attack, this is the first step of authentication, and it is named as A-situation in the paper. After digital signature extraction, hash values of matrix $U_{\mathrm{w}}$ and $V^{\mathrm{T}}{}_{\mathrm{w}}$ obtained from the embedding side and the values entered by user are compared. The matrices $U_{\mathrm{w}}$ and $V^{\mathrm{T}}{}_{\mathrm{w}}$ are components of SVD decomposition of watermark image (4). This test is called as B-test. The authentication process during embedding and extraction is shown in Fig. 1a, b.

$$\text{watermark\_image} = U_{\mathrm{w}} \times S_{\mathrm{w}} \times V_{\mathrm{w}}^{\mathrm{T}} \tag{4}$$

## 3.1 Analysis and Discussion

The authentication in [14] depends on results of situation $A$ and situation $B$. As mentioned in the paper, if at least 5 bits are correctly extracted from digital signature, $A$-situation is considered as passed. In case if, less than 5 bits are recovered correctly the $A$-situation is considered as failed. When $A$-situation fails to establish the authenticity, then extraction depends on result of B-test only. The authentication situation is expressed in Table 1, here *1* denotes authentication passed, 0 denotes authentication failed, and $X$ denotes don't care condition.

Following are the analysis of first stage ($A$-situation) of authentication:

- It is clear from Table 1 that $A$-situation has no impact on extraction and the extraction process is solely governed by $B$-Test.
- Table 1 establishes the fact that $A$-situation is an unnecessary step and can be omitted without any impact on the algorithm.

The $B$-test is based on comparison of hashed values of $U_{\mathrm{w}}$ and $V^{\mathrm{T}}{}_{\mathrm{w}}$ obtained from embedding side and present with the user at extraction side. Therefore, during extraction, an image will be considered authentic only when the hash values provided by extracting user and supplied by embedding side matches and same watermark

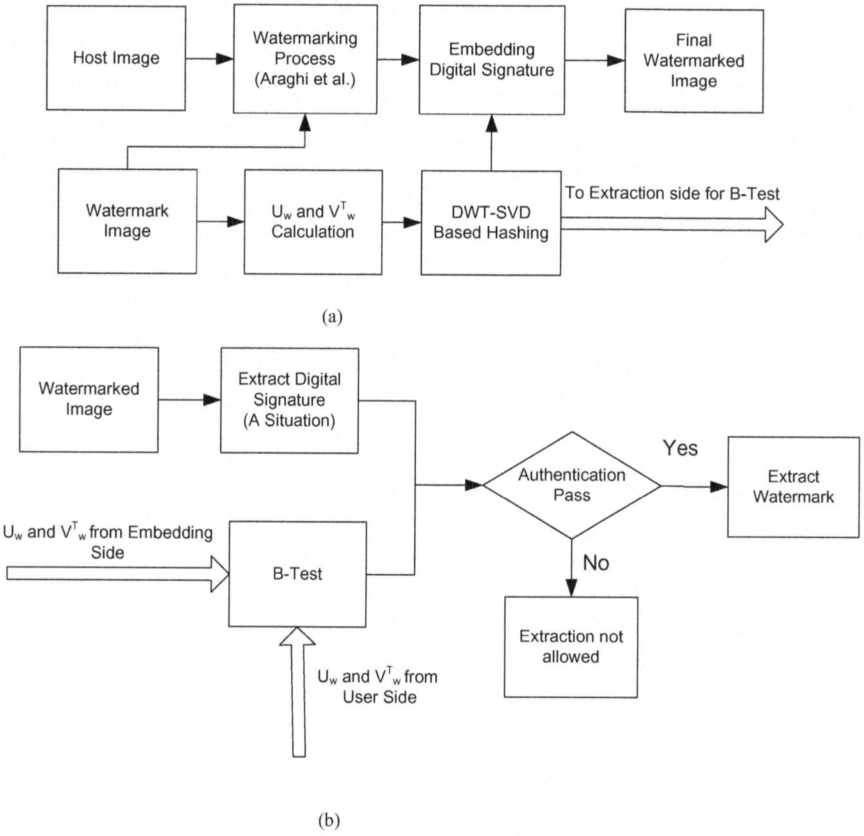


**Fig. 1** **a** Embedding process, **b** extraction process [14]

**Table 1** Extraction status based on authentication

| *A*-situation | *B*-test | Extraction |
|---|---|---|
| 1 | 1 | Allowed |
| 0 | 1 | Allowed |
| *X* | 0 | Not allowed |

image is extracted as embedded watermark using the extraction algorithm. For B-test either hashed values or watermark image must be present during extraction. There is only one way by which extracting user will get this data, and it will be either provided by embedding side or obtained from some network storage (this makes the algorithm a non-blind one). In case, when watermark image is present during extraction $U_w$ and $V^T{}_w$ are calculated from DWT-SVD-based hashing mentioned in [15]. The problem with B-test is that, it could be cheated where a third party could claim a forged image as an authentic image watermarked by same system. The scenario is explained in detail in following paragraph through an example.

The procedure for deceiving the B-test is explained with an example in Fig. 2. Consider Lena image as host image, which is watermarked by method proposed in [14]. The watermark is cameraman image (W1). The PSNR of the watermarked image is 51.73. Post watermarking, the digital signature is generated from watermark, and watermarked image is signed. The resultant image is digitally signed watermarked image (called W_Lena). The extraction side has watermarked/attacked image and hash values of watermark parameters obtained from embedding side. The extraction side will be able to extract the watermark after successful authentication which requires hash values of watermark parameter from extracting user. It is assumed that, if an attacker gets hold of a watermarked image, he/she will not be able to extract watermark from it as the hashed values would not be present with attacker. The authorization would fail if hash value of different watermark image is used by attacker.

In this scenario, the attacker will get hold of watermarked image W_Lena. Attacker could again watermark the W_Lena by procedure mentioned in [14] using the watermark of choice (W2) and will generate a new image Lena1. The Lena1 image and the watermark image of coins (W2) are also present in Fig. 2. The PSNR of repeated watermarked image with respect to original watermarked and host image is 51.82 and 42.8, respectively. The PSNR of forged watermarked image is good enough to pass it as originally watermarked image. The attacker would also digitally sign the image after watermarking.

During extraction, attacker could claim that it is the actual image watermarked by system legally. The attacker's image will fail *A*-test/situation but it will pass the

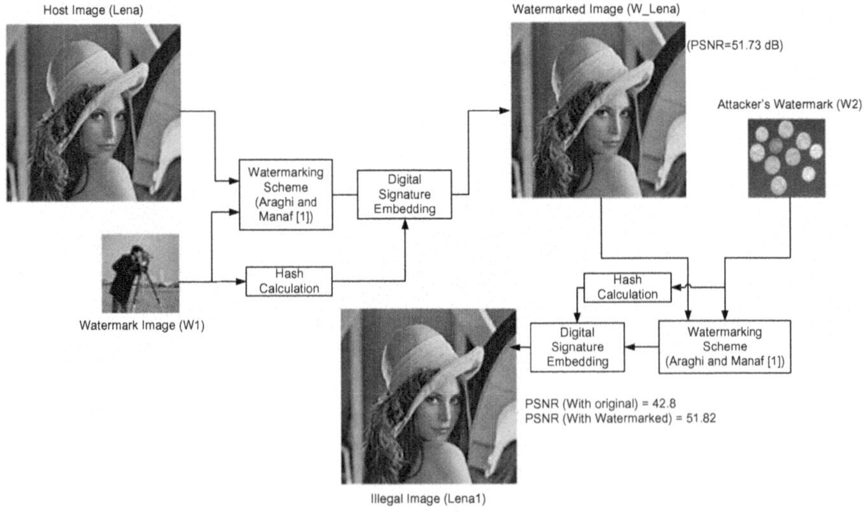


**Fig. 2** False image generation

*B*-test. The *B*-test requires the hashed values of embedded watermark parameter. As the hashing scheme is publically available, attacker will generate these parameters from watermark image. The attacker will be authorized to extract the watermark, and the extracted watermark will be same as used for *B*-test. Thus, attacker will claim that the image with attacker is real watermarked image and will claim ownership of image. There will be absolutely no evidence to denounce the claim made by attacker. Therefore, in the scenario explained here the authorization step mentioned in [14] can be deceived.

Following are the analysis of second stage (*B*-test) of authentication:

- Attacker could claim ownership of an image by fooling authentication step by watermarking the watermarked image again.
- The requirement of $U_w$ and $V^T{}_w$ or watermark image during extraction makes this scheme a non-blind scheme.
- The system will not be able to decide legally watermarked image from illegal image in case of multiple watermarking.

**Possible Solution** The two-stage authentication in [14] is introduced as a solution for false negative and false positive problems. The A-situation mentioned in [14] is actually redundant. It could be made meaningful by increasing the number of digital signature bits; this will also strengthen the authorization. In [14], only 8 bits are embedded as signature; therefore, there is always a chance that even in the presence of attacks the bits could be correctly extracted. Thus, by increasing the digital signature size to 16 bits gives a better chance for authentication of first step. Although there is always a chance that even with 16 bits digital signature, there is a possibility of incorrect extraction. This could be eradicated by embedding more than one copy of digital signature in the image, as an authentication step.

The *B*-test as discussed above will always be the Achilles heel of this algorithm, in case of multiple watermarking on image. Therefore, the *B*-test of authentication could be eliminated and replaced by a better method for false positive detection. There are various strategies for false positive problems mentioned in [12]. In [9, 16–19], the false positive problem is eliminated by embedding watermark in U or V component of singular value decomposition instead of *S*. In [20], complete watermark image is embedded in host image to prevent FP problem. One of the methods to eradicate false positive problem is embedding principal components in place of *S* matrix. This strategy is discussed and implemented in [10, 11].

## 4 Conclusion

In this paper, security analysis was presented on SVD-based methods for robust image watermarking. Two different SVD-based approaches were analyzed. The first scheme was examined in light of false positive problem, and possible solutions were discussed. The second scheme was a two-level authorization, it was found out that, the primary authorization stage, which was digital signature embedding, was redundant. The second stage was shown insufficient, as it could be cheated by showing an attacker as an actual owner of the image. For both the schemes, strategies for improving the algorithm were also suggested.

**Acknowledgements** This work was supported by Faculty Initiation Grant of PDPM Indian Institute of Information Technology Design and Manufacturing Jabalpur, India.

## References

1. Potdar, V.M., Han, S., Chang, E.: A survey of digital image watermarking techniques. In: INDIN'05. 2005 3rd IEEE International Conference on Industrial Informatics, pp. 709–716. IEEE (2005)
2. Sinhal, R., Ansari, I.A.: A source and ownership identification framework for mobile-based messenger applications. In: Pant, M., Sharma, T., Verma, O., Singla, R., Sikander, A. (eds.) Soft Computing: Theories and Applications. Advances in Intelligent Systems and Computing, vol 1053. Springer, Singapore (2020)
3. Agarwal, N., Singh, A.K., Singh, P.K.: Survey of robust and imperceptible watermarking. Multimedia Tools Appl. **78**(7), 8603–8633 (2019)
4. Arya, D.: A survey of frequency and wavelet domain digital watermarking techniques. Int. J. Sci. Eng. Res. **1**(2), 286–292 (2010)
5. Ganic, E., Eskicioglu, A.M.: Robust DWT-SVD domain image watermarking: embedding data in all frequencies. In: Proceedings of the 2004 Workshop on Multimedia and Security, pp. 166–174 (2004).
6. Mishra, A., Agarwal, C., Sharma, A., Bedi, P.: Optimized gray-scale image watermarking using DWT–SVD and Firefly algorithm. Expert Syst. Appl. **41**(17), 7858–7867 (2014)
7. Araghi, T.K., Manaf, A.A., Alarood, A., Zainol, A.B.: Host Feasibility investigation to improve robustness in hybrid DWT + SVD based image watermarking schemes. In: Advances in Multimedia (2018)
8. Zhang, X.P., Li, K.: Comments on" An SVD-based watermarking scheme for protecting rightful oOwnership . IEEE Trans. Multimedia **7**(3), 593–594 (2005)
9. Thakkar, F.N., Srivastava, V.K.: A blind medical image watermarking: DWT-SVD based robust and secure approach for telemedicine applications. Multimedia Tools Appl. **76**(3), 3669–3697 (2017)
10. Ali, M., Ahn, C.W.: An optimized watermarking technique based on self-adaptive DE in DWT–SVD transform domain. Signal Process. **94**, 545–556 (2014)
11. Ansari, I.A., Pant, M.: Multipurpose image watermarking in the domain of DWT based on SVD and ABC. Pattern Recogn. Lett. **94**, 228–236 (2017)
12. Makbol, N.M., Khoo, B.E., Rassem, T.H.: Security analyses of false positive problem for the SVD-based hybrid digital image watermarking techniques in the wavelet transform domain. Multimedia Tools Appl. **77**(20), 26845–26879 (2018)
13. Zear, A., Singh, A.K., Kumar, P.: A proposed secure multiple watermarking technique based on DWT, DCT and SVD for application in medicine. Multimedia Tools Appl. **77**(4), 4863–4882 (2018)
14. Araghi, T.K., Manaf, A.A.: An enhanced hybrid image watermarking scheme for security of medical and non-medical images based on DWT and 2-D SVD. Future Gener. Comput. Syst. **101**, 1223–1246 (2019)
15. Kozat, S.S., Venkatesan, R., & Mihçak, M.K. Robust perceptual image hashing via matrix invariants. In: 2004 International Conference on Image Processing, ICIP'04, vol. 5, pp. 3443–3446. IEEE (2004)

16. Chang, C.C., Tsai, P., Lin, C.C.: SVD-based digital image watermarking scheme. Pattern Recogn. Lett. **26**(10), 1577–1586 (2005)
17. Chung, K.L., Yang, W.N., Huang, Y.H., Wu, S.T., Hsu, Y.C.: On SVD-based watermarking algorithm. Appl. Math. Comput. **188**(1), 54–57 (2007)
18. Fan, M.Q., Wang, H.X., Li, S.K.: Restudy on SVD-based watermarking scheme. Appl. Math. Comput. **203**(2), 926–930 (2008)
19. Lai, C.C.: An improved SVD-based watermarking scheme using human visual characteristics. Opt. Commun. **284**(4), 938–944 (2011)
20. Yadav, B., Kumar, A., Kumar, Y.: A robust digital image watermarking algorithm using DWT and SVD. In: Pant, M., Ray, K., Sharma, T., Rawat, S., Bandyopadhyay, A. (eds.) Soft Computing: Theories and Applications. Advances in Intelligent Systems and Computing, vol 583. Springer, Singapore (2018)

# Real-Time Wildfire Detection via Image-Based Deep Learning Algorithm

**Parul Jindal, Himanshu Gupta, Nikhil Pachauri, Varun Sharma, and Om Prakash Verma**

**Abstract** Forests are the major source of oxygen that keep our ecosystem balanced. However, with rapid economic development, the colossal threat of wildfire also enhances globally. Uncontrolled wildfires can cause billions of losses. Therefore, the need of early fire detection with high accuracy is required to reduce these losses. This paper presents an algorithm based on the advanced object detection CNN models (YOLOv3 and YOLOv4) for the detection of forest smoke. Evaluation of both the algorithms has been performed based on five parameters that are precision, recall, F1-score, average IOU, and mAP. A comparison of both the algorithms reveals that the accuracy of fire detection using smoke is higher in YOLOv3 than YOLOv4. The increment of 28.1%, 3.4%, 15.1%, 11.7%, and 31.8% in precision, recall, F1-score, mAP, and IOU, respectively, has been observed in YOLOv3 as compared to YOLOv4.

**Keywords** CNN · Forest safety · Object detection · Real-time forest fire detection · YOLO

P. Jindal · H. Gupta · O. P. Verma (✉)
Department of Instrumentation and Control Engineering, Dr. B. R. Ambedkar National Institute of Technology, Jalandhar, India
e-mail: vermaop@nitj.ac.in

P. Jindal
e-mail: parul.ic.19@nitj.ac.in

N. Pachauri
School of Electrical and Electronics Engineering, SASTRA University, Tamil Nadu, India
e-mail: nikhilpachauri@eie.sastra.edu

V. Sharma
Department of Industrial and Production Engineering, Dr. B. R. Ambedkar National Institute of Technology, Jalandhar, India
e-mail: sharmav@nitj.ac.in

 
T. K. Sharma et al. (eds.), *Soft Computing: Theories and Applications*,
Advances in Intelligent Systems and Computing 1381,
https://doi.org/10.1007/978-981-16-1696-9_51

## 1 Introduction

Forest fires are responsible for massive global damage every year. In 2020 inspection, it has been observed that there is 13% increase in wildfires than in 2019 [1]. There has been a constant threat to the existence of the Amazon rainforest (lungs of earth) as wildfires can spread extremely fast if they are not controlled in time. The best way to minimize wildfire loss is to be able to detect it in its early stages for rapid containment and suppression.

Therefore, lots of work has been found in the literature. Zhang et al. [2, 3] proposed the cluster-based wireless sensor network for detection of forest fire. Cheon et al. [4] introduced single-chip CMOS-based sensor which measures smoke density and temperature to detect fire. However, traditional detection technologies based on sensors [5–7], like smoke and heat detectors, can only detect fire in nearby places where they are installed. Moreover, it encounters difficulty in the distribution of sensors in complex outdoor environments and is also not suitable for large spaces, complex buildings, or spaces with many disturbances. These systems are plagued by frequent false alarms from non-fire sources. These detection technologies lead to false alarms, missed or delayed detections and have battery charging problems, making it even more difficult to achieve early fire warnings. Due to these limitations, nowadays visual/image-based methods have been employed for detection which can monitor the forest 24-h and detect fire as early as possible.

Fire can be detected by detecting smoke or flames. But forest flames can easily obscure by obstacles such as mountains or may not be visible if present far away from the monitoring camera. Therefore, smoke (easily visible from long distance) detection based on computer vision techniques has been commonly utilized to determine the presence of fire.

Image fire detection has many advantages such as early fire detection, high accuracy, flexible system installation, and the capability to effectively detect fires in large spaces. In this technique, real-time image data is processed to extract features of smoke. However, identifying smoke in images is difficult because it lacks a specific shape or colour pattern and often misleads by clouds or heavy fog.

Muhammad et al. [8] proposed GoogleNet architecture to detect fire in surveillance videos. Filonenko et al. [9] implemented AlexNet, VGG, Inception, Xception, and ResNet networks for fire detection. Yin et al. [10] introduced DNCNN which uses normalization and convolutional layers to increase the performance of smoke detection. Though these detection algorithms have higher accuracy than sensor-based technologies, but are not preferable because these models can only classify the image into one class (fire) and not able to locate the contaminated region. However, in the early stage, fire only covered a small area of the image. Thus, using the entire image without region proposals would decrease the accuracy and cause a delay in detection. Therefore, in this study, YOLO [11] algorithm has been proposed which improves the ability to detect and locate the fire in early stage and thus reduces fire accidents.

This study demonstrates two types of advanced CNN algorithms, i.e. YOLOv3 and YOLOv4 applied for fire detection in real-time images. The theoretical overview of these networks and their training algorithm has been illustrated in Sect. 2. To evaluate the performance, five parameters have been used, as described in Sect. 3. In Sect. 4, results including a comparison between the performance of both the algorithms have been presented, and finally, the present work has been concluded in Sect. 5. The results provide useful information that can be further used in the future for modification of detection algorithms for preventing fire accidents and associated losses.

## 2 Proposed Framework

The following subsection depicts CNN architecture and the proposed network for smoke detection. The training algorithm and its parameters have been further discussed in the following subsections.

### *2.1 CNN Architecture*

The advancement of CNNs brought remarkable improvements in the performance of many computer vision tasks, particularly in the detection of an object from real-time image or video [12–17]. For object detection, two independent steps are required that are image classification and localization. Image classification is a process of identifying an object and assigning a label to an entire image. Object localization estimates the location of the detected object by subdividing the process into region proposal, feature extraction, and classification. The performance of classification crucially relies on the features extracted from the image. The convolution kernel slides on the input images and generate a feature map that reflects the features of the original image [18]. Feature map can be computed using Eq. (1).

$$y = \sum_{j=0}^{J-1} \sum_{i=0}^{I-1} w_{ij} x_{m+i,n+j} + b; \ (0 \le m \le M, 0 \le n \le N) \tag{1}$$

where

$w$ Convolution kernel of a size J × I.
$x$ Input image of a size M × N.
$b$ Bias.
$y$ Output feature map.

Object detection techniques (based on region proposals) usually offer poor computational speed due to their pipeline architecture, thus making unsuitable for real-time detection. In recent years, YOLO has been a highly appreciated object detection technique which offers the advantage of accurate and fast detection and can be used in real-time applications. YOLO (single neural network) transforms the object detection task into a regression model and performs training on the entire image. Therefore, it eliminates the limitations of classical object detection techniques.

YOLO splits the image into S × S grid, and for each grid cell, it predicts five bounding boxes and calculates their confidence score. Each bounding box consists of five predictions: *x*, *y*, *w*, *h*, and confidence score. Equation (2) represents the loss function of YOLO algorithm which is minimized during the training process.

$$L = \lambda\text{coord}\sum_{i=0}^{s^2}\sum_{j=0}^{B} 1_{ij}^{\text{obj}}\left[(xi - \hat{x}i)^2 + (yi - \hat{y}i)^2\right]$$
$$+ \lambda\text{coord}\sum_{i=0}^{s^2}\sum_{j=0}^{B} 1_{ij}^{\text{obj}}\left[\left(\sqrt{wi} - \sqrt{\hat{w}i}\right)^2 + \left(\sqrt{hi} - \sqrt{\hat{h}i}\right)^2\right]$$
$$+ \sum_{i=0}^{s^2}\sum_{j=0}^{B} 1_{ij}^{\text{obj}}\left[\left(Ci - \hat{C}i\right)^2\right] + \lambda\text{noobj}\sum_{i=0}^{s^2}\sum_{j=0}^{B} 1_{ij}^{\text{noobj}}\left[\left(Ci - \hat{C}i\right)^2\right]$$
$$+ \sum_{i=0}^{s^2} 1_{i}^{\text{obj}} \sum_{c\in\text{classes}} (pi(c) - \hat{p}i(c))^2 \tag{2}$$

where

| | |
|---|---|
| $L$ | Loss function. |
| $x$, $y$ | Coordinates of the centre of bounding box relative to the bounds of the grid cell. |
| $w$, $h$ | Width and height of predicted box. |
| $C$ | Confidence score. |
| $1_{\text{i}}^{\text{obj}}$ | Object appears in cell *i*. |
| $1_{\text{ij}}^{\text{obj}}$ | *j*th bounding box predictor is responsible for detection in *i*th cell. |
| $\lambda$ | Parameter used to increase model stability. |

Confidence score for each predicted bounding box can be calculated using Eq. (3)

$$C = \text{pr(Object)} \times \text{IoU}_{\text{pred}}^{\text{truth}},\ \text{pr(Object)} \in \{0, 1\} \tag{3}$$

here

| | |
|---|---|
| $\text{IoU}_{\text{pred}}^{\text{truth}}$ | Intersection ratio of the ground truth box to the predicted bounding box. |
| pr(Object) | Probability that the predicted bounding box contains object. |

Finally, bounding boxes having a higher confidence score than the threshold value are selected for object localization.

**YOLOv3**

YOLOv3 [19] has been evolved from YOLO and YOLOv2 [20] networks. As compared to faster RCNN, YOLO detects and locates object faster, but still it has large detection error. Therefore, to improve the accuracy, YOLOv2 introduces the idea of the "anchor box" in faster R-CNN and generates priori bounding box using k-means clustering method. YOLOv2 improves the network structure by replacing fully connected layer (output layer) to convolution layer. To further enhance the accuracy, YOLOv3 (advancement in YOLOv2) has been introduced. Its architecture (shown in Fig. 1) is more complex than YOLOv2, having 106 layers containing residual blocks, detection blocks, and 75 convolutional layers together with upsampling layers. YOLOv3 predicts bounding boxes on three different scales, thus detects multi-size objects. Due to three-scale method, it has achieved commendable detection performance particularly for small objects in the image. YOLOv3 predicts boxes at three different scales (e.g. 13 × 13, 26 × 26, 52 × 52).

**YOLOv4**

Alexey et al. [22] introduce YOLOv4 with new features having cross-Stage partial connections (CSP), weighted residual connections (WRC), mosaic data augmentation, self-adversarial training (SAT), and Mish activation. YOLOv4 architecture (shown in Fig. 2) consists of CSPDarknet53 classifier (backbone) to extract features

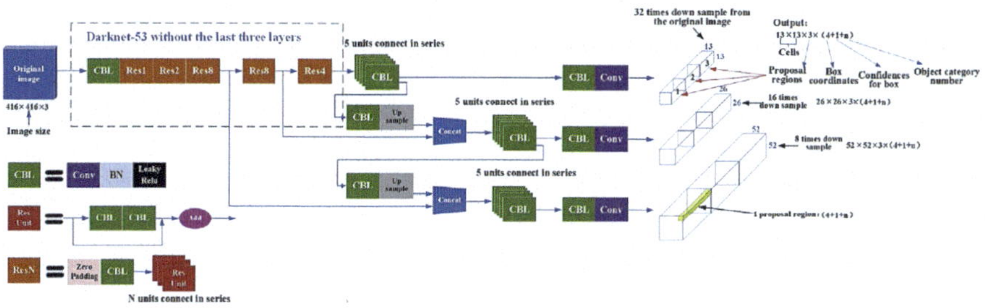


**Fig. 1** YOLOv3 architecture [21]

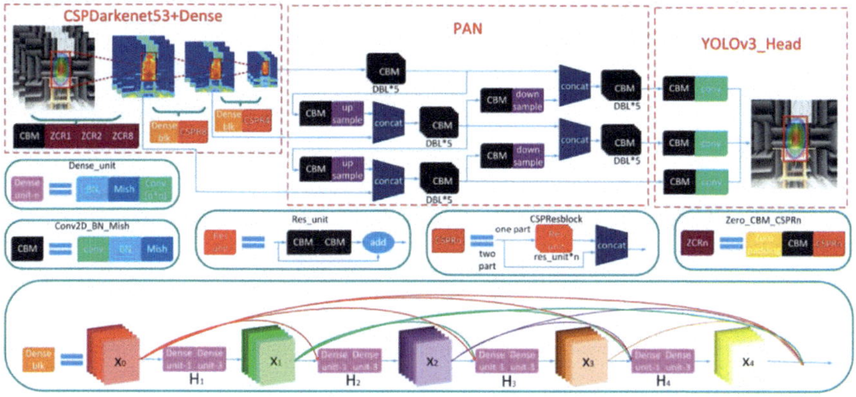


**Fig. 2** YOLOv4 architecture [23]

from image, spatial pyramid pooling, and path aggregation network (neck) to improve extracted feature map and YOLOv3 (head) for detection. Several other improvements were also introduced in the training process such as data augmentation, known as "bag of freebies." It improves accuracy and does not alter inference speed. Further, "bag of specials," which refers to post-processing methods, increases the inference cost but returns good performance.

## 2.2 *Training Algorithm*

This subsection describes the data set, hardware, and software tools used to train the network. Performance parameters used to evaluate both the models are also illustrated in subsequent sections.

**Image Data set**

Image-based deep learning methods often require learning of millions of parameters. Additionally, this process requires the collection of a huge number of images. In this study, public available data set [24] released by "AI for mankind" and "HPWREN" has been used which contains 737 images of size 480 × 640. Images were captured before and after fire ignition using HPWREN cameras. To prevent the model from overfitting and to obtain accurate results, data set has been divided into train (70% of total data set, i.e. 516 images), validation (20%, i.e. 147 images), and test data (10%, i.e. 74 images).

**Learning Parameters**

For fire detection, model was trained to recognize only one class, i.e. "smoke." The parameters changed in YOLO configuration file to customize the network are specified in Table 1.

**Table 1** Parameters used for training the model

| Parameters | YOLOv3 | YOLOv4 |
|---|---|---|
| Width | 608 | 608 |
| Height | 608 | 608 |
| Subdivisions | 32 | 64 |
| Batch | 64 | 64 |
| Filters | 18 | 18 |
| Momentum | 0.9 | 0.949 |
| Learning rate | 0.001 | 0.0013 |
| Weight decay | 0.005 | 0.005 |
| Threshold | 0.7 | 0.7 |
| Maximum no. of batches | 10,000 | 10,000 |
| Steps | 8000, 9000 | 8000, 9000 |

**Table 2** Computer configurations

| Name | Related configuration |
|---|---|
| Operating system | Windows10 |
| CPU | Intel Core i7-10750H (six cores–twelve threads, Comet Lake architecture) |
| RAM | 16 GB |
| Graphic card | Nvidia RTX 2060, 6 GB GDDR6 |

### Computer Specifications

The computational process presented in this study was performed on a computer configuration illustrated in Table 2.

## 3 Performance Evaluation Parameters

Five fundamental parameters have been used to evaluate the performance of proposed algorithms. These were calculated after every 1000 iterations during training process as shown in Table 3. The parameters description is as follows:

1. Precision: It identifies the correctness of detected bounding box, by matching it with actual ground truth box and represented as Eq. (4)

$$\text{Precision} = \frac{\text{total no. of objects detected correctly}}{\text{total objects detected}} = \frac{N_{\text{TP}}}{N_{\text{TP}} + N_{\text{FP}}} \tag{4}$$

Here

**Table 3** Detection performance of YOLOv3 and YOLOv4 algorithm

| No. of Iteration | Precision | | Recall | | F1-score | | mAP% | | Avg IOU% | |
|---|---|---|---|---|---|---|---|---|---|---|
| | v3 | v4 | v3 | v4 | v3 | v4 | v3 | v4 | v3 | v4 |
| 1000 | 0.71 | 0.23 | 0.52 | 0.15 | 0.60 | 0.18 | 64.15 | 13.94 | 46.95 | 14.21 |
| 2000 | 0.86 | 0.80 | 0.83 | 0.88 | 0.84 | 0.84 | 84.87 | 87.07 | 62.64 | 56.37 |
| 3000 | 0.90 | 0.66 | 0.88 | 0.85 | 0.89 | 0.75 | 87.89 | 81.61 | 65.17 | 48.57 |
| 4000 | 0.89 | 0.61 | 0.89 | 0.81 | 0.89 | 0.70 | 88.13 | 78.23 | 65.95 | 42.29 |
| 5000 | 0.91 | 0.75 | 0.91 | 0.89 | 0.91 | 0.82 | 90.21 | 87.59 | 66.88 | 55.94 |
| 6000 | 0.92 | 0.70 | 0.90 | 0.86 | 0.91 | 0.77 | 88.54 | 80.59 | 67.33 | 50.45 |
| 7000 | 0.92 | 0.66 | 0.90 | 0.86 | 0.91 | 0.75 | 90.16 | 77.34 | 69.88 | 47.99 |
| 8000 | 0.91 | 0.73 | 0.89 | 0.85 | 0.90 | 0.79 | 89.23 | 77.86 | 68.00 | 54.30 |
| 9000 | 0.90 | 0.73 | 0.90 | 0.90 | 0.90 | 0.80 | 87.72 | 81.82 | 68.52 | 53.81 |
| 10,000 | 0.91 | 0.71 | 0.91 | 0.88 | 0.91 | 0.79 | 88.73 | 79.40 | 68.92 | 52.27 |

$N_{\text{TP}}$ Number of correctly detected bounding box (true positive).
$N_{\text{FP}}$ Number of false detections which do not resemble with ground truth box (false positives).

2. Recall: It is defined as true positive rate of a given class in classification. It is represented in Eq. (5)

$$\text{recall} = \frac{\text{no. of objects correctly detected}}{\text{no. of ground truth objects}} = \frac{N_{\text{TP}}}{N_{\text{TP}} + N_{\text{FN}}} \quad (5)$$

where
$N_{\text{FN}}$ = Number of ground truth objects that could not be detected (false negatives).
3. F1-score: It is the weighted average of recall and precision, which can be evaluated as in Eq. (6)

$$\text{F}_1\text{ score} = \frac{2 \times \text{Precision} \times \text{Recall}}{\text{Precision} + \text{Recall}} \quad (6)$$

4. mAP: Average precision encapsulates both precision and recall, and it is represented as region covered by their curve, expressed as Eq. (7):

$$\text{AP} = \int_0^1 p(r)\text{d}r \quad (7)$$

mAP is the mean of AP which can be represented by Eq. (8)

$$\text{mAP} = \frac{\sum_{i=1}^{N} \text{AP}_i}{N} \quad (8)$$

where $N$ = total object classes.
Intersection over union (IOU): It measures closeness of predicted bounding box to the ground truth box. It is represented as Eq. (9)

$$\text{IoU} = \frac{A \cap B}{A \cup B} \quad (9)$$

where $A$ denotes area covered by ground truth box in the image, and $B$ represents the area of detected box.

## 4 Result and Discussion

The results obtained by training YOLOv3 and YOLOv4 networks were based on the analysis of 737 image data set (described in Sect. 2), trained for 10 k iterations. Figure 3 depicts the rise in mAP% with increase in epochs during training of

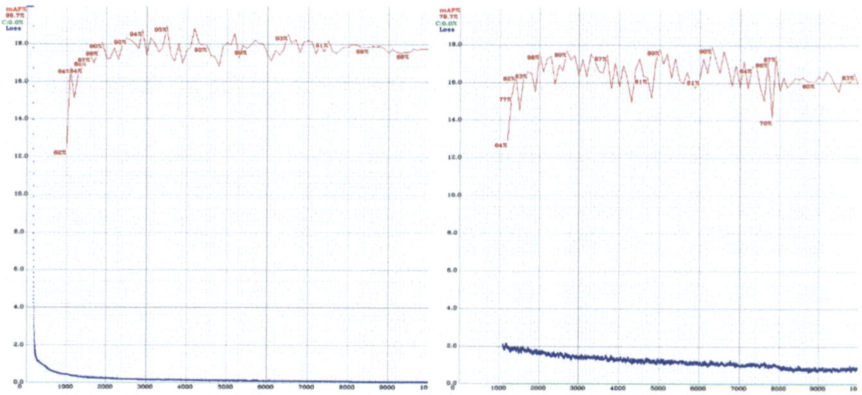

(a) (b)

**Fig. 3** mAP% and loss function during the training phase **a** YOLOv3 **b** YOLOv4

networks. Both networks reached maximum mAP% while successfully reducing the values of their respective loss functions. YOLOv3 presents higher precision (~10% more) and minimal loss value (~90% less) than YOLOv4.

Figure 4 shows the sample of test data and their detection on both the networks. From this figure, it has been found that YOLOv4 detects multiple bounding boxes for one ground truth, which cannot be acceptable. The performance of these networks has been compared on the basis of five parameters (described in Sect. 3), and their results have been illustrated in Table 3. The graphical representation of these parameters is shown in Fig. 5.

From Table 3 and Fig. 5, it is evident that YOLOv3 overtakes YOLOv4 in all the listed parameters. In last epoch, precision, recall, F1-score, mAP, and IOU of YOLOv3 have increment of 28.1%, 3.4%, 15.1%, 11.7%, and 31.8%, respectively, than YOLOv4. This might be due to the limited computation capability because of which 32 and 64 subdivision has been chosen for YOLOv3 and YOLOv4, respectively.

Table 4 demonstrates execution time of both networks to detect smoke on a sample test image. From this table, it is clear that YOLOv4 detects fast (~80% faster) than YOLOv3.

## 5 Conclusion

To monitor wildfire, image fire detection based on deep learning, i.e. YOLOv3 and YOLOv4, has been proposed. These algorithms extract features of smoke from images and successfully detect fire in different scenes. The experimental results served as guidance for monitoring and controlling of forest fire. To train the network, public available data set (released by "AI for mankind" and "HPWREN") has been

**Fig. 4** Experimental results tested on sample test images **a** original image **b** detection by YOLOv3 **c** Detection by YOLOv4

**Fig. 5** Comparison between different performance parameters

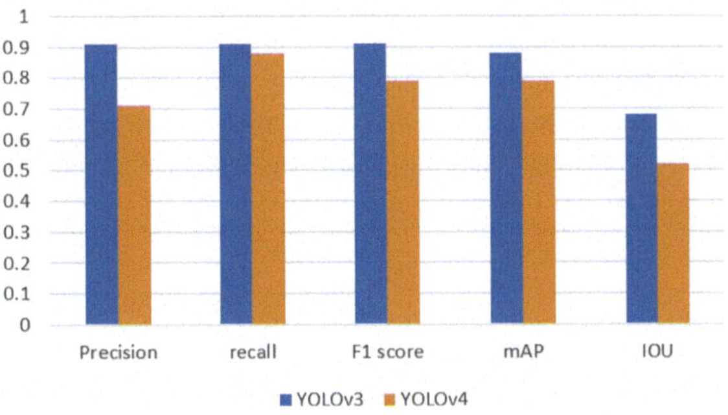

**Table 4** Detection on sample test image

| Test image | Predicted in time (ms) | |
|---|---|---|
| | YOLOv3 | YOLOv4 |
| 1 | 376.19 | 53.31 |
| 2 | 313.73 | 53.03 |
| 3 | 333.03 | 53.14 |

used. YOLOv3 performs better than YOLOv4 as YOLOv3 has an increment of mAP (11.7%), IOU (31.8%), precision (28.1%), recall (3.4%), and F1-score (15.1%) than YOLOv4. It has also been observed that YOLOv4 detects multiple bounding boxes for single ground truth. Therefore, the present investigation concluded the supremacy of YOLOv3 over YOLOv4 in fire detection. Further, to increase the robustness, the network can be trained on multiple object classes as flame or fire-like objects with a greater number of images and iterations which will be addressed in future works.

**Acknowledgements** The first and second authors would like to thank the Ministry of Human Resource Development, New Delhi, India, for providing the Research Fellowship for carrying out this work.

## References

1. WWF: Fires, forest and the future: a crisis raging out of control? World Wide Fund for Nature, pp. 1–20 (2020). www.swim2birds.co.uk
2. Zhang, J., Li, W., Yin, Z., Liu, S., Guo, X.: Forest fire detection system based on wireless sensor network. In: 2009 4th IEEE Conference on Industrial Electronics and Applications, Xi'an, pp. 520–523 (2009). https://doi.org/10.1109/ICIEA.2009.5138260
3. Zhang, J., Li, W., Han, N., et al.: Forest fire detection system based on a ZigBee wireless sensor network. Front. For. China **3**, 369–374 (2008). https://doi.org/10.1007/s11461-008-0054-3
4. Cheon, J., Lee, J., Lee, I., Chae, Y., Yoo, Y., Han, G.: A single-chip CMOS smoke and temperature sensor for an intelligent fire detector. IEEE Sens. J. **9**(8), 914–921 (2009). https://doi.org/10.1109/JSEN.2009.2024703
5. Chen, S.-J., Hovde, D.C., Peterson, K.A., Marshall, A.W.: Fire detection using smoke and gas sensors. Fire Saf. J. **42**(8) (2007). https://doi.org/10.1016/j.firesaf.2007.01.006
6. Chaturvedi, P., Daniel, A.K: A novel approach for target coverage in wireless sensor networks based on network coding. In: Ray, K., Sharma, T.K., Rawat, S., Saini, R.K, Bandyopadhyay, A. (eds.) Soft Computing: Theories and Applications, pp. 303–310. Springer Singapore, Singapore (2019)
7. Zhang, L. et al.: Design and implementation of automatic fire alarm system based on wireless sensor networks. In: Proceedings of the International Symposium on Information Processing. Microcontroller. Retrieved 16 Mar 2006
8. Muhammad, K., Ahmad, J., Mehmood, I., et al.: Convolutional neural networks-based fire detection in surveillance videos. IEEE Access **6**, 18174–18183 (2018)
9. Filonenko, A., Kurnianggoro, L., Jo, K.: Comparative study of modern convolutional neural networks for smoke detection on image data. In: 2017 10th International Conference on Human System Interactions (HSI), pp. 64–68 (2017)

10. Yin, Z., Wan, B., Yuan, F., Xia, X., Shi, J.: A deep normalization and convolutional neural network for image smoke detection. IEEE Access **5**, 18429–18438 (2017). https://doi.org/10.1109/ACCESS.2017.2747399
11. Redmon, J., Divvala, S., Girshick, R., Farhadi, A.: You only look once: unified, real-time object detection. In: Proceedings of the IEEE Conference on Computer Vision and Pattern Recognition, Las Vegas, 27–30 June 2016, pp. 779–788 (2016). https://doi.org/10.1109/CVPR.2016.91https://doi.org/10.1109/CVPR.2016.91
12. Goel, R., Sharma, A., Kapoor, R.: State-of-the-art object recognition techniques: a comparative study. In: Pant, M., Sharma, T.K., Verma, O.P., Singla, R., Sikander, A. (eds.) Soft Computing: Theories and Applications, pp. 925–932. Springer Singapore, Singapore (2020)
13. Kumar, P., Gupta, M., Gupta, M., Sharma, A.: Profession identification using handwritten text images. In: Nain, N., Vipparthi, S.K., Raman, B. (eds.) Computer Vision and Image Processing, pp. 25–35. Springer Singapore, Singapore (2020)
14. Gupta, K.K., Vijay, R., Pahadiya, P.: A review paper on feature selection techniques and artificial neural networks architectures used in thermography for early stage detection of breast cancer. In: Pant, M., Sharma, T.K., Arya, R., Sahana, B.C., Zolfagharinia, H. (eds.) Soft Computing: Theories and Applications, pp. 455–465. Singapore: Springer Singapore (2020)
15. Bordia, B., Nishanth, N., Patel, S., Anand Kumar, M., Rudra, B.: Automated traffic light signal violation detection system using convolutional neural network. In: Pant, M., Sharma, T.K., Arya, R., Sahana, B.C., Zolfagharinia, H. (eds.) Soft Computing: Theories and Applications, pp. 579–592. Springer Singapore, Singapore (2020)
16. Dahiya, S., Tyagi, R., Gaba, N.: Streamlining choice of CNNs and structure framing of convolution layer. In: Pant, M., Sharma, T.K., Arya, R., Sahana, B.C., Zolfagharinia, H. (eds.) Soft Computing: Theories and Applications, pp. 705–718. Springer Singapore, Singapore (2020)
17. Pathak, K.C., Kundaram, S.S.: Accuracy-based performance analysis of Alzheimer's disease classification using deep convolution neural network. In: Pant, M., Sharma, T.K., Arya, R., Sahana, B.C., Zolfagharinia, H. (eds.)*Soft Computing: Theories and Applications*, edited by Millie Pant, Tarun Kumar Sharma, Rajeev Arya, B C Sahana, and Hossein Zolfagharinia, 731–44. Singapore: Springer Singapore, 2020.
18. Li, P., Zhao, W.: Image fire detection algorithms based on convolutional neural networks. Case Stud. Therm. Eng. **19** (2020). https://doi.org/10.1016/j.csite.2020.100625
19. Redmon, J., Farhadi, A.: YOLOv3: An incremental improvement (2018). arXiv preprint arXiv: 1804.02767
20. Redmon, J., Farhadi, A. YOLO9000: better, faster, stronger (2017). arXiv preprint arXiv:1612.0824
21. Yin, X., Chen, Y., Bouferguene, A., Zaman, H., Al-Hussein, M., Kurach, L.: A deep learning-based framework for an automated defect detection system for sewer pipes. Autom. Constr. **109**, 102967 (2020). https://doi.org/10.1016/j.autcon.2019.102967
22. Bochkovskiy, A., Wang, C.-Y., Liao, H.-Y.M.: YOLOv4: optimal speed and accuracy of object detection (2020). arXiv preprint arXiv: 2004.10934
23. Zhu, Q., Zheng, H., Wang, Y., Cao, Y., Guo, S.: Study on the evaluation method of sound phase cloud maps based on an improved YOLOv4 algorithm. Sensors **20**(15), 4314 (2020). https://doi.org/10.3390/s20154314
24. Roboflow Homepage, https://public.roboflow.com/object-detection/wildfire-smoke. Last accessed 09 Nov 2020

# Automatic Vehicle Detection from Satellite Images Using Deep Learning Algorithm

**Himanshu Gupta, Parul Jindal, and Om Prakash Verma**

**Abstract** Attributable to global economic development and rapid urbanization, traffic explosion has been observed. This situation becomes very prominent as the traditional traffic control systems are incapable to efficiently monitor and control the traffic. Therefore, this paper presents automatic vehicle detection from satellite images using deep learning approaches. For this purpose, two most renowned and widely used detection algorithms (YOLOv4 and YOLOv3) have been employed to develop satellite image vehicle detector using publicly available DOTA dataset. This work confirms the supremacy of YOLOv4 over YOLOv3 by large improvements in mAP, IOU, precision, recall, F1-score with increase of 45%, 20%, 11.1%, 45.9%, and 29.5%, respectively. Therefore, these investigational results verify the strength of YOLOv4 algorithm for satellite image vehicle detection and recommend its use for the development of an intelligent traffic control system.

**Keywords** Deep learning · Object detection · Intelligent traffic control · Vehicle detection · YOLOv3 · YOLOv4 algorithm

## 1 Introduction

With the growth of urban traffic and the necessity to control it, attention has been paid to intelligent traffic control systems. Nowadays, urban traffic is controlled by the cameras which are installed on highways located at a long distance from each other. With the advancement in technology, this procedure seems inefficient. However, remote sensing images seem much more proficient in terms of both accuracy and fast response. Typical applications include monitoring vegetation growth, urban planning,

H. Gupta · P. Jindal · O. P. Verma (✉)
Department of Instrumentation and Control Engineering, Dr. B R Ambedkar National Institute of Technology, Jalandhar, India
e-mail: vermaop@nitj.ac.in

P. Jindal
e-mail: parul.ic.19@nitj.ac.in

 
T. K. Sharma et al. (eds.), *Soft Computing: Theories and Applications*,
Advances in Intelligent Systems and Computing 1381,
https://doi.org/10.1007/978-981-16-1696-9_52

fire detection, glacial observations, natural and man-made disaster management, land cover mapping, etc.

In the intelligent transportation era, traffic accidents or vehicle flows are controlled by using high-resolution remote sensing images. As the key technology of intelligent transportation system, vehicle detection is the basis for measurement of traffic parameters such as vehicle location, traffic flow, and density (i.e., detecting the number of vehicles present on road at a certain period of time), tracking and traffic data mining, etc. [1]. More recently, the analysis of aerial videos taken from the aerial vehicle has become an important way of managing traffic and can also be used in surveillance. For monitoring fast-moving targets, aerial surveillance is more suitable as compared to ground-plane surveillance systems as it covers a much larger spatial area. Therefore, ground-plane surveillance systems have now been replaced by aerial surveillance systems.

Further, a big concern in intelligent aerial surveillance is the tracking of vehicles. Vehicle detection from real-time images/videos is the first step for tracking. Therefore, a lot of research is going on to extract valuable information from aerial images. For accurate detection, a large number of aerial images are required to train the network. Remote sensing satellite and unmanned aerial vehicle (UAV) technologies offer better quality and high-resolution aerial images. Compared to traditional methods of aerial data collection, UAVs have drawn much attention due to their easy operation, good storage capacity, less resource-intensive, and large coverage area. These are more affordable and can provide high-resolution imagery from difficult-to-reach places [2].

To obtain acceptable results and gets benefitted from remote sensing image data, the data should be processed and analyzed carefully. Object detection is a pivotal job in computer vision, which aims to extract features from the image. However, object detection from satellite images is a challenging task because object appearance may vary due to scale variation, various illumination conditions, images may have vague, and cluttered backgrounds. In addition, objects are relatively small and always express an overhead appearance, which causes the difficulty in extracting features and thus leads to false detection.

Zheng [3] proposed a model to detect vehicles on highways using GIS road vector map. Grayscale transformation is used to identify hypothesis vehicles by changing the background color to white or black. Zhou [4] introduces adaptive background estimation to detect moving vehicles. The image is divided into many small nonoverlapped blocks. Vehicle is detected from the blocks by calculating gray-level change between the current image and the background. Cheng et al. [5] proposed a model based on Bayesian networks to detect vehicles which can be used in aerial surveillance system. For detection, vehicle color was extracted by utilizing color transform techniques which aids in distinguishing vehicles from its background. These studies were based on background elimination, which are not accurate due to change in illuminations.

Chen et al. [6] designed an aerial object detector based on region proposal algorithm. The author proposed two models MORPN and ODN to detect small size and oriented objects in aerial images. Guirado et al. [7] trained faster R-CNN on satellite images to automatically detect and count the number of whales. Han et al. [8] run real-time object detection on embedded hardware using R-CNN and kernelized correlation filter (KCF). All these experiments use regional-based algorithms to localize the object within the image. Drawback of these networks is that it does not work on the complete image. It only focuses on the parts of the image which have high probability of containing the object. Therefore, in this research YOLO [9] (based on regression algorithms) is used. YOLO does not select the interested regions in the image, instead it predicts classes and bounding boxes for the entire image in one go.

The rest of the paper is structured as follows: Sect. 2 provides a theoretical overview of YOLOv3 and YOLOv4 model. Experimental setup and comparison between them have been discussed in Sect. 3. Section 4 discusses the obtained results, and finally, Sect. 5 concludes the paper.

## 2 Theoretical Overview of YOLOv3 and YOLOv4

This section illustrates the architecture of YOLOv3 and YOLOv4. Modifications in YOLOv3 to attain YOLOv4 model have been discussed.

### *2.1 Network Architecture*

You Only Look Once (YOLO) is a single stage detector and one of the fastest detectors which detects object in a image. It divides the image into S $\times$ S grid and detects m bounding box for each grid. Bounding box which has high class probability than threshold value is selected and used to locate the object. Though it is not that much accurate, still it is acceptable for real-time detection. As objects in aerial images are small, YOLOv2 [10] is not a good choice as it often struggles with detection of small objects. Therefore, in this work, aerial image detector based on YOLOv3 [11] and YOLOv4 [12] has been implemented and their performance has been evaluated on public available DOTA dataset [13].

**YOLOv3**

YOLOv3 uses DarkNet as the backbone network, which originally has 53-layer network trained on ImageNet. In this architecture, 53 more layers are stacked onto darknet and thus make 106 fully convolutional layer. The architecture of YOLOv3 is shown in Fig. 1.

YOLOv3 extracts features and detects multiple size objects at three different scales. Detection kernels of different sizes are applied on a feature map at three different places in the network. The algorithm downsamples the dimensions of the image by the stride of 32, 16, 8 at each scale. It reduces the spatial resolution while

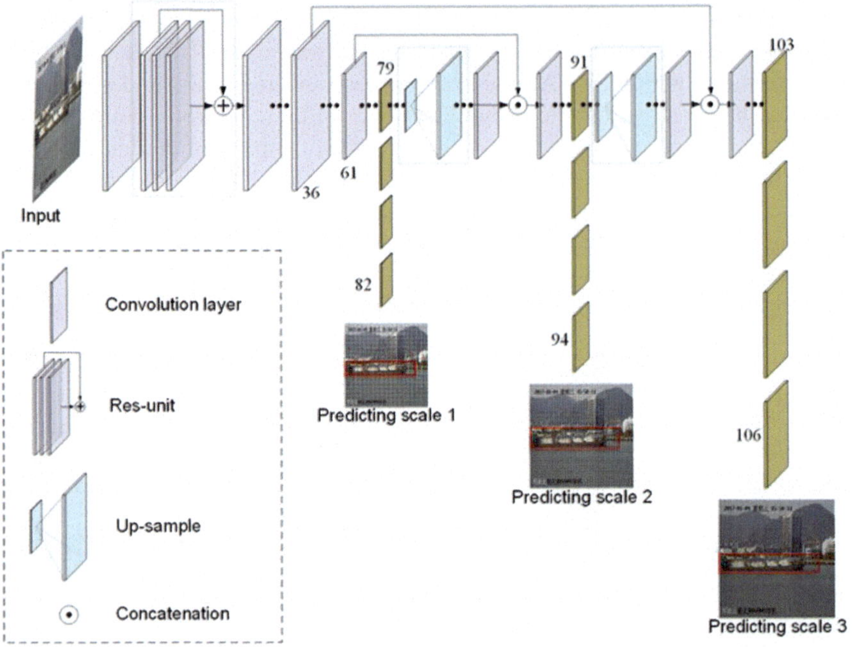


**Fig. 1** YOLOv3 architecture [14]

keeping the image representation unchanged. Result of YOLO is represented in 3-D tensor that encodes bounding box, objectness score, and prediction over classes.

The shape of the detection kernel has been represented in Eq. (1):

$$S \times S \times (B \times (1 + 4 + C)) \tag{1}$$

Here

| | |
|---|---|
| S × S | number of the grid cells in input image. |
| $B$ | number of bounding boxes predicted in each cell on the feature map. |
| $1 + 4$ | 1 for object confidence and 4 for bounding box attributes. |
| $C$ | number of classes. |

The first scale (82nd layer) detects large objects and gives a detection feature map of 13 × 13 × 255. Medium size objects are detected at the second scale (94th layer) and yields detection feature map of 26 × 26 × 255 and the last scale (106th layer) detects small objects, generates feature map of size 52 × 52 × 255. Due to the concatenation of unsampled layers with previous layers, fine-grained features are preserved which improves the detection of small objects [15].

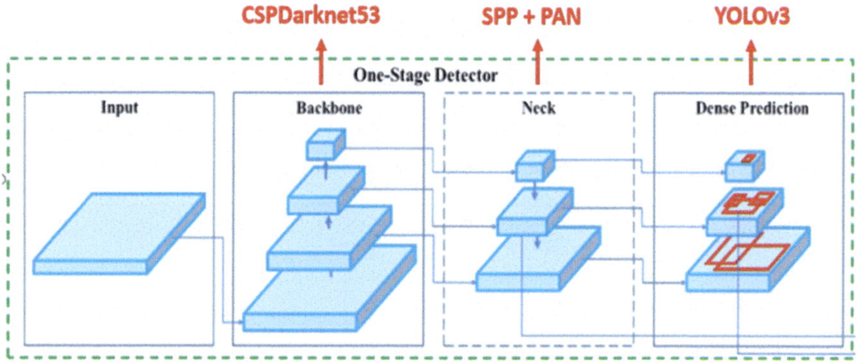


**Fig. 2** YOLOv4 architecture [12]

### YOLOv4

Alexey et al. [12] introduce YOLOv4, and the architecture (shown in Fig. 2) is a logical evolutionary step from the YOLOv3 architecture which utilizes the CSPDarknet53 (backbone of the network) as a classifier. The main objective of the backbone is to extract the essential features. Often pre-trained neural networks are used to train the backbone. The architecture uses spatial pyramid pooling (SPP) and path aggregation network (PAN) for the neck that connects it to the YOLOv3 head. This combination creates a balanced network [represented by Eq. (2)] that can deliver a perfect balance in speed and accuracy.

$$\text{YOLOV4} = \text{YOLOV3} + \text{CSPDarknet53} + \text{SPP} + \text{PAN} + \text{BoF} + \text{BoS} \quad (2)$$

Augmentation techniques are modified in YOLOv4. The techniques are classified into bag of freebies and bag of specials. 'Bag of Freebies' improves the precision up to 5% compared to state-of-the-art baselines. As there is no variation in model's architecture, inference cost will remain unchanged [16]. Data augmentations that fall under the 'bag of freebies' are meant to increase the variability of the input images to increase robustness and generalization of the model. The most used methods are photometric distortion (adjusts brightness, contrast, hue, saturation, and noise), geometric distortion (rotate the image, random scaling, cropping or flipping), MixUp (weighted linear interpolation of two existing images). These methods display more varieties of the same image.

'Bag of Specials' refers to post-processing methods which increase inference cost but improves the accuracy of object detection. Bag of specials includes: spatial pyramid pooling (SPP), receptive field block (RFB), spatial attention module, squeeze and excitation, feature integration.

# 3 Experimental Setup and Comparison Between YOLOv3 and YOLOv4

This section illustrates the specifications of hardware and software tools and dataset used to train the model. The experimental results and comparison between both the algorithms based on different performance parameters are further discussed in the following subsections.

## *3.1 Dataset*

To perform the experimental part of our study, DOTA dataset has been used which contains 15 object categories like bridge, basketball court, ground track field, plane, ship, baseball diamond, tennis court, helicopter, roundabout, large and small vehicle. In this study, only vehicles are selected to train the model. This dataset was collected from images taken from satellite JL-1 and GF-2. The dataset includes 2806 high-resolution (~4000 × 4000) images. From these images, 2130 images have been selected which covers vehicle. But detection performance on high-resolution images turns out to be poor due to the presence of multi-scale and small objects in the image. To address this problem, images are spit into 9888 of size 1920 × 1080. After image processing, dataset further splits into training set, validation set, and test set. The training set consists of 7476 images (~75% of images), the validation set contains 2390 images, and 22 images has been used to test the model.

## *3.2 Hardware and Software Tools*

For training the YOLO model, DarkNet, OpenCV 3.3.0, and CUDA 10.0 have been installed. The model was trained to recognize only one class, i.e., 'vehicles.' During training, modified YOLO configuration file for both the networks has been utilized. The parameters used to customize the network are specified in Table 1.

The number of filters used in the last layer of both networks were set according to Eq. (3):

$$\text{filters} = (\text{class} + 5) \times 3 = (1 + 5) \times 3 = 18 \tag{3}$$

The details of the Computer configurations used in the present analysis have been presented in Table 2.

**Table 1** Parameters used for training the model

| Parameters | YOLOv3 | YOLOv4 |
|---|---|---|
| Width | 608 | 608 |
| Height | 608 | 608 |
| Subdivisions | 32 | 64 |
| Batch | 64 | 64 |
| Momentum | 0.9 | 0.949 |
| Learning rate | 0.001 | 0.0013 |
| Weight decay | 0.005 | 0.005 |
| Threshold | 0.7 | 0.7 |
| Maximum no. of batches | 20,000 | 20,000 |
| Steps | 10,000, 15,000 | 16,000, 18,000 |

**Table 2** Computer configurations

| Name | Related configuration |
|---|---|
| Operating system | Windows10 |
| CPU | Intel Core i7-10750H (six cores-twelve threads, Comet Lake architecture) |
| RAM | 16 GB |
| Graphic card | Nvidia RTX 2060, 6 GB GDDR6 |

### *3.3 Performance Evaluation and Metrics*

To evaluate the performance, some fundamental parameters were calculated at specific intervals during the training process. The parameters are:

1. Precision: It is defined as the probability of detected bounding box that matches with actual ground truth box and represented as Eq. (4)

$$\text{Precision} = \frac{\text{total no. of objects detected correctly}}{\text{total objects detected}} = \frac{N_{\text{TP}}}{N_{\text{TP}} + N_{\text{FP}}} \quad (4)$$

   Here,
   $N_{\text{TP}}$ = Number of objects detected correctly (true positives).
   $N_{\text{FP}}$ = Number of false detections which could not correspond to the ground truth objects (false positives)

2. Recall: Recall measures the probability of correct detection of ground truth objects. It is calculated using Eq. (5)

$$\text{recall} = \frac{\text{no. of objects correctly detected}}{\text{no. of ground truth objects}} = \frac{N_{\text{TP}}}{N_{\text{TP}} + N_{\text{FN}}} \quad (5)$$

   where

$N_{FN}$ = Number of ground truth objects that could not be detected (false negatives)

3. F1-score: This is another performance parameter used to measure the accuracy of model. It is evaluated as Eq. (6)

$$F_1 \text{ score} = \frac{2 \times \text{Precision} \times \text{Recall}}{\text{Precision} + \text{Recall}} \quad (6)$$

4. mAP: Average precision is a single number which encapsulate both precision and recall, and it is represented as region covered by their curve, expressed as Eq. (7):

$$\text{AP} = \int_0^1 p(r)\text{d}r \quad (7)$$

mAP is the mean of AP of all $N$ classes which can be represented by Eq. (8)

$$\text{mAP} = \frac{\sum_{i=1}^{N} \text{AP}_i}{N} \quad (8)$$

where $N$ = total object classes.

5. Intersection Over Union (IOU): It is used to evaluate the closeness of predicted bounding box to the ground truth box. It is represented as Eq. (9)

$$\text{IoU} = \frac{A \cap B}{A \cup B} \quad (9)$$

where $A$ represents area of ground truth box and $B$ represents the area of detected box.

## 4 Result and Discussion

As mentioned earlier, comparison of YOLOv3 and YOLOv4 on vehicle detection is done based on performance parameters as described in Sect. 3. Table 3 shows the results obtained from both the algorithms after every 1000 iteration. For better visualization of results, graphical comparison between two models is shown in Fig. 3.

From Fig. 3, it is clear that YOLOv4 outperforms YOLOv3 in all the listed performance parameters. The training continues till 20 k iterations, and after the training process, significant improvement has been found in the performance of YOLOv4 as compared with YOLOv3. Scientifically in this work, the increment of 45%, 20%, 11.1%, 45.9%, and 29.5% for mAP, IOU, precision, recall, and F1-score, respectively, has been evaluated for YOLOv4 as compared to YOLOv3. Screenshots of sample test images on both the network have been presented in Fig. 4, and for better clarification, processing time required for detection and number of vehicles detected

**Table 3** Detection performance of YOLOv3 and YOLOv4 algorithm on DOTA dataset

| No. of Iteration | Precision | | Recall | | F1-score | | mAP% | | Avg IOU% | |
|---|---|---|---|---|---|---|---|---|---|---|
| | v3 | v4 | v3 | v4 | v3 | v4 | v3 | v4 | v3 | v4 |
| 1000 | 0.26 | 0.21 | 0.14 | 0.23 | 0.18 | 0.22 | 7.36 | 9.46 | 15.94 | 12.73 |
| 2000 | 0.45 | 0.43 | 0.23 | 0.38 | 0.31 | 0.40 | 18.40 | 24.12 | 29.22 | 28.67 |
| 3000 | 0.45 | 0.50 | 0.34 | 0.44 | 0.39 | 0.47 | 24.85 | 32.83 | 29.82 | 34.35 |
| 4000 | 0.50 | 0.51 | 0.33 | 0.45 | 0.40 | 0.48 | 25.88 | 32.04 | 33.29 | 35.02 |
| 5000 | 0.50 | 0.53 | 0.32 | 0.48 | 0.39 | 0.50 | 24.84 | 35.54 | 33.24 | 37.11 |
| 6000 | 0.51 | 0.59 | 0.36 | 0.48 | 0.42 | 0.53 | 28.64 | 38.68 | 34.60 | 41.88 |
| 7000 | 0.52 | 0.59 | 0.35 | 0.45 | 0.42 | 0.51 | 27.31 | 36.01 | 35.61 | 41.74 |
| 8000 | 0.48 | 0.55 | 0.38 | 0.52 | 0.42 | 0.54 | 27.44 | 38.66 | 32.08 | 38.53 |
| 9000 | 0.51 | 0.56 | 0.37 | 0.51 | 0.43 | 0.54 | 28.62 | 38.97 | 35.08 | 40.18 |
| 10,000 | 0.52 | 0.54 | 0.37 | 0.53 | 0.43 | 0.53 | 28.73 | 39.44 | 35.52 | 38.30 |
| 11,000 | 0.53 | 0.56 | 0.38 | 0.54 | 0.44 | 0.55 | 29.82 | 40.87 | 36.61 | 40.56 |
| 12,000 | 0.54 | 0.57 | 0.38 | 0.55 | 0.44 | 0.56 | 30.20 | 41.40 | 37.27 | 41.28 |
| 13,000 | 0.54 | 0.56 | 0.38 | 0.50 | 0.44 | 0.53 | 30.24 | 37.15 | 37.34 | 39.13 |
| 14,000 | 0.55 | 0.57 | 0.36 | 0.52 | 0.44 | 0.54 | 29.91 | 40.21 | 38.56 | 40.88 |
| 15,000 | 0.53 | 0.57 | 0.38 | 0.50 | 0.44 | 0.53 | 29.99 | 38.10 | 36.34 | 40.39 |
| 16,000 | 0.51 | 0.63 | 0.39 | 0.49 | 0.44 | 0.55 | 29.88 | 41.95 | 35.55 | 44.60 |
| 17,000 | 0.53 | 0.61 | 0.38 | 0.53 | 0.44 | 0.57 | 29.93 | 43.30 | 36.82 | 44.68 |
| 18,000 | 0.51 | 0.59 | 0.39 | 0.54 | 0.44 | 0.56 | 29.92 | 42.93 | 35.49 | 43.70 |
| 19,000 | 0.52 | 0.62 | 0.38 | 0.52 | 0.44 | 0.57 | 29.94 | 43.48 | 36.11 | 46.30 |
| 20,000 | 0.54 | 0.60 | 0.37 | 0.54 | 0.44 | 0.57 | 30.00 | 43.56 | 37.18 | 44.66 |

**Fig. 3** Performance comparison between YOLOv3 and YOLOv4

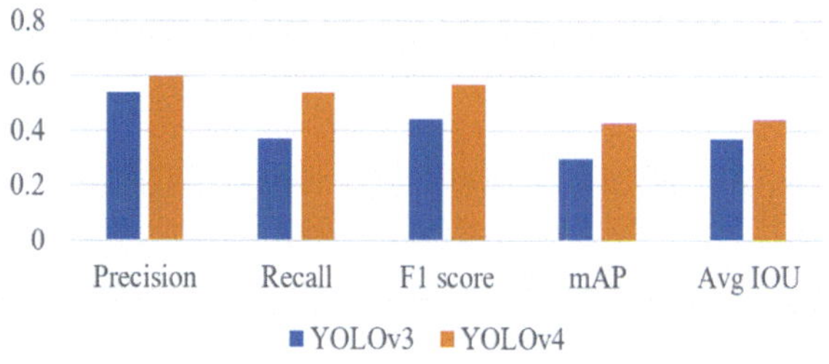


per sample image has been illustrated in Table 4. Further, by the evaluation of trained models (YOLOv3 and YOLOv4) on test images, significantly high accuracy (>8%) of YOLOv4 as compared to YOLOv3 has been observed. Moreover, as contrast to YOLOv4, YOLOv3 predicts more number of FN. However, due to very small size and cluttered background, even YOLOv4 is not able to predict all objects in the sample test images as depicted in Table 4 and Fig. 4.

Furthermore, to process each test image, the time required for YOLOv3 ranges between 33 and 42 ms with an average of 37.5 ms, whereas for YOLOv4, that of

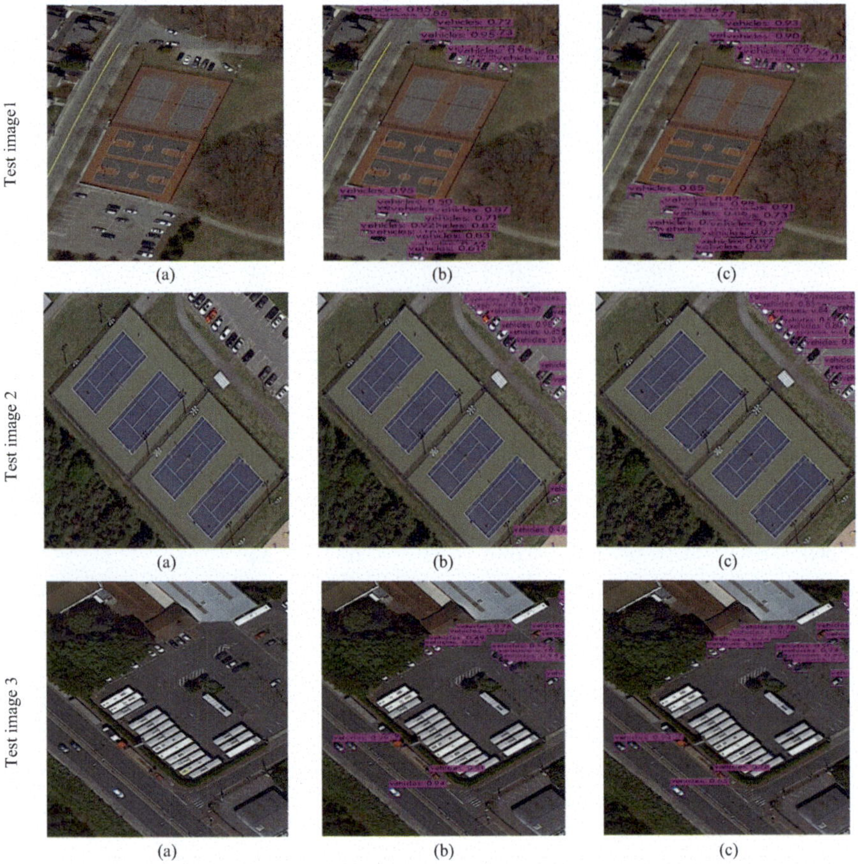


**Fig. 4** Experimental results tested on sample test images **a** Original image **b** Detection by YOLOv3 **c** Detection by YOLOv4

**Table 4** Comparison of YOLOv3 and YOLOv4 based on detection

| Test image | Total no. of vehicles | No. of vehicles detected | | Predicted in time (ms) | |
|---|---|---|---|---|---|
| | | YOLOv3 | YOLOv4 | YOLOv3 | YOLOv4 |
| 1 | 25 | 22 | 25 | 121.70 | 144.61 |
| 2 | 18 | 19 | 18 | 122.50 | 145.57 |
| 3 | 19 | 17 | 18 | 122.32 | 148.01 |

35 ms, 50 ms, and 42.5 ms, respectively, has been computed. However, for testing single image at a time, average prediction time for YOLOv3 and YOLOv4 has been evaluated as 122.17 and 146.06, respectively, as demonstrated in Table 4. The 19.55% more average prediction time is due to the large network size of YOLOv4 (161 layers) as compared to YOLOv3 (106 layers).

## 5 Conclusion

The present investigation served as guidance for the assessment and control of road traffic and accidents from satellite images. For this purpose, publicly available DOTA has been employed. Further, this dataset contains satellite images of very high resolution; therefore, each image in this dataset has been divided into several parts to extract a high-quality feature map. Although this dataset contains 15 object classes, in this work, for comparison and visualization only one class (vehicle) have been utilized. Therefore, in spite of 2806 original images, 9888 images have been used for training, validation, and testing. The considerable improvement of mAP (45%), IOU (20%), precision (11.1%), recall (45.9%), and F1-score (29.5%) for YOLOv4 than to YOLOv3 authorizes the efficiency of YOLOv4 over YOLOv3 specifically for vehicle detection using satellite images. These obtained results clearly recommend the development of satellite image object detection using YOLOv4. Further, to prove the robustness, the proposed methodology needs to be tested on a higher dataset with multiple object classes which will be addressed in future works.

**Acknowledgements** The first and second authors would like to thank the Ministry of Human Resource Development, New Delhi, India, for providing the Research Fellowship for carrying out this work.

## References

1. Lu, J., Ma, C., Li, L., Xing, X., Zhang, Y., Wang, Z., Xu, J.: A vehicle detection method for aerial image based on YOLO. J. Comput. Commun. **6**, 98–107 (2018). https://doi.org/10.4236/jcc.2018.611009
2. Pi, Y., Nath, N.D., Behzadan, A.H.: Convolutional neural networks for object detection in aerial imagery for disaster response and recovery. Adv. Eng. Inform. **43**, 101009 (2020)
3. Zheng, Z., et al.: A novel vehicle detection method with high resolution highway aerial image. IEEE J. Sel. Top. Appl. Earth Obs. Remote Sens. **6**(6), 2338–2343 (2013). https://doi.org/10.1109/JSTARS.2013.2266131
4. Zhou, J., Gao, D., Zhang, D.: Moving vehicle detection for automatic traffic monitoring. IEEE Trans. Veh. Technol. **56**(1), 51–59 (2007). https://doi.org/10.1109/TVT.2006.883735
5. Cheng, H.-Y., Weng, C.-C., Chen, Y.-Y.: Vehicle detection in aerial surveillance using dynamic Bayesian networks. IEEE Trans. Image Process **21**(4), 2152–2159 (2012). https://doi.org/10.1109/tip.2011.2172798
6. Chen, C., Zhong, J., Tan, Y.: Multiple-oriented and small object detection with convolutional neural networks for aerial image, MDPI (2019)
7. Guirado, E., Tabik, S., Rivas, M.L., Alcaraz-Segura, D., Herrera, F.: Automatic whale counting in satellite images with deep learning. bioRxiv (2018)
8. Han, S., Shen, W., Liu, Z.: Deep drone: object detection and tracking for smart drones on embedded system. Stanford University (2012)
9. Redmon, J., Divvala, S., Girshick, R., Farhadi, A.: You only look once: unified, real-time object detection. In: Proceedings of the IEEE Conference on Computer Vision and Pattern Recognition, Las Vegas, 27–30 June 2016, 779–788 (2016). https://doi.org/10.1109/CVPR.2016.91

10. Redmon, J., Farhadi, A.: YOLO9000: better, faster, stronger (2017). arXiv preprint arXiv:1612.08242
11. Redmon, J., Farhadi, A.: YOLOv3: An incremental improvement (2018). arXiv preprint arXiv:1804.02767
12. Bochkovskiy, A., Wang, C.-Y., Mark Liao, H.-Y. YOLOv4: optimal speed and accuracy of object detection (2020). arXiv preprint arXiv: 2004.10934
13. Xia, G., et al.: DOTA: a large-scale dataset for object detection in aerial images. In: 2018 IEEE/CVF Conference on Computer Vision and Pattern Recognition, Salt Lake City, UT, pp. 3974–3983 (2018). https://doi.org/10.1109/CVPR.2018.00418
14. Nie, X., Yang, M., Liu, R.W.: Deep neural network-based robust ship detection under different weather conditions. In: IEEE International Conference on Intelligent Transportation Systems, Auckland, New Zealand (2019)
15. Benjdira, B., Khursheed, T., Koubaa, A., Ammar, A., Ouni, K.: Car detection using unmanned aerial vehicles: comparison between faster R-CNN and YOLOv3. In: 2019 1st International Conference on Unmanned Vehicle Systems-Oman (UVS) (2019). https://doi.org/10.1109/uvs.2019.8658300
16. Zhang, Z., He, T., Zhang, H., Zhang, Z., Xie, J., Li, M.: Bag of freebies for training object detection neural networks (2019). ArXiv, abs/1902.04103

# Semantic Segmentation of Retinal Blood Vessel with Autoencoders

**Hirdesh Varshney, Utpal Kant, Himanshu Gupta, Om Prakash Verma, Tarun K. Sharma, and Irshad Ahmad Ansari**

**Abstract** In the medical imaging, early and precise segmentation of retina blood vessel (RBV) has been considered as one of the most key factors to diagnose the ophthalmologic diseases such as diabetic retinopathy, hypertension, arteriosclerosis, cardiovascular disease, and age-related macular degeneration. However, owing to very complex anatomy of the fundus, manual segmentation has been found as troublesome and tedious task along with lots of required knowledge and skills. Therefore, in the proposed work incorporating autoencoders has been proposed. To investigate the effectiveness of proposed methodology, DRIVE dataset has been employed. Further, the available dataset images have been converted into patches of 10 × 10 to encounter the very small size of utilized dataset. The developed model achieves more than 90% classification accuracy in most of the cases which validates the effectiveness of the proposed methodology.

H. Varshney
Department of Computer Science and Engineering, Babu Banarsi Das University, Lucknow, India
e-mail: hrideshvarshney@bbdu.ac.in

U. Kant
Doctoral School SISEO, Grenoble, France
e-mail: utpal.kant@univ-savoie.fr

H. Gupta · O. P. Verma
Department of Instrumentation and Control Engineering, Dr B R Ambedkar National Institute of Technology, Jalandhar, India
e-mail: guptah.nitj@gmail.com

O. P. Verma
e-mail: vermaop@nitj.ac.in

T. K. Sharma
Department of Computer Science and Engineering, Shobhit University, Gangoh, Saharanpur, India
e-mail: taruniitr1@gmail.com

I. A. Ansari (✉)
Electronics and Communication Engineering, PDPM Indian Institute of Information Technology Design and Manufacturing, Jabalpur, India
e-mail: irshad@iiitdmj.ac.in

 
T. K. Sharma et al. (eds.), *Soft Computing: Theories and Applications*,
Advances in Intelligent Systems and Computing 1381,
https://doi.org/10.1007/978-981-16-1696-9_53

**Keywords** Autoencoders · BCE loss · MSE loss · Retina blood vessel · Semantic segmentation · Unsupervised learning

## 1 Introduction

In this modern era, medical imaging has become a vital tool for medical practitioners which in turn provides very crucial information and helps in the exact diagnosis of various diseases. Segmentation of RBV from fundus vascular imaging is one such technique that helps in the diagnosis of many ophthalmic and non-ophthalmic diseases like arteriosclerosis, diabetes mellitus, cardiovascular disease, age-related macular degeneration, and hypertension [1, 2]. Further, diabetes mellitus has been found as the most common cause of diabetic retinopathy which may lead to blindness [3]. A recent study projected more than 150 million diabetic people by 2040 in China only [4]. Another study reveals that early-stage diagnosis with timely treatment and follow-up can prevent blindness by 95% in the reported cases [5] which strengthens early detection and diagnosis of diabetic retinopathy.

Retina has been considered as one of the most sensitive and inner parts of the human body that contains all the photosensory neurons, which has been composed of very thin layer and placed back to the eyeball. It converts light into signals and sends them to the brain which translates it into sight. If, because of any reason, the retina has been damaged, it may lead to complete blindness or any other type of serious problems such as blurry vision, retinal vein occlusion, and eye stroke [1]. Therefore, early assessment of RBV has been considered as the key to avoid these complications. These assessment techniques can be broadly classified into traditional and modern techniques. Traditional methods require a manual assessment of RBV which is very time-consuming and depends on the knowledge and belief of medical experts. Further, the very complex structure of RBV (diameter, tortuosity, and optic disk shape) and a large number of images often lead to many unidentified human errors [6]. Therefore, there is an urgent need of automatic RBV segmentation (RBVS) for early and accurate diagnosis of retinopathy and macular degeneration.

Modern techniques utilize automated RBVS which not only enhances the analyzing capability of vasculature for various pathological conditions or biometric identification [7] but also provides a more accurate and precise assessment of disease even at a very early stage. This motivated many researchers across the globe for the design and development of automatic RBVS methodologies. Typically, these mechanisms can be divided into rule-based and machine learning (ML) or deep learning (DL)-based techniques. Rule-based techniques employ pixel-wise labeling of blood vessels based on pre-defined criteria, whereas ML- or DL-based techniques utilizes automatic feature extraction mechanism. Further, depending upon the type of segmentation strategy utilized, they may be further classified into (i) Kernel filtering [8] (ii) vascular tracking [9] (iii) model-based [10] (iv) supervised and unsupervised approaches [11]. Kernel filtering incorporates tuned 2D kernels for efficient mapping of blood vessels within the retinal image [12]. However, it often leads to non-blood vessel regions, therefore resulting in unacceptable performance [13]. Vessel tracking methods capture the central lines of blood vessels; however, they have been suffered from drifts due to the complex structure of RBV and therefore resulting in degraded performance [14]. Model-based methodologies incorporate energy minimization to extricate RBV profiles [15]. Supervised methods rely on prior labeling and proper training under the ground truth values, whereas unsupervised methods classify pixels without prior knowledge. Therefore, supervised methods are usually more accurate; however, unsupervised methods have proved their dominance in the classification of large unlabeled data, particularly for small blood vessels.

Supervised learning-based methods subdivide the task of RBVS into three main parts: (i) image acquisition, (ii) feature extraction, and (iii) classification. Out of these parts, feature extraction has been considered as the most crucial part because the quality of extracted features directly affects classification accuracy. Therefore, reported literature utilized Gabor filter [16], Gaussian filter [17], K-nearest neighbor (K-NN) [18], support vector machine (SVM) [19], SVM with Adaboost [20], and convolutional neural network (CNN) [21] to train the classifier. However, the classification accuracy in these approaches greatly depends upon the quality and volume of labelled data which is not always possible in RBVS. To overcome this problem, denoising autoencoders has been proposed [22]. Further, a fusion of deep and ensemble learning has been proposed for RBVS and more than 93% accuracy has been achieved. Furthermore, a combination of 12 CNN has been utilized in a group of three layers where each layer has been separately trained on randomly generated 60,000 window patches of size 31 × 31 × 31 and has obtained more than 94% classification accuracy.

In light of the above discussion, it may be concluded that though, supervised learning-based ML and DL models have shown tremendous performance in most of the classification task, their performance greatly reduced in RBVS due to limited available data, inherent camera losses, and very complex nature of RBV. On the other hand, unsupervised techniques have shown promising RBV classification capabilities. Therefore, based on the above framework the novel contribution of the presented work has been sketched as:

1. Unsupervised learning using autoencoders has been employed to prevent overfitting.
2. To create a decent dataset, and therefore effective training, random patches of size 10 × 10 have been utilized instead of whole image at once.

The remainder of this paper is organized as follows: Sect. 2 presents the dataset utilized for the present investigation. The detail description of methodology used is shown in Sect. 3. In Sect. 4, the experimental results are presented and discussed. And finally, Sect. 5 gives the concluding remarks of the present work.

## 2 Dataset

Retinal imaging requires a specialized low power microscope that is capable of simultaneously illuminating and imaging the retina with the help of an attached camera. It has been designed to efficiently capture the inner surface of the retina which includes retina, optic disk, macula, etc. [7]. Further, the vasculature has been composed of various arteries and veins with varying widths (from 1 to 20 pixel). The traces of these vasculature along with inner surfaces of retina have been depicted in Fig. 1 for better understanding.

In this work, open-source digital retinal image for image extraction (DRIVE) has been utilized for feature extraction and therefore segmentation of RBV [23]. DRIVE dataset contains 40 retinal images of diabetic patients aged between 25 to 90 years in which 33 didn't have any sign of diabetic retinopathy, and the rest 7 images have shown mild symptoms of early diabetic retinopathy as shown in Fig. 2. Further, the dataset utilizes JPEG compression techniques for images and incorporates a 50–50 strategy for training and testing. Therefore, both the training and testing datasets contain 20 images. Further, single and two sets of manual segmentation have been adopted for training and testing datasets, where manual segmentation is referred to as ground truth values. Furthermore, these images have been captured by a canon CR5 non-mydriatic 3CCD camera with a 45° field of view with a size of 768 × 584 pixels and clustered into an original, gold standard (ground truth), and mask (indicating region of interest).

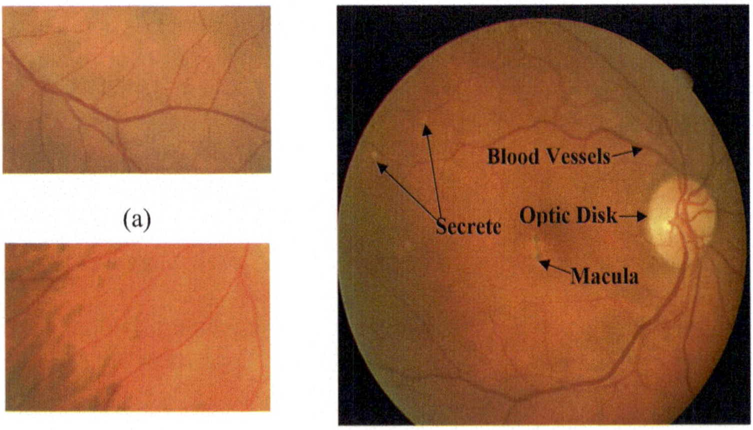


(b) (c)

**Fig. 1** Retinal image anatomy **a** central vessel reflex **b** hard secrete **c** morphology

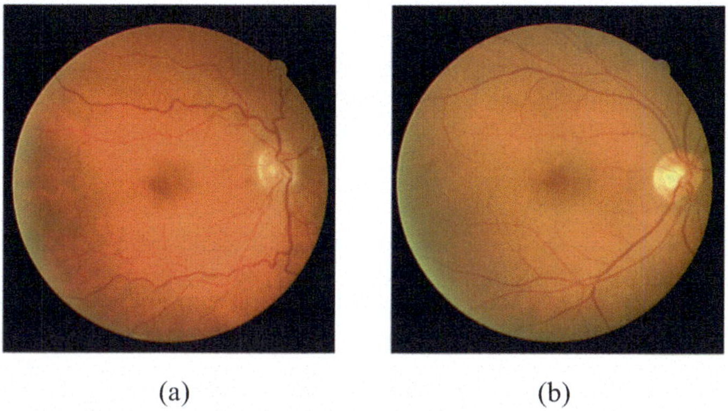


**Fig. 2** Sample retinal images from DRIVE dataset **a** healthy **b** diabetic retinopathy

## *2.1 Image Preprocessing*

The RBVS typically depends upon the attributes of the retinal vessel such as length, width, branching pattern, structure, tortuosity, and angles. However, it requires a lot of features to be learned for accurate classification. Further, image preprocessing techniques will help the model to learn more complex features. Therefore, in the reported work, various image preprocessing techniques including gray scale, red channel, blue channel, and green channel have been utilized as shown in Fig. 3.

These sampled images reveal that red, blue, and green channel images have better contrast than to grayscale image which helps in extracting enrich features. Further, because of non-uniform illumination, the background intensity of retinal images has

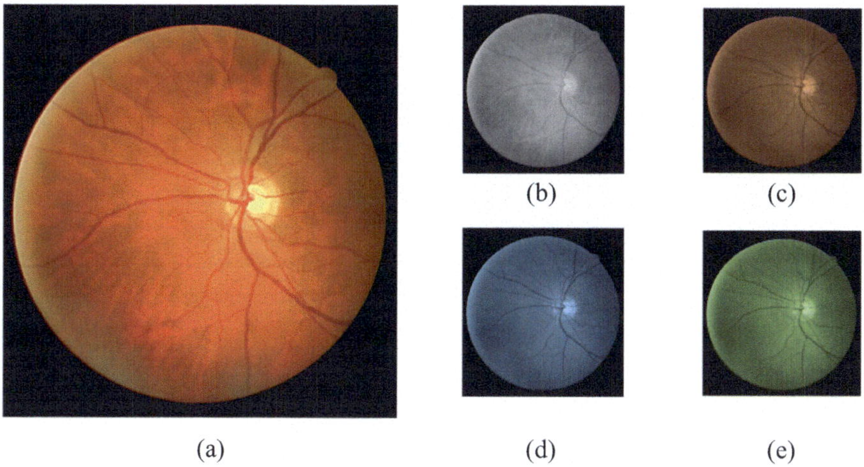


**Fig. 3** Sample image **a** original **b** gray scale **c** red **d** blue and **e** green channel

been drastically affected which significantly downgrades the RBVS performance. To effectively handle this issue, a contrast limited adaptive histogram equalization (CLAHE) algorithm has been also incorporated.

## 3 Methodology

Semantic segmentation has been most widely used in digital image processing for slicing the digital image into several pieces. These divisions helped in more precise visualization and therefore better feature extraction. Therefore, in this proposed work semantic segmentation approach has been employed for segmenting RBV images.

The retina blood vessel semantic segmentation (RBVSS) labelled each and every pixel of the retina image and therefore produces a highly precise fundus image which will boost the early-stage diagnosis capabilities. The flowchart of the proposed methodology has been presented in Fig. 4. This approach can be classified into two parts (i) feature extraction and (ii) patching of the image. The complex feature that

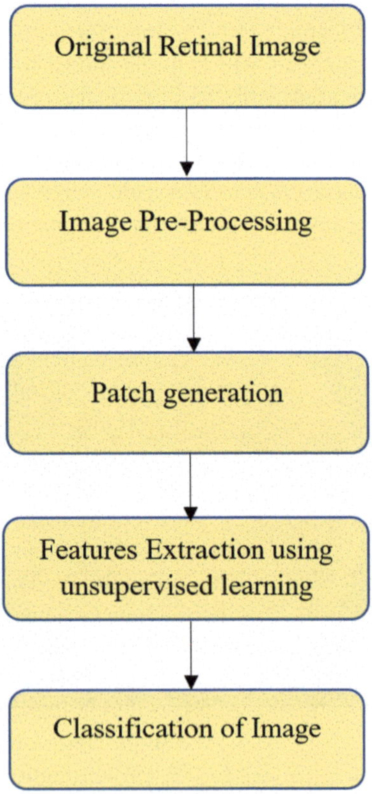


**Fig. 4** Flowchart of proposed unsupervised methodology

may results in very good classification cannot be extricated directly from the DRIVE dataset because of the very small size of the training dataset. Therefore, the images in the dataset have been converted into patches of small size.

Although the incorporated dataset contains three types of images, namely actual images, ground truth, and mask images, only two subsets (actual and ground truth) have been utilized for the present analysis. Further, instead of taking only image as input, pixel-wise labelled image using autoencoder has been considered as an input. These labelled images after proper preprocessing have been converted into grayscale images. These grayscale images have been converted into patches of 10 × 10, and out of these patches, 1000 patches have been randomly selected from each image, and therefore, in totality 20,000 patches have been employed for training. Thereafter, a neural network (NN) consisting of one input layer, two hidden layers, and one output layer has been incorporated to train the end-to-end model. Further, 300 neurons for the input layer, 100 neurons for each hidden layer, and 100 neurons for the output layer have been used. Furthermore, to prevent the model from overfitting, autoencoders utilizing tanh for encoding, and a combination of tanh and sigmoid activation functions have been employed. Therefore, autoencoders play a vital role in extracting features responsible for classification. Further, during the entire learning of autoencoders as feature extractor, Adam optimizer with default parameter settings (learning rate = 0.001, beta1 = 0.9, beta2 = 0.999, epsilon = 1e−08) has been incorporated to reduce the mean square loss (MSE) [24] as represented by Eq. (1).

$$L(y, \hat{y}) = \frac{1}{N} \sum_{(i=0)}^{N} (y - \hat{y}_i)^2 \tag{1}$$

where $L$, $N$, $y$, $\hat{y}$, and $\hat{y}_i$ represent loss function, total number of iterations, ground truth value, predicted value, and prediction at $i$th iteration, respectively. Further, after proper learning of autoencoders, the updated weights have been utilized for the classification task. For this purpose, binary cross-entropy (BCE) [25] loss has been employed along with Adam optimizer. Mathematically this loss function has been represented by Eq. (2).

$$H_p(q) = -\frac{1}{N} \sum_{(i=0)}^{N} y_i \log(p(y_i)) + (1 - y_i) \log(1 - p(y_i)) \tag{2}$$

where $y$ and $p(y_i)$ represent labels and prediction probability for all $N$ epochs.

## 4 Result and Discussion

The entire training and evaluation of the proposed technique have been done on the simulation platform as specified in Table 1. Further, python 3.7.9 installed

**Table 1** Description of simulation platform

| Name | Related configuration |
|---|---|
| Operating system | Ubuntu 18.04 |
| CPU | Intel(R) Core (TM) i7-9750H CPU @ 2.60 GHz |
| RAM | 8 GB |
| GPU | MSI Gaming GeForce GTX 1650 (Max-Q) |
| GPU acceleration library | CUDA10.0, CUDNN7.4 |

over anaconda platform has been used as programming environment for all the programming for training and evaluation.

As mentioned earlier, the overall task of classification has been subdivided into two parts (i) feature extraction using autoencoders (representative learning) and (ii) utilization of weights learned through autoencoders for further training of classifier. The training begins with random weight initialization for representative learning which with each passing iteration modifies a bit until the end of training. A sample of these weights has been shown in Fig. 5 for better visualization.

Both representative learning and classifier training have been done for 1000 iterations for a batch size of 1000. Further, during the whole training process MSE and BCE losses have been regularly monitored and presented in Table 2.

For better visualization, these losses have been also demonstrated in Fig. 6. Further, Fig. 6 and Table 2 reveal comparatively small value of MSE loss than BCE loss at every step. Furthermore, a sudden and sharp fall in MSE loss has been observed as compared to BCE loss. This might be because of sequential strategy in which MSE loss also enhances BCE loss.

After accomplishment of training, the trained model has been evaluated on test images and it has been found that the model able to classify RBV in most of the test images with adequate accuracy of more than 90%. For demonstration, the sample of classifier output along with original and ground truth image has been shown in Fig. 7.

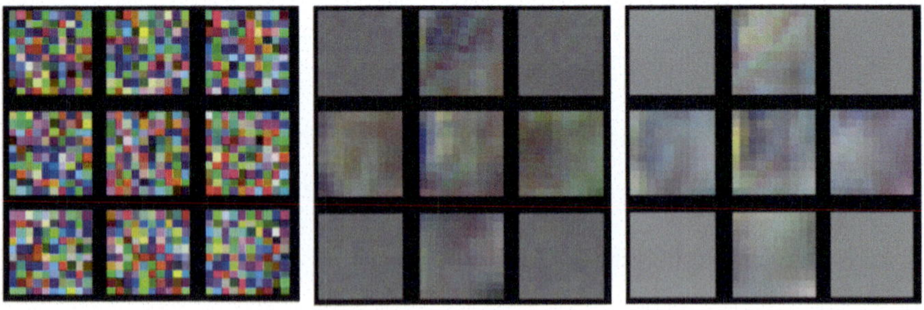

(a) (b) (c)

**Fig. 5** Training weights **a** initial weights **b** final weights **c** updated weights

**Table 2** Training losses: MSE (Representative learning) and BCE (Classifier)

| Iteration number | MSE loss | BCE loss |
|---|---|---|
| 1 | 0.086436 | 0.536363 |
| 101 | 0.001967 | 0.244587 |
| 201 | 0.001109 | 0.199314 |
| 301 | 0.000720 | 0.178213 |
| 401 | 0.000574 | 0.169834 |
| 501 | 0.000462 | 0.163858 |
| 601 | 0.000380 | 0.159562 |
| 701 | 0.000443 | 0.152836 |
| 801 | 0.000369 | 0.149130 |
| 901 | 0.000370 | 0.148650 |
| 1000 | 0.000365 | 0.139587 |

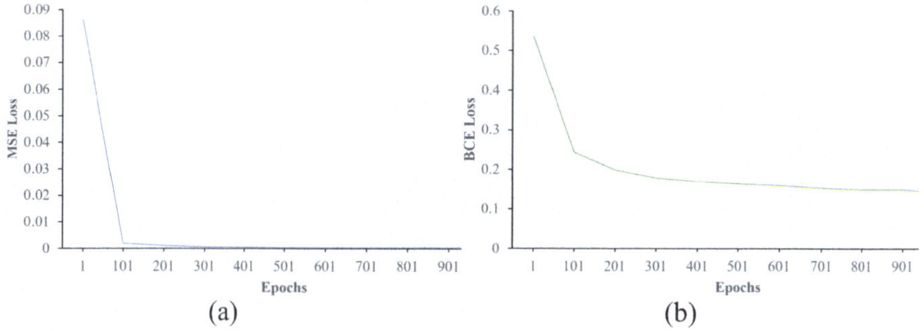


**Fig. 6** Training losses **a** representative learning **b** classifier

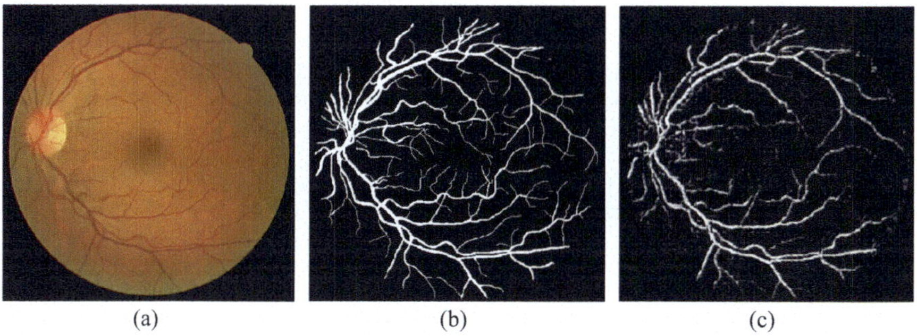


**Fig. 7** Classification results **a** original fundus image **b** ground truth **c** classifier output

## 5 Conclusion

In this paper, RBVSS method has been presented for classifying the retinal image by using unsupervised learning technique. For this purpose, the publically available DRIVE dataset has been utilized in the present investigation. The dataset contains only 40 images (20 each for both training and testing) which are not sufficient to train a classifier. Therefore, training images have been converted into small patches of size $10 \times 10$. Further, 1000 patches from each training image have been randomly selected to extract features via autoencoders. During entire training, Adam optimizer with default parameter settings has been incorporated to reduce the MSE loss and BCE loss for representative learning and classifier, respectively. The obtained classifier results have been compared against ground truth. It has been found that the developed model achieved more than 90% classification accuracy which validates the effectiveness of proposed technique. In the future, other dataset of retinal images will be used to validate the robustness of developed model.

## References


1. Trucco, E., Azegrouz, H., Dhillon, B.: Modeling the tortuosity of retinal vessels: does caliber play a role? IEEE Trans. Biomed. Eng. **57**, 2239–2247 (2010). https://doi.org/10.1109/TBME.2010.2050771
2. Wang, X., Jiang, X., Ren, J.: Blood vessel segmentation from fundus image by a cascade classification framework. Pattern Recogn. **88**, 331–341 (2019). https://doi.org/10.1016/j.patcog.2018.11.030
3. Nayak, C., Kaur, L.: Retinal blood vessel segmentation for diabetic retinopathy using multilayered thresholding (2013)
4. Luo, L., Chen, D., Xue, D.: Retinal blood vessels semantic segmentation method based on modified U-Net. In: Proceedings of the 30th Chinese Control and Decision Conference CCDC 2018, pp. 1892–1895 (2018). https://doi.org/10.1109/CCDC.2018.8407435
5. Jiang, Y., Zhang, H., Tan, N., Chen, L.: Automatic retinal blood vessel segmentation based on fully convolutional neural networks. Symmetry (Basel) **11** (2019). https://doi.org/10.3390/sym11091112
6. Hassan, G., El-Bendary, N., Hassanien, A.E., et al.: Retinal blood vessel segmentation approach based on mathematical morphology. Procedia Comput. Sci. Elsevier, 612–622 (2015)
7. Fraz, M.M., Remagnino, P., Hoppe, A., et al.: Blood vessel segmentation methodologies in retinal images—a survey. Comput. Meth. Programs Biomed. **108**, 407–433 (2012). https://doi.org/10.1016/j.cmpb.2012.03.009
8. Zolfagharnasab, H., Naghsh-Nilchi, A.R.: Cauchy based matched filter for retinal vessels detection. J. Med. Signals Sens. **4**, 1–9 (2014)
9. Vlachos, M., Dermatas, E.: Multi-scale retinal vessel segmentation using line tracking. Comput. Med. Imaging Graph **34**, 213–227 (2010). https://doi.org/10.1016/j.compmedimag.2009.09.006
10. Espona Pernas, L., Carreira, M., Penedo, M.G., Ortega, M.: Retinal vessel tree segmentation using a deformable contour model. In: 19th International Conference on Pattern Recognition, pp. 1–4 (2009)
11. Ganesan, K., Naik, G., Adapa, D., et al.: A supervised blood vessel segmentation technique for digital Fundus images using Zernike Moment based features. PLoS One **15** (2020)

12. Al-Rawi, M., Qutaishat, M., Arrar, M.: An improved matched filter for blood vessel detection of digital retinal images. Comput. Biol. Med. **37**, 262–267 (2007). https://doi.org/10.1016/j.compbiomed.2006.03.003
13. Almotiri, J., Elleithy, K., Elleithy, A.: Retinal vessels segmentation techniques and algorithms: a survey. Appl. Sci. **8** (2018). https://doi.org/10.3390/app8020155
14. Ben Abdallah, M., Malek, J., Azar, A.T., et al.: Automatic extraction of blood vessels in the retinal vascular tree using multiscale medialness. Int. J. Biomed. Imaging (2015). https://doi.org/10.1155/2015/519024
15. Espona, L., Carreira, M.J., Ortega, M., Penedo, M.G.: A snake for retinal vessel segmentation. In: Lecture Notes in Computer Science (including subseries Lecture Notes in Artificial Intelligence and Lecture Notes in Bioinformatics), pp. 178–185. Springer Verlag (2007)
16. Uchimura, S.: Optimization of Gabor filter parameters for pattern recognition. LibTkkFi, 49–52 (2001)
17. Nguyen, V., Blumenstein, M.: An application of the 2D Gaussian filter for enhancing feature extraction in off-line signature verification. In: Proceedings of the International Conference on Document Analysis Recognition, ICDAR, pp. 339–343 (2011). https://doi.org/10.1109/ICDAR.2011.76
18. Korotkova, O., Salem, M., Dogariu, A., Wolf, E.: Changes in the polarization ellipse of random electromagnetic beams propagating through the turbulent atmosphere. Waves Random Complex Media **15**, 353–364 (2005). https://doi.org/10.1080/17455030500184511
19. Li, X., Wang, L., Sung, E.: AdaBoost with SVM-based component classifiers. Eng. Appl. Artif. Intell. **21**, 785–795 (2008). https://doi.org/10.1016/j.engappai.2007.07.001
20. Ricci, E., Perfetti, R.: Retinal blood vessel segmentation using line operators and support vector classification. IEEE Trans. Med. Imaging **26**, 1357–1365 (2007). https://doi.org/10.1109/TMI.2007.898551
21. Tan, J.H., Fujita, H., Sivaprasad, S., et al.: Automated segmentation of exudates, haemorrhages, microaneurysms using single convolutional neural network. Inf. Sci. (Ny) **420**, 66–76 (2017). https://doi.org/10.1016/j.ins.2017.08.050
22. Roy, A.G., Sheet, D.: DASA: Domain adaptation in stacked autoencoders using systematic dropout. In: Proceedings—3rd IAPR Asian Conference on Pattern Recognition, ACPR 2015, pp. 735–739. Institute of Electrical and Electronics Engineers Inc. (2016)
23. DRIVE: Digital Retinal Images for Vessel Extraction,https://drive.grand-challenge.org/
24. Ghosh, A., Kumar, H., Sastry, P.S.: Robust loss functions under label noise for deep neural networks. In: 31st AAAI Conference on Artificial Intelligence, AAAI 2017, pp. 1919–1925 (2017)
25. Jadon, S. (2020, October). A survey of loss functions for semantic segmentation. In 2020 IEEE Conference on Computational Intelligence in Bioinformatics and Computational Biology (CIBCB) (pp. 1–7). IEEE

# Vehicle Trajectory Outlier Detection for Road Safety

**Hanan S. AlShanbari, Heba T. Al-Qadi, Ashwaq M. Al-Hassani, and Dalya M. Mitwally**

**Abstract** This work is making use of trajectory analytics for vehicular traffic study through developing and testing a system to identify outlier trajectories of vehicles. Real traffic data is analyzed to provide useful insights of traffic and generate traffic alerts to identify violating behavior, with the aim to facilitate traffic safety. Road safety is very critical issue, and this problem is faced by most people on the road. We developed a computer vision system that analyzes the trajectories of vehicles on the road to identify outlier trajectories. Outlier trajectories can be many types (non-typical/irregular behavior) on the road, such as backing up on a highway or changing lanes, turning, and other abnormal behavior. The work involves detection of vehicles and then analyzing its behavior on road. Yolov3, a TensorFlow model, is used for vehicle detection. Learning models are used to distinguish between normal and abnormal behavior. Finally, a system is developed using machine learning techniques, which is further trained using sufficient dataset examples. The performance of the developed system is tested in real-life situation, and it is found to be working quite comparable to human level detection of abnormal vehicle behavior.

**Keywords** Computer vision · Yolov3 · Machine learning · Trajectory outlier detection · Bayes theorem

## 1 Introduction

With the development of technology and communication tools, lots of things are now possible, which was earlier very tuff to do. Monitoring & surveillance is a tedious and complex task, which used to require human checking. Today, machine learning and computer vision system have provided us a way to replace the human

H. S. AlShanbari (✉) · H. T. Al-Qadi · A. M. Al-Hassani · D. M. Mitwally
College of Computer Science and Information System, Umm Al-Qura University, Makkah, Saudi Arabia
e-mail: Hsshanbari@uqu.edu.sa

D. M. Mitwally
e-mail: Dalya1417@windowslive.com

 
T. K. Sharma et al. (eds.), *Soft Computing: Theories and Applications*,
Advances in Intelligent Systems and Computing 1381,
https://doi.org/10.1007/978-981-16-1696-9_54

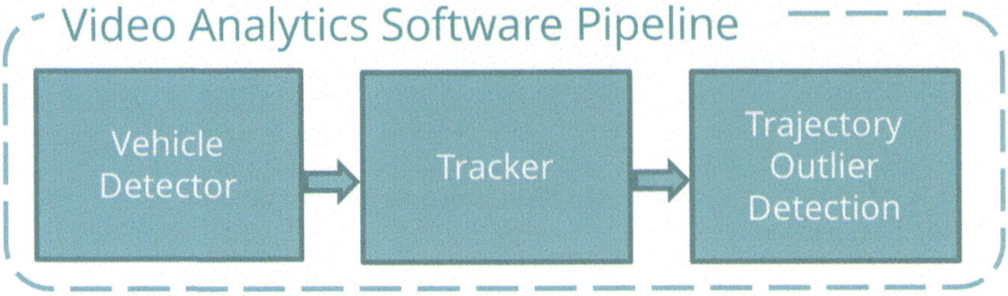


**Fig. 1** Main components of proposed system

monitoring to automatic (AI—artificial intelligence) monitoring. Mostly video data is used for surveillance, for example, vision-based transportation processing [1, 2] and traffic engineering applications [3, 4]. These vision systems, when combined with machine learning algorithm can do wonders like vehicle tracing, flow control, vehicle counting, trajectory detection and prediction, abnormal activity detection, etc. [1–4]. Earlier days, automated toll tax collection was only the prime use of automated vehicle detection and surveillance but with the inclusion of vision-based intelligence, the application got extended to vehicle type, color, driver behavior, and collision avoidance analysis. One such prime area of research that got a lot of attention from the researchers is vehicle trajectories outlier detection and classification. The performance (detection and classification) of such a system depends highly on the good quality video recording (input data) and intelligent decision of classification algorithm (decision making) [5, 6].

This paper is providing an insight of the test (vehicle trajectories outlier detection and classification) on the self-collected high-quality video recording (in Saudi Arabia) using the yolov3 for vehicle detection and Bayes theorem for vehicle classifications. Figure 1 shows the main components of the proposed system.

The main idea of our project is based on trajectory analytics for vehicular traffic through develop and test a system to identify outlier trajectories of vehicles on a road, in real traffic data.

We developed a computer vision system which analyzes the trajectories of vehicles on the road to identify outlier trajectories. Outlier trajectories can be any type of non-typical behavior on the road, such as backing up on a highway or changing lanes, turning, and other abnormal behavior. Proposed system is not a vehicle detection or tracking system, whereas proposed system uses existing model for vehicle detection and tracking (yolo3) for outlier detection. Trajectory analysis (on Saudi Arabia traffic dataset) and decision making on this is the main contribution of this study. The Naive Bayes algorithm is used for trajectory analysis and outlier detection. The vehicle detection & behavior analysis are achieved with various stages. Figure 2 shows the various levels on on-road online monitoring.

In this work, we have developed a computer vision system that has the ability to monitor the vehicle movement & detect patterns of motion. The system is designed with following objectives:

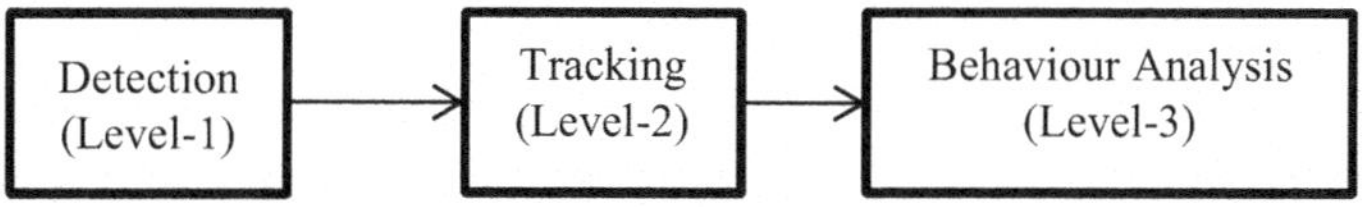


**Fig. 2** Vision levels of on-road online monitoring

- Detection of abnormal vehicle trajectories (abnormal behavior) in an efficient way.
- Use of computer vision to recognize outlier vehicle detection in reliable manner with real-time performance ability.

## 2 Related Work

Surveillance and positioning data collection has seen a rapid growth in the last decade due to the advancement in Internet, wireless sensor network (WSN) sensor & related technologies. Global positioning system, Wi-Fi, and video-based surveillance become very common these days. The detection of vehicle and its tracking has also seen a lot of advancement [7, 8]. Surveillance generates a lot of data but machine learning techniques are needs to make sense (extract useful information) of this data. For example, motion tracking, color, model of car, other variants detection, and outlier detection (after learning patterns) [9, 10]. Vehicle trajectory outlier detection is one of the prime areas of vehicle surveillance [11].

Vehicle trajectory detection can be made possible with machine learning technique such as yolov3 [12], which can detect the objects as per requirements. If we train it to detect vehicles of different type, it can be used for different type of vehicle classification such as company, model, color [13]. As we are interested in tracking rather than classification, we can use series of image frames (video) and feed them to yolov3 for vehicle detection. Eventually, it will generate a tracking on vehicle. This trajectory metadata can be used with any learning model such as Naive Bayesian [14] to differentiate between normal and abnormal trajectory (Fig. 3).

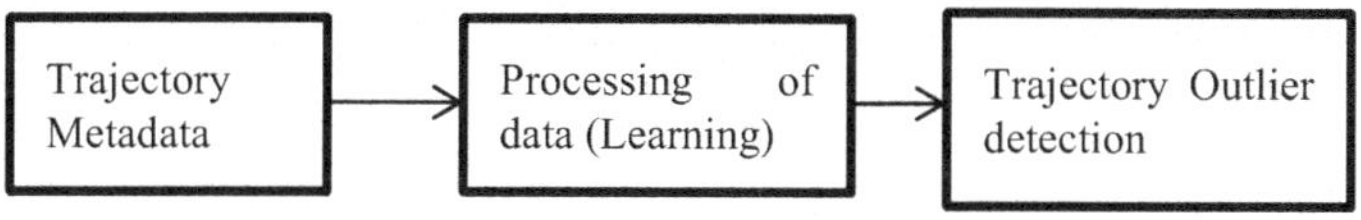


**Fig. 3** Steps to detect trajectory outlier

## 3 Methodology

Machine learning provides computers the ability to work like human being by learning from example. The analysis of video recording via machine learning techniques can provide useful information. In the proposed work, we have used machine learning algorithms to identify and analyze outlier trajectories of vehicles from a video. First, our model detects and tracks all vehicles to extract their trajectories and then create a learning model to identify outlier trajectories of vehicles. These outlier trajectories indicate some abnormal behavior on part of the driver, which cause an alert to be generated. The purpose of the proposed system is that it will learn "normal" behavior from clustering trajectories together over time and then identify outliers as those deviating significantly from this norm. In this system, we built a computer vision (CV) system that analyzes the trajectories of vehicles on the road to identify outlier trajectories. Outlier trajectories can be any type of non-typical behavior on the road.

For vehicle detection, yolov3 TensorFlow model is employed. yolov3 uses deep neural network-based supervised learning approach for detection of object in almost real time. When a set of frames are provide to yolov3, it analyzes the same and detects the vehicle presence by separating the background from the foreground. This detection on number of frames results into vehicle tracking, which ultimately generates the trajectory metadata (location $(x, y)$ & speed/direction $(dx, dy)$ for further analysis.

The tracking is possible with location coordinate $(x, y)$ in 2-D/with height & location in 3-D. In proposed work, we have used 2-D tracking as height of the vehicle remains constant and it does not provide trajectory information. "Kalman filter" along with "yolov3 TensorFlow" model is used to track the vehicle trajectory.

After that, we calculated the direction and speed for vehicles from each frame by represented on the histogram to learn the normal and abnormal behaviors, by using probability density function (PDF) for direction and speed. For continuous distributions values, precisely the area under its PDF and the probability is high so that considered normal behavior. For discrete distributions values, precisely values in the different areas its PDF, the probability is low and that is called outlier (abnormal behavior). We use Bayes Theorem to predict the likelihood of an occasion happening given the likelihood of another occasion that has just happened. Bayes theorem makes use of PDF and following formula (Eq. 1):

$$P\left(\frac{\text{event}-1}{\text{event}-2}\right)=\frac{P\left(\frac{\text{event}-2}{\text{event}-1}\right)P(\text{event}-1)}{P(\text{event}-2)} \quad (1)$$

Here:

$P\left(\frac{\text{event}-1}{\text{event}-2}\right)$ is conditional probability; likelihood of event-1, given event-2 is true.
$P\left(\frac{\text{event}-2}{\text{event}-1}\right)$ is conditional probability; likelihood of event-2, given event-1 is true.
$P(\text{event}-1)$ and $P(\text{event}-2)$ are probability of event-1 and event-2, respectively.
Also, event-1 and event-2 are independent events.

In our proposed model, Bayes theorem is used. Exactly what we are doing is that we are computing the class conditional probabilities (from the histograms), and multiplying them together. This is what is done in naive Bayes classifier. However, since we are doing outlier detection, we just look at the value of the product, and if it is too small, we say it is an outlier.

## 4 Implemented System and Results

The dataset used in this work was provided by one of the companies incubated in "Wadi Makkah," and this dataset consists of real traffic videos collected in Makkah and Riyadh. The video was captured by road-side cameras mounted all a pole, and it was stored in H-264 standard video format like (.avi &.mov &.mp4) format. The complete algorithm implementation involves total 6 phases, which are as follows:

***Phase 1 (Collection of original videos)***: Firstly, stationary traffic cameras are used to capture the video data in H-264 standard video format. Then, some preprocessing steps are applied to make this recorded video suitable for further processing.

***Phase 2 (Vehicle detection)***: To detect the presence of vehicle in an image, yolov3 TensorFlow Model is used. The model used deep neural networks to detect the vehicle presence in the image.

***Phase 3 (Vehicle tracking)***: The location $(x, y)$ in each frame and its speed with direction $(\mathrm{d}x, \mathrm{d}y)$ is recorded. This data provided the trajectory of the vehicle and tracking become possible. Kalman filter-based Gaussian model is also developed to find future position using the past positions.

***Phase 4 (Extract vehicle trajectories)***: The location $(x, y)$ in each frame and its speed with direction $(\mathrm{d}x, \mathrm{d}y)$ is extracted for every vehicle (for entire dataset).

***Phase 5 (Learn a model of normal behavior)***: Vehicle outlier detection is achieved by learning from trajectories metadata. Trajectories of moving objects are the reflection of their activities in the physical world. Therefore, we calculated the direction and speed for vehicles from each frame by represented it on histogram to learn the normal and abnormal behaviors, by using probability density function (PDF) of direction and speed. For continuous distributions values, precisely the area under its PDF and the probability is high so that considered as a normal behavior. For discrete distributions values, precisely values in the different areas of PDF, the probability is low and that is called outlier (abnormal behavior).

So basically, we build a model to work on learning the common behaviors and determining outliers (unusual behavior) as shown in Fig. 4.

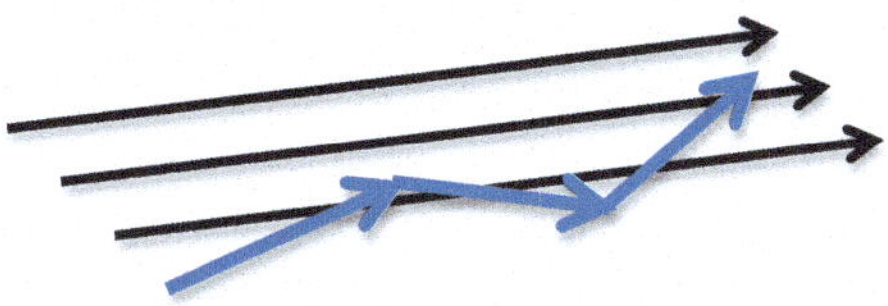

**Fig. 4** An example of outlaying (unusual) behavior

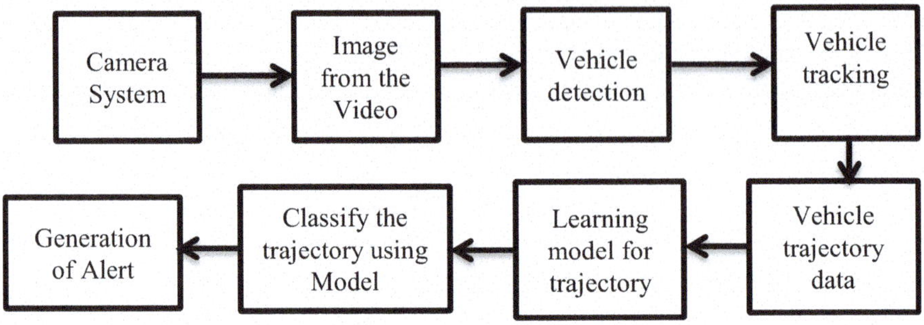


**Fig. 5** Complete system representation and alert generation

***Phase 6 (Generate an alert)***: Fig. 5 shows the complete process of trajectory tracking. After an abnormal behavior is detected by the learned model, it will generate an alert (red box around the vehicle).

## 5 Results

Firstly, the input frame is divided into 4 blocks (Fig. 6). The trajectory for every block is found using yolov3. For each block, histogram of trajectories are calculated & plotted as shown in Fig. 7 (for one of the blocks).

Entire dataset is divided into training and testing dataset. Then, naive Bayes algorithm is used to predict the conditional probabilities. What we exactly do it to calculate the conditional probabilities (Fig. 8) of the class (from the graphs) and multiply them together. This is what is done in the naive Bayes workbook. However, since we are

**Fig. 6** Dividing the frame into 4 blocks

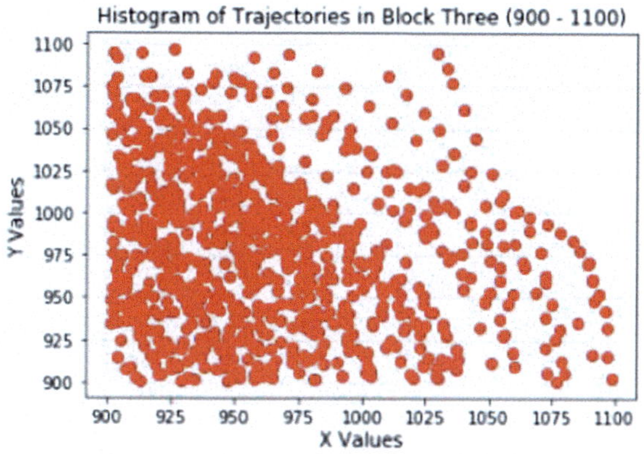


**Fig. 7** Histogram of trajectories in block 3

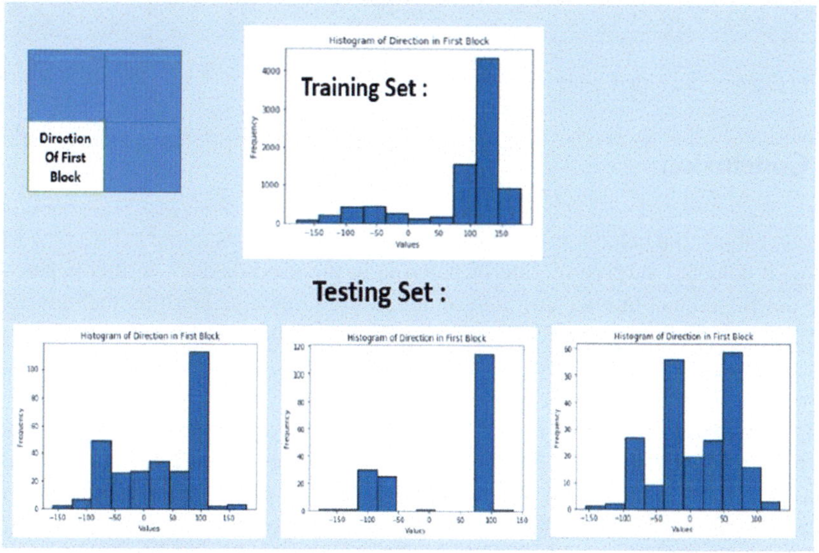


**Fig. 8** Comparative vehicle behavior analysis in testing and training phases

doing external discovery, we are only looking at the value of the product, and if it is too small, we say it is an anomaly.

Then, we repeat all these steps taking some other videos (testing set), then comparing the results between them and the (training set) for each block.

Finally, after finishing all the analysis to detect the outlier using vehicle trajectory, an alert will be generated in real time. This alert will change the color of tracking box to red and display the on the screen. Example of an alert generation in actual scenario is shown in Fig. 9.

**Fig. 9** Example of an alert generation in actual scenario

## 6 Conclusion

To detect abnormal behavior of vehicles, outlier detection was found to be a powerful tool as it detected surprising data or patterns in the used dataset. In this paper, we have developed a computer vision system that used machine learning techniques to monitors vehicle behavior over time & detect abnormal movement by comparing it with exiting set of normal movements. Both, the tracking using yolov3 and PDF-based outlier detection was very efficient.

As future work, we would like to continue this work and improving outlier detection by extending it to detect different abnormal vehicle behavior and classify those behaviors in different groups. We would also like to explore computer vision to try other more powerful technologies, and planning to develop our own TensorFlow model and consider our problem as a detection & tracking-related problem. Also, we would like to add notification messages and linked it with the identity of the vehicle driver.

## References

1. Guerreiro, G., Figueiras, P., Silva, R., Costa, R., Jardim-Goncalves, R.: An architecture for big data processing on intelligent transportation systems. An application scenario on highway traffic flows. In 2016 IEEE 8th International Conference on Intelligent Systems (IS), pp. 65–72. IEEE (2016)

2. Babar, M., Arif, F.: Real-time data processing scheme using big data analytics in internet of things based smart transportation environment. J. Ambient. Intell. Humaniz. Comput. **10**(10), 4167–4177 (2019)
3. Kar, K., Kodialam, M., Lakshman, T.V.: Minimum interference routing of bandwidth guaranteed tunnels with MPLS traffic engineering applications. IEEE J. Sel. Areas Commun. **18**(12), 2566–2579 (2000)
4. Trimponias, G., Xiao, Y., Wu, X., Xu, H., Geng, Y.: Node-constrained traffic engineering: theory and applications. IEEE/ACM Trans. Netw. **27**(4), 1344–1358 (2019)
5. Petrovskaya, A., Thrun, S.: Model based vehicle detection and tracking for autonomous urban driving. Auton. Robot. **26**(2–3), 123–139 (2009)
6. Stauffer, C., Grimson, W.E.L.: Learning patterns of activity using real-time tracking. IEEE Trans. Pattern Anal. Mach. Intell. **22**(8), 747–757 (2000)
7. Sivaraman, S., Trivedi, M.M.: Looking at vehicles on the road: A survey of vision-based vehicle detection, tracking, and behavior analysis. IEEE Trans. Intell. Transp. Syst. **14**(4), 1773–1795 (2013)
8. Meng, F., Yuan, G., Lv, S., Wang, Z., Xia, S.: An overview on trajectory outlier detection. Artif. Intell. Rev. **52**(4), 2437–2456 (2019)
9. Lee, J.G., Han, J., Li, X.: Trajectory outlier detection: a partition-and-detect framework. In: 2008 IEEE 24th International Conference on Data Engineering, pp. 140–149. IEEE (2008)
10. Yuan, G., Xia, S., Zhang, L., Zhou, Y., Ji, C.: Trajectory outlier detection algorithm based on structural features. J. Comput. Inf. Syst. **7**(11), 4137–4144 (2011)
11. Wu, H., Sun, W., Zheng, B.: A fast trajectory outlier detection approach via driving behavior modeling. In: Proceedings of the 2017 ACM on Conference on Information and Knowledge Management, pp. 837–846 (2017)
12. Redmon, J., Farhadi, A.: Yolov3: an incremental improvement (2018). arXiv preprint arXiv: 1804.02767
13. Mao, J., Wang, T., Jin, C., Zhou, A.: Feature grouping-based outlier detection upon streaming trajectories. IEEE Trans. Knowl. Data Eng. **29**(12), 2696–2709 (2017)
14. Flach, P.A., Lachiche, N.: Naive Bayesian classification of structured data. Mach. Learn. **57**(3), 233–269 (2004)

# Author Index





T. K. Sharma et al. (eds.), *Soft Computing: Theories and Applications*,
Advances in Intelligent Systems and Computing 1381,
https://doi.org/10.1007/978-981-16-1696-9

The manufacturer's authorised representative in the EU is Springer Nature Customer Service Centre GmbH, Europaplatz 3, 69115 Heidelberg, Germany. If you have any concerns regarding our products, please contact ProductSafety@springernature.com

Printed and bound by CPI Group (UK) Ltd, Croydon, CR0 4YY
30/09/2026
02225232-0010